工业和信息化部“十四五”规划教材

数值分析双语教程

Numerical Analysis: A Bilingual Course

檀结庆　主　编
张　莉　副主编

科 学 出 版 社
北　京

内 容 简 介

本书系双语教材，主体部分用英语撰写，延伸阅读部分用汉语撰写. 主体部分主要内容包括：常见数学公式和数学表达式的英语读法、解线性方程组的直接法、矩阵代数迭代技术、一元方程求根、多项式插值、逼近论、数值微分与数值积分、常微分方程初值问题等. 延伸阅读部分内容包括：数学家传记、求解非线性方程组的最小二乘法、非线性方程组的不动点迭代法、牛顿迭代法及拟牛顿迭代法、有理函数插值、Thiele 型连分式插值、Padé 逼近、最佳一致逼近、高斯求积公式的收敛性、Radau 求积公式与 Lobatto 求积公式、Euler-Maclaurin 展开、常微分方程边值问题等.

本书适用于高等学校数学类各专业本科生、工科各专业感兴趣的本科生和研究生.

图书在版编目(CIP)数据

数值分析双语教程 = Numerical Analysis: A Bilingual Course: 汉、英/檀结庆主编. —北京：科学出版社，2023.9

工业和信息化部“十四五”规划教材

ISBN 978-7-03-076331-0

Ⅰ. ①数…　Ⅱ. ①檀…　Ⅲ. ①数值分析–高等学校–教材–汉、英　Ⅳ. ①O241

中国国家版本馆 CIP 数据核字(2023)第 169685 号

责任编辑：蒋　芳/责任校对：杨聪敏

责任印制：赵　博/封面设计：许　瑞

科学出版社 出版

北京东黄城根北街 16 号

邮政编码：100717

http://www.sciencep.com

北京凌奇印刷有限责任公司印刷

科学出版社发行　各地新华书店经销

*

2023 年 9 月第 一 版　开本：787×1092　1/16

2024 年 1 月第二次印刷　印张：25 3/4

字数：608 000

定价：119.00 元

(如有印装质量问题，我社负责调换)

前　言

“数值分析”是一门重要的基础课程. 十余年来，我们一直采用双语方式为合肥工业大学数学学院的本科生讲授“数值分析”课程，选用的教材是由 Richard L. Burden & J. Douglas Faires 编写的同名英文教材 *Numerical Analysis*(*Seventh Edition*) 的影印版，由高等教育出版社出版. 该影印版教材内容较丰富，而实际课堂教学课时有限，没有足够的时间讲授完教材中的所有内容，教学内容的取舍成为一个不得不面对的现实问题. 通过十余年的双语教学实践，我们深深感受到，编写一本切合实际需要的数值分析双语教材，显得很有必要.

考虑到数值分析双语教材应尽可能体现出双语特色，因此，本教材既不能编撰成国外出版的英文版教材的注释版，也不能编写成像某些英语文学作品那样的英汉对照版，而要对教材的组织架构重新进行设计，使其达到既“利于学”，又“便于教”的双重效果.

本书的编写逻辑体现在以下两个方面：一是使学生能够在规定的学时内比较轻松地完成该教程的系统学习；二是完成本教程的学习后，学生的学科专业知识将得到进一步夯实，专业英语应用能力也将有较大程度的提高.

本书的编写思路是：尽力促使学生在学习该课程时，能心无旁骛地将注意力集中在对课程内容的理解、消化和吸收上，而不是面对专业英语，一脸茫然、一筹莫展，将大量时间和精力花费在对英语语言的理解上.

本书的主要特色体现在以下几个方面：一是每个章节标题均有中英文对照；二是在每章具体内容之前，均用中文列出每章的学习目标，使学生怀着明确的目的去学习；三是正文内容用英语陈述，对关键词或关键术语用中文进行标注，从而减少语言阻尼时间；四是每章后面均用中文概括出该章的要点，帮助学生强化所学内容，达到去粗取精的目的；五是每章后均配置若干与该章内容相关的问题供学生思考与讨论，进一步激发学生的学习兴趣，培养学生的创新能力，实现内涵与外延的有机统一；六是每章后均以英汉对照的方式列出了该章出现的关键术语，使学生轻松熟悉专业英语词汇；七是每章最后都设计了“延伸阅读”模块，该模块用中文撰写，作为对课堂教学内容的自然延伸和补充，供学生课外阅读，借此拓展学生的视野.

本书共分 8 章. 第 1 章是数学基础知识，内容包括：常见数学公式和数学表达式的英语读法、微积分中的一些定义与定理、绝对误差与相对误差的概念等. 第 2 章是解线性方程组的直接法，内容包括：Gauss 消元法、主元法、矩阵分解法、特殊矩阵等. 第 3 章是矩阵代数迭代技术，内容包括：向量范数与矩阵范数、特征值与特征向量、解线性方程组的迭代法、收敛性分析与 SOR 迭代法、误差分析与迭代优化等. 第 4 章是一元方程求根，内容包括：二分法、不动点迭代及误差分析、牛顿法、弦截法等. 第 5 章是多项式插值，内容包括：Lagrange 插值、Neville 插值、Newton 插值、Hermite 插值、三次样条插值等. 第 6 章是逼近论，内容包括：离散最小二乘逼近、正交多项式与最小平方逼近、多项式与幂级数约化等. 第 7 章是数值微分与数值积分，内容包括：数值微分、Richardson 外推、数值积分、复化数值积分、Romberg 积分、Gauss 求积等. 第 8 章是常微分方程初值问题，内容包括：初值问题基本理论、Euler 方法、Runge-Kutta 方法、多步方法等.

本书第 1 章、第 3 章由张莉执笔，第 2 章由江平执笔，“延伸阅读”版块的内容由檀结庆撰写，第 4 章由朱晓临执笔，第 5 章至第 8 章主要由檀结庆执笔，其中，第 7 章前两节的部分内容由刘植撰写，第 2 章至第 8 章的“算法程序及实例”模块内容全部由彭凯军撰写，全书由檀结庆统稿.

本书已被工业与信息化部列入“十四五”规划教材进行立项建设，衷心感谢各位匿名的评审专家对出版本教材的支持. 本教材编写团队成员，既是国家级双语示范课程和国家级一流本科课程“数值分析”的主讲教师，也是纳入质量工程建设的安徽省线下课程“数值分析”省级教学团队的成员，教材中的一些内容正是源自团队成员的教学研究成果. 感谢合肥工业大学对本教材出版所给予的大力支持. 感谢科学出版社蒋芳编辑为本教材出版所付出的辛勤劳动.

我们深知，双语类教材的结构设计和编写逻辑见仁见智，本书的编写是教材建设方面的一种尝试和探索. 囿于学识和水平，本书难免存在不足之处，敬请读者批评指正.

编　者

2023 年 1 月于合肥

目　　录

Chapter 1 Mathematical Preliminaries

(数学基础知识)

本章内容 (Contents)

学习目标 (Learning Objectives)

- 熟悉常见数学公式及数学表达式的读法
- 掌握微积分学中的一些基本定义与引理
- 掌握极限与连续的定义
- 熟悉微分的定义并熟练运用其推广的重要定理
- 熟悉积分的定义并熟练运用积分中值定理
- 掌握泰勒多项式与级数的定义
- 了解误差的来源
- 熟悉绝对误差与相对误差的计算
- 了解有效数字的概念

This chapter contains a short review of those topics from single-variable calculus that will be needed in later chapters. A solid knowledge of calculus is essential for the understanding of numerical techniques and their analysis, and more thorough review might be needed if you have been away from this subject for a while. In addition, there is an introduction to convergence, error analysis, the machine representation of numbers, and some techniques for categorizing and minimizing computational error.

1.1 Mathematics English (数学英语)

$1/2$	is read	one half
$1/3$	is read	one third

$2/3$	is read	two thirds
$1/4$	is read	one quarter
$3/4$	is read	three quarters
$5/14$	is read	five over fourteen
$25\frac{7}{9}$	is read	twenty five and seven ninths
35%	is read	thirty five percent
-3.14	is read	minus three point one four
$\pm a$	is read	plus or minus a
$\mp a$	is read	minus or plus a
$\|a\|$	is read	the absolute value of a
$[a]$	is read	the greatest integer not exceeding a
$n!$	is read	factorial n
$\mathrm{Re}(z)$	is read	the real part of (complex number) z
$\mathrm{Im}(z)$	is read	the imaginary part of (complex number) z
$a+b=c$	is read	a plus b equals (is) c
$a-b=c$	is read	a minus b equals (is) c
$a\times b=c$	is read	a times b equals (is) c
$a\div b=c$	is read	a divided by b equals (is) c
a^2	is read	a square
a^3	is read	a cube
a^n	is read	a to the power of n
a^{-n}	is read	a to the power of minus n
$\sqrt{a}$	is read	the square root of a
$\sqrt[3]{a}$	is read	the cube root of a
$\sqrt[n]{a}$	is read	the nth root of a
$\sqrt[n]{a^m}$	is read	the nth root of a to the power of m
e^x	is read	the exponential of x (to the base e)
a^x	is read	the exponential of x (to the base a)
$\ln x$	is read	the natural logarithm of x (to the base e)
$\lg x$	is read	the (common) logarithm of x (to the base 10)
$\log_a x$	is read	the logarithm of x (to the base a)
$a=b$	is read	a is equal to b
$a\neq b$	is read	a is not equal to b
$a\equiv b$	is read	a is identically equal to b
$a\approx b$	is read	a is approximately equal to b
$a>b$	is read	a is greater / larger than b
$a<b$	is read	a is less than b

$a \geqslant b$	is read	a is greater than or equal to b
$a \leqslant b$	is read	a is less than or equal to b
$a \gg b$	is read	a is much greater than b
$a \ll b$	is read	a is much less than b
$a \parallel b$	is read	a is parallel to b
$a \perp b$	is read	a is perpendicular to b
(a,b)	is read	open interval
$[a,b]$	is read	closed interval
$(a,b]$	is read	half-open interval open at the left
$[a,b)$	is read	half-open interval open at the right
$A \cap B$	is read	the intersection of A and B
$A \cup B$	is read	the union of A and B
$A \subseteq B$	is read	A is a subset of B
$A \subset B$	is read	A is a proper subset of B
$A \not\subset B$	is read	A is not a proper subset of B
$A - B$	is read	the difference of A and B
A^c	is read	the complement of A
$\varnothing$	is read	the empty set
$a \in A$	is read	a belongs to A
$a \notin A$	is read	a is not in A
$\boldsymbol{a} \cdot \boldsymbol{b}$	is read	the scalar product of vectors $\boldsymbol{a}$ and $\boldsymbol{b}$
$\boldsymbol{a} \times \boldsymbol{b}$	is read	the vector product of vectors $\boldsymbol{a}$ and $\boldsymbol{b}$
$\Rightarrow$	is read	imply
$\Leftrightarrow$	is read	be equivalent to
$\lim\limits_{n \to \infty} a_n = a$	is read	the limit of the sequence $\{a_n\}$ is a as n tends to infinity
$\lim\limits_{x \to x_0} f(x) = A$	is read	the limit of f of x is A as x tends to x_0
$\lim\limits_{x \to x_0^-} f(x) = A$	is read	the limit of f of x is A as x tends to x_0 from the left
$\lim\limits_{x \to x_0^+} f(x) = A$	is read	the limit of f of x is A as x tends to x_0 from the right
Δ	is read	delta
$\dfrac{dy}{dx}$	is read	the first derivative of y with respect to x
$\dfrac{d^2y}{dx^2}$	is read	the second derivative of y with respect to x
$\dfrac{d^ny}{dx^n}$	is read	the nth derivative of y with respect to x
$f'(x), f''(x)$	is read	f prime of x, f double prime of x

$\dfrac{\partial z}{\partial x}$	is read	the first partial derivative of z with respect to x
$\dfrac{\partial^2 z}{\partial x^2}$	is read	the second partial derivative of z with respect to x
$\dfrac{\partial^n y}{\partial x^n}$	is read	the nth partial derivative of y with respect to x
$\int f(x)\mathrm{d}x$	is read	the (indefinite) integral of f of x, dx
$\int_a^b f(x)\mathrm{d}x$	is read	the (definite) integral from a to b of f of x, dx
$\sum_{i=1}^{n} a_i$	is read	sigma of a_i as i runs from 1 to n
$\sum_{n=1}^{\infty} a_n$	is read	sigma of a_n as n runs from 1 to infinity
$\det \boldsymbol{A}$	is read	the determinant of (the matrix) $\boldsymbol{A}$
$\boldsymbol{A}^{\mathrm{T}}$	is read	the transpose of (the matrix) $\boldsymbol{A}$
$\boldsymbol{A}^{-1}$	is read	the inverse (matrix) of (the matrix) $\boldsymbol{A}$
$\boldsymbol{A}^{*}$	is read	the adjoint matrix of (the matrix) $\boldsymbol{A}$
$\overline{\boldsymbol{A}}$	is read	the conjugate matrix of (the matrix) $\boldsymbol{A}$
$\operatorname{rank} \boldsymbol{A}$	is read	the rank of (the matrix) $\boldsymbol{A}$
$\operatorname{tr} \boldsymbol{A}$	is read	the trace of (the matrix) $\boldsymbol{A}$

1.2 Review of calculus (微积分回顾)

- Limits and continuity (极限和连续性)
- Differentiability (可微性)
- Integration (积分)
- Taylor polynomials and series (泰勒多项式和级数)
- Examples (例题)

1.2.1 Limits and continuity (极限和连续性)

The concepts of limit (极限) and continuity (连续性) of a function are fundamental to the study of calculus, and form the basis for the analysis of numerical techniques.

Definition 1.1 A function f defined on a set X of real numbers has the limit (极限) A at x_0, written

$$\lim_{x \to x_0} f(x) = A, \tag{1.2.1}$$

if given any real number $\varepsilon > 0$, there exists a real number $\delta > 0$ such that

$$|f(x) - A| < \varepsilon, \text{ whenever } x \in X \text{ and } 0 < |x - x_0| < \delta.$$

This is the definition of the limit of a function in ε, δ-terminology. It can also be written without the limit symbol as follows:

$$f(x) \to A, \quad x \to x_0. \tag{1.2.2}$$

Definition of "one-sided" limits (单侧极限): right (left)-hand limit (右(左)极限)

If $f(x) \to A$ as $x \to x_0$ through values greater (less) than x_0, then we say that A is the right (left)-hand limit (右(左)极限) of f at x_0, and we indicate this by writing

$$\lim_{x \to x_0^+} f(x) = A \left(\lim_{x \to x_0^-} f(x) = A \right). \tag{1.2.3}$$

Definition of neighborhood (邻域) of a point:

Any open interval containing a point x as its midpoint is called neighborhood (邻域) of x, denoted by $U(x)$ or $U(x, \varepsilon)$, if we wish to specify its radius $\varepsilon > 0$.

In neighborhood (邻域) terminology, the symbol

$$\lim_{x \to x_0} f(x) = A$$

means that for every neighborhood $U(A, \varepsilon)$ there is some neighborhood $U(x_0, \delta)$ such that $f(x) \in U(A, \varepsilon)$, whenever $x \in U(x_0, \delta)$ and $x \neq x_0$.

Definition 1.2 (The definition of the continuity of a function) Let f be a function defined on a set X of real numbers and $x_0 \in X$. Then f is said to be continuous (连续的) at x_0, if

$$\lim_{x \to x_0} f(x) = f(x_0). \tag{1.2.4}$$

The function f is said to be continuous on the set X (在集合 X 上是连续的) if it is continuous at each number in X.

In ε, δ-terminology, the continuity of the function f at x_0 can be formulated as: if given any real number $\varepsilon > 0$, there exists a real number $\delta > 0$ such that $|f(x) - f(x_0)| < \varepsilon$, whenever $x \in X$ and $|x - x_0| < \delta$.

The limit of a sequence (序列的极限) of real or complex numbers is defined in a similar manner.

Definition 1.3 (The definition of the limit of a sequence) Let $\{x_n\}_{n=1}^{\infty}$ be an infinite sequence of real or complex numbers. The sequence $\{x_n\}_{n=1}^{\infty}$ has the limit x (极限为 x) (converges to x (收敛至 x)), if for any $\varepsilon > 0$, there exists a positive integer $N(\varepsilon)$ such that $|x_n - x| < \varepsilon$, whenever $n > N(\varepsilon)$. The notation

$$\lim_{n \to \infty} x_n = x, \text{ or } x_n \to x \text{ as } n \to \infty$$

means that the sequence $\{x_n\}_{n=1}^{\infty}$ converges to x.

Theorem 1.1 If f is a function defined on the set X of real numbers and $x_0 \in X$, then the following statements are equivalent:

(a) f is continuous at x_0;

(b) If $\{x_n\}_{n=1}^{\infty}$ is any sequence in X converging to x_0, then

$$\lim_{n\to\infty} f(x_n) = f(x_0). \tag{1.2.5}$$

1.2.2 Differentiability (可微性)

More sophisticated assumptions about a function generally lead to better approximation results. For example, a function with a smooth graph will normally behave more predictably than one with numerous jagged features. The smoothness condition relies on the concept of the derivative.

Definition 1.4 (The definition of the derivative of a function) Let f be defined in an open interval containing x_0. The function f is said to be differentiable(可微的) at x_0, if

$$f'(x_0) = \lim_{x\to x_0} \frac{f(x) - f(x_0)}{x - x_0} \tag{1.2.6}$$

exists. The number $f'(x_0)$ (read as " f prime of x_0") is called the derivative (导数) of f at x_0. A function that has a derivative at each number in a set X is said to be differentiable (可微的) on X.

Theorem 1.2 If the function f is differentiable at x_0, then f is continuous at x_0.

Theorem 1.3 (Rolle's Theorem) Suppose $f \in C[a,b]$ and f is differentiable on (a,b). If $f(a) = f(b)$, then a number ξ in (a,b) exists with $f'(\xi) = 0$.

Theorem 1.4 (Mean Value Theorem) If $f \in C[a,b]$ and f is differentiable on (a,b), then a number ξ in (a,b) exists with

$$f'(\xi) = \frac{f(b) - f(a)}{b - a}.$$

Theorem 1.5 (Extreme Value Theorem) If $f \in C[a,b]$, then there exist c_1, c_2 with $f(c_1) \leqslant f(x) \leqslant f(c_2)$, for all $x \in [a,b]$. In addition, if f is differentiable on (a,b), then the numbers c_1 and c_2 occur either at the endpoints of $[a,b]$ or where $f'(x)$ is zero.

Theorem 1.6 (Generalized Rolle's Theorem) Suppose $f \in C[a,b]$ is n times differentiable on (a,b) and $f(x)$ is zero at $n+1$ distinct numbers $x_0, x_1, \cdots, x_n$ in $[a,b]$, then a number ξ in (a,b) exists with $f^{(n)}(\xi) = 0$.

Theorem 1.7 (Intermediate Value Theorem) If $f \in C[a,b]$ and c is any number between $f(a)$ and $f(b)$, then there exists a number ξ in (a,b) for which $f(\xi) = c$.

1.2.3 Integration (积分)

The other basic concept of calculus that will be used extensively is the Riemann integral (黎曼积分).

Definition 1.5 (Riemann integral) The Riemann integral (黎曼积分) of the function f

on the interval $[a,b]$ is the following limit, provided it exists

$$\int_a^b f(x)\mathrm{d}x = \lim_{\max \Delta_{x_i} \to 0} \sum_{i=1}^{n} f(z_i)\Delta_{x_i}, \tag{1.2.7}$$

where the numbers $x_0, x_1, \cdots, x_n$ satisfy condition of $a = x_0 \leqslant x_1 \leqslant \cdots \leqslant x_n = b$, $\Delta_{x_i} = x_i - x_{i-1}$, for each $i = 1,2,\cdots,n$, and z_i is arbitrarily chosen in the interval $[x_{i-1}, x_i]$.

A function f that is continuous on an interval $[a,b]$ is also Riemann integrable on $[a,b]$. This permits us to choose, for computational convenience, the points x_i to be equally spaced in $[a,b]$, and for each $i = 1,2,\cdots,n$, to choose $z_i = x_i$. In this case,

$$\int_a^b f(x)\mathrm{d}x = \lim_{n \to \infty} \frac{b-a}{n} \sum_{i=1}^{n} f(x_i), \tag{1.2.8}$$

where the numbers x_i are $x_i = a + i(b-a)/n$.

Theorem 1.8 (Weighted Mean Value Theorem for Integrals) Suppose $f \in C[a,b]$, the Riemann integral of g exists on $[a,b]$, and $g(x)$ does not change sign on $[a,b]$. Then there exists a number ξ in (a,b) with

$$\int_a^b f(x)g(x)\mathrm{d}x = f(\xi)\int_a^b g(x)\mathrm{d}x. \tag{1.2.9}$$

When $g(x) \equiv 1$, the above formula is the usual Mean Value Theorem for Integrals. It gives the average value of the function f over the interval $[a,b]$ as

$$f(\xi) = \frac{1}{b-a}\int_a^b f(x)\mathrm{d}x. \tag{1.2.10}$$

1.2.4 Taylor polynomials and series (泰勒多项式和级数)

The final contents in this review of calculus describes the Taylor polynomials. These polynomials are used extensively in numerical analysis.

Theorem 1.9 (Taylor's Theorem) Suppose $f \in C^n[a,b]$, $f^{(n+1)}$ exists on (a,b) and $x_0 \in [a,b]$. For every $x \in [a,b]$, there exists a number $\xi(x)$ between x_0 and x with

$$f(x) = P_n(x) + R_n(x), \tag{1.2.11}$$

where

$$\begin{aligned} P_n(x) &= f(x_0) + f'(x_0)(x-x_0) + \frac{f''(x_0)}{2!}(x-x_0)^2 + \cdots + \frac{f^{(n)}(x_0)}{n!}(x-x_0)^n \\ &= \sum_{k=0}^{n} \frac{f^{(k)}(x_0)}{k!}(x-x_0)^k \end{aligned} \tag{1.2.12}$$

and

$$R_n(x)=\frac{f^{(n+1)}(\xi(x))}{(n+1)!}(x-x_0)^{n+1}. \tag{1.2.13}$$

Here

- $P_n(x)$ is called the nth Taylor polynomials (泰勒多项式) for f about x_0, and $R_n(x)$ is called the remainder term (余项) (or truncation error (截断误差)), associated with $P_n(x)$.
- The infinite series obtained by taking the limit of $P_n(x)$ as $n\to\infty$ is called the Taylor series (泰勒级数) for f about x_0.
- In the case $x_0=0$, the Taylor polynomials is often called a Maclaurin polynomial (麦克劳林多项式), and the Taylor series is called a Maclaurin series (麦克劳林级数).
- The term truncation error (截断误差) refers to the error involved in using a truncated, or finite, summation to approximate the sum of an infinite series (无穷级数).

1.2.5 Examples (例题)

Example 1.1 For the function below, determine whether $f'(1)$ exists.

$$f(x)=\begin{cases}2x^2+1, & \text{if } x\leqslant 1;\\ 4x-1, & \text{if } x>1.\end{cases}$$

Solution. Since the rule for f changes at 1, we calculate that one-side derivatives of f at 1:

$$\lim_{x\to 1^-}\frac{f(x)-f(1)}{x-1}=\lim_{x\to 1^-}\frac{(2x^2+1)-3}{x-1}=\lim_{x\to 1^-}\frac{2(x^2-1)}{x-1}=\lim_{x\to 1^-}2(x+1)=4,$$

$$\lim_{x\to 1^+}\frac{f(x)-f(1)}{x-1}=\lim_{x\to 1^+}\frac{(4x-1)-3}{x-1}=\lim_{x\to 1^+}\frac{4(x-1)}{x-1}=4.$$

Thus, $f'(1)=4$. The graph of f is smooth at 1; that is, the graph of f has a tangent line at 1 and its slope is $f'(1)=4$.

Example 1.2 Show that $x^5-2x^3+3x^2-1=0$ has a solution in the interval $[0,1]$.

Solution. Consider the function defined by $f(x)=x^5-2x^3+3x^2-1=0$. The function f is continuous on $[0,1]$. In addition,

$$f(0)=-1<0 \quad\text{and}\quad 0<1=f(1).$$

The Intermediate Value Theorem (介值定理) implies that a number x exists, with $0<x<1$, for which $x^5-2x^3+3x^2-1=0$.

Example 1.3 Let $f(x)=\cos x$ and $x_0=0$. Determine

(a) the second Taylor polynomials for f about x_0; and

(b) the third Taylor polynomials for f about x_0.

Solution. Since $f\in C^\infty(\mathbf{R})$, Taylor's Theorem can be applied for any $n\geqslant 0$. Also,

$$f'(x)=-\sin x,\ f''(x)=-\cos x,\ f'''(x)=\sin x, \text{ and } f^{(4)}(x)=\cos x$$

so

$$f(0)=1,\ f'(0)=0,\ f''(0)=-1, \text{ and } f'''(0)=0.$$

(a) For $x \in (-\infty,\infty)$ and $x_0 = 0$, we have

$$\cos x = f(0)+f'(0)x+\frac{f''(0)}{2!}x^2+\frac{f'''(\xi(x))}{3!}x^3$$
$$=1-\frac{1}{2}x^2+\frac{1}{6}x^3\sin\xi(x),$$

where $\xi(x)$ is some (generally unknown) number between 0 and x.

(b) Since $f'''(0)=0$, the third Taylor polynomials with remainder term about $x_0 = 0$ is

$$\cos x = 1-\frac{1}{2}x^2+\frac{1}{24}x^4\cos\tilde{\xi}(x),$$

where $\tilde{\xi}(x)$ is some number between 0 and x.

1.3 Errors and significant digits (误差和有效数字)

- Source of errors (误差的来源)
- Absolute error and relative error (绝对误差和相对误差)
- Significant digit (or figure) (有效数字)
- How to avoid the loss of accuracy (如何避免精度的丢失)
- Examples (例题)

Analysis of the error (误差) involved in calculations is an important topic in numerical analysis. In this section, the contents of errors and significant digits (有效数字) will be reviewed.

1.3.1 Source of errors (误差的来源)

- Model error (systematic error) (模型误差(系统误差))
- Observation error (measure error) (观测误差(测量误差))
- Truncation error (截断误差)
- Round-off error (舍入误差)

Model error(systematic error): To solve scientific computing problems with computers, the first step is to establish mathematical models, which are approximately obtained by abstracting and simplifying the actual problems. This error between the mathematical model and the actual problem is called the model error (模型误差).

Observation error(measure error): In mathematical models, there are often some physical quantities based on observation, which obviously also contain errors. This error due to observation is called observation error (观测误差).

Truncation error: Truncation error (截断误差) results when approximations are used to represent exact mathematical procedures.

For example,

$$\mathrm{e}=1+1+\frac{1}{2!}+\frac{1}{3!}+\cdots+\frac{1}{n!}+\cdots,$$

but we have to use its finite items to approximate exp, say as

$$\mathrm{e}\approx 1+1+\frac{1}{2!}+\frac{1}{3!}+\cdots+\frac{1}{n!},$$

and the error

$$R_n(1)=\frac{\mathrm{e}^{\xi}}{(n+1)!},\ \text{where}\ \ 0<\xi<1$$

is the truncation error (截断误差).

Round-off error: Computer can only represent a number with a finite number of digits. Round-off error (舍入误差) is produced when a calculator or computer is used to perform real number calculations. The error that results from replacing a number with its floating-point form is called round-off error(regardless of whether the rounding or chopping method is used).

In our traditional mathematical world, we permit numbers with an infinite number of digits. The arithmetic we use in this world defines $\sqrt{3}$ as the unique positive number that when multiplied by itself produces the integer 3. In the computational world, however, each representable number has only a fixed and finite number of digits. This means, for example, that only rational numbers and not even all of these can be represented exactly. Since $\sqrt{3}$ is not rational, it is given an approximate representation, one whose square will not be precisely 3, although it will likely be sufficiently close to 3, which is acceptable in most situation. In most cases, then, this machine arithmetic is satisfactory and passes without notice or concern, but at times problems arise because of this discrepancy.

Machine numbers (机器数) are represented in the normalized decimal floating-point form

$$\pm 0.\,d_1 d_2\cdots d_k\times 10^n,\ 1\leqslant d_1\leqslant 9,\ 0\leqslant d_i\leqslant 9,\ 2\leqslant i\leqslant k$$

which are called k- digit decimal machine numbers.

For example, the number π has an infinite decimal expansion of the form $\pi=3.141592653589793$, written in normalized decimal form, we have

$$\pi=0.3141592653589793\times 10^1,$$

then the floating-point form of π using five-digit chopping is

$$fl(\pi)=0.31415\times 10^1=3.1415,$$

and the floating-point form of π using five-digit rounding is

$$fl_1(\pi)=(0.31415+0.00001)\times 10^1=3.1416\ \text{(Rounding up)},$$

and the floating-point form of π using six-digit rounding is

$$fl_1(\pi)=0.314159\times 10^1=3.14159\ \text{(Rounding down)}.$$

1.3.2 Absolute error and relative error (绝对误差和相对误差)

Definition 1.6 (Absolute error and relative error) If p^* is an approximation to p, the absolute error is

$$\left|p-p^*\right|$$

and the relative error is

$$\frac{\left|p-p^*\right|}{\left|p\right|} \quad \text{or} \quad \frac{\left|p-p^*\right|}{\left|p^*\right|}$$

provided that $p \neq 0$ or $p^* \neq 0$.

Suppose that you have the task of measuring the lengths of a bridge and a rivet and come up with 9999 cm and 9 cm, respectively. If the true values are 10000 cm and 10 cm, respectively, then the error for measuring the bridge is

$$10000-9999=1(\text{cm})$$

and for the rivet it is

$$10-9=1(\text{cm}).$$

However, the relative error for the bridge is

$$\frac{1}{10000}\times 100\% = 0.01\%$$

and for the rivet it is

$$\frac{1}{10}\times 100\% = 10\%.$$

Thus, although both measurements have an error of 1 cm, the relative error for the rivet is much greater. We would conclude that we have done a good job of measuring the bridge, but our estimate for the rivet is far from enough.

1.3.3 Significant digit (or figure) (有效数字)

Definition 1.7 (Significant digit (or figure)) Suppose

$$p^* = \pm 0.\, d_1 d_2 \cdots d_k \times 10^m,\ 1 \leqslant d_1 \leqslant 9,\ 0 \leqslant d_i \leqslant 9,\ 2 < i \leqslant k$$

is the approximation to p. If

$$\left|p-p^*\right| \leqslant 0.5\times 10^{m-t}, 0 \leqslant t \leqslant k,$$

then the number p^* is said to be approximate p to t significant digits (or figures), or p^* retains t significant digits as an approximation to p.

We know that the unit magnitude in the position of digit d_t is 10^{m-t}; therefore, we can define significant digits as follows:

If the absolute error $|p-p^*|$ is less than or equal to half unit magnitude of some digit position, then all the digits from this digit till the first leftmost nonzero digit are called significant digits.

1.3.4 How to avoid the loss of accuracy (如何避免精度的丢失)

1. Avoid the subtraction of nearly equal numbers

One of the most common error-producing calculations involves the cancelation of significant digits due to the subtraction of nearly equal numbers. The loss of accuracy due to round-off error can be avoided by a reformulation of the problem, illustrated in the next example.

2. Avoid big numbers "swallowing" important small numbers

The quadratic formula states that the roots of $ax^2+bx+c=0$, when $a\neq 0$, are

$$x_1=\frac{-b+\sqrt{b^2-4ac}}{2a} \text{ and } x_2=\frac{-b-\sqrt{b^2-4ac}}{2a}.$$

If b is much larger than $4ac$, then $\sqrt{b^2-4ac}\approx b$, and a poor approximation to x_1 may yield with large relative error.

3. Reduce the number of computations

Another way to reduce round-off error is to reduce the number of error-producing computations, above all, the multiplications.

For example, polynomials should always be expressed in nested form before performing an evaluation, because this form minimizes the number of arithmetic calculations.

4. Avoid dividing by a number with small absolute value

This principle mainly refers to trying to avoid division where the absolute value of the divisor is much smaller than the absolute value of the dividend.

5. Use the stable algorithm

If all the round-off errors of an algorithm do not exceed a given thresholding value, then the algorithm is said to be well behaved or numerically stable.

1.3.5 Examples 例题

Example 1.4 Suppose $p=\pi=3.141592653589793\cdots$, and

$$p_1^*=3.14,\ p_2^*=3.141,\ p_3^*=3.142$$

are the approximations to π respectively. Try to find how many significant digits they have respectively.

Solution. Since

$$p_1^* = 0.314\times 10^1,\ p_2^* = 0.3141\times 10^1,\ p_3^* = 0.3142\times 10^1$$

which yields $m = 1$, and

$$\left|p - p_1^*\right| = 0.00159265\cdots = 0.159265\cdots\times 10^{-2} < 0.5\times 10^{1-3}$$

$$\left|p - p_2^*\right| = 0.00059265\cdots = 0.059265\cdots\times 10^{-2} < 0.5\times 10^{1-3}$$

$$\left|p - p_3^*\right| = 0.00040734\cdots = 0.407346\cdots\times 10^{-3} < 0.5\times 10^{1-4}$$

then p_1^*, p_2^* and p_3^* have 3, 3 and 4 significant digits, respectively.

Example 1.5 Using six-digit floating-point rounding arithmetic, calculate the approximation to $\lg 1.99999 - \lg 1.99998$, whose true value is 2.1714886×10^{-6}.

Solution. Using six-digit floating-point rounding arithmetic, we have

$$\lg 1.99999 - \lg 1.99998 \approx 0.301028 - 0.301026 = 0.000002$$

which has only one significant digit.

To obtain a more accurate six-digit rounding approximation for $\lg 1.99999 - \lg 1.99998$, we change the form of this expression by the property of the logarithm function

$$\lg 1.99999 - \lg 1.99998 = \lg\frac{1.99999}{1.99998} \approx 2.17149\times 10^{-6}$$

which has 6 significant digits and is very close to the true value 2.1714886×10^{-6}.

Example 1.6 Using four-digit rounding arithmetic, consider this formula applied to the equation $x^2 + 1000x + 1 = 0$, whose roots are

$$x_1 = -0.001001\cdots \quad \text{and} \quad x_2 = -999.998.$$

Solution. In this equation, b^2 is much larger than $4ac$, which induces

$$b^2 - 4ac = 1000^2 - 4.000 = 0.1000\times 10^7 - 0.4000\times 10^1 \approx 0.1000\times 10^7.$$

Consequently,

$$\sqrt{b^2 - 4ac} = \sqrt{1000^2 - 4.000} \approx 1000.0,$$

and we have

$$fl(x_1) = \frac{-b + \sqrt{b^2 - 4ac}}{2a} = \frac{-1000 + 1000}{2.000} = 0.000,$$

a poor approximation to the true value $x_1 = -0.00100\cdots$.

Example 1.7 We re-express the following polynomial of degree three in nested form

$$a_0 + a_1x + a_2x^2 + a_3x^3 = a_0 + x\left(a_1 + x\left(a_2 + xa_3\right)\right)$$

which not only reduces the calculation but also is easier to program:

$$\begin{cases} s_3 = a_3; \\ s_n = a_n + xs_{n+1} \quad \text{for} \quad n = 2,1,0. \end{cases}$$

1.4 本章要点 (Highlights)

1. 极限是指无限趋近于一个固定的数值. 极限可分为数列极限和函数极限.
2. 左 (右) 极限就是函数从一个点的左 (右) 侧无限靠近该点时所取到的极限值，且误差可以小到任意指定的程度.
3. 左右极限存在且相等时，极限存在.
4. 对于一元函数，可导必可微，可微必可导，可导必连续，连续必可积.
5. 泰勒公式可以用函数在某一点的各阶导数值做系数构建一个多项式来近似该函数.
6. 常见的误差来源包括：模型误差、观测误差、截断误差和舍入误差.
7. 机器数以归一化的十进制浮点形式表示.
8. 绝对误差是测量值与真实值之差的绝对值，即绝对误差=|测量值–真实值|.
9. 相对误差是绝对误差所占真实值的百分比，即相对误差=|测量值–真实值|/真实值.
10. 有效数字是指在分析工作中实际能够测量到的数字，能够测量到的是包括最后一位估计的不确定的数字.
11. 如果绝对误差小于或等于某个数字位置单位大小的一半，那么从这个数字到最左边的第一个非零数字的所有数字都称为有效数字.
12. 合理地运用计算方法可以避免精度的丢失.

1.5 问题讨论 (Questions for discussion)

1. 极限的定性描述与定量描述有什么区别？
2. 如何理解连续的概念？连续函数有什么应用？
3. 导数是如何定义的？举例说明导数的应用.
4. 如何理解微分的概念？
5. 微分学中的中值定理有何作用？如何运用中值定理解决实际问题？
6. 积分中值定理与微分中值定理有什么区别与联系？
7. 举例说明泰勒公式的实际运用.
8. 举例说明各种误差的区别与联系.
9. 举例说明绝对误差、相对误差以及有效数字的概念.
10. 尝试说明在实际中有哪些方法可以减少误差.

1.6 关键术语 (Key terms)

limit (极限)

continuity (连续性)
one-sided limits (单侧极限)
right (left)-hand limit (右 (左) 极限)
neighborhood (邻域)
limit of a sequence (数列极限)
differentiability (可微性)
derivative (导数)
integration (积分)
Taylor polynomials (泰勒多项式)
remainder term (余项)
Maclaurin polynomial (麦克劳林多项式)
infinite series (无穷级数)
error (误差)
significant digits (有效数字)
model error (模型误差)
truncation error (截断误差)
observation error (观测误差)
round-off error (舍入误差)
machine numbers (机器数)
absolute error (绝对误差)
relative error (相对误差)

1.7 延伸阅读 (Extending reading)

1.7.1 背景知识

1. 微积分的基本概念及其应用是在 17 世纪末到 18 世纪初发展起来的，但是极限和连续性的数学概念直到 19 世纪后半叶，才由奥古斯丁 • 路易斯 • 柯西 (Augustin Louis Cauchy)(1789~1857)、海因里希 • 爱德华 • 海涅 (Heinrich Edward Heine) (1821~1881)和卡尔 • 魏尔斯特拉斯 (Karl Weierstrass) (1815~1897) 给出精确的描述.

2. 如果某个函数在某一点可微，那么函数在该点是连续的. 这一定理是由米歇尔 • 罗尔 (Michel Rolle)(1652~1719)提出的，出现在 1691 年的一篇论文中，题目是 *Méthode pour résoundre les égalites*. 最初罗尔批评由艾萨克 • 牛顿 (Isaac Newton) (1643~1727)和戈特弗里德·莱布尼茨 (Gottfried Wilhelm Leibniz) (1646~1716)发展的微积分，但后来成为它的支持者之一.

3. 符号数学算法和系统设计的研究工作始于 20 世纪 60 年代. 第一个投入使用的系统是 20 世纪 70 年代的一个基于 lisp 的系统，叫做 MACSYMA.

4. 乔治 • 弗里德里希 • 伯恩哈德 • 黎曼 (Georg Friedrich Bernhard Riemann)

(1826~1866)对可积函数进行分类，取得了许多重要的发现. 他还在几何和复函数理论方面做了基础工作，被认为是 19 世纪最杰出的数学家之一.

5. 科林·麦克劳林 (Colin Maclaurin) (1698~1746) 以牛顿微积分的捍卫者而闻名，当时牛顿微积分受到了爱尔兰哲学家乔治·伯克利 (George Berkeley) (1685~1753) 的猛烈抨击.

1.7.2 数学家传记：泰勒 (Taylor)

布鲁克·泰勒 (Brook Taylor，1685 年 8 月 18 日~1731 年 12 月 29 日) 出生于英格兰埃德蒙顿 (前米德尔塞克斯)，逝世于伦敦，是一名数学家，主要以泰勒公式和泰勒级数出名.

1701 年，布鲁克·泰勒进入剑桥大学圣约翰学院，1709 年获得法学学士，1714 年获法学博士学位. 1715 年他发表的 *Methodus Incrementorum Directa et Inversa* (《正反增量方法》)为高等数学添加了一个新的分支，今天这个方法被称为有限差分方法. 在同一著作中他还提出了著名的泰勒公式. 直到 1772 年，约瑟夫·拉格朗日才认识到这个公式的重要性，并称之为微分学基本定理，但泰勒于证明当中并没有考虑级数的收敛性，因而证明不严谨，这项工作直至 19 世纪 20 年代才由柯西完成. 泰勒定理开创了有限差分理论，使任何单变量函数都可展成幂级数，同时使泰勒成了有限差分理论的奠基者.

1.7.3 数学家传记：黎曼 (Riemann)

乔治·弗里德里希·伯恩哈德·黎曼(Georg Friedrich Bernhard Riemann，1826 年 9 月 17 日~1866 年 7 月 20 日)，是德国著名的数学家，在数学分析和微分几何方面做出过重要贡献，他开创了黎曼几何，并且为后来爱因斯坦的广义相对论提供了数学基础. 1846 年，黎曼进入哥廷根大学学习哲学和神学. 1847 年春，黎曼转到柏林大学，投入卡尔·古斯塔夫·雅各布·雅可比 (Carl Gustav Jacob Jacobi) (1804~1851)、约翰·彼得·古斯塔夫·勒热纳·狄利克雷 (Johann Peter Gustav Lejeune Dirichlet) (1805~1859)门下，两年后他回到哥廷根. 1851 年，黎曼在哥廷根大学获博士学位，并论证了复变函数可导的充分必要条件(即柯西-黎曼方程)，借助狄利克雷原理阐述了黎曼映射定理，奠定了函数的几何理论基础. 1853 年，他定义了黎曼积分并研究了三角级数收敛的准则. 1854 年，他发展了高斯关于曲面的微分几何理论，提出用流形的概念理解空间的实质，用微分弧长的平方所确定的正定二次型理解度量，建立了黎曼空间的概念. 1854 年，黎曼在哥廷根大学做了题为“论作为几何学基础的假设”的演讲，创立了黎曼几何学. 黎曼将曲面本身看成一个独立的几何实体，而不是仅仅把它看作欧几里得空间中的一个几何实体. 1915 年，爱因斯坦运用黎曼几何和张量分析工具创立了新的引力理论——广义相对论. 1857 年，黎曼引出黎曼曲面的概念，将阿贝尔积分与阿贝尔函数的理论带到新的转折点，其中对黎曼曲面从拓扑、分析、代数几何各角度做了深入研究，创造了一系列对代数拓扑发展影响深远的概念. 1857 年，他升为哥廷根大学的编外教授. 1859 年，他接替狄利克雷成为教授，并发表论文《论小于某给定值的素数的个数》，研究了黎曼 ζ 函数，给出了 ζ 函数的积分表示和它满足的函数方程，指出素数的分布与黎曼 ζ 函数之间存在深刻的联系，提出了著名的黎曼猜想. 黎曼猜想是大卫 • 希尔伯特 (David Hilbert) (1862~1943)在 1900 年提出的 23 个问题中的第 8 个问题，后来被列为千禧年七大难题之一. 黎曼猜想可以表述为：黎曼 $\zeta(s)$函数的所有非平凡零点都位于复平面内 $\mathrm{Re}(s)=1/2$ 这条临界线上. 2018 年 9 月，英国数学家迈克尔 • 弗朗西斯 • 阿蒂亚 (Michael Francis Atiyah) (1929~2019)声称证明了黎曼猜想.

黎曼的系列工作直接影响了 19 世纪后半叶的数学发展，许多杰出的数学家重新论证黎曼断言过的定理，在黎曼思想的影响下，许多数学分支取得了辉煌成就. 首先黎曼提出用复变函数论特别是用 ζ 函数研究数论的新思想和新方法，开创了解析数论的新时期，并对单复变函数论的发展有深刻的影响. 他是世界数学史上最具独创精神的数学家之一，黎曼的著作不多，但却异常深刻，极富对概念的创造与想象. 他的名字出现在黎曼 ζ 函数、黎曼积分、黎曼引理、黎曼流形、黎曼空间、黎曼映照定理、黎曼-希尔伯特问题、柯西-黎曼方程、黎曼思路回环矩阵中.

1.8 习题 (Exercises)

1. Use the Intermediate Value Theorem and Rolle's Theorem to show that the graph of $f(x)=x^3+2x+k$ crosses the x- axis exactly once, regardless of the value of the constant k.
2. Suppose $f\in C[a,b]$ and $f'(x)$ exists on (a,b). Show that if $f'(x)\neq 0$ for all x in (a,b), then there can exist at most one number p in $[a,b]$ with $f(p)=0$.
3. Let $f(x)=x^3$.

   a. Find the second Taylor polynomials $P_2(x)$ about $x_0=0$.

   b. Find $R_2(0.5)$ and the actual error in using $P_2(0.5)$ to approximate $f(0.5)$.

   c. Repeat part (a) using $x_0=1$.

   d. Repeat part (b) using the polynomial from part (c).
4. Find the third Taylor polynomials $P_3(x)$ for the function $f(x)=\sqrt{x+1}$ about $x_0=0$. Approximate $\sqrt{0.5},\sqrt{0.75},\sqrt{1.25}$ and $\sqrt{1.5}$ using $P_3(x)$, and find the actual errors.
5. Find the second Taylor polynomials $P_2(x)$ for the function $f(x)=\mathrm{e}^x\cos x$ about $x_0=0$.

   a. Use $P_2(0.5)$ to approximate $f(0.5)$. Find an upper bound for error $|f(0.5)-P_2(0.5)|$ using the error formula, and compare it to the actual error.

   b. Find a bound for the error $|f(x)-P_2(x)|$ in using $P_2(x)$ to approximate $f(x)$ on the interval $[0,1]$.

   c. Approximate $\int_0^1 f(x)\,\mathrm{d}x$ using $\int_0^1 P_2(x)\,\mathrm{d}x$.

   d. Find an upper bound for the error in (c) using $\int_0^1 |R_2(x)|\,\mathrm{d}x$, and compare the bound to the actual error.
6. Let $f(x)=2x\cos(2x)-(x-2)^2$ and $x_0=0$.

   a. Find the third Taylor polynomials $P_3(x)$, and use it to approximate $f(0.4)$.

   b. Use the error formula in Taylor's Theorem to find an upper bound for the error $|f(0.4)-P_3(0.4)|$. Compute the actual error.

   c. Find the fourth Taylor polynomials $P_4(x)$, and use it to approximate $f(0.4)$.

   d. Use the error formula in Taylor's Theorem to find an upper bound for the error $|f(0.4)-P_3(0.4)|$. Compute the actual error.
7. Compute the absolute error and relative error in approximations of p by p^*.

   a. $p=\pi, p^*=22/7$

   b. $p=\pi, p^*=3.1416$

c. $p=\mathrm{e}, p^*=2.718$

d. $p=\sqrt{2}, p^*=1.414$

e. $p=\mathrm{e}^{10}, p^*=22000$

f. $p=10^{\pi}, p^*=1400$

g. $p=8!, p^*=39900$

h. $p=9!, p^*=\sqrt{18\pi}(9/\mathrm{e})^9$

8. Find the largest interval in which p^* must lie to approximate p with relative error at most 10^{-4} for each value of p.

a. π

b. e

c. $\sqrt{2}$

d. $\sqrt[3]{7}$

9. Suppose p^* must approximate p with relative error at most 10^{-3}. Find the largest interval in which p^* must lie for each value of p.

a. 150

b. 900

c. 1500

d. 90

10. The number e can be defined by $\mathrm{e}=\sum_{n=0}^{\infty}(1/n!)$, where $n!=n(n-1)\cdots 2\cdot 1$ for $n\neq 0$ and $0!=1$. Compute the absolute error and relative error in the following approximations of e:

a. $\sum_{n=0}^{5}\frac{1}{n!}$

b. $\sum_{n=0}^{10}\frac{1}{n!}$

11. Show that the following equations have at least one solution in the given intervals.

a. $x\cos x-2x^2+3x-1=0$, $[0.2,0.3]$ and $[1.2,1.3]$

b. $(x-2)^2-\ln x$, $[1,2]$ and $[\mathrm{e},4]$

c. $2x\cos(2x)-(x-2)^2=0$, $[2,3]$ and $[3,4]$

d. $x-(\ln x)^x=0$, $[4,5]$

12. Find intervals containing solutions to the following equations.

a. $x-3^{-x}=0$

b. $4x^2-\mathrm{e}^x=0$

c. $x^3-2x^2-4x+2=0$

d. $x^3 + 4.001x^2 + 4.002x + 1.101 = 0$

13. Show that $f'(x)$ is 0 at least once in the given intervals.

a. $f(x) = 1 - e^x + (e-1)\sin\left(\frac{\pi}{2}x\right)$, $[0,1]$

b. $f(x) = (x-1)\tan x + x\sin \pi x$, $[0,1]$

c. $f(x) = x\sin \pi x - (x-2)\ln x$, $[1,2]$

d. $f(x) = (x-2)\sin x \ln(x+2)$, $[-1,3]$

Chapter 2 Direct Methods for Solving Linear Systems (解线性方程组的直接法)

本章内容 (Contents)

2.1 Gauss elimination method (Gauss 消元法)
2.2 Pivoting strategies (选主元策略)
2.3 Matrix factorization (矩阵分解法)
2.4 Special types of matrices (特殊形式矩阵的三角分解)

学习目标 (Learning Objectives)

- 掌握消元法的基本思想
- 掌握矩阵三角分解法
- 了解选主元策略
- 掌握 Gauss 列主元消元法
- 掌握 Gauss 按比例消元法
- 掌握 Gauss 全主元消元法
- 熟练掌握 LU 分解 (Doolittle 分解、Crout 分解)
- 熟悉 GG^T 分解 (Cholesky 分解或平方根法)
- 了解 LDL^T 分解(改进的平方根法)

2.1 Gauss elimination method (Gauss 消元法)

- Some preliminaries (预备知识)
- Gauss elimination with backward-substitution process (可回代的 Gauss 消元法)

2.1.1 Some preliminaries (预备知识)

Let linear system of equations of order n (n 阶线性方程组) be as follows

$$\begin{cases} a_{11}x_1 + a_{12}x_2 + \cdots + a_{1n}x_n = b_1, \\ a_{21}x_1 + a_{22}x_2 + \cdots + a_{2n}x_n = b_2, \\ \qquad \vdots \\ a_{n1}x_1 + a_{n2}x_2 + \cdots + a_{nn}x_n = b_n. \end{cases} \tag{2.1.1}$$

For $x_1, x_2, \cdots, x_n$ given the constants a_{ij}, for each $i, j = 1, 2, \cdots, n$, and b_i for each $i = 1, 2, \cdots, n$, Eq. (2.1.1) can be rewritten into a matrix form

$$\boldsymbol{Ax} = \boldsymbol{b}, \tag{2.1.2}$$

where

$$\boldsymbol{A} = \left(a_{ij}\right)_{n\times n} = \begin{bmatrix} a_{11} & a_{12} & \cdots & a_{1n} \\ a_{21} & a_{22} & \cdots & a_{2n} \\ \vdots & \vdots & & \vdots \\ a_{n1} & a_{n2} & \cdots & a_{nn} \end{bmatrix}, \quad \boldsymbol{x} = \begin{bmatrix} x_1 \\ x_2 \\ \vdots \\ x_n \end{bmatrix}, \quad \boldsymbol{b} = \begin{bmatrix} b_1 \\ b_2 \\ \vdots \\ b_n \end{bmatrix}, \tag{2.1.3}$$

and $\boldsymbol{A}$ is called the coefficient matrix (系数矩阵). By Cramer's rule (Cramer 法则), the solution to the linear system of equations exists and is unique while $\det \boldsymbol{A} \neq 0$. Throughout chapter 2, we assume that $\det \boldsymbol{A} \neq 0$ or $|\boldsymbol{A}| \neq 0$.

In this chapter, direct techniques are considered to solve the linear system of equations $\boldsymbol{Ax} = \boldsymbol{b}$. The basic idea of the direct method (直接法) is to equivalently transform a complex system of equations into a simple form that is easy to solve, and then the simple system is solved to obtain the solution of the original system of equations, i.e.

$$\boldsymbol{Ax} = \boldsymbol{b} \rightleftarrows \boldsymbol{Gx} = \boldsymbol{c},$$

where the coefficient matrix $\boldsymbol{G}$ is usually a triangular matrix or some simple matrix. The direct method can find the exact solution of the system of equations by a finite number of arithmetic operations, if there are no rounding errors (舍入误差) in the calculation process; however, in practice, due to the existence and influence of rounding errors, this method can only find the approximate solution of the linear system of equations. In this chapter several common direct methods for solving linear systems of equations are introduced, including the Gauss elimination method (Gauss 消元法) and the matrix triangular decomposition method (矩阵三角分解法).

The Gauss elimination method is to transform a linear system of equations into an equivalent system of upper triangular equations (上三角方程组) using elementary linear transformations (初等线性变换) and then back-substitute (回代) to find the solution to the system of triangular equations.

For the sake of convenience, we use the following three operations to simplify the linear system given in Eq. (2.1.1):

1. Equation E_i can be multiplied by any nonzero constant λ with the resulting equation used in place of E_i. This operation is denoted $(\lambda E_i) \to (E_i)$.

2. Equation E_j can be multiplied by any constant λ and added to equation E_i with the resulting equation used in place of E_i. This operation is denoted $\left(E_i+\lambda E_j\right)\to\left(E_i\right)$.

3. Equations E_i and E_j can be transposed in order. This operation is denoted $\left(E_i\right)\leftrightarrow\left(E_j\right)$.

2.1.2 Gauss elimination with backward-substitution process(可回代的 Gauss 消元法)

An example of the Gauss elimination method is given below to illustrate the main process of solving a linear system of equations.

Example 2.1 Suppose we are given the linear system

$$\begin{cases} E_1: x_1+4x_2+7x_3=1, \\ E_2: 2x_1+5x_2+8x_3=1, \\ E_3: 3x_1+6x_2+11x_3=1. \end{cases}$$

Then the Gauss elimination process to solve the linear system proceeds as follows.

Step 1 Gauss elimination

Firstly, the operations corresponding to $\left(E_2-2E_1\right)\to\left(E_2\right)$, $\left(E_3-3E_1\right)\to\left(E_3\right)$ are performed to eliminate the coefficient of x_1 in E_2, E_3, which results in the new system

$$\begin{cases} E_1: x_1 & +4x_2 & +7x_3 & =1, \\ E_2: & -3x_2 & -6x_3 & =-1, \\ E_3: & -6x_2 & -10x_3 & =-2. \end{cases}$$

Similarly, the operation corresponding to $\left(E_3-2E_2\right)\to\left(E_3\right)$ is performed to eliminate the coefficient of x_2 in E_3, which gives

$$\begin{cases} E_1: x_1 & +4x_2 & +7x_3 & =1, \\ E_2: & -3x_2 & -6x_3 & =-1, \\ E_3: & & 2x_3 & =0, \end{cases} \tag{2.1.4}$$

where the coefficient matrix of system (2.1.4) is an upper triangular matrix.

We also can write the above process into a matrix form as follows. The augmented matrix (增广矩阵) of the linear systems is

$$\tilde{A}=[A, b]=\begin{bmatrix} 1 & 4 & 7 & \vdots & 1 \\ 2 & 5 & 8 & \vdots & 1 \\ 3 & 6 & 11 & \vdots & 1 \end{bmatrix}.$$

Performing the operations to eliminate the coefficient of x_i in each of these rows results in the new matrices:

$$\tilde{A}=\begin{bmatrix} 1 & 4 & 7 & \vdots & 1 \\ 2 & 5 & 8 & \vdots & 1 \\ 3 & 6 & 11 & \vdots & 1 \end{bmatrix} \xrightarrow[-3r_1+r_3\to r_3]{-2r_1+r_2\to r_2} \begin{bmatrix} 1 & 4 & 7 & \vdots & 1 \\ 0 & -3 & -6 & \vdots & -1 \\ 0 & -6 & -10 & \vdots & -2 \end{bmatrix}$$

$$\xrightarrow{-2r_2+r_3\to r_3}\begin{bmatrix}1 & 4 & 7 & \vdots & 1\\ 0 & -3 & -6 & \vdots & -1\\ 0 & 0 & 2 & \vdots & 0\end{bmatrix}.$$

Step 2 Backward substitution

By the system of equations (2.1.4), E_3 implies $x_3=0$. Substituting the $x_3=0$ into E_2 obtains

$$x_2=-\frac{1}{3}(-1+6x_3)=\frac{1}{3}.$$

Similarly, we can solve E_1 for x_1 by substituting the $x_3=0, x_2=1/3$, then

$$x_1=1-4x_2-7x_3=-\frac{1}{3}.$$

Hence the solution of the original linear system of equations is

$$x_1=-\frac{1}{3},\ x_2=\frac{1}{3},\ x_3=0.$$

In general, using the Gauss elimination method to solve the linear system of equations (2.1.1) consists of two processes: elimination and backward substitution.

1. Gauss elimination (高斯消元)

We denote $\boldsymbol{A}=\boldsymbol{A}^{(0)}=\left(a_{ij}^{(0)}\right)_{n\times n}$, $\boldsymbol{b}=\boldsymbol{b}^{(0)}=\left(a_{1,n+1}^{(0)},a_{2,n+1}^{(0)},\cdots,a_{n,n+1}^{(0)}\right)^{\mathrm{T}}$, and then the system (2.1.1) can be written as

$$\begin{cases}a_{11}^{(0)}x_1+a_{12}^{(0)}x_2+\cdots+a_{1n}^{(0)}x_n=a_{1,n+1}^{(0)},\\ a_{21}^{(0)}x_1+a_{22}^{(0)}x_2+\cdots+a_{2n}^{(0)}x_n=a_{2,n+1}^{(0)},\\ \qquad\vdots\\ a_{n1}^{(0)}x_1+a_{n2}^{(0)}x_2+\cdots+a_{nn}^{(0)}x_n=a_{n,n+1}^{(0)}.\end{cases}\tag{2.1.5}$$

(1) Suppose $a_{11}^{(0)}\neq 0$; otherwise, this can be done by interchanging rows. Let $l_{i1}=a_{i1}^{(0)}/a_{11}^{(0)}$, $i=2,3,\cdots,n$. The operations corresponding to $(E_i-l_{i1}E_1)\to(E_i)$ for $i=2,3,\cdots,n$ are performed to eliminate the coefficient of x_i in each of these rows, which results in the first new system

$$\begin{cases}a_{11}^{(0)}x_1+a_{12}^{(0)}x_2+\cdots+a_{1n}^{(0)}x_n=a_{1,n+1}^{(0)},\\ \qquad a_{22}^{(1)}x_2+\cdots+a_{2n}^{(1)}x_n=a_{2,n+1}^{(1)},\\ \qquad\qquad\vdots\\ \qquad a_{n2}^{(1)}x_2+\cdots+a_{nn}^{(1)}x_n=a_{n,n+1}^{(1)},\end{cases}\tag{2.1.6}$$

where $a_{ij}^{(1)}=a_{ij}^{(0)}-l_{i1}a_{1j}^{(0)}$, $i=2,3,\cdots,n$; $j=2,3,\cdots,n+1$.

(2) Suppose $a_{22}^{(1)} \neq 0$, and let $l_{i2} = a_{i2}^{(1)} / a_{22}^{(1)}$, $i = 3,4,\cdots,n$. The operations corresponding to $(E_i - l_{i2}E_2) \to (E_i)$ for $i = 3,4,\cdots,n$ are performed to eliminate the coefficient of x_i in each of these rows, which results in the second new system

$$\begin{cases} a_{11}^{(0)}x_1 + a_{12}^{(0)}x_2 + a_{13}^{(0)}x_3 + \cdots + a_{1n}^{(0)}x_n = a_{1,n+1}^{(0)}, \\ \qquad a_{22}^{(1)}x_2 + a_{23}^{(1)}x_3 + \cdots + a_{2n}^{(1)}x_n = a_{2,n+1}^{(1)}, \\ \qquad\qquad a_{33}^{(2)}x_3 + \cdots + a_{3n}^{(2)}x_n = a_{3,n+1}^{(2)}, \\ \qquad\qquad\qquad \vdots \\ \qquad\qquad a_{n3}^{(2)}x_3 + \cdots + a_{nn}^{(2)}x_n = a_{n,n+1}^{(2)}, \end{cases} \tag{2.1.7}$$

where $a_{ij}^{(2)} = a_{ij}^{(1)} - l_{i2}a_{2j}^{(1)}, i = 3, 4, \cdots, n; \ j = 3, 4, \cdots, n+1$.

(3) Suppose $a_{kk}^{(k-1)} \neq 0$, and let $l_{ik} = a_{ik}^{(k-1)} \big/ a_{kk}^{(k-1)}$, $i = k+1,\ k+2,\ \cdots,\ n$. The operations corresponding to $(E_i - l_{ik}E_k) \to (E_i)$ for $i = k+1, k+2, \cdots, n$ are performed to eliminate the coefficient of x_i in each of these rows below the kth for $k = 3, 4, \cdots, n$, which results in the kth new system

$$\begin{cases} a_{11}^{(0)}x_1 + a_{12}^{(0)}x_2 + \cdots + a_{1,k+1}^{(0)}x_{k+1} + \cdots + a_{1n}^{(0)}x_n = a_{1,n+1}^{(0)}, \\ \qquad a_{22}^{(1)}x_2 + \cdots + a_{2,k+1}^{(0)}x_{k+1} + \cdots + a_{2n}^{(1)}x_n = a_{2,n+1}^{(1)}, \\ \qquad\qquad\qquad \vdots \\ \qquad\qquad a_{k+1,k+1}^{(k)}x_{k+1} + \cdots + a_{k+1,n}^{(k)}x_n = a_{k+1,n+1}^{(k)}, \\ \qquad\qquad\qquad \vdots \\ \qquad\qquad a_{n,k+1}^{(k)}x_{k+1} + \cdots + a_{nn}^{(k)}x_n = a_{n,n+1}^{(k)}, \end{cases} \tag{2.1.8}$$

where $a_{ij}^{(k)} = a_{ij}^{(k-1)} - l_{ik}a_{kj}^{(k-1)}$, $i = k+1, \cdots, n; \ j = k+1, \cdots, n+1$.

Keeping on the above processes $n-1$ times produces the $(n-1)$th new system

$$\begin{cases} a_{11}^{(0)}x_1 + a_{12}^{(0)}x_2 + a_{13}^{(0)}x_3 + \cdots + a_{1n}^{(0)}x_n = a_{1,n+1}^{(0)}, \\ \qquad a_{22}^{(1)}x_2 + a_{23}^{(1)}x_3 + \cdots + a_{2n}^{(1)}x_n = a_{2,n+1}^{(1)}, \\ \qquad\qquad a_{33}^{(2)}x_3 + \cdots + a_{3n}^{(2)}x_n = a_{3,n+1}^{(2)}, \\ \qquad\qquad\qquad \vdots \\ \qquad\qquad\qquad a_{nn}^{(n-1)}x_n = a_{n,n+1}^{(n-1)}, \end{cases} \tag{2.1.9}$$

where for $k = 1, 2, \cdots, n-1; \ i = k+1, \cdots, n; \ j = k+1, \cdots, n+1$,

$$l_{ik} = a_{ik}^{(k-1)} \big/ a_{kk}^{(k-1)}, \quad a_{ij}^{(k)} = a_{ij}^{(k-1)} - l_{ik}a_{kj}^{(k-1)}. \tag{2.1.10}$$

The procedure involved in is called the **Gauss Elimination Process**, where entries $a_{kk}^{(k-1)} \neq 0, k = 1,2,\cdots,n$ are called the pivot elements (主元素), and $l_{ik}\ (1 \leqslant k < i \leqslant n)$ are called multipliers (乘数). Thus, by the Gauss Elimination, the original linear system (2.1.1) is transformed into the equivalent upper triangular linear system (2.1.9).

2. Backward-substitution process (回代过程)

Backward-substitution process is used to solve the solution of the above upper triangular linear system (2.1.9). Supposing $a_{nn}^{(n-1)} \neq 0$，the operations corresponding to E_n (2.1.9) gives

$$x_n = a_{n,n+1}^{(n-1)} / a_{n,n}^{(n-1)},$$

and from $E_k, k = n-1, \cdots, 1$ one derives

$$x_i = \frac{1}{a_{ii}^{(i-1)}}\left(a_{i,n+1}^{(i-1)} - \sum_{j=i+1}^{n} a_{ij}^{(i-1)} x_j\right), \quad i = n-1, n-2, \cdots, 1 \tag{2.1.11}$$

for each $k = n-1, \cdots, 1$.

Definition 2.1 The method to solve the solution of linear system of equations by Eq. (2.1.5)~(2.1.11) is called Gauss elimination method (Gauss 消元法)，including two processes elimination (消元) and backward substitution (回代).

Consider the amount of arithmetic operations in Gauss elimination method.

The involved operations for triangularization in Gauss elimination are as following: multiplications/divisions (乘法/除法)

$$\sum_{k=1}^{n-1}(n-k)(n-k+1) + \sum_{k=1}^{n-1}(n-k) = \frac{2n^3+3n^2-5n}{6},$$

additions/subtractions (加法/减法)

$$\sum_{k=1}^{n-1}(n-k)(n-k+1) = \frac{n^3-n}{3}.$$

During the process for backward substitution, the operations are counted as follows: multiplications/divisions

$$\sum_{k=1}^{n}(n-k+1) = \frac{n(n+1)}{2},$$

additions/subtractions

$$\sum_{k=1}^{n}(n-k) = \frac{n(n-1)}{2}.$$

Hence the total number of arithmetic operations in Gauss elimination is as follows: multiplications/divisions

$$\frac{n(n-1)(2n+5)}{6} + \frac{n(n+1)}{2} = \frac{n^3+3n^2-n}{3},$$

additions/subtractions

$$\frac{n^3-n}{3}+\frac{n(n-1)}{2}=\frac{2n^3+3n^2-5n}{6}.$$

2.2 Pivoting strategies (选主元策略)

- Partial pivoting (maximal column pivoting) (最大列主元)
- Scaled partial pivoting (scaled-column pivoting) (按比例列主元)

In the Gauss elimination method, the assumption of the pivot elements $a_{kk}^{(k-1)} \neq 0$ is crucial for implementing the algorithm. However, it will increase round-off error even if $a_{kk}^{(k-1)} \neq 0$, but it is very small. If $a_{kk}^{(k-1)}$ is small in magnitude compared with $a_{ik}^{(k-1)}\left(k+1 \leqslant i \leqslant n\right)$, then the magnitude of the multiplier $l_{ik}=a_{ik}^{(k-1)}/a_{kk}^{(k-1)}$ will be much larger than 1.

Since

$$a_{ij}^{(k)}=a_{ij}^{(k-1)}-l_{ik}a_{kj}^{(k-1)}, \quad i=k+1,\cdots,n,\ j=k+1,\cdots,n+1,$$

round-off error introduced in the computation of one of the terms $a_{kj}^{(k-1)}$ is multiplied by l_{ik} when computing $a_{ij}^{(k)}$, which compounds the original error. Also, when performing the backward substitution

$$x_i=\frac{1}{a_{ii}^{(i-1)}}(a_{i,n+1}^{(i-1)}-\sum_{j=i+1}^{n}a_{ij}^{(i-1)}x_j), \quad i=n-1,n-2,\cdots,1$$

a small value of $a_{ii}^{(i-1)}$ can dramatically increase any error in the numerator because of the division by $a_{ii}^{(i-1)}$.

Example 2.2 Use Gauss elimination with backward substitution and four-digit rounding arithmetic to solve the following linear system

$$\begin{cases}0.0020x_1+52.88x_2=52.90,\\ 4.573x_1\ \ -7.29x_2\ \ =38.44.\end{cases}$$

Solution. The exact solution is $x_1^*=10.00, x_2^*=1.000$.

The first pivot element is $a_{11}=0.002000$, and its associated multiplier $l_{21}=\dfrac{4.573}{a_{11}}=2286.5$ rounds to the large number $l_{21}=2286.5\approx 2287$. Performing $(E_2-l_{21}E_1)\to(E_2)$ and the approximate rounding gives

$$\begin{cases}0.0020x_1+52.88x_2=52.90,\\ \qquad\quad -120900x_2=-121000,\end{cases}$$

instead of the precise values

$$\begin{cases} 0.0020x_1 + 52.88x_2 = 52.90, \\ \qquad -120917.41x_2 = -120917.41. \end{cases}$$

Backward substitution yields $x_2 \approx 1.001$, which is a close approximation to the actual value $x_2^* = 1.000$, and

$$x_1 = \frac{52.90 - 52.88 \times 1.001}{0.002} \approx \frac{52.90 - 52.93}{0.002} = -15.00,$$

this ruins the approximation to the actual value $x_1^* = 10.00$, since the small error 0.001 in x_2 is now multiplied by $\frac{52.88}{0.002} = 26440$ which is generated from the small pivot $a_{11} = 0.002000$.

To reduce round-off error (舍入误差), it is often necessary to perform row interchanges even when the pivot elements are not zero.

To avoid the problem such as that arising in Example 2.2, pivoting strategies (选主元策略) should be performed by selecting a larger element $a_{pk}^{(k-1)}$ for the pivot element, including partial pivoting (maximal column pivoting, 最大列主元) and scaled partial pivoting (scaled-column pivoting, 按比例列主元).

2.2.1 Partial pivoting (maximal column pivoting) (最大列主元)

The simplest pivoting strategy is to select an element for the pivot in the same column that is below the diagonal and has the largest absolute value. Specifically, we determine the smallest $p \geqslant k$ such that $\left|a_{pk}^{(k-1)}\right| = \max_{k \leqslant i \leqslant n} \left|a_{ik}^{(k-1)}\right|$, and perform $E_p \leftrightarrow E_k$ to transform matrix into a linear system of upper triangular equations with Gauss elimination, which is called Gauss elimination with partial pivoting (Gauss 列主元消元法). In this case no interchanging of columns is used.

Example 2.3 Reconsider the system in Example 2.2

$$\begin{cases} 0.0020x_1 + 52.88x_2 = 52.90, \\ 4.573x_1 \quad - 7.29x_2 \quad = 38.44. \end{cases}$$

Solution. The pivoting procedure just described results in first finding

$$\max\{a_{11}, a_{21}\} = \max\{0.002, 4.573\} = 4.573.$$

Then a_{21} is taken as the pivot element, and the operation $(E_1 \leftrightarrow E_2)$ is then performed to give the system

$$\begin{cases} 4.573x_1 \quad - 7.29x_2 = 38.44, \\ 0.0020x_1 + 52.88x_2 = 52.90. \end{cases}$$

The multiplier for this system is $l_{21} = \frac{0.002000}{4.573} \approx 0.0004373$, and the operation $(E_2 - l_{21}E_1 \to E_2)$ reduces the system to

$$\begin{cases}4.573x_1 - 7.29x_2 = 38.44, \\ \qquad\qquad 52.88x_2 = 52.88.\end{cases}$$

The four-digit answers resulting from the backward substitution are the correct values $x_2 = 1.000, x_1 = 10.00$.

2.2.2 Scaled partial pivoting (scaled-column pivoting) (按比例列主元)

Each multiplier l_{ij} in the partial pivoting algorithm has a magnitude less than or equal to 1. Although this strategy is sufficient for most linear systems, it is not always the case.

Example 2.4 Consider the linear system

$$\begin{cases}20.000x_1 + 528800x_2 = 529000, \\ 4.573x_1 \quad -7.290x_2 \quad = 38.44,\end{cases}$$

which is the same as that in Examples 2.2 and 2.3 except that all the entries in the first equation have been multiplied by 10^4.

Using partial pivoting, the maximal value in the first column is

$$\max\left\{a_{11}, a_{21}\right\} = \max\{20.00, 4.5731\} = 20.00 = a_{11},$$

and the multiplier $l_{21} = \dfrac{4.573}{20.00} \approx 0.2287$ leads to the system

$$\begin{cases}20.00x_1 + 528800x_2 = 529000, \\ \qquad\qquad -120900x_2 = -121000,\end{cases}$$

which has the same inaccurate solutions as in Example 2.2

$$x_2 \approx 1.001,\ x_1 \approx -15.00.$$

Scaled partial pivoting, also called scaled-column pivoting, is approximate for the system in Example 2.4. It places the element in the pivot position that is the largest relative to the entries in its row. The first step in this procedure is to define a scale factor s_i for each row as

$$s_i = \max_{1 \leqslant j \leqslant n} \left|a_{ij}\right|.$$

If, for some i, we have $s_i = 0$, then the system has no unique solution since all entries in the i th row are 0. Assuming that this is not the case, the approximate row interchanging to place zeros in the first column is determined by choosing the least integer $p\left(p \geqslant 1\right)$ with

$$\frac{a_{p1}^{(0)}}{s_p} = \max_{1 \leqslant k \leqslant n} \frac{a_{k1}^{(0)}}{s_k},$$

and performing $E_1 \leftrightarrow E_p$.

The effect of scaling is to ensure that the largest element in each row has a relative magnitude of 1 before the comparison for row interchanging is performed.

In a similar manner, before eliminating the variable x_i using the operations $\left(E_k - l_{ki}E_i \to E_k\right)$ for $k = i+1, \cdots, n$, we select the smallest integer $p\left(p \geqslant k\right)$ with

$$\frac{a_{pk}^{(k-1)}}{s_p}=\max_{1\leqslant i\leqslant n}\frac{a_{ik}^{(k-1)}}{s_i},$$

and perform the row interchange $(E_k\leftrightarrow E_p)$ if $p\neq k$.

The scale factors $s_1,s_2,\cdots,s_n$ are computed only once, at the start of the procedure, and must also be interchanged when row interchanges are performed.

Solution. Applying scaled partial pivoting to Example 2.4 gives

$$s_1=\max\{20.00,528800\}=528800,$$
$$s_2=\max\{4.573,|-7.290|\}=7.290.$$

Consequently,

$$\frac{a_{11}}{s_1}=\frac{20.00}{528800}\approx 0.3782\times 10^{-4},\quad \frac{a_{21}}{s_2}=\frac{4.573}{7.290}\approx 0.6273.$$

Since $0.6273>0.3782\times 10^{-4}$, a_{21} is taken as the pivot element, and the interchange $(E_1\leftrightarrow E_2)$ is made. Applying Gauss elimination to the new system

$$\begin{cases}4.573x_1-7.290x_2=38.44,\\20.00x_1+528800x_2=529000,\end{cases}$$

produces the correct results $x_1=10.00, x_2=1.000$.

The method performed as above process to solve Example 2.3 is called Gauss elimination with scaled partial pivoting (Gauss 按比例主元消元法).

Example 2.5 Solve the linear system using three-digit rounding arithmetic with scaled partial pivoting

$$\begin{cases}x_1+4x_2+7x_3=1,\\2x_1+5x_2+8x_3=1,\\3x_1+6x_2+11x_3=1.\end{cases}$$

Solution. Applying scaled partial pivoting

$$s_1=\max\{1,4,7\}=7,\quad s_2=\max\{2,5,8\}=8,\quad s_3=\max\{3,6,11\}=11,$$

and

$$\frac{a_{11}^{(0)}}{s_1}=\frac{1}{7},\quad \frac{a_{21}^{(0)}}{s_2}=\frac{1}{4},\quad \frac{a_{31}^{(0)}}{s_3}=\frac{3}{11}.$$

Since $\frac{3}{11}>\frac{1}{4}>\frac{1}{7}$, take $a_{31}^{(0)}$ as the first pivot element, and the row interchange $(E_1\leftrightarrow E_3)$ is made. Applying Gauss elimination to the new system

$$\begin{cases}3x_1+6x_2+11x_3=1,\\2x_1+5x_2+8x_3=1,\\x_1+4x_2+7x_3=1,\end{cases}$$

produces

$$\begin{cases} 3x_1 + 6x_2 + 11x_3 = 1, \\ x_2 + \dfrac{2}{3}x_3 = \dfrac{1}{3}, \\ 2x_2 + \dfrac{10}{3}x_3 = \dfrac{2}{3}. \end{cases}$$

Similarly, since

$$\frac{a_{22}^{(1)}}{s_2} = \frac{1}{8}, \quad \frac{a_{32}^{(1)}}{s_1} = \frac{2}{7}, \quad \frac{2}{7} > \frac{1}{8},$$

$a_{32}^{(1)}$ is chosen as the second pivot element, the row interchange $(E_2 \leftrightarrow E_3)$ is made to generate the new system

$$\begin{cases} 3x_1 + 6x_2 + 11x_3 = 1, \\ 2x_2 + \dfrac{10}{3}x_3 = \dfrac{2}{3}, \\ x_2 + \dfrac{2}{3}x_3 = \dfrac{1}{3}. \end{cases}$$

Applying Gauss elimination produces

$$\begin{cases} 3x_1 + 6x_2 + 11x_3 = 1, \\ 2x_2 + \dfrac{10}{3}x_3 = \dfrac{2}{3}, \\ -x_3 = 0. \end{cases}$$

Backward substitution gives $x_3 = 0$, $x_2 = \frac{1}{3}$, $x_1 = -\frac{1}{3}$.

2.3 Matrix factorization (矩阵分解法)

- Doolittle factorization (Doolittle 分解)
- Crout factorization (Crout 分解)
- Permutation matrix (置换矩阵)

Solving linear system $\boldsymbol{Ax} = \boldsymbol{b}$, if $\boldsymbol{A}$ has been factored into the triangular form $\boldsymbol{A} = \boldsymbol{LU}$, then we can solve for $\boldsymbol{x}$ more easily by using a two-step process, where $\boldsymbol{L}$ is a lower triangular and $\boldsymbol{U}$ is an upper triangular. First we let $\boldsymbol{Ux} = \boldsymbol{y}$, and solve the system $\boldsymbol{Ly} = \boldsymbol{b}$ for $\boldsymbol{y}$, and then solve the system $\boldsymbol{Ux} = \boldsymbol{y}$ for $\boldsymbol{x}$. Suppose that Gauss elimination can be performed on the system $\boldsymbol{Ax} = \boldsymbol{b}$ without row interchanges. With the introduced notation this is equivalent to having nonzero pivot elements $a_{ii}^{(i-1)}$ for each $i = 1, 2, \cdots, n$.

2.3.1 Doolittle factorization (Doolittle 分解)

Definition 2.2 Let $A \in \mathbf{R}^{n \times n}$. $A = LU$ is called Doolittle factorization or Doolittle decomposition of matrix A (Doolittle 分解), where L is a lower-triangular matrix and U is an upper-triangular matrix, and

$$L = \begin{bmatrix} 1 & & & & \\ l_{21} & 1 & & & \\ l_{31} & l_{32} & 1 & & \\ \vdots & \vdots & \vdots & \ddots & \\ l_{n1} & l_{n2} & \cdots & l_{n,n-1} & 1 \end{bmatrix}, \quad U = \begin{bmatrix} u_{11} & u_{12} & \cdots & u_{1n} \\ & u_{22} & \cdots & u_{2n} \\ & & \ddots & \vdots \\ & & & u_{nn} \end{bmatrix}.$$

In general, any given matrix does not always have an LU decomposition (LU 分解), and sufficient conditions for a matrix to have an LU decomposition are given below.

Theorem 2.1 For any given matrix $A \in \mathbf{R}^{n \times n}$ $(n \geqslant 2)$, the Doolittle factorization of matrix A is unique if each of its leading principal submatrices has a nonzero determinant (各阶顺序主子式非零).

Proof. We conduct the proof by induction on the order n of matrix. First we verify that the statement of the theorem is true for $n = 2$. Suppose matrix A is a 2×2 matrix of the form

$$A = \begin{pmatrix} a & b \\ c & d \end{pmatrix},$$

where $a \neq 0$, $\det A \neq 0$. Let

$$L = \begin{pmatrix} 1 & 0 \\ l & 1 \end{pmatrix}, \quad U = \begin{pmatrix} \alpha & u \\ 0 & \beta \end{pmatrix}.$$

Then

$$LU = \begin{pmatrix} \alpha & u \\ l\alpha & lu + \beta \end{pmatrix}.$$

Equating

$$\begin{pmatrix} \alpha & u \\ l\alpha & lu + \beta \end{pmatrix} = \begin{pmatrix} a & b \\ c & d \end{pmatrix}$$

yields

$$\alpha = a, \quad u = b, \quad l\alpha = c, \quad lu + \beta = d.$$

Since $a \neq 0$, $\det A \neq 0$, it follows

$$\alpha = a \neq 0, \ u = b, \ l = \frac{c}{a}, \ \beta = d - lu = d - \frac{cb}{a} = \frac{ad - cb}{a} = \frac{\det A}{a} \neq 0.$$

Therefore, A can be written as the product of a unit lower triangular matrix L of order 2 and an upper triangular matrix U of order 2.

Remark In the case of $n = 2$, one can prove that A can also be written as the product of a lower triangular matrix $\boldsymbol{L}$ of order 2 and a unit upper triangular matrix $\boldsymbol{U}$ of order 2. In fact, let

$$\boldsymbol{L}=\begin{pmatrix}\alpha & 0\\ l & \beta\end{pmatrix},\quad \boldsymbol{U}=\begin{pmatrix}1 & u\\ 0 & 1\end{pmatrix}.$$

Then

$$\boldsymbol{LU}=\begin{pmatrix}\alpha & \alpha u\\ l & lu+\beta\end{pmatrix}.$$

Equating

$$\begin{pmatrix}\alpha & \alpha u\\ l & lu+\beta\end{pmatrix}=\begin{pmatrix}a & b\\ c & d\end{pmatrix}$$

yields

$$\alpha=a,\quad \alpha u=b,\quad l=c,\quad lu+\beta=d.$$

Since $a\neq 0$, $\det \boldsymbol{A}\neq 0$, it follows

$$\alpha=a\neq 0,\quad u=\frac{b}{a},\quad l=c,\quad \beta=d-lu=d-\frac{cb}{a}=\frac{ad-cb}{a}=\frac{\det \boldsymbol{A}}{a}\neq 0.$$

Now assume that the statement of the theorem is true for $n=k$; that is, if $\boldsymbol{A}^{(k)}$ is a matrix of order k, and each of its leading principal submatrices has a nonzero determinant, then $\boldsymbol{A}^{(k)}=\boldsymbol{L}^{(k)}\boldsymbol{U}^{(k)}$, where $\boldsymbol{L}^{(k)}$ is a unit lower triangular matrix of order k, and $\boldsymbol{U}^{(k)}$ is an upper triangular matrix of order k.

Let matrix $\boldsymbol{A}$ be a $(k+1)\times(k+1)$ matrix with all its leading principal submatrices being nonsingular. Then $\boldsymbol{A}$ can be written as the following form

$$\boldsymbol{A}=\begin{pmatrix}\boldsymbol{A}^{(k)} & \boldsymbol{b}\\ \boldsymbol{c}^{\mathrm{T}} & d\end{pmatrix},$$

where $\boldsymbol{A}^{(k)}=\boldsymbol{L}^{(k)}\boldsymbol{U}^{(k)}$, with $\boldsymbol{L}^{(k)}$ being a unit lower triangular matrix of order k, and $\boldsymbol{U}^{(k)}$ an upper triangular matrix of order k, $\boldsymbol{b}$ and $\boldsymbol{c}$ are column vectors of size k, and d is a real number. Let

$$\boldsymbol{L}=\begin{pmatrix}\boldsymbol{L}^{(k)} & \boldsymbol{0}\\ \boldsymbol{l}^{\mathrm{T}} & 1\end{pmatrix},\quad \boldsymbol{U}=\begin{pmatrix}\boldsymbol{U}^{(k)} & \boldsymbol{u}\\ \boldsymbol{0}^{\mathrm{T}} & \alpha\end{pmatrix},$$

where $\boldsymbol{u}$, $\boldsymbol{l}$ and $\boldsymbol{0}$ are column vectors of size k, and α is a real number. Then

$$\boldsymbol{LU}=\begin{pmatrix}\boldsymbol{L}^{(k)}\boldsymbol{U}^{(k)} & \boldsymbol{L}^{(k)}\boldsymbol{u}\\ \boldsymbol{l}^{\mathrm{T}}\boldsymbol{U}^{(k)} & \boldsymbol{l}^{\mathrm{T}}\boldsymbol{u}+\alpha\end{pmatrix}.$$

Equating

$$\boldsymbol{LU}=\begin{pmatrix}\boldsymbol{L}^{(k)}\boldsymbol{U}^{(k)} & \boldsymbol{L}^{(k)}\boldsymbol{u}\\ \boldsymbol{l}^{\mathrm{T}}\boldsymbol{U}^{(k)} & \boldsymbol{l}^{\mathrm{T}}\boldsymbol{u}+\alpha\end{pmatrix}=\boldsymbol{A}=\begin{pmatrix}\boldsymbol{A}^{(k)} & \boldsymbol{b}\\ \boldsymbol{c}^{\mathrm{T}} & d\end{pmatrix}$$

produces

$$A^{(k)} = L^{(k)}U^{(k)},\ L^{(k)}u = b,\ l^{T}U^{(k)} = c^{T},\ l^{T}u + \alpha = d.$$

The first equality $A^{(k)} = L^{(k)}U^{(k)}$ is based on the hypothesis of induction, and A^k is nonsingular implies that both L^k and U^k are nonsingular.

From the second equality, it follows $u = (L^{(k)})^{-1}b$. The third equality implies $l^{T} = c^{T}(U^{(k)})^{-1}$. And solving the last equality gives

$$\alpha = d - l^{T}u = d - c^{T}(U^{(k)})^{-1}(L^{(k)})^{-1}b = d - c^{T}(L^{(k)}U^{(k)})^{-1}b = d - c^{T}(A^{(k)})^{-1}b .$$

Consequently, we proved that there exist column vectors u and l of size k, and real number α such that the unit lower triangular matrix L of order $k+1$ and the upper triangular matrix U of order $k+1$ of the forms

$$L = \begin{pmatrix} L^{(k)} & 0 \\ l^{T} & 1 \end{pmatrix},\quad U = \begin{pmatrix} U^{(k)} & u \\ 0^{T} & \alpha \end{pmatrix}$$

satisfy

$$A = \begin{pmatrix} A^{(k)} & b \\ c^{T} & d \end{pmatrix} = LU.$$

Therefore the proof of the theorem is completed by induction.

Theorem 2.2 If Gauss elimination can be performed on the linear system $Ax = b$ without row interchanges, then the matrix A can be factored into the product of a lower-triangular matrix L and an upper-triangular matrix U, that is, $A = LU$, where

$$L = \begin{bmatrix} 1 & & & & \\ l_{21} & 1 & & & \\ l_{31} & l_{32} & 1 & & \\ \vdots & \vdots & \vdots & \ddots & \\ l_{n1} & l_{n2} & \cdots & l_{n,n-1} & 1 \end{bmatrix},\quad U = \begin{bmatrix} a_{11}^{(0)} & a_{12}^{(0)} & a_{13}^{(0)} & \cdots & a_{1n}^{(0)} \\ & a_{22}^{(1)} & a_{23}^{(1)} & \cdots & a_{2n}^{(1)} \\ & & \ddots & \vdots & \vdots \\ & & & a_{n-1,n-1}^{(n-2)} & a_{n-1,n}^{(n-2)} \\ & & & & a_{nn}^{(n-1)} \end{bmatrix}, \tag{2.3.1}$$

and for $k = 1, 2, \cdots, n-1;\ i = k+1, \cdots, n;\ j = k+1, \cdots, n+1$,

$$l_{ik} = a_{ik}^{(k-1)} \big/ a_{kk}^{(k-1)},\quad a_{ij}^{(k)} = a_{ij}^{(k-1)} - l_{ik}a_{kj}^{(k-1)}.$$

Proof. The Doolittle decomposition of a matrix A can be obtained by Gauss elimination.

Suppose each leading principal determinants of A is nonzero.

The first step in the Gauss elimination process consists of performing the equations $(E_i - l_{i1}E_1 \to E_i)$ for each $i = 2,3,\cdots,n$, which is equivalent to the following matrix multiplication (all the following l_{ik} satisfy Eq.(2.1.10))

$$Ax = A^{(0)}x = b \ \Rightarrow\ A^{(1)}x = b^{(1)},\quad L_1A^{(0)} = A^{(1)}, \tag{2.3.2}$$

where

$$L_1 = \begin{bmatrix} 1 & & & & \\ -l_{21} & 1 & & & \\ -l_{31} & 0 & 1 & & \\ \vdots & \vdots & \vdots & \ddots & \\ -l_{n1} & 0 & \cdots & 0 & 1 \end{bmatrix}$$

is called the first Gauss transformation matrix.

The second step in the Gauss elimination process consists of performing the operations $(E_i - l_{i2}E_2 \to E_i)$ for each $i = 3,4,\cdots,n$, which is equivalent to the following multiplication operations for matrices

$$A^{(1)}x = b^{(1)} \quad \Rightarrow \quad A^{(2)}x = b^{(2)}, \quad L_2A^{(1)} = A^{(2)}, \tag{2.3.3}$$

where

$$L_2 = \begin{bmatrix} 1 & & & & \\ 0 & 1 & & & \\ 0 & -l_{32} & 1 & & \\ \vdots & \vdots & \vdots & \ddots & \\ 0 & -l_{n2} & 0 & \cdots & 1 \end{bmatrix}$$

is the second Gauss transformation matrix.

Repeating the process ends with the formation of $A^{(n-1)}x = b^{(n-1)}$, where $A^{(n-1)} = L_{n-1}A^{(n-2)}$, and

$$L_{n-1} = \begin{bmatrix} 1 & & & & \\ 0 & 1 & & & \\ 0 & 0 & \ddots & & \\ \vdots & \vdots & \vdots & 1 & \\ 0 & 0 & \cdots & -l_{n,n-1} & 1 \end{bmatrix}.$$

Then

$$A^{(n)} = L_{n-1}A^{(n-2)} = L_{n-1}L_{n-2}A^{(n-3)} = \cdots = L_{n-1}L_{n-2}\cdots L_1A^{(0)} \tag{2.3.4}$$

is the upper triangular matrix

$$A^{(n)} = \begin{bmatrix} a_{11}^{(0)} & a_{12}^{(0)} & a_{13}^{(0)} & \cdots & a_{1n}^{(0)} \\ & a_{22}^{(1)} & a_{23}^{(1)} & \cdots & a_{2n}^{(1)} \\ & & \ddots & \vdots & \vdots \\ & & & a_{n-1,n-1}^{(n-2)} & a_{n-1,n}^{(n-2)} \\ & & & & a_{nn}^{(n-1)} \end{bmatrix}. \tag{2.3.5}$$

And let $U = A^{(n)}$, $L = L_1^{-1}L_2^{-1}\cdots L_{n-1}^{-1}$. Then

$$L=\begin{bmatrix}1 & & & & \\ l_{21} & 1 & & & \\ l_{31} & l_{32} & 1 & & \\ \vdots & \vdots & \vdots & \ddots & \\ l_{n1} & l_{n2} & \cdots & l_{n,n-1} & 1\end{bmatrix} \tag{2.3.6}$$

is a lower triangular matrix. Therefore we have

$$A=A^{(0)}=LA^{(n)}=LU. \tag{2.3.7}$$

The Doolittle decomposition $A=LU$ can also be obtained directly by multiplication of the matrix. Supposing

$$\begin{bmatrix}a_{11} & a_{12} & \cdots & a_{1n}\\ a_{21} & a_{22} & \cdots & a_{2n}\\ \vdots & \vdots & & \vdots\\ a_{n1} & a_{n2} & \cdots & a_{nn}\end{bmatrix}=\begin{bmatrix}1 & & & \\ l_{21} & 1 & & \\ \vdots & \vdots & \ddots & \\ l_{n1} & l_{n2} & \cdots & 1\end{bmatrix}\begin{bmatrix}u_{11} & u_{12} & \cdots & u_{1n}\\ & u_{22} & \cdots & u_{2n}\\ & & \ddots & \vdots\\ & & & u_{nn}\end{bmatrix}, \tag{2.3.8}$$

by multiplication of the matrix， we have the recursive formula for L and U as follows：

(1) $u_{1j}=a_{1j}$, $j=1,2,\cdots,n$； $l_{i1}=\dfrac{a_{i1}}{a_{11}}$, $i=2,3,\cdots,n$; (2.3.9)

(2) for $k=2,3,\cdots,n$,

$$\begin{cases}u_{kj}=a_{kj}-\sum\limits_{s=1}^{k-1}l_{ks}u_{sj}, & j=k,k+1,\cdots,n,\\ l_{ik}=\dfrac{1}{u_{kk}}\left(a_{ik}-\sum\limits_{s=1}^{k-1}l_{is}u_{sk}\right), & i=k+1,\cdots,n.\end{cases} \tag{2.3.10}$$

Theorem 2.2 implies that finding the solution of system $Ax=b$ is to solve the system $Ly=b$ for y, and then solve the system $Ux=y$ for x, which obtain

$$\begin{cases}y_1=b_1,\\ y_i=b_i-\sum\limits_{j=1}^{i-1}l_{ij}y_j, \quad i=2,3,\cdots,n,\end{cases} \tag{2.3.11}$$

and

$$\begin{cases}x_n=y_n/u_{nn},\\ x_i=\left(y_i-\sum\limits_{j=i+1}^{n}u_{ij}x_j\right)\Big/u_{ii} \quad i=n-1,n-2,\cdots,1.\end{cases} \tag{2.3.12}$$

Example 2.6 Solve the following linear system using Doolittle factorization

$$\begin{cases}x_1+4x_2+7x_3=1,\\ 2x_1+5x_2+8x_3=1,\\ 3x_1+6x_2+11x_3=1.\end{cases}$$

Solution 1. First use Gauss elimination to factor A into LU. Let

$$\boldsymbol{A}=\begin{bmatrix}1&4&7\\2&5&8\\3&6&11\end{bmatrix},\quad \boldsymbol{b}=\begin{bmatrix}1\\1\\1\end{bmatrix}.$$

By Example 2.2 in Section 2.2, multipliers for each Gauss elimination process are

$$l_{21}=2,\ l_{31}=3,\ l_{32}=2,$$

then the lower-triangular matrix $\boldsymbol{L}$ is given

$$\boldsymbol{L}=\begin{bmatrix}1&&\\l_{21}&1&\\l_{31}&l_{32}&1\end{bmatrix}=\begin{bmatrix}1&&\\2&1&\\3&2&1\end{bmatrix},$$

and the Gauss elimination ends with an upper-triangular coefficient matrix, that is

$$\begin{bmatrix}1&4&7\\&-3&-6\\&&2\end{bmatrix}=\boldsymbol{U}.$$

Then we solve the system $\boldsymbol{Ly}=\boldsymbol{b}$ for $\boldsymbol{y}$ and $\boldsymbol{Ux}=\boldsymbol{y}$ for $\boldsymbol{x}$ respectively. Solving $\boldsymbol{Ly}=\boldsymbol{b}$, i.e.

$$\begin{bmatrix}1&&\\2&1&\\3&2&1\end{bmatrix}\begin{bmatrix}y_1\\y_2\\y_3\end{bmatrix}=\begin{bmatrix}1\\1\\1\end{bmatrix}$$

for $y_i\,(i=1,2,3)$ by forward substitution gives

$$\begin{bmatrix}y_1\\y_2\\y_3\end{bmatrix}=\begin{bmatrix}1\\1-2y_1\\1-3y_1-2y_2\end{bmatrix}=\begin{bmatrix}1\\-1\\0\end{bmatrix}.$$

Solving $\boldsymbol{Ux}=\boldsymbol{y}$, i.e.

$$\begin{bmatrix}1&4&7\\&-3&-6\\&&2\end{bmatrix}\begin{bmatrix}x_1\\x_2\\x_3\end{bmatrix}=\begin{bmatrix}1\\-1\\0\end{bmatrix}$$

for $x_i\,(i=1,2,3)$ by backward substitution gives

$$x_3=0,\ x_2=-\left(-1+6x_3\right)/3=1/3,\ x_1=1-4x_2-7x_3=-1/3.$$

Therefore the solution to the given linear system is $\boldsymbol{x}=\frac{1}{3}(-1,\ 1,\ 0)^{\mathrm{T}}$. This is the same as the Gauss elimination.

Solution 2. Using the multiplication of matrix leads to

(1) $u_{11}=a_{11}=1, u_{12}=a_{12}=4, u_{13}=a_{13}=7,$

$$l_{21}=\frac{a_{21}}{a_{11}}=2,\quad l_{31}=\frac{a_{31}}{a_{11}}=3.$$

(2) For $k=2$, we have

$$u_{22}=a_{22}-l_{21}u_{12}=-3,\ u_{23}=a_{23}-l_{21}u_{13}=-6,\quad l_{32}=\frac{1}{u_{22}}\left(a_{32}-l_{31}u_{12}\right)=2,$$

and for $k=3$, we have $u_{33}=a_{33}-\left(l_{31}u_{13}+l_{32}u_{23}\right)=2$.

2.3.2 Crout factorization (Crout 分解)

Definition 2.3 Let $A\in\mathbf{R}^{n\times n}$. $\boldsymbol{A}=\boldsymbol{L}\boldsymbol{U}$ is called Crout factorization or Crout decomposition of matrix $\boldsymbol{A}$ (Crout 分解), where $\boldsymbol{L}$ is a lower-triangular matrix and $\boldsymbol{U}$ is an upper-triangular matrix, and

$$\boldsymbol{L}=\begin{bmatrix} l_{11} & & & \\ l_{21} & l_{22} & & \\ \vdots & \vdots & \ddots & \\ l_{n1} & l_{n2} & \cdots & l_{nn}\end{bmatrix},\quad \boldsymbol{U}=\begin{bmatrix} 1 & u_{12} & u_{13} & \cdots & u_{1n} \\ & 1 & u_{23} & \cdots & u_{2n} \\ & & \ddots & \vdots & \vdots \\ & & & 1 & u_{n-1,n} \\ & & & & 1\end{bmatrix}. \tag{2.3.13}$$

By multiplication of the matrix, the recursive formula for $\boldsymbol{L}$ and $\boldsymbol{U}$ are as follows:

(1) The first column entries l_{i1} of $\boldsymbol{L}$ and first rows entries u_{1j} of $\boldsymbol{U}$ are

$$l_{i1}=a_{i1},\ i=1,2,\cdots,n;\ u_{1j}=\frac{a_{1j}}{l_{11}},\ j=2,3,\cdots,n; \tag{2.3.14}$$

(2) For $k=2,3,\cdots,n$ (in order: the second column of $\boldsymbol{L}$, the second row of $\boldsymbol{U}$, the third column of $\boldsymbol{L}$, the third row of $\boldsymbol{U}$, $\cdots$)

$$\begin{cases} l_{ik}=a_{ik}-\sum_{s=1}^{k-1}l_{is}u_{sk}, & i=k,k+1,\cdots,n, \\ u_{kj}=\frac{1}{l_{kk}}\left(a_{kj}-\sum_{s=1}^{k-1}l_{ks}u_{sj}\right), & j=k+1,\cdots,n. \end{cases} \tag{2.3.15}$$

Therefore to find the solution of system $\boldsymbol{Ax}=\boldsymbol{b}$ is to solve the system $\boldsymbol{Ly}=\boldsymbol{b}$ for $\boldsymbol{y}$, and then solve the system $\boldsymbol{Ux}=\boldsymbol{y}$ for $\boldsymbol{x}$, which leads to

$$\begin{cases} y_1=b_1/l_{11}, \\ y_i=\left(b_i-\sum_{j=1}^{i-1}l_{ij}y_j\right), & i=2,3,\cdots,n, \end{cases} \tag{2.3.16}$$

and

$$\begin{cases} x_n=y_n, \\ x_i=y_i-\sum_{j=i+1}^{n}u_{ij}x_j, & i=n-1,n-2,\cdots,1. \end{cases} \tag{2.3.17}$$

2.3.3 Permutation matrix (置换矩阵)

Definition 2.4 If an $n\times n$ matrix $\boldsymbol{P}$ is obtained by rearranging the rows of $\boldsymbol{I}_n$, then the $n\times n$ identity matrix $\boldsymbol{P}$ is called permutation matrix.

For any nonsingular matrix $\boldsymbol{A}$, $\boldsymbol{Ax}=\boldsymbol{b}$ has a unique solution, which implies that we can arrange the original equations $\boldsymbol{Ax}=\boldsymbol{b}$ so that its equivalent linear system $\overline{\boldsymbol{A}}\boldsymbol{x}=\overline{\boldsymbol{b}}$ can be solved by Gauss elimination without row interchanges.

Theorem 2.3 For any nonsingular matrix $\boldsymbol{A}$, there exists permutation matrix (置换矩阵) $\boldsymbol{P}$, such that the matrix $\boldsymbol{PA}$ can be factored into $\boldsymbol{PA}=\boldsymbol{LU}$, where $\boldsymbol{L}$ is lower triangular with 1's on its diagonal (主对角线元素为 1) and $\boldsymbol{U}$ is upper triangular.

Theorem 2.3 implies that for any nonsingular matrix $\boldsymbol{A}$, a permutation matrix $\boldsymbol{P}$ exists for which $\boldsymbol{PAx}=\boldsymbol{Pb}$ can be solved without row interchanges, and then solving the systems $\boldsymbol{Ux}=\boldsymbol{y}, \boldsymbol{Ly}=\boldsymbol{Pb}$ obtains the solution $\boldsymbol{x}$ of original linear system of equations.

Example 2.7 Suppose we are given the linear system

$$\begin{cases} E_1: & x_1 & -x_2 & +2x_3 & -x_4 & = & -8, \\ E_2: & 2x_1 & -2x_2 & +3x_3 & -x_4 & = & -20, \\ E_3: & x_1 & +x_2 & +x_3 & & = & -2, \\ E_4: & x_1 & -x_2 & +4x_3 & +3x_4 & = & 4. \end{cases}$$

Performing the operation $(E_2-2E_1\to E_2),(E_3-E_1\to E_3)$, and $(E_4-E_1\to E_4)$ gives

$$A^{(1)}=\begin{pmatrix} 1 & -1 & 2 & -1 \\ 0 & 0 & -1 & -1 \\ 0 & 2 & -1 & 1 \\ 0 & 0 & 2 & 4 \end{pmatrix}.$$

Then the Permutation matrix is

$$\boldsymbol{P}=\begin{pmatrix} 1 & 0 & 0 & 0 \\ 0 & 0 & 1 & 0 \\ 0 & 1 & 0 & 0 \\ 0 & 0 & 0 & 1 \end{pmatrix}.$$

2.4 Special types of matrices (特殊形式矩阵的三角分解)

- Strictly diagonally dominant matrix (严格对角占优矩阵)
- Positive definite matrix (正定矩阵)
- Strictly diagonally dominant tridiagonal matrix (严格对角占优三对角矩阵)

2.4.1 Strictly diagonally dominant matrix (严格对角占优矩阵)

Definition 2.5 (Strictly diagonally dominant matrix) The $n\times n$ matrix $\boldsymbol{A}=\left(a_{ij}\right)_{n\times n}$ is said to be strictly diagonally dominant if $|a_{ii}|>\sum\limits_{j=1,j\neq i}^{n}|a_{ij}|$ holds for each $i=1,2,\cdots,n$.

Theorem 2.4 A strictly diagonally dominant matrix $\boldsymbol{A}$ is nonsingular. Moreover, in this case, Gauss elimination can be performed on any linear system of the form $\boldsymbol{Ax}=\boldsymbol{b}$ to obtain its unique solution without row or column interchanges, and the computations will be stable with respect to the growth of round-off errors.

Proof. We use proof by contradiction (reduction to absurdity).

To prove that $\boldsymbol{A}$ is nonsingular is to prove that $\boldsymbol{Ax}=\boldsymbol{0}$ has only zero solution. Suppose that $\boldsymbol{x}=(x_i)$ is a nonzero solution to this system. Let $|x_k|=\max\limits_{1\leqslant i\leqslant n}|x_i|>0$, then from $\sum\limits_{j=1}^{n}a_{kj}x_j=0$, it follows $a_{kk}x_k=-\sum\limits_{j=1,j\neq k}^{n}a_{kj}x_j$, which implies that

$$|a_{kk}||x_k|\leqslant\sum_{j=1,j\neq k}^{n}|a_{kj}||x_j|,\ \text{and}\ |a_{kk}|\leqslant\sum_{j=1,j\neq k}^{n}|a_{kj}|\left|\frac{x_j}{x_k}\right|\leqslant\sum_{j=1,j\neq k}^{n}|a_{kj}|.$$

This contradicts the strict diagonal dominance of $\boldsymbol{A}$. As a result, the only solution to $\boldsymbol{Ax}=\boldsymbol{0}$ is $\boldsymbol{x}=\boldsymbol{0}$.

Theorem 2.5 A strictly diagonally dominant matrix $\boldsymbol{A}$ permits to be factored into the product of a lower-triangular matrix $\boldsymbol{L}$ and an upper-triangular matrix $\boldsymbol{U}$, $\boldsymbol{A}=\boldsymbol{LU}$.

Proof. It suffices for us only to show that Gauss elimination can be performed on $\boldsymbol{A}$ without row interchanges since we know that Doolittle factorization ensures that $\boldsymbol{A}=\boldsymbol{LU}$ if Gauss elimination can be performed on $\boldsymbol{A}$ without row interchanges.

We prove that each of the matrices $\boldsymbol{A}^{(1)},\boldsymbol{A}^{(2)},\cdots,\boldsymbol{A}^{(n-1)}$ generated by the Gauss elimination process is strictly diagonally dominant. This implies that all the diagonal entries are nonzero, so Gauss elimination can be performed without row interchanges.

Since $\boldsymbol{A}^{(0)}=\boldsymbol{A}$ is strictly diagonally dominant, $a_{11}^{(0)}\neq 0$, and $\boldsymbol{A}^{(1)}$ can be formed

$$\boldsymbol{A}^{(1)}=\begin{pmatrix} a_{11}^{(0)} & a_{12}^{(0)} & \cdots & a_{1n}^{(0)} \\ & a_{22}^{(1)} & \cdots & a_{2n}^{(1)} \\ & \vdots & & \vdots \\ & a_{n2}^{(1)} & \cdots & a_{nn}^{(1)} \end{pmatrix},$$

where $a_{ij}^{(1)}=a_{ij}^{(0)}-a_{1j}^{(0)}a_{i1}^{(0)}/a_{11}^{(0)}, i=2,3,\cdots,n;\ j=2,3,\cdots,n+1$. Then

$$\begin{aligned}\sum_{j=2,j\neq i}^{n}\left|a_{ij}^{(1)}\right| &= \sum_{j=2,j\neq i}^{n}\left|a_{ij}^{(0)}-a_{1j}^{(0)}a_{i1}^{(0)}/a_{11}^{(0)}\right| \\ &\leqslant \sum_{j=2,j\neq i}^{n}\left|a_{ij}^{(0)}\right|+\left|\frac{a_{i1}^{(0)}}{a_{11}^{(0)}}\right|\sum_{j=2,j\neq i}^{n}\left|a_{1j}^{(0)}\right|<\left|a_{ii}^{(0)}\right|-\left|\frac{a_{i1}^{(0)}a_{1i}^{(0)}}{a_{11}^{(0)}}\right|\leqslant\left|a_{ii}^{(1)}\right|.\end{aligned}$$

This establishes the strict diagonal dominance for rows $2,3,\cdots,n$. Since the first row of $\boldsymbol{A}^{(1)}$ and $\boldsymbol{A}^{(0)}$ are the same, $\boldsymbol{A}^{(1)}$ is strictly diagonally dominant.

Similarly, each of $\boldsymbol{A}^{(2)},\boldsymbol{A}^{(3)},\cdots,\boldsymbol{A}^{(n-1)}$ is strictly diagonally dominant.

2.4.2 Positive definite matrix (正定矩阵)

Definition 2.6 A matrix $\boldsymbol{A}$ is said to be positive definite (正定) if it is symmetric and if $\boldsymbol{x}^{\mathrm{T}}\boldsymbol{A}\boldsymbol{x} > \boldsymbol{0}$ for every n-dimensional vector $\boldsymbol{x} \neq \boldsymbol{0}$.

Theorem 2.6 If $\boldsymbol{A} = (a_{ij})$ is an $n \times n$ positive definite matrix, then

(1) $\boldsymbol{A}$ is nonsingular;

(2) $a_{ii} > 0$ for each $i = 1, 2, \cdots, n$;

(3) $\max\limits_{1 \leqslant i \leqslant n} |a_{ii}| \geqslant \max\limits_{1 \leqslant k, j \leqslant n} |a_{kj}|$;

(4) $(a_{ij})^2 < a_{ii}a_{jj}$ for each $i \neq j$.

Theorem 2.7 A symmetric matrix $\boldsymbol{A}$ is positive definite if and only if each of its leading principal submatrices

$$\boldsymbol{A}_k = \begin{pmatrix} a_{11} & a_{12} & \cdots & a_{1k} \\ a_{21} & a_{22} & \cdots & a_{2k} \\ \vdots & \vdots & & \vdots \\ a_{k1} & a_{k2} & \cdots & a_{kk} \end{pmatrix}, k = 1, 2, \cdots, n$$

has a positive determinant.

Theorem 2.8 The symmetric matrix $\boldsymbol{A}$ is positive definite if and only if Gauss elimination without row interchanges can be performed on the linear system $\boldsymbol{Ax} = \boldsymbol{b}$ with all pivot elements positive.

Theorem 2.9 ($\boldsymbol{LDL}^{\mathrm{T}}$ Factorization) If the symmetric matrix $\boldsymbol{A}$ is positive definite, then $\boldsymbol{A}$ can be factored in the form $\boldsymbol{LDL}^{\mathrm{T}}$, where $\boldsymbol{L}$ is lower triangular with 1's on its diagonal, $\boldsymbol{D}$ is a diagonal matrix (对角矩阵) with positive diagonal entries, and $\boldsymbol{L}^{\mathrm{T}}$ is transpose matrix of $\boldsymbol{L}$ (转置矩阵).

Proof. By Theorem 2.2, if the symmetric matrix $\boldsymbol{A}$ is positive definite, then $\boldsymbol{A}$ can be factored in the form $\boldsymbol{A} = \boldsymbol{LU}$, where $\boldsymbol{L}$ is lower triangular, $\boldsymbol{U}$ is an upper- triangular matrix, and

$$\det \boldsymbol{A} = \prod_{i=1}^{n} u_{ii} \neq 0 \quad \Rightarrow \quad u_{ii} \neq 0 \, (i = 1, 2, \cdots, n),$$

then there exists a diagonal matrix $\boldsymbol{D}$, such that $\boldsymbol{U} = \boldsymbol{D}\widetilde{\boldsymbol{U}}$, where

$$\boldsymbol{D} = \begin{bmatrix} u_{11} & & & \\ & u_{22} & & \\ & & \ddots & \\ & & & u_{nn} \end{bmatrix}, \quad \widetilde{\boldsymbol{U}} = \begin{bmatrix} 1 & u_{12}/u_{11} & u_{13}/u_{11} & \cdots & u_{1n}/u_{11} \\ & 1 & u_{23}/u_{22} & \cdots & u_{2n}/u_{22} \\ & & \ddots & \cdots & \vdots \\ & & & 1 & u_{n-1,n}/u_{n-1,n-1} \\ & & & & 1 \end{bmatrix}.$$

The symmetric matrix $\boldsymbol{A}$ leads to

$$\left(\boldsymbol{L}\boldsymbol{D}\widetilde{\boldsymbol{U}}\right)^{\mathrm{T}}=\widetilde{\boldsymbol{U}}^{\mathrm{T}}\boldsymbol{D}\boldsymbol{L}^{\mathrm{T}}=\boldsymbol{L}\boldsymbol{D}\widetilde{\boldsymbol{U}},$$

from the uniqueness of Doolittle factorization, we have $\widetilde{\boldsymbol{U}}^{\mathrm{T}}=\boldsymbol{L}$ or $\boldsymbol{U}=\boldsymbol{L}^{\mathrm{T}}$. Therefore $\boldsymbol{A}=\boldsymbol{L}\boldsymbol{D}\boldsymbol{L}^{\mathrm{T}}$.

Let $\boldsymbol{D}=\begin{pmatrix} d_1 & & \\ & \ddots & \\ & & d_n \end{pmatrix}$. Positive definite matrix $\boldsymbol{A}$ leads to $d_i>0\,(i=1,2,\cdots,n)$, then $\boldsymbol{D}=\boldsymbol{D}^{\frac{1}{2}}\boldsymbol{D}^{\frac{1}{2}}$, where

$$\boldsymbol{D}^{\frac{1}{2}}=\begin{pmatrix} \sqrt{d_1} & & \\ & \ddots & \\ & & \sqrt{d_n} \end{pmatrix},$$

which follows

$$\boldsymbol{A}=\boldsymbol{L}\boldsymbol{D}\boldsymbol{L}^{\mathrm{T}}=\boldsymbol{L}\boldsymbol{D}^{\frac{1}{2}}\boldsymbol{D}^{\frac{1}{2}}\boldsymbol{L}^{\mathrm{T}}=\left(\boldsymbol{L}\boldsymbol{D}^{\frac{1}{2}}\right)\left(\boldsymbol{L}\boldsymbol{D}^{\frac{1}{2}}\right)^{\mathrm{T}}.$$

Denote $\boldsymbol{L}\boldsymbol{D}^{\frac{1}{2}}=\boldsymbol{G}$, then we have the following theorem.

Theorem 2.10 (Cholesky Factorization) If the matrix $\boldsymbol{A}$ is positive definite, then $\boldsymbol{A}$ can be factored in the form $\boldsymbol{A}=\boldsymbol{G}\boldsymbol{G}^{\mathrm{T}}$, which is called Cholesky factorization (Cholesky 分解), where $\boldsymbol{G}=\boldsymbol{L}\boldsymbol{D}^{1/2}$, $\boldsymbol{L}$ is lower triangular with 1's on its diagonal and $\boldsymbol{D}$ is a diagonal matrix.

2.4.3 Strictly diagonally dominant tridiagonal matrix (严格对角占优三对角矩阵)

In scientific engineering calculations, such as edge value problems for second order ordinary differential equations, the solution of spline interpolation, and heat conduction equations, tridiagonal linear system of equations $\boldsymbol{Ax}=\boldsymbol{d}$ of the following form are often solved

$$\begin{bmatrix} b_1 & c_1 & & & \\ a_2 & b_2 & c_2 & & \\ & \ddots & \ddots & \ddots & \\ & & a_{n-1} & b_{n-1} & c_{n-1} \\ & & & a_n & b_n \end{bmatrix}\begin{bmatrix} x_1 \\ x_2 \\ \vdots \\ x_{n-1} \\ x_n \end{bmatrix}=\begin{bmatrix} d_1 \\ d_2 \\ \vdots \\ d_{n-1} \\ d_n \end{bmatrix}. \tag{2.4.1}$$

Definition 2.7 Suppose

$$\boldsymbol{A}=\begin{bmatrix} b_1 & c_1 & & & \\ a_2 & b_2 & c_2 & & \\ & \ddots & \ddots & \ddots & \\ & & a_{n-1} & b_{n-1} & c_{n-1} \\ & & & a_n & b_n \end{bmatrix}$$

is a tridiagonal matrix, it is strictly diagonally dominant, if

$$\begin{cases} |b_1| > |c_1| > 0, \\ |b_n| > |a_n| > 0, \\ |b_i| > |a_i| + |c_i|, \quad a_i c_i \neq 0, \quad i = 2,3,\cdots,n-1. \end{cases}$$

Theorem 2.11 Suppose

$$\boldsymbol{A} = \begin{bmatrix} b_1 & c_1 & & & \\ a_2 & b_2 & c_2 & & \\ & \ddots & \ddots & \ddots & \\ & & a_{n-1} & b_{n-1} & c_{n-1} \\ & & & a_n & b_n \end{bmatrix}$$

is strictly diagonally dominant, then $\boldsymbol{A}$ can be uniquely factored in the form $\boldsymbol{A} = \boldsymbol{LU}$, where

$$\boldsymbol{L} = \begin{bmatrix} 1 & & & \\ l_2 & 1 & & \\ & \ddots & \ddots & \\ & & l_n & 1 \end{bmatrix}, \quad \boldsymbol{U} = \begin{bmatrix} u_1 & c_1 & & & \\ & u_2 & c_2 & & \\ & & \ddots & \ddots & \\ & & & u_{n-1} & c_{n-1} \\ & & & & u_n \end{bmatrix}, \tag{2.4.2}$$

and

$$\begin{cases} u_1 = b_1, \\ l_i = a_i / u_{i-1}, \qquad i = 2,3,\cdots,n. \\ u_i = b_i - l_i c_{i-1}, \end{cases} \tag{2.4.3}$$

By the Doolittle factorization of the coefficient matrix $\boldsymbol{A}$, solving $\boldsymbol{Ax} = \boldsymbol{d}$ is to solve $\boldsymbol{Ly} = \boldsymbol{d}$ for $\boldsymbol{y}$ and $\boldsymbol{Ux} = \boldsymbol{y}$ for $\boldsymbol{x}$. From Eqs. (2.4.2) and (2.4.3) it is obtained

$$\begin{cases} y_1 = d_1, \\ y_i = d_i - l_i y_{i-1}, \quad i = 2,3,\cdots,n, \end{cases} \tag{2.4.4}$$

and

$$\begin{cases} x_n = y_n / u_n, \\ x_i = \left(y_i - c_i x_{i+1}\right) / u_i, \quad i = n-1, n-2, \cdots, 1. \end{cases} \tag{2.4.5}$$

Example 2.8 Solve the following linear system

$$\begin{bmatrix} 2 & -1 & & & \\ -1 & 2 & -1 & & \\ & -1 & 2 & -1 & \\ & & -1 & 2 & -1 \\ & & & -1 & 2 \end{bmatrix} \begin{bmatrix} x_1 \\ x_2 \\ x_3 \\ x_4 \\ x_5 \end{bmatrix} = \begin{bmatrix} 1 \\ 0 \\ 0 \\ 0 \\ 0 \end{bmatrix}. \tag{2.4.6}$$

Solution. Let

$$
\boldsymbol{A}=\begin{bmatrix} 2 & -1 & & & \\ -1 & 2 & -1 & & \\ & -1 & 2 & -1 & \\ & & -1 & 2 & -1 \\ & & & -1 & 2 \end{bmatrix}.
$$

Then $\boldsymbol{A}$ is a strictly diagonally dominant tridiagonal matrix. Let

$$
\boldsymbol{A}=\boldsymbol{LU}=\begin{bmatrix} 1 & & & & \\ l_1 & 1 & & & \\ & l_2 & 1 & & \\ & & l_3 & 1 & \\ & & & l_4 & 1 \end{bmatrix}\begin{bmatrix} u_1 & -1 & & & \\ & u_2 & -1 & & \\ & & u_3 & -1 & \\ & & & u_4 & -1 \\ & & & & u_5 \end{bmatrix}.
$$

From Eq.(2.4.3)， we have

$$
u_1=2,\ l_1=1/u_1=-\frac{1}{2},\ u_2=\frac{3}{2},\ l_2=1/u_2=-\frac{2}{3},
$$

$$
u_3=\frac{4}{3},\ l_3=-\frac{3}{4},\ u_4=\frac{5}{4},\ l_4=-\frac{4}{5},\ u_5=\frac{6}{5}.
$$

Solving $\boldsymbol{Ly}=\boldsymbol{d}$ for $\boldsymbol{y}$, i.e.

$$
\begin{bmatrix} 1 & & & & \\ -1/2 & 1 & & & \\ & -2/3 & 1 & & \\ & & -3/4 & 1 & \\ & & & -4/5 & 1 \end{bmatrix}\begin{bmatrix} y_1 \\ y_2 \\ y_3 \\ y_4 \\ y_5 \end{bmatrix}=\begin{bmatrix} 1 \\ 0 \\ 0 \\ 0 \\ 0 \end{bmatrix},
$$

gives $\boldsymbol{y}=\left(1,\ \frac{1}{2},\ \frac{1}{3},\ \frac{1}{4},\ \frac{1}{5}\right)^{\mathrm{T}}$. And solving $\boldsymbol{Ux}=\boldsymbol{y}$ for $\boldsymbol{x}$, i.e.

$$
\begin{bmatrix} 2 & -1 & & & \\ & 3/2 & -1 & & \\ & & 4/3 & -1 & \\ & & & 5/4 & -1 \\ & & & & 6/5 \end{bmatrix}\begin{bmatrix} x_1 \\ x_2 \\ x_3 \\ x_4 \\ x_5 \end{bmatrix}=\begin{bmatrix} 1 \\ 1/2 \\ 1/3 \\ 1/4 \\ 1/5 \end{bmatrix},
$$

results in

$$
x_5=\frac{1}{6},\ x_4=\frac{1}{3},\ x_3=\frac{1}{2},\ x_2=\frac{2}{3},\ x_1=\frac{5}{6}.
$$

Therefore the solution of original system (2.4.6) is

$$
\boldsymbol{x}=\left(\frac{5}{6},\ \frac{2}{3},\ \frac{1}{2},\ \frac{1}{3},\ \frac{1}{6}\right)^{\mathrm{T}}.
$$

2.5 本章算法程序及实例 (Algorithms and examples)

2.5.1 Gauss 消元法 (Gauss elimination method)

例 2.9 采用 Gauss 消元法求解方程组

$$\begin{pmatrix} 6 & 2 & 1 & -2 \\ 2 & 5 & 0 & -2 \\ -2 & 0 & 8 & 5 \\ 1 & 3 & 2 & 7 \end{pmatrix}\begin{pmatrix} x_1 \\ x_2 \\ x_3 \\ x_4 \end{pmatrix}=\begin{pmatrix} 4 \\ 7 \\ -1 \\ 0 \end{pmatrix}.$$

```
%高斯消元法的MATLAB程序
clc; clear; close all;
A=[6 2 1 -2;2 5 0 -2;-2 0 8 5;1 3 2 7];    %输入系数矩阵
B=[4;7;-1;0];
n=length(B);    %计算未知数个数
x=zeros(n,1);   %未知数向量
for k=1:n-1
    if A(k,k)==0;
    error('Error');
end
for i=k+1:n
    temp=A(i,k)/A(k,k);
    for j=k:n
        A(i,j)=A(i,j)-temp*A(k,j);
    end
    B(i)=B(i)-temp*B(k);
end
x(n)=B(n)/A(n,n);
for k=n-1:-1:1
   S=B(k);
   for j=k+1:n
       S=S-A(k,j)*x(j);
   end
   x(k)=S/A(k,k);
end
disp(x);
```

结果：0.0520；1.1509；0.2446；-0.5706.

2.5.2 选主元策略 (Pivoting strategies)

1. Gauss 列主元消元法求解线性方程组

例 2.10 采用 Gauss 列主元消元法求解方程组

$$\begin{pmatrix} 6 & 2 & 1 & -2 \\ 2 & 5 & 0 & -2 \\ -2 & 0 & 8 & 5 \\ 1 & 3 & 2 & 7 \end{pmatrix}\begin{pmatrix} x_1 \\ x_2 \\ x_3 \\ x_4 \end{pmatrix}=\begin{pmatrix} 4 \\ 7 \\ -1 \\ 0 \end{pmatrix}.$$

```
%高斯列主元消元法的MATLAB程序
clc; clear; close all;
A=[6 2 1 -2;2 5 0 -2;-2 0 8 5;1 3 2 7];      %输入矩阵系数
B=[4;7;-1;0];
C=[A B];
[m,n]=size(C);  %计算出增广矩阵的行数和列数
H=m;L=n;
for i=1:n-2       %第i列
   max=0;
   for j=i:m%第j行
      if(abs(C(j,i))>max)
        max=abs(C(j,i));
        a=i; %记录i列中最大在第a列
        b=j; %记录i列中最大在第b行
       end
   end
  C([i,b],:)=C([b,i],:);%交换第i行和最大值所在的第b行,使列最大元素在C(i,j);
  for k=i+1:m
     temp=C(i,i)/C( k,i);
     for d=1:n
       C(k,d)=temp*C(k,d)-C(i,d);   %消元
     end
  end
end
x=zeros(1,m);   %生成1行m列的0矩阵
for s=m:-1:1;
    x(s)=C(s,n);
    for t=1:m-s;
        x(s)=x(s)-C(s,n-t)*x(m+1-t);    %反带回方程
```

```
    end
    x(s)=x(s)/C(s,s);
end
disp(x);
```

结果：0.0520；1.1509；0.2446；-0.5706.

2. Gauss 全主元消元法

例 2.11 采用 Gauss 全主元消元法求解方程组

$$\begin{pmatrix} 6 & 2 & 1 & -2 \\ 2 & 5 & 0 & -2 \\ -2 & 0 & 8 & 5 \\ 1 & 3 & 2 & 7 \end{pmatrix}\begin{pmatrix} x_1 \\ x_2 \\ x_3 \\ x_4 \end{pmatrix}=\begin{pmatrix} 4 \\ 7 \\ -1 \\ 0 \end{pmatrix}.$$

```
%Gauss全主元消元法的MATLAB程序
clc;clear all;
A=[6 2 1 -2;2 5 0 -2;-2 0 8 5;1 3 2 7];          %输入矩阵系数
B=[4;7;-1;0];
ZG=[A B];                %增广矩阵
n=length(B);             %维度
X=zeros (n,1);           %n维列向量
x=(1:1:n)';              %n个解的顺序
C=zeros(1,n+1);          %n+1维行向量+初始化
for p=1:n-1
    max_v=max(max(ZG(p:n,p:n)));
    [i,j]=find(ZG(:,1:n)==max_v);    %矩阵中最大元素的下标
    H=ZG(p,:);             %H为ZG的第p行元素
    ZG(p,:)=ZG(i,:);
    ZG(i,:)=H;             %交换第p行和最大的行
    L=ZG(:,p);             %L为ZG的第p列元素
    ZG(:,p)=ZG(:,j);ZG(:,j)=L;          %交换第p列和最大的列
    xx=x(j,1);
    x(j,1)=x(p,1);x(p,1)=xx;
    for k=p+1:n
        m=ZG(k,p)/ZG(p,p);
        ZG(k,p:n+1)=ZG(k,p:n+1)-m*ZG(p,p:n+1);          %消元
    end
end
B=ZG(1:n,n+1);A=ZG(1:n,1:n);          %上三角化后的新增广矩阵[A b]
X(n)=B(n)/A(n,n);                     %计算最后一行的X(n)
```

```
for q=n-1:-1:1            %以-1为步长递增(实际是减少)
    X(q)=(B(q)-sum(A(q,q+1:n)*X(q+1:n)))/A(q,q);
end
ans_=[x X];        %乱序的x及其对应的值
for i=1:n
    [o,~]=find(ans_==i);value=ans_(o,2);
    fprintf('x%d=%.4f; ',i,value)
end
fprintf('\n')
```

结果 x1=0.0520; x2=1.1509; x3=0.2446; x4=-0.5706.

2.5.3 LU 分解法 (LU decomposition)

例 2.12 采用 LU 分解法求解方程组

$$\begin{pmatrix} 6 & 2 & 1 & -2 \\ 2 & 5 & 0 & -2 \\ -2 & 0 & 8 & 5 \\ 1 & 3 & 2 & 7 \end{pmatrix}\begin{pmatrix} x_1 \\ x_2 \\ x_3 \\ x_4 \end{pmatrix}=\begin{pmatrix} 4 \\ 7 \\ -1 \\ 0 \end{pmatrix}.$$

```
% LU分解法的MATLAB程序
clc;clear;close all;
A=[6 2 1 -2;2 5 0 -2;-2 0 8 5;1 3 2 7];     %输入矩阵系数
B=[4;7;-1;0];
[n,n]=size(A);  %实现对矩阵A的LU分解, L为下三角矩阵
L=zeros(n,n);
U=zeros(n,n);
for i=1:n
    L(i,i)=1;
end
for k=1:n
    for j=k:n
        U(k,j)=A(k,j)-sum(L(k,1:k-1).*U(1:k-1,j)');
    end
    for i=k+1:n
        L(i,k)=(A(i,k)-sum(L(i,1:k-1).*U(1:k-1,k)'))/U(k,k);
    end
end
y(1)=B(1);     %LU分解法求解方程组
for i=2:n
     for j=1:i-1
```

```
        B(i)=B(i)-L(i,j)*y(j);
       end
       y(i)=B(i);
   end
   x(n)=y(n)/U(n,n);
   for i=(n-1):-1:1
        for j=n:-1:i+1
            y(i)=y(i)-U(i,j)*x(j);
        end
        x(i)=y(i)/U(i,i);
   end
   disp(x);
   结果: 0.0520; 1.1509; 0.2446; -0.5706.
```

2.6 本章要点 (Hightlights)

1. 线性方程组的直接解法：Gauss 消元法与矩阵三角分解法. 消元法包括 Gauss 消元法、Gauss 主元消元法.
2. 若方程组系数矩阵的各阶顺序主子式非零，则可直接运用 Gauss 消元法.
3. 当方程组系数矩阵的顺序主子式为零时，则需要考虑采用 Gauss 主元消元法. Gauss 主元消元法包括 Gauss 列主元消元法、Gauss 按比例消元法和 Gauss 全主元消元法.
4. 直接法的优点是运算量小，精度高，但在用计算机实现时程序较复杂，一般适用求解中小型 (阶数 $n<1000$) 线性方程组.
5. 若系数矩阵为三对角矩阵，特别是严格对角占优矩阵，用特殊形式的 Doolittle 分解法，即追赶法，既快速又稳定.
6. 当系数矩阵为对称正定阵时，采用平方根法或改进的平方根法比较有效.

2.7 问题讨论 (Questions for discussion)

1. 用 Cramer 法则求解线性方程组需要进行多少次乘除法运算？
2. 用 Doolittle 分解求解线性方程组需要进行多少次乘除法运算？
3. 求解 n 阶三对角方程组需要进行多少次乘除法运算？
4. Doolittle 分解与 Crout 分解有何异同？
5. 矩阵满足哪些条件时可进行三角分解？
6. 非奇异矩阵 $\boldsymbol{A}$ 的条件数 (详见第 3 章定义 3.12) 定义为

$$\text{Cond}\ (\boldsymbol{A})\ =\ \|\boldsymbol{A}\|\|\boldsymbol{A}^{-1}\|$$

其中，$\|\cdot\|$表示矩阵范数 (详见 3.1 节). 按照范数的性质，有 $\text{Cond}(\boldsymbol{A})\geqslant 1$. 当 $\text{Cond}(\boldsymbol{A})$ 接近 1 时，矩阵 $\boldsymbol{A}$ 称作是良态的 (well-conditioned) ，当 $\text{Cond}(\boldsymbol{A})\gg 1$时，矩阵 $\boldsymbol{A}$

称作病态的(ill-conditioned). 直接法求解方程组的一般步骤如下：先求方程组系数矩阵的条件数，判断方程组的性态，若为良态，则根据系数矩阵的特性，选取相应有效可靠的直接法求解；若为病态，则要采取其他方法求解，例如，本章延伸阅读部分的 $\boldsymbol{QR}$ 分解方法. 还有哪些方法可以用来求解病态方程组？

2.8 关键术语 (Key terms)

linear system of equations of order n (n 阶线性方程组)
coefficient matrix (系数矩阵)
augmented matrix (增广矩阵)
Cramer's rule (Cramer 法则)
direct method (直接解法)
rounding error (舍入误差)
Gauss elimination method (Gauss 消元法)
matrix triangular decomposition method (矩阵三角分解法)
system of upper triangular equations (上三角方程组)
elementary linear transformations (初等线性变换)
backward-substitute (回代)
backward-substitution process (回代过程)
Gauss elimination process (Gauss 消元过程)
pivot elements (主元素)
multiplier (乘数)
maximal column pivoting (最大列主元)
scaled-column pivoting (按比例列主元)
Gauss elimination with partial pivoting (Gauss 列主元消元法)
Gauss elimination with scaled partial pivoting (Gauss 按比例主元消元法)
leading principal submatrices (顺序主子矩阵)
each of its leading principal submatrices has a nonzero determinant (各阶顺序主子式非零)
Doolittle factorization (Doolittle 分解)
Doolittle decomposition (Doolittle 分解)
LU decomposition (LU 分解)
Crout factorization (Crout 分解)
permutation matrix (置换矩阵)
strictly diagonally dominant matrix (严格对角占优矩阵)
strictly diagonally dominant tridiagonal matrix (严格对角占优三对角矩阵)
positive definite matrix (正定矩阵)
Cholesky factorization (Cholesky 分解)

2.9 延伸阅读 (Extending reading)

求解线性方程组的最小二乘方法 (Least squares method for solving the system of linear equations)

在本章中，我们详细介绍了线性方程组 $\boldsymbol{Ax}=\boldsymbol{b}$ 的求解问题，其中 $\boldsymbol{A}\in\mathbf{R}^{n\times n}$，$\boldsymbol{x}\in\mathbf{R}^{n}$，$\boldsymbol{b}\in\mathbf{R}^{n}$. 然而，在诸如数据拟合 (data fitting) 等某些实际问题中，涉及的线性方程组形如：$\boldsymbol{Ax}=\boldsymbol{b}$，其中，$\boldsymbol{A}\in\mathbf{R}^{m\times n}$，$\boldsymbol{x}\in\mathbf{R}^{n}$，$\boldsymbol{b}\in\mathbf{R}^{m}$，$m>n$，亦即，方程的个数多于未知数的个数. 此时，系数矩阵不再是方阵，这样的方程组属于超定方程组，通常无解，或无法求出精确解. 此时，研究如何近似求解 $\boldsymbol{Ax}=\boldsymbol{b}$，即，求解向量 $\boldsymbol{x}\in\mathbf{R}^{n}$，使得：$\boldsymbol{Ax}-\boldsymbol{b}\approx\boldsymbol{0}$ 就显得很有必要. 令 $\boldsymbol{r}=\boldsymbol{Ax}-\boldsymbol{b}$，称 $\boldsymbol{r}$ 为残差向量 (residual vector)，求解向量 $\boldsymbol{x}\in\mathbf{R}^{n}$，使残差向量 $\boldsymbol{r}$ 达到最小，是很自然的指导思想. 为此，考虑如下最小二乘问题 (least squares problem)：

$$\min_{x\in\mathbf{R}^{n}}\|\boldsymbol{Ax}-\boldsymbol{b}\|_2, \tag{2.9.1}$$

其中，$\|\cdot\|_2$ 表示向量的 2-范数(2-norm of vector)，定义为

$$\|\boldsymbol{x}\|_2=\left(\sum_{i}^{n}x_i^2\right)^{1/2},\quad \boldsymbol{x}\in\mathbf{R}^{n}. \tag{2.9.2}$$

$\|\boldsymbol{Ax}-\boldsymbol{b}\|_2$ 的极小化问题显然等价于 $\|\boldsymbol{Ax}-\boldsymbol{b}\|_2^2$ 的极小化问题. 注意到，

$$\|\boldsymbol{Ax}-\boldsymbol{b}\|_2^2=(\boldsymbol{Ax}-\boldsymbol{b})^{\mathrm{T}}(\boldsymbol{Ax}-\boldsymbol{b})=\boldsymbol{x}^{\mathrm{T}}\boldsymbol{A}^{\mathrm{T}}\boldsymbol{Ax}-2\boldsymbol{x}^{\mathrm{T}}\boldsymbol{A}^{\mathrm{T}}\boldsymbol{b}+\boldsymbol{b}^{\mathrm{T}}\boldsymbol{b},$$

是一个关于 $\boldsymbol{x}$ 的 n 个分量 (components) $x_1,x_2,\cdots,x_n$ 的非负二次函数(nonnegative quadratic function). 记

$$f(x_1,x_2,\cdots,x_n)=\|\boldsymbol{Ax}-\boldsymbol{b}\|_2^2=\boldsymbol{x}^{\mathrm{T}}\boldsymbol{A}^{\mathrm{T}}\boldsymbol{Ax}-2\boldsymbol{x}^{\mathrm{T}}\boldsymbol{A}^{\mathrm{T}}\boldsymbol{b}+\boldsymbol{b}^{\mathrm{T}}\boldsymbol{b}. \tag{2.9.3}$$

令 $\dfrac{\partial f}{\partial x_i}=0,\quad i=1,2,\cdots,n$，推得

$$\boldsymbol{A}^{\mathrm{T}}\boldsymbol{Ax}=\boldsymbol{A}^{\mathrm{T}}\boldsymbol{b}. \tag{2.9.4}$$

令 $\boldsymbol{B}=\boldsymbol{A}^{\mathrm{T}}\boldsymbol{A}$，则 $\boldsymbol{B}\in\mathbf{R}^{n\times n}$，是一个对称的 n 阶方阵，如果矩阵 $\boldsymbol{A}\in\mathbf{R}^{m\times n}$，$m>n$，是满秩的，即，$\operatorname{rank}(\boldsymbol{A})=n$，此时 $\boldsymbol{B}$ 是一个非奇异矩阵，称 $\boldsymbol{B}$ 为规范矩阵 (normal matrix)，而方程组 $\boldsymbol{Bx}=\boldsymbol{A}^{\mathrm{T}}\boldsymbol{b}$ 称作规范方程组 (the system of normal equations).

值得注意的是，由最小二乘问题诱导出的规范方程组的数值求解会遇到一些困难. 如果矩阵 $\boldsymbol{A}\in\mathbf{R}^{m\times n}$ $(m\geqslant n)$ 是严重病态的 (quite ill-conditioned)，则规范矩阵 $\boldsymbol{B}=\boldsymbol{A}^{\mathrm{T}}\boldsymbol{A}$，就可能是极端病态的 (extremely ill-conditioned) (关于病态矩阵的定义，参见 Chapter 3). 例如，如果

$$\boldsymbol{A}=\begin{pmatrix}\varepsilon & 0\\ 0 & 1\end{pmatrix},\quad 0<\varepsilon<1,$$

则有

$$A^{-1}=\begin{pmatrix}\varepsilon^{-1} & 0\\ 0 & 1\end{pmatrix},\quad B=A^{\mathrm T}A=\begin{pmatrix}\varepsilon^{2} & 0\\ 0 & 1\end{pmatrix},\quad B^{-1}=\begin{pmatrix}\varepsilon^{-2} & 0\\ 0 & 1\end{pmatrix},$$

于是

$$\mathrm{Cond}(A)=\|A\|_\infty\|A^{-1}\|_\infty=\varepsilon^{-1},\quad \mathrm{Cond}(B)=\|B\|_\infty\|B^{-1}\|_\infty=\varepsilon^{-2}.$$

对正整数 $N\gg 1$，当 $\varepsilon\in(0,\frac{1}{N})$ 时，有

$$\mathrm{Cond}(A)=\varepsilon^{-1}>N\gg 1,\ \mathrm{Cond}(B)=\varepsilon^{-2}=\varepsilon^{-1}\varepsilon^{-1}>N\varepsilon^{-1}=N\,\mathrm{Cond}(A).$$

上例表明，矩阵 A 是严重病态的，而矩阵 $B=A^{\mathrm T}A$ 则是极端病态的.

为了避免求解病态方程组，介绍一种矩阵正交分解算法.

定理 2.12　设 $A\in\mathbf{R}^{m\times n}, m\geqslant n$，则矩阵 A 可以分解为如下形式：

$$A=\hat{Q}\hat{R},\tag{2.9.5}$$

其中，$\hat{R}$ 是一个上三角 $n\times n$ 矩阵，$\hat{Q}$ 是一个 $m\times n$ 矩阵，满足：

$$\hat{Q}^{\mathrm T}\hat{Q}=I_n,\tag{2.9.6}$$

其中，I_n 为 $n\times n$ 单位矩阵. 如果 $\mathrm{rank}(A)=n$，则 $\hat{R}$ 非奇异.

证明　对 n 采用归纳法证明. 当 $n=1$ 时，矩阵 A 实际上是一个列向量，记作 c，假设 $c\neq 0$，此时，令 $\hat{Q}=c/\|c\|_2$，$\hat{R}=\|c\|_2$，则 $A=\hat{Q}\hat{R}$，$\hat{Q}^{\mathrm T}\hat{Q}=1=I_1$. 如果 $c=0$，则令 $\hat{R}=0$，而取 $\hat{Q}$ 为 $\mathbf{R}^m$ 中任意 2-范数为 1 的列向量，即可. 现设定理的结论对 $n=k, 1\leqslant k<m$，为真. 现设 A_{k+1} 是一个 m 行 $k+1$ 列的矩阵，将其写成

$$A_{k+1}=(A_k\ a),\quad A_k\in\mathbf{R}^{m\times k},\quad a\in\mathbf{R}^m,$$

因为 $A_k\in\mathbf{R}^{m\times k}$，由归纳法假设，存在 $\hat{Q}_k\in\mathbf{R}^{m\times k}$，$\hat{R}_k\in\mathbf{R}^{k\times k}$，使得 $A_k=\hat{Q}_k\hat{R}_k$，其中，$\hat{R}_k$ 是上三角矩阵，$\hat{Q}_k$ 满足：$\hat{Q}_k^{\mathrm T}\hat{Q}_k=I_k$. 现在证明：存在 $\hat{Q}_{k+1}\in\mathbf{R}^{m\times(k+1)}$，$\hat{R}_{k+1}\in\mathbf{R}^{(k+1)\times(k+1)}$，使得 $A_{k+1}=\hat{Q}_{k+1}\hat{R}_{k+1}$，其中，$\hat{R}_{k+1}$ 是上三角矩阵，$\hat{Q}_{k+1}$ 满足：$\hat{Q}_{k+1}^{\mathrm T}\hat{Q}_{k+1}=I_{k+1}$. 为此，设

$$\hat{Q}_{k+1}=(\hat{Q}_k\ b),\ \hat{R}_{k+1}=\begin{pmatrix}\hat{R}_k & c\\ 0 & 1\end{pmatrix},\tag{2.9.7}$$

其中，$b\in\mathbf{R}^m$，$c\in\mathbf{R}^k$，则由矩阵分块乘法，有

$$\hat{Q}_{k+1}\hat{R}_{k+1}=(\hat{Q}_k\hat{R}_k\ \hat{Q}_kc+b),$$

由于 $A_k=\hat{Q}_k\hat{R}_k$，因此，欲证 $A_{k+1}=\hat{Q}_{k+1}\hat{R}_{k+1}$，只需证明：$(A_k\ \hat{Q}_kc+b)=A_{k+1}=(A_k\ a)$，从而只需证明存在 $b\in\mathbf{R}^m$，$c\in\mathbf{R}^k$，使得 $\hat{Q}_kc+b=a$，即可. 因 $\hat{Q}_{k+1}^{\mathrm T}\hat{Q}_{k+1}=I_{k+1}$，有

$$\begin{pmatrix}\hat{Q}_k^{\mathrm T}\\ b^{\mathrm T}\end{pmatrix}(\hat{Q}_k\ b)=\begin{pmatrix}\hat{Q}_k^{\mathrm T}\hat{Q}_k & \hat{Q}_k^{\mathrm T}b\\ b^{\mathrm T}\hat{Q}_k & b^{\mathrm T}b\end{pmatrix}=\begin{pmatrix}I_k & 0\\ 0 & 1\end{pmatrix}.\tag{2.9.8}$$

因此，向量 $b\in\mathbf{R}^m$，可通过求解 $\hat{Q}_k^{\mathrm T}b=0$ 和 $b^{\mathrm T}b=1$ 得到. 对 $x\in\mathbf{R}^m$，因 $\hat{Q}_k^{\mathrm T}\in\mathbf{R}^{k\times m}$，

且 $k<m$, 故线性方程组 $\hat{\boldsymbol{Q}}_k^{\mathrm{T}}\boldsymbol{x}=0$ 有无穷多个非零解，任意取其中一个非零解 $\boldsymbol{x}_0\in\mathbf{R}^m$，并令 $\boldsymbol{b}=\boldsymbol{x}_0/\|\boldsymbol{x}_0\|_2$，则 $\boldsymbol{b}\in\mathbf{R}^m$，且 $\hat{\boldsymbol{Q}}_k^{\mathrm{T}}\boldsymbol{b}=0$，$\boldsymbol{b}^{\mathrm{T}}\boldsymbol{b}=1$. 求出 $\boldsymbol{b}=\boldsymbol{x}_0/\|\boldsymbol{x}_0\|_2$ 后，由 $\hat{\boldsymbol{Q}}_k\boldsymbol{c}+\boldsymbol{b}=\boldsymbol{a}$，即可求解出 $\boldsymbol{c}$. 实际上，由 $\hat{\boldsymbol{Q}}_k\boldsymbol{c}+\boldsymbol{b}=\boldsymbol{a}$ 得：$\hat{\boldsymbol{Q}}_k\boldsymbol{c}=\boldsymbol{a}-\boldsymbol{b}$，于是，$\hat{\boldsymbol{Q}}_k^{\mathrm{T}}\hat{\boldsymbol{Q}}_k\boldsymbol{c}=\hat{\boldsymbol{Q}}_k^{\mathrm{T}}(\boldsymbol{a}-\boldsymbol{b})=\hat{\boldsymbol{Q}}_k^{\mathrm{T}}\boldsymbol{a}$，因 $\hat{\boldsymbol{Q}}_k^{\mathrm{T}}\hat{\boldsymbol{Q}}_k=\boldsymbol{I}_k$， 故有 $\boldsymbol{c}=\hat{\boldsymbol{Q}}_k^{\mathrm{T}}\boldsymbol{a}$. 至此，由归纳法原理，即证得定理 2.12 的矩阵分解结论. 最后证明，如果 $\operatorname{rank}(\boldsymbol{A})=n$，则 $\hat{\boldsymbol{R}}$ 非奇异. 用反证法. 假设 $\hat{\boldsymbol{R}}\in\mathbf{R}^{n\times n}$ 奇异，则存在非零向量 $\boldsymbol{p}\in\mathbf{R}^n$，使得 $\hat{\boldsymbol{R}}\boldsymbol{p}=\boldsymbol{0}$，从而由 $\boldsymbol{A}=\hat{\boldsymbol{Q}}\hat{\boldsymbol{R}}$ 推得 $\boldsymbol{A}\boldsymbol{p}=0$，这说明矩阵 $\boldsymbol{A}\in\mathbf{R}^{m\times n}$ 的 n 个列向量线性相关，因此，$\operatorname{rank}(\boldsymbol{A})<n$，与 $\operatorname{rank}(\boldsymbol{A})=n$ 矛盾. 定理 2.12 证毕.

定理 2.12 中描述的矩阵分解也称为 $\boldsymbol{QR}$ 分解，属于一种正交分解.

定理 2.13 设 $A\in\mathbf{R}^{m\times n}, m\geqslant n$，且 $\operatorname{rank}(A)=n$. 则对于 $\boldsymbol{b}\in\mathbf{R}^m$，方程组 $\boldsymbol{A}\boldsymbol{x}=\boldsymbol{b}$ 存在唯一的最小二乘解 (unique least squares solution) $\boldsymbol{x}^*\in\mathbf{R}^n$, 即，

$$\min_{x\in\mathbf{R}^n}\|\boldsymbol{A}\boldsymbol{x}-\boldsymbol{b}\|_2=\|\boldsymbol{A}\boldsymbol{x}^*-\boldsymbol{b}\|_2, \tag{2.9.9}$$

利用定理 2.12 关于矩阵 $\boldsymbol{A}$ 的分解 $\boldsymbol{A}=\hat{\boldsymbol{Q}}\hat{\boldsymbol{R}}$，通过求解非奇异上三角方程组 (nonsingular upper triangular system) $\hat{\boldsymbol{R}}\boldsymbol{x}=\hat{\boldsymbol{Q}}^{\mathrm{T}}\boldsymbol{b}$, 即可求得最小二乘解 $\boldsymbol{x}^*$.

证明 由定理 2.12，矩阵 $\boldsymbol{A}$ 可以分解为如下形式：

$$\boldsymbol{A}=\hat{\boldsymbol{Q}}\hat{\boldsymbol{R}},$$

其中，$\hat{\boldsymbol{R}}$ 是一个上三角 $n\times n$ 矩阵，$\hat{\boldsymbol{Q}}$ 是一个 $m\times n$ 矩阵，满足：

$$\hat{\boldsymbol{Q}}^{\mathrm{T}}\hat{\boldsymbol{Q}}=\boldsymbol{I}_n.$$

当 $m=n$ 时，由 $\operatorname{rank}(\boldsymbol{A})=n$, 推得

$$\boldsymbol{x}=\boldsymbol{A}^{-1}\boldsymbol{b}=\left(\hat{\boldsymbol{Q}}\hat{\boldsymbol{R}}\right)^{-1}\boldsymbol{b}=\hat{\boldsymbol{R}}^{-1}\hat{\boldsymbol{Q}}^{-1}\boldsymbol{b}=\hat{\boldsymbol{R}}^{-1}\hat{\boldsymbol{Q}}^{\mathrm{T}}\boldsymbol{b},$$

因此，$\hat{\boldsymbol{R}}\boldsymbol{x}=\hat{\boldsymbol{Q}}^{\mathrm{T}}\boldsymbol{b}$，定理的结论成立. 接下来，只要论证 $m>n$ 情形即可. 记 $\hat{\boldsymbol{Q}}=(\boldsymbol{q}_1\ \boldsymbol{q}_2\cdots\boldsymbol{q}_n)$，其中，$\boldsymbol{q}_i\in\mathbf{R}^m, i=1,2,\cdots,n$, 由 $\hat{\boldsymbol{Q}}^{\mathrm{T}}\hat{\boldsymbol{Q}}=\boldsymbol{I}_n$ 知，$\operatorname{rank}(\hat{\boldsymbol{Q}})=n$， 且 $\hat{\boldsymbol{Q}}$ 的 n 列 $\boldsymbol{q}_1,\boldsymbol{q}_2,\cdots,\boldsymbol{q}_n$ 形成了 m 维空间 $\mathbf{R}^m$ 中的一组正交向量. 令

$$S_n=\operatorname{Span}\{\boldsymbol{q}_1,\boldsymbol{q}_2,\cdots,\boldsymbol{q}_n\}$$

表示正交向量 $\boldsymbol{q}_1,\boldsymbol{q}_2,\cdots,\boldsymbol{q}_n$ 张成的空间，则 S_n 是 $\mathbf{R}^m$ 的一个 n 维子空间，令

$$S_n^*=\mathbf{R}^m-S_n,$$

则 S_n^* 是 S_n 在 $\mathbf{R}^m$ 中的正交补 (直交补) 空间 (the orthogonal complement of the space). 因此，对于 $\boldsymbol{b}\in\mathbf{R}^m$，$\boldsymbol{b}$ 可以表示成 $\boldsymbol{b}=\boldsymbol{b}_q+\boldsymbol{b}_o$，其中，$\boldsymbol{b}_q\in S_n, \boldsymbol{b}_o\in S_n^*$. $\boldsymbol{b}_q\in S_n$ 意味着存在 $\boldsymbol{c}\in\mathbf{R}^n$, 使得 $\boldsymbol{b}_q=\hat{\boldsymbol{Q}}\boldsymbol{c}$； $\boldsymbol{b}_o\in S_n^*$ 意味着 $\hat{\boldsymbol{Q}}^{\mathrm{T}}\boldsymbol{b}_o=\boldsymbol{0}$.

现设 $\boldsymbol{x}^*\in\mathbf{R}^n$, 是 $\hat{\boldsymbol{R}}\boldsymbol{x}=\hat{\boldsymbol{Q}}^{\mathrm{T}}\boldsymbol{b}$ 的解，即，$\hat{\boldsymbol{R}}\boldsymbol{x}^*=\hat{\boldsymbol{Q}}^{\mathrm{T}}\boldsymbol{b}$. $\forall\boldsymbol{y}\in\mathbf{R}^n$, 反复利用 $\boldsymbol{b}_q=\hat{\boldsymbol{Q}}\boldsymbol{c}$, $\hat{\boldsymbol{Q}}^{\mathrm{T}}\boldsymbol{b}_o=\boldsymbol{0}$, 有

$$\boldsymbol{A}\boldsymbol{y}-\boldsymbol{b}=\hat{\boldsymbol{Q}}\hat{\boldsymbol{R}}\boldsymbol{y}-\boldsymbol{b}$$

$$
\begin{aligned}
&= \hat{\boldsymbol{Q}}\hat{\boldsymbol{R}}(\boldsymbol{y}-\boldsymbol{x}^*)+\hat{\boldsymbol{Q}}\hat{\boldsymbol{R}}\boldsymbol{x}^*-\boldsymbol{b} \\
&= \hat{\boldsymbol{Q}}\hat{\boldsymbol{R}}(\boldsymbol{y}-\boldsymbol{x}^*)+\hat{\boldsymbol{Q}}\hat{\boldsymbol{Q}}^{\mathrm{T}}\boldsymbol{b}-\boldsymbol{b} \\
&= \hat{\boldsymbol{Q}}\hat{\boldsymbol{R}}(\boldsymbol{y}-\boldsymbol{x}^*)+\hat{\boldsymbol{Q}}\hat{\boldsymbol{Q}}^{\mathrm{T}}(\boldsymbol{b}_q+\boldsymbol{b}_o)-(\boldsymbol{b}_q+\boldsymbol{b}_o) \\
&= \hat{\boldsymbol{Q}}\hat{\boldsymbol{R}}(\boldsymbol{y}-\boldsymbol{x}^*)+\hat{\boldsymbol{Q}}\hat{\boldsymbol{Q}}^{\mathrm{T}}\boldsymbol{b}_q-\boldsymbol{b}_q+\hat{\boldsymbol{Q}}\hat{\boldsymbol{Q}}^{\mathrm{T}}\boldsymbol{b}_o-\boldsymbol{b}_o \\
&= \hat{\boldsymbol{Q}}\hat{\boldsymbol{R}}(\boldsymbol{y}-\boldsymbol{x}^*)+\hat{\boldsymbol{Q}}\hat{\boldsymbol{Q}}^{\mathrm{T}}\hat{\boldsymbol{Q}}\boldsymbol{c}-\boldsymbol{b}_q-\boldsymbol{b}_o \\
&= \hat{\boldsymbol{Q}}\hat{\boldsymbol{R}}(\boldsymbol{y}-\boldsymbol{x}^*)+\hat{\boldsymbol{Q}}\boldsymbol{c}-\boldsymbol{b}_q-\boldsymbol{b}_o \\
&= \hat{\boldsymbol{Q}}\hat{\boldsymbol{R}}(\boldsymbol{y}-\boldsymbol{x}^*)-\boldsymbol{b}_o,
\end{aligned}
$$

因此，利用 $\boldsymbol{b}_q=\hat{\boldsymbol{Q}}\boldsymbol{c}$，$\hat{\boldsymbol{Q}}^{\mathrm{T}}\hat{\boldsymbol{Q}}=\boldsymbol{I}_n$，推得

$$
\begin{aligned}
\|\boldsymbol{A}\boldsymbol{y}-\boldsymbol{b}\|_2^2 &= (\hat{\boldsymbol{Q}}\hat{\boldsymbol{R}}(\boldsymbol{y}-\boldsymbol{x}^*)-\boldsymbol{b}_o)^{\mathrm{T}}(\hat{\boldsymbol{Q}}\hat{\boldsymbol{R}}(\boldsymbol{y}-\boldsymbol{x}^*)-\boldsymbol{b}_o) \\
&= (\boldsymbol{y}-\boldsymbol{x}^*)^{\mathrm{T}}\hat{\boldsymbol{R}}^{\mathrm{T}}\hat{\boldsymbol{Q}}^{\mathrm{T}}\hat{\boldsymbol{Q}}\hat{\boldsymbol{R}}(\boldsymbol{y}-\boldsymbol{x}^*)+\boldsymbol{b}_o^{\mathrm{T}}\boldsymbol{b}_o-2(\boldsymbol{y}-\boldsymbol{x}^*)^{\mathrm{T}}\hat{\boldsymbol{R}}^{\mathrm{T}}\hat{\boldsymbol{Q}}^{\mathrm{T}}\boldsymbol{b}_o \\
&= (\boldsymbol{y}-\boldsymbol{x}^*)^{\mathrm{T}}\hat{\boldsymbol{R}}^{\mathrm{T}}\hat{\boldsymbol{R}}(\boldsymbol{y}-\boldsymbol{x}^*)+\boldsymbol{b}_o^{\mathrm{T}}\boldsymbol{b}_o \\
&= \left\|\hat{\boldsymbol{R}}(\boldsymbol{y}-\boldsymbol{x}^*)\right\|_2^2+\|\boldsymbol{b}_o\|_2^2 \\
&\geqslant \|\boldsymbol{b}_o\|_2^2,
\end{aligned}
$$

其中，等式成立，当且仅当 $\boldsymbol{y}=\boldsymbol{x}^*$．因此，由方程组 $\hat{\boldsymbol{R}}\boldsymbol{x}^*=\hat{\boldsymbol{Q}}^{\mathrm{T}}\boldsymbol{b}$ 解得的 $\boldsymbol{x}^*$ 满足：

$$
\min_{\boldsymbol{y}\in\mathbb{R}^n}\|\boldsymbol{A}\boldsymbol{y}-\boldsymbol{b}\|_2=\left\|\boldsymbol{A}\boldsymbol{x}^*-\boldsymbol{b}\right\|_2,
$$

即，方程组 $\boldsymbol{A}\boldsymbol{x}=\boldsymbol{b}$ 存在唯一的最小二乘解 $\boldsymbol{x}^*\in\mathbf{R}^n$，且 $\boldsymbol{x}^*$ 可以从上三角方程组 $\hat{\boldsymbol{R}}\boldsymbol{x}=\hat{\boldsymbol{Q}}^{\mathrm{T}}\boldsymbol{b}$ 求解得到.

定理 2.13 证毕.

2.10　习题 (Exercises)

1. Use Gauss elimination with backward substitution and four-digit rounding arithmetic (使用 4 位浮点数计算) to solve the following linear system.

a. $$
\begin{cases}
1.19x_1+2.11x_2-100x_3+x_4=1.12, \\
14.2x_1-0.122x_2+12.2x_3-x_4=3.44, \\
100x_2-99.9x_3+x_4=2.15, \\
15.3x_1+0.110x_2-13.1x_3-x_4=4.16.
\end{cases}
$$

b. $\begin{cases} 4t_1 - t_2 - t_4 = 30, \\ -t_1 + 4t_2 - t_3 - t_5 = 20, \\ -t_2 + 4t_3 - t_6 = 60, \\ -t_1 + 4t_4 - t_5 = 40, \\ -t_2 - t_4 + 4t_5 - t_6 = 30, \\ -t_3 - t_5 + 4t_6 = 70. \end{cases}$

2. Solve the linear system using three-digit rounding arithmetic with partial pivoting scaled partial pivoting.

(1) $\begin{cases} -3x_1 + 2x_2 + 6x_3 = 4, \\ 10x_1 - 7x_2 = 7, \\ 5x_1 - x_2 + 5x_3 = 6. \end{cases}$ (2) $\begin{cases} -3.01x_1 + 0.921x_2 + 2.16x_3 = 4.01, \\ 0.210x_1 - 4.27x_2 + 1.73x_3 = 7.32, \\ 2.25x_1 + 0.982x_2 - 5.87x_3 = 6.28. \end{cases}$

3. Suppose

$$A = \begin{pmatrix} 4 & -1 & 0 & -1 & 0 & 0 \\ -1 & 4 & -1 & 0 & -1 & 0 \\ 0 & -1 & 4 & 0 & 0 & -1 \\ -1 & 0 & 0 & 4 & -1 & 0 \\ 0 & -1 & 0 & -1 & 4 & -1 \\ 0 & 0 & -1 & 0 & -1 & 4 \end{pmatrix}.$$

(1) Use Doolittle factorization and Crout factorization to factor matrix $\boldsymbol{A}$ into form $\boldsymbol{A} = \boldsymbol{LU}$ respectively;

(2) Whether can matrix $\boldsymbol{A}$ be factored into form $\boldsymbol{A} = \boldsymbol{LDL}^{\mathrm{T}}, \boldsymbol{A} = \boldsymbol{LL}^{\mathrm{T}}$? If yes, please factor $\boldsymbol{A}$ to corresponding form.

4. Using Cholesky factorization to solve the following linear system of equations

$$\begin{cases} 4x_1 - x_2 + x_3 = 6, \\ -x_1 + 4.25x_2 + 2.75x_3 = -0.5, \\ x_1 + 2.75x_2 + 3.5x_3 = 1.25. \end{cases}$$

5. Using $\boldsymbol{LDL}^{\mathrm{T}}$ factorization to solve the following linear system of equations

$$\begin{bmatrix} 3 & 3 & 5 \\ 3 & 5 & 9 \\ 5 & 9 & 17 \end{bmatrix} \begin{bmatrix} x_1 \\ x_2 \\ x_3 \end{bmatrix} = \begin{bmatrix} 10 \\ 16 \\ 30 \end{bmatrix}.$$

6. Solve the following linear system of equations

$$\begin{bmatrix} 2 & 1 & & \\ 1 & 3 & 1 & \\ & 1 & 1 & 1 \\ & & 2 & 1 \end{bmatrix} \begin{bmatrix} x_1 \\ x_2 \\ x_3 \\ x_4 \end{bmatrix} = \begin{bmatrix} 1 \\ 2 \\ 2 \\ 0 \end{bmatrix}.$$

7. Let $\boldsymbol{A}$ be n order nonsingular matrix, and $\boldsymbol{A}$ can be factored in the form $\boldsymbol{A} = \boldsymbol{LU}$. Show that

the determinants of all the leading principal submatrices (顺序主子式) of matrix $\boldsymbol{A}$ are nonzero.

8. Using Crout factorization to factor tridiagonal matrix (三对角矩阵)

$$\boldsymbol{A}=\begin{bmatrix} b_1 & c_1 & & & \\ a_2 & b_2 & c_2 & & \\ & \ddots & \ddots & \ddots & \\ & & a_{n-1} & b_{n-1} & c_{n-1} \\ & & & a_n & b_n \end{bmatrix}.$$

9. Show that the total number of multiplications/divisions operations in Gauss elimination to solve n order linear system of equations is $\frac{1}{3}n^3+n^2-\frac{1}{3}n$.

Chapter 3 Iterative Techniques in Matrix Algebra

(矩阵代数迭代技术)

本章内容 (Contents)

3.1 Norms of vectors and matrices (向量范数与矩阵范数)
3.2 Eigenvalues and eigenvectors (特征值和特征向量)
3.3 Iterative techniques for solving linear systems (解线性方程组的迭代法)
3.4 Convergence analysis and SOR iterative method (收敛性分析与 SOR 迭代法)
3.5 Condition number and iterative refinement (条件数和迭代优化)

学习目标 (Learning Objectives)

- 了解向量范数和矩阵范数的概念
- 了解向量距离和矩阵距离的概念
- 掌握向量范数和矩阵范数的性质
- 会求基本的向量范数和矩阵范数
- 掌握特征值和特征向量的概念
- 会求矩阵的特征值和特征向量
- 掌握迭代的基本思想
- 熟悉迭代矩阵的概念
- 掌握并熟练应用雅可比迭代方法
- 掌握并熟练应用高斯-赛德尔迭代方法
- 掌握迭代松弛方法
- 掌握并熟练应用 SOR 迭代方法
- 熟悉各迭代方法的收敛条件
- 了解残差向量的概念
- 了解条件数的概念
- 掌握迭代优化的思想

3.1 Norms of vectors and matrices (向量范数与矩阵范数)

- Vector norm (向量范数)
- Distance between vectors (向量之间的距离)
- Matrix norm and distance (矩阵范数和距离)
- Examples (例题)

3.1.1 Vector norm (向量范数)

Let $\mathbf{R}^n$ denote the set of all n-dimensional column vectors with real-number components. To define a distance in $\mathbf{R}^n$, we use the notion of a norm, which is the generalization of the absolute value on $\mathbf{R}$, the set of real numbers.

Definition 3.1 A vector norm (向量范数) on $\mathbf{R}^n$ is a function, $\|\cdot\|$, from $\mathbf{R}^n$ into $\mathbf{R}$ with the following properties:

(a) $\|\boldsymbol{x}\| \geqslant 0$ for all $\boldsymbol{x} \in \mathbf{R}^n$;

(b) $\|\boldsymbol{x}\| = 0$ if and only if $\boldsymbol{x} = 0$; (Positivity, 正定性)

(c) $\|\alpha \boldsymbol{x}\| = |\alpha| \|\boldsymbol{x}\|$ for all $\alpha \in \mathbf{R}$ and $\boldsymbol{x} \in \mathbf{R}^n$; (Homogeneity, 齐次性)

(d) $\|\boldsymbol{x}+\boldsymbol{y}\| \leqslant \|\boldsymbol{x}\| + \|\boldsymbol{y}\|$ for all $\boldsymbol{x}, \boldsymbol{y} \in \mathbf{R}^n$. (Triangle inequality, 三角不等式)

Definition 3.2 For the vector $\boldsymbol{x} = (x_1, x_2, \cdots, x_n)^{\mathrm{T}}$, the l_1, l_2 and l_∞ norms are defined by

$$\|\boldsymbol{x}\|_1 = \sum_{i=1}^{n} |x_i|, \quad \|\boldsymbol{x}\|_2 = \left[\sum_{i=1}^{n} x_i^2\right]^{1/2} \text{ and } \|\boldsymbol{x}\|_\infty = \max_{1 \leqslant i \leqslant n} |x_i|. \tag{3.1.1}$$

The l_2 norm is also called the Euclidean norm (欧几里得范数) of the vector $\boldsymbol{x}$ since it represents the usual notion of distance from the origin in case $\boldsymbol{x}$ is in $\mathbf{R}^1 \equiv \mathbf{R}$ and $\mathbf{R}^2$ or $\mathbf{R}^3$.

For example, the l_2 norm of the vector $\boldsymbol{x} = (x_1, x_2, x_3)^{\mathrm{T}}$ gives the length of the line segment joining the points $(0,0,0)$ and (x_1, x_2, x_3).

Since vectors in $\mathbf{R}^n$ are column vectors, it is convenient to use the transpose notation when a vector is represented in terms of its components. For example, the vector

$$\boldsymbol{x} = \begin{bmatrix} x_1 \\ x_2 \\ \vdots \\ x_n \end{bmatrix}$$

will be written as

$$\boldsymbol{x} = (x_1, x_2, \cdots, x_n)^{\mathrm{T}}, \quad \boldsymbol{x} = (x_1, x_2, \cdots, x_n)^{\mathrm{t}} \quad \text{or} \quad \boldsymbol{x} = (x_1, x_2, \cdots, x_n)'.$$

It is easy to show that the properties in Definition 3.1 hold for the l_∞ norm because they follow from similar results for absolute values. The only property that requires much demonstration is Definition 3.1(d), and in this case if $\boldsymbol{x}=(x_1,x_2,\cdots,x_n)^{\mathrm{T}}$ and $\boldsymbol{y}=(y_1,y_2,\cdots,y_n)^{\mathrm{T}}$, then

$$\|\boldsymbol{x}+\boldsymbol{y}\|_\infty=\max_{1\leqslant i\leqslant n}|x_i+y_i|\leqslant\max_{1\leqslant i\leqslant n}(|x_i|+|y_i|)\leqslant\max_{1\leqslant i\leqslant n}|x_i|+\max_{1\leqslant i\leqslant n}|y_i|=\|\boldsymbol{x}\|_\infty+\|\boldsymbol{y}\|_\infty.$$

The first three conditions are also easy to show for the l_2 norm. But to show that

$$\|\boldsymbol{x}+\boldsymbol{y}\|_2\leqslant\|\boldsymbol{x}\|_2+\|\boldsymbol{y}\|_2,\ \text{for each } \boldsymbol{x},\boldsymbol{y}\in\mathbf{R}^n,$$

we need the following famous inequality.

Theorem 3.1 (Cauchy-Bunyakovsky-Schwarz Inequality for Sums, 柯西-施瓦兹不等式)

For each $\boldsymbol{x}=(x_1,x_2,\cdots,x_n)^{\mathrm{T}}$ and $\boldsymbol{y}=(y_1,y_2,\cdots,y_n)^{\mathrm{T}}$ in $\mathbf{R}^n$,

$$\boldsymbol{x}^{\mathrm{T}}\boldsymbol{y}=\sum_{i=1}^n x_iy_i\leqslant\left\{\sum_{i=1}^n x_i^2\right\}^{1/2}\left\{\sum_{i=1}^n y_i^2\right\}^{1/2}=\|\boldsymbol{x}\|_2\cdot\|\boldsymbol{y}\|_2.$$

3.1.2 Distance between vectors (向量之间的距离)

The norm of a vector gives a measure for the distance between an arbitrary vector and the zero vector, just as the absolute value of a real number describes its distance from 0. Similarly, the distance between two vectors is defined as the norm of the difference of the vectors just as distance between two real numbers is the absolute value of their difference.

Definition 3.3 If $\boldsymbol{x}=(x_1,x_2,\cdots,x_n)^{\mathrm{T}}$ and $\boldsymbol{y}=(y_1,y_2,\cdots,y_n)^{\mathrm{T}}$ are vectors in the $\mathbf{R}^n$, l_2 and l_∞ distances (距离) between $\boldsymbol{x}$ and $\boldsymbol{y}$ are defined by

$$\|\boldsymbol{x}-\boldsymbol{y}\|_2=\left[\sum_{i=1}^n(x_i-y_i)^2\right]^{1/2}\quad\text{and}\quad\|\boldsymbol{x}-\boldsymbol{y}\|_\infty=\max_{1\leqslant i\leqslant n}|x_i-y_i|.$$

Definition 3.4 A sequence $\left\{\boldsymbol{x}^{(k)}\right\}_{k=1}^\infty$ of vectors in $\mathbf{R}^n$ is said to converge to (收敛至) $\boldsymbol{x}$ with respect to the norm $\|\cdot\|$ if given any $\varepsilon>0$, there exists an integer $N(\varepsilon)>0$ such that

$$\|\boldsymbol{x}^{(k)}-\boldsymbol{x}\|<\varepsilon\quad\text{for any}\quad k\geqslant N(\varepsilon).$$

Theorem 3.2 The sequence of vector $\left\{\boldsymbol{x}^{(k)}\right\}_{k=1}^\infty$ converges to $\boldsymbol{x}$ in $\mathbf{R}^n$ with respect to norm $\|\cdot\|_\infty$ if and only if

$$\lim_{k\to\infty}x_i^{(k)}=x_i,\ \text{for any } i=1,2,\cdots,n.$$

Proof. Suppose $\left\{\boldsymbol{x}^{(k)}\right\}_{k=1}^\infty$ converges to $\boldsymbol{x}$ with respect to l_∞ norm. Given any $\varepsilon>0$, there exists an integer $N(\varepsilon)$ such that for all $k\geqslant N(\varepsilon)$,

$$\max_{i=1,2,\ldots,n}\left|x_i^{(k)}-x_i\right| = \|\boldsymbol{x}^{(k)}-\boldsymbol{x}\|_\infty < \varepsilon.$$

This result implies that $\left|x_i^{(k)}-x_i\right|<\varepsilon$, for each $i=1,2,\cdots,n$, so, $\lim_{k\to\infty} x_i^{(k)}=x_i$ for each i. Conversely, suppose that $\lim_{k\to\infty} x_i^{(k)}=x_i$, for every $i=1,2,\cdots,n$. For a given $\varepsilon>0$, let $N_i(\varepsilon)$ for each i represent an integer with the property that

$$\left|x_i^{(k)}-x_i\right|<\varepsilon$$

whenever $k\geqslant N_i(\varepsilon)$.

Define $N(\varepsilon)=\max_{i=1,2,\cdots,n}\left|N_i(\varepsilon)\right|$. If $k\geqslant N(\varepsilon)$, then

$$\max_{i=1,2,\cdots,n}\left|x_i^{(k)}-x_i\right| = \|\boldsymbol{x}^{(k)}-\boldsymbol{x}\|_\infty < \varepsilon.$$

This implies that $\boldsymbol{x}^{(k)}$ converges to $\boldsymbol{x}$ with respect to the l_∞ norm.

Theorem 3.3 For each $\boldsymbol{x}\in\mathbf{R}^n$, $\|\boldsymbol{x}\|_\infty\leqslant \|\boldsymbol{x}\|_2 \leqslant\sqrt{n}\|\boldsymbol{x}\|_\infty$.

Proof. Let x_j be a coordinate of $\boldsymbol{x}$ such that $\|\boldsymbol{x}\|_\infty=\max_{1\leqslant i\leqslant n}|x_i|=|x_j|$. Then on the one hand

$$\|\boldsymbol{x}\|_\infty^2=\max_{1\leqslant i\leqslant n}|x_i|^2=x_j^2\leqslant\sum_{i=1}^n x_i^2 = \|\boldsymbol{x}\|_2^2,$$

namely

$$\|\boldsymbol{x}\|_\infty\leqslant \|\boldsymbol{x}\|_2.$$

On the other hand

$$\|\boldsymbol{x}\|_2^2=\sum_{i=1}^n x_i^2\leqslant\sum_{i=1}^n x_j^2=nx_j^2=n\|\boldsymbol{x}\|_\infty^2,$$

which implies

$$\|\boldsymbol{x}\|_2\leqslant\sqrt{n}\|\boldsymbol{x}\|_\infty.$$

It can be shown that all norms on $\mathbf{R}^n$ are equivalent with respect to convergence; that is, if $\|\cdot\|$ and $\|\cdot\|'$ are any two norms on $\mathbf{R}^n$ and $\left\{\boldsymbol{x}^{(k)}\right\}_{k=1}^\infty$ has the limit $\boldsymbol{x}$ with respect to the norm $\|\cdot\|$, then also has the limit $\boldsymbol{x}$ with respect to the norm $\|\cdot\|'$: The case for the norms $\|\cdot\|'$ and $\|\cdot\|$ follows from Theorem 3.3.

3.1.3 Matrix norm and distance (矩阵范数和距离)

In the subsequent sections of this and later chapters, we will need methods for determining the distance between $n\times n$ matrices. This again requires the use of a norm.

Theorem 3.4 A matrix norm $\|\cdot\|$ on the set of all $n\times n$ matrices is a real-valued function defined on this set, satisfying for all $n\times n$ matrices $\boldsymbol{A}$ and $\boldsymbol{B}$ and all real numbers α:

(a) $\| A \| \geqslant 0$;

(b) $\| A \| = 0$, if and only if A is O, the matrix with all 0 entries;

(c) $\| \alpha A \| = |\alpha| \| A \|$;

(d) $\| A+B \| \leqslant \|A\| + \|B\|$;

(e) $\| AB \| \leqslant \|A\| \|B\|$. (Compatibility 相容性)

The distance between $n \times n$ matrices $\boldsymbol{A}$ and $\boldsymbol{B}$ with respect to this matrix norm is

$$\| A - B \| .$$

Theorem 3.5 If $\| \cdot \|$ is a vector norm on $\mathbf{R}^n$, then

$$\| A \| = \max_{\|x\|=1} \| Ax \| \tag{3.1.2}$$

is a matrix norm, where $\| A \|$ defined in Eq.(3.1.2) is called the natural, or induced, matrix norm associated with the vector norm. In this text, all matrix norms will be assumed to be natural matrix norms unless specified otherwise.

For any $\| \mathbf{z} \| \neq 0$, we have $x = \mathbf{z} / \|\mathbf{z}\|$ as a unit vector. Hence,

$$\max_{\|x\|=1} \|Ax\| = \max_{\|\mathbf{z}\|\neq 0} \left\| A\left(\frac{\mathbf{z}}{\|\mathbf{z}\|}\right) \right\| = \max_{\|\mathbf{z}\|\neq 0} \frac{\| A\mathbf{z} \|}{\| \mathbf{z} \|}$$

and we can alternatively write

$$\| A \| = \max_{\|\mathbf{z}\|\neq 0} \frac{\| A\mathbf{z} \|}{\| \mathbf{z} \|}. \tag{3.1.3}$$

Corollary 3.6 For $\| \mathbf{z} \| \neq 0$, matrix A, and any natural norm $\| \cdot \|$, we have

$$\| A\mathbf{z} \| \leqslant \| A \| \| \mathbf{z} \|. \tag{3.1.4}$$

The measure given to a matrix under a natural norm describes how the matrix stretches unit vectors relative to that norm. The maximum stretch is the norm of the matrix. The matrix norms we will consider have the forms.

$$\| A \|_\infty = \max_{\|x\|_\infty=1} \| Ax \|_\infty, \text{ the } l_\infty \text{ norm} \tag{3.1.5}$$

and

$$\| A \|_2 = \max_{\|x\|_2=1} \| Ax \|_2, \text{ the } l_2 \text{ norm.} \tag{3.1.6}$$

Theorem 3.7 If $A = (a_{ij})$ is an $n \times n$ matrix, then

$$\| A \|_\infty = \max_{1\leqslant i\leqslant n} \sum_{j=1}^{n} |a_{ij}|. \tag{3.1.7}$$

3.1.4 Examples (例题)

Example 3.1 Determine the l_2 norm and the l_∞ norm of the vector $x = (-1, 1, -2)^{\mathrm{T}}$.

Solution. The vector $\boldsymbol{x}=(-1,\ 1,-2)^{\mathrm{T}}$ in $\mathbf{R}^3$ has norms

$$\|\boldsymbol{x}\|_2=\sqrt{(-1)^2+1^2+(-2)^2}=\sqrt{6}$$

and

$$\|\boldsymbol{x}\|_\infty=\max\{|-1|,\ |1|,\ |-2|\}=2.$$

Example 3.2 Show that

$$\boldsymbol{x}^{(k)}=\left(x_1^{(k)},x_2^{(k)},x_3^{(k)},x_4^{(k)}\right)^{\mathrm{T}}=\left(1,2+\frac{1}{k},\frac{3}{k^2},\mathrm{e}^{-k}\sin k\right)^{\mathrm{T}}$$

converges to $\boldsymbol{x}=(1,\ 2,\ 0,\ 0)^{\mathrm{T}}$ with respect to the l_∞ norm.

Solution. Since

$$\lim_{k\to\infty}1=1,\ \lim_{k\to\infty}(2+1/k)=2,\ \lim_{k\to\infty}3/k^2=0\ \text{ and }\ \lim_{k\to\infty}\mathrm{e}^{-k}\sin k=0,$$

then Theorem 3.2 implies that the sequence $\left\{\boldsymbol{x}^{(k)}\right\}$ converges to $(1,\ 2,\ 0,\ 0)^{\mathrm{T}}$ with respect to the l_∞ norm.

Example 3.3 Determine $\|\boldsymbol{A}\|_\infty$ for the matrix

$$\boldsymbol{A}=\begin{bmatrix}1 & 2 & -1\\ 0 & 3 & -1\\ 5 & -1 & 1\end{bmatrix}.$$

Solution. We have

$$\sum_{j=1}^{3}|a_{1j}|=|1|+|2|+|-1|=4,\ \sum_{j=1}^{3}|a_{2j}|=|0|+|3|+|-1|=4$$

and

$$\sum_{j=1}^{3}|a_{3j}|=|5|+|-1|+|1|=7.$$

So, Theorem 3.7 implies that the $\|\boldsymbol{A}\|_\infty=\max\limits_{1\leqslant i\leqslant n}\{4,4,7\}=7.$

3.2 Eigenvalues and eigenvectors (特征值和特征向量)

- Eigenvalues and eigenvectors (特征值和特征向量)
- Spectral radius (谱半径)
- Convergent matrices (收敛矩阵)
- Examples (例题)

An $n\times m$ matrix can be considered as a function that uses matrix multiplication to take m-dimensional column vectors into n-dimensional column vectors. So, an $n\times m$ matrix is actually a linear function from $\mathbf{R}^n$ to $\mathbf{R}^n$. A square matrix $\boldsymbol{A}$ takes the set of n-dimensional vectors into itself, which gives a linear function from $\mathbf{R}^n$ to $\mathbf{R}^n$. In this case, certain nonzero vectors $\boldsymbol{x}$ might be parallel to $\boldsymbol{Ax}$, which means that a constant λ exists with $\boldsymbol{Ax}=\lambda\boldsymbol{x}$. For these vectors, we have $(\boldsymbol{A}-\lambda\boldsymbol{I})\boldsymbol{x}=0$. There is a close connection between these numbers λ and the likelihood that an iterative method will converge. We will consider this connection in this section.

3.2.1 Eigenvalues and eigenvectors (特征值和特征向量)

Definition 3.5 If $\boldsymbol{A}$ is a square matrix, then the polynomial defined by

$$p(\lambda)=\det(\boldsymbol{A}-\lambda\boldsymbol{I})$$

is called the characteristic polynomial (特征多项式) of $\boldsymbol{A}$.

It is not difficult to show that p is an nth-degree polynomial, and, consequently, has at most n distinct zeros, some of which might be complex. If λ is a zero of p, then, $\det(\boldsymbol{A}-\lambda\boldsymbol{I})=\boldsymbol{0}$, the linear system defined by $(\boldsymbol{A}-\lambda\boldsymbol{I})\boldsymbol{x}=\boldsymbol{0}$ has a solution with $\boldsymbol{x}\neq 0$. We wish to study the zeros of p and the nonzero solutions corresponding to these systems.

Definition 3.6 If p is the characteristic polynomial of the matrix $\boldsymbol{A}$, then the zeros of p are called eigenvalues (特征值), or characteristic values, of the matrix $\boldsymbol{A}$. If λ is an eigenvalue of $\boldsymbol{A}$ and $\boldsymbol{x}\neq 0$ satisfies $(\boldsymbol{A}-\lambda\boldsymbol{I})\boldsymbol{x}=\boldsymbol{0}$, then $\boldsymbol{x}$ is an eigenvector, or characteristic vector of $\boldsymbol{A}$ corresponding to the eigenvalue λ.

To determine the eigenvalues of a matrix, we can use the fact that

- λ is an eigenvalue of $\boldsymbol{A}$ if and only if $\det(\boldsymbol{A}-\lambda\boldsymbol{I})=\boldsymbol{0}$.

Once an eigenvalue has been found a corresponding eigenvector $\boldsymbol{x}\neq 0$ is determined by solving the system.

- $(\boldsymbol{A}-\lambda\boldsymbol{I})\boldsymbol{x}=\boldsymbol{0}$.

3.2.2 Spectral radius (谱半径)

Definition 3.7 The spectral radius $\rho(\boldsymbol{A})$ of a matrix $\boldsymbol{A}$ is defined by

$$\rho(\boldsymbol{A})=\max|\lambda|$$

where λ is an eigenvalue of $\boldsymbol{A}$.

Recall that for complex $\lambda=\alpha+\beta i$, we have $|\lambda|=\left(\alpha^2+\beta^2\right)^{1/2}$.

The spectral radius is closely related to the norm of a matrix, as shown in the following theorem.

Theorem 3.8 If $\boldsymbol{A}$ is an $n\times n$ matrix, then

(a) $\|\boldsymbol{A}\|_2=\left[\rho\left(\boldsymbol{A}^{\mathrm{T}}\boldsymbol{A}\right)\right]^{1/2}$;

(b) $\rho(\boldsymbol{A})\leqslant\|\boldsymbol{A}\|$, for any natural norm $\|\cdot\|$.

Part (a) of Theorem 3.8 implies that if $\boldsymbol{A}$ is symmetric, then

$$\|\boldsymbol{A}\|_2=\sqrt{\rho\left(\boldsymbol{A}^{\mathrm{T}}\boldsymbol{A}\right)}=\sqrt{\rho\left(\boldsymbol{A}^2\right)}=\rho(\boldsymbol{A}),$$

since if λ is an eigenvalue of the matrix A, then λ^2 is an eigenvalue of the matrix A^2.

An interesting and useful result, which is similar to part (b) of Theorem 3.8, is that for any matrix A and any $\varepsilon > 0$, there exists a natural norm $\|\cdot\|$ with property that

$$\rho(A) < \|A\| < \rho(A) + \varepsilon.$$

Consequently, $\rho(A)$ is the greatest lower bound for the natural norms on A.

3.2.3 Convergent matrices (收敛矩阵)

In studying iterative matrix techniques, it is of particular importance to know when powers of a matrix become small (that is, when all the entries approach zero). Matrices of this type are called convergent (收敛的).

Definition 3.8 We call an $n \times n$ matrix A convergent (收敛的) if

$$\lim_{k\to\infty}\left(A^k\right)_{ij} = 0$$

for each $i = 1, 2, \cdots, n$ and $j = 1, 2, \cdots, n$.

3.2.4 Examples (例题)

Example 3.4 Show that there are no nonzero vectors x in $\mathbf{R}^2$ with Ax parallel to x if

$$A = \begin{bmatrix} 0 & 1 \\ -1 & 0 \end{bmatrix}.$$

Solution. The eigenvalues of A are the solutions to the characteristic polynomial

$$0 = \det(A - \lambda I) = \det\begin{bmatrix} -\lambda & 1 \\ -1 & -\lambda \end{bmatrix} = \lambda^2 + 1,$$

so, the eigenvalues of A are the complex numbers $\lambda_1 = i$ and $\lambda_2 = -i$. A corresponding eigenvector x for λ_1 needs to satisfy

$$\begin{bmatrix} 0 \\ 0 \end{bmatrix} = \begin{bmatrix} -i & 1 \\ -1 & -i \end{bmatrix}\begin{bmatrix} x_1 \\ x_2 \end{bmatrix} = \begin{bmatrix} -ix_1 + x_2 \\ -x_1 - ix_2 \end{bmatrix}.$$

That is, $0 = -ix_1 + x_2$, so $x_2 = ix_1$, and $0 = -ix_1 - x_2$. Hence, if x is an eigenvector of A, then exactly one of its components is real and the other is complex. As a consequence, there are no nonzero vectors x in $\mathbf{R}^2$ with Ax parallel to x.

Example 3.5 Determine the eigenvalues and eigenvectors for the matrix

$$A = \begin{bmatrix} 2 & 0 & 0 \\ 1 & 1 & 2 \\ 1 & -1 & 4 \end{bmatrix}.$$

Solution. The characteristic polynomial of A is

$$p(\lambda)=\det(\boldsymbol{A}-\lambda\boldsymbol{I})=\det\begin{bmatrix}2-\lambda & 0 & 0\\ 1 & 1-\lambda & 2\\ 1 & -1 & 4-\lambda\end{bmatrix}$$
$$=-(\lambda^3-7\lambda^2+16\lambda-12)=-(\lambda-3)(\lambda-2)^2,$$

so there are two eigenvalues of $\boldsymbol{A}$: $\lambda_1=3$ and $\lambda_2=2$.

The eigenvector $\boldsymbol{x}_1$ corresponding to the eigenvalue $\lambda_1=3$ is a solution to the vector-matrix equation $(\boldsymbol{A}-3\boldsymbol{I})\boldsymbol{x}_1=0$, so

$$\begin{bmatrix}-1 & 0 & 0\\ 1 & -2 & 2\\ 1 & -1 & 1\end{bmatrix}\begin{bmatrix}x_1\\ x_2\\ x_3\end{bmatrix}=\begin{bmatrix}0\\0\\0\end{bmatrix}$$

which implies that $x_1=0$ and $x_2=x_3$.

Any nonzero value of x_3 produces an eigenvector for the eigenvalue $\lambda_1=3$. For example, when $x_3=1$ we have the eigenvector $\boldsymbol{x}_1=(0,1,1)^{\mathrm{T}}$, and any eigenvector of $\boldsymbol{A}$ corresponding to $\lambda=3$ is a nonzero multiple of $\boldsymbol{x}_1$.

An eigenvector $\boldsymbol{x}_1\neq 0$ of $\boldsymbol{A}$ associated with $\lambda_2=2$ is a solution of the system $(\boldsymbol{A}-2\boldsymbol{I})\boldsymbol{x}_1=0$, so

$$\begin{bmatrix}0 & 0 & 0\\ 1 & -1 & 2\\ 1 & -1 & 2\end{bmatrix}\begin{bmatrix}x_1\\ x_2\\ x_3\end{bmatrix}=\begin{bmatrix}0\\0\\0\end{bmatrix}.$$

In this case the eigenvector has only to satisfy the equation

$$x_1-x_2+2x_3=0$$

which can be done in various ways. For example, when $x_1=0$ we have $x_2=2x_3$, so one choice would be $\boldsymbol{x}_2=(0,2,1)^{\mathrm{T}}$. We could also choose $x_2=0$, which requires that $x_1=-2x_3$. Hence, $\boldsymbol{x}_3=(-2,0,1)^{\mathrm{T}}$ gives a second eigenvector for the eigenvalue $\lambda_2=2$ that is not a multiple of $\boldsymbol{x}_2$. The eigenvectors of $\boldsymbol{A}$ corresponding to the eigenvalue $\lambda_2=2$ generate an entire plane. This plane is described by all vectors of the form

$$\alpha\boldsymbol{x}_2+\beta\boldsymbol{x}_3=(-2\beta,2\alpha,\alpha+\beta)^{\mathrm{T}}$$

for arbitrary constants α and β, provided that at least one of the constants is nonzero.

Example 3.6 Let

$$\boldsymbol{A}=\begin{bmatrix}1 & 0 & 2\\ 0 & 1 & -1\\ -1 & 1 & 1\end{bmatrix}.$$

Compute the eigenvalues of $\boldsymbol{A}$ and the spectral radius $\rho(\boldsymbol{A})$.

Solution. To compute the eigenvalues of $\boldsymbol{A}$, consider

$$p(\lambda)=\det(\boldsymbol{A}-\lambda\boldsymbol{I})=\begin{vmatrix}1-\lambda & 0 & 2\\ 0 & 1-\lambda & -1\\ -1 & 1 & 1-\lambda\end{vmatrix}=(1-\lambda)(\lambda^2-2\lambda+4).$$

The eigenvalues of $\boldsymbol{A}$ are the solution of $p(\lambda)=0$, which are

$$\lambda_1=1,\quad \lambda_2=1+\sqrt{3}i,\quad \lambda_2=1-\sqrt{3}i$$

and the spectral radius of $\boldsymbol{A}$ is

$$\rho(A)=\max\left\{1,\left|1+\sqrt{3}i\right|,\left|1-\sqrt{3}i\right|\right\}=\max\{1,2,2\}=2.$$

Example 3.7　Compute $\|\boldsymbol{A}\|_2$ for the matrix

$$\boldsymbol{A}=\begin{bmatrix}1 & 1 & 0\\ 1 & 2 & 1\\ -1 & 1 & 2\end{bmatrix}.$$

Solution.　Since $\|\boldsymbol{A}\|_2=\sqrt{\rho(\boldsymbol{A}^{\mathrm{T}}\boldsymbol{A})}$, from Theorem 3.8, we need to compute the eigenvalues of $\boldsymbol{A}^{\mathrm{T}}\boldsymbol{A}$. We first compute

$$\boldsymbol{A}^{\mathrm{T}}\boldsymbol{A}=\begin{bmatrix}1 & 1 & -1\\ 1 & 2 & 1\\ 0 & 1 & 2\end{bmatrix}\begin{bmatrix}1 & 1 & 0\\ 1 & 2 & 1\\ -1 & 1 & 2\end{bmatrix}=\begin{bmatrix}3 & 2 & -1\\ 2 & 6 & 4\\ -1 & 4 & 5\end{bmatrix}.$$

If

$$0=\det(\boldsymbol{A}^{\mathrm{T}}\boldsymbol{A}-\lambda\boldsymbol{I})=\begin{vmatrix}3-\lambda & 2 & -1\\ 2 & 6-\lambda & 4\\ -1 & 4 & 5-\lambda\end{vmatrix}$$
$$=-\lambda^3+14\lambda^2-42\lambda=-\lambda(-\lambda^2+14\lambda-42),$$

then

$$\lambda=0 \quad\text{or}\quad \lambda=7\pm\sqrt{7},$$

so

$$\|\boldsymbol{A}\|_2=\sqrt{\rho(\boldsymbol{A}^{\mathrm{T}}\boldsymbol{A})}=\sqrt{\max\left\{0,7-\sqrt{7},7+\sqrt{7}\right\}}=\sqrt{7+\sqrt{7}}\approx 3.106.$$

3.3　Iterative techniques for solving linear systems (解线性方程组的迭代法)

- Jacobi iterative method (Jacobi 迭代法)
- Gauss-Seidel iterative method (Gauss-Seidel 迭代法)
- General iteration method (一般迭代法)
- Examples (例题)

In this section, we describe the Jacobi and the Gauss-Seidel iterative methods (Jacobi 和 Gauss-Seidel 迭代法), classic methods that dated to the late eighteenth century. Iterative techniques are seldom used for solving linear systems of small dimension since the time required for sufficient accuracy exceeds that required for direct techniques such as Gauss elimination. For large systems with a high percentage of 0 entries, however, these techniques are efficient in terms of both computer storage and computation. Systems of this type arise frequently in the numerical solution of boundary-value problems and partial-differential equations.

An iterative technique to solve the $n \times n$ linear system $\boldsymbol{Ax} = \boldsymbol{b}$ starts with an initial approximation $\boldsymbol{x}^{(0)}$ to the solution $\boldsymbol{x}$ and generates a sequence of vectors $\left\{\boldsymbol{x}^{(k)}\right\}_{k=0}^{\infty}$ that converges to $\boldsymbol{x}$. Iterative techniques involve a process that converts the system $\boldsymbol{Ax} = \boldsymbol{b}$ into an equivalent system of the form $\boldsymbol{x} = \boldsymbol{Mx} + \boldsymbol{d}$ for some fixed matrix $\boldsymbol{M}$ and vector $\boldsymbol{d}$. After the initial vector $\boldsymbol{x}^{(0)}$ is selected, the sequence of approximation solution vectors is generated by computing

$$\boldsymbol{x}^{(k)} = \boldsymbol{M}\boldsymbol{x}^{(k-1)} + \boldsymbol{d}$$

for each $k = 1,2,3,\cdots$, and the matrix $\boldsymbol{M}$ is called the iterative matrix (迭代矩阵).

3.3.1 Jacobi iterative method (Jacobi 迭代法)

Definition 3.9 Given the linear system

$$\begin{cases} E_1 : a_{11}x_1 + a_{12}x_2 + \cdots + a_{1n}x_n = b_1 \\ E_2 : a_{21}x_1 + a_{22}x_2 + \cdots + a_{2n}x_n = b_2 \\ \quad \vdots \qquad\qquad \vdots \\ E_n : a_{n1}x_1 + a_{n2}x_2 + \cdots + a_{nn}x_n = b_n \end{cases} \tag{3.3.1}$$

for $x_1, x_2, \cdots, x_n$,with the constants a_{ij} ,for each $i, j = 1,2,\cdots,n$.

If $a_{ii} \neq 0$,we obtain by solving E_i for x_i ,for each $i = 1,2,\cdots,n$

$$x_i = \sum_{\substack{j=1 \\ j \neq i}}^{n} \left(-\frac{a_{ij}x_j}{a_{ii}} \right) + \frac{b_i}{a_{ii}} \tag{3.3.2}$$

and generate each $x_i^{(k)}$ from components of $x^{(k-1)}$ for $k \geqslant 1$ by

$$x_i^{(k)} = \frac{1}{a_{ii}} \left[-\sum_{\substack{j=1 \\ j \neq i}}^{n} a_{ij} x_j^{(k-1)} + b_i \right]. \tag{3.3.3}$$

The method defined in Eq. (3.3.3) is called Jacobi iterative method (Jacobi 迭代法).

In general, iterative techniques for solving linear systems involve a process that converts the system $\boldsymbol{Ax} = \boldsymbol{b}$ into an equivalent system of the form $\boldsymbol{x} = \boldsymbol{Mx} + \boldsymbol{d}$ for some fixed matrix $\boldsymbol{T}$ and vector $\boldsymbol{c}$. After the initial vector $\boldsymbol{x}^{(0)}$ is selected, the sequence of approximation solution vectors is generated by computing

$$x^{(k)} = Mx^{(k-1)} + d$$

for each $k = 1,2,3,\cdots$

The Jacobi method can be written in the form $x^{(k)} = Mx^{(k-1)} + d$ by splitting A into its diagonal and off-diagonal parts. To see this, let D be the diagonal matrix whose diagonal entries are those of A, L be the strictly lower-triangular part (严格下三角部分) of A. With this notion,

$$A = \begin{bmatrix} a_{11} & a_{12} & \cdots & a_{1n} \\ a_{21} & a_{22} & \cdots & a_{2n} \\ \vdots & \vdots & & \vdots \\ a_{n1} & a_{n2} & \cdots & a_{nn} \end{bmatrix}$$

is split into

$$A = \begin{bmatrix} a_{11} & 0 & \cdots & 0 \\ 0 & a_{22} & \ddots & \vdots \\ \vdots & \ddots & \ddots & 0 \\ 0 & \cdots & 0 & a_{nn} \end{bmatrix} + \begin{bmatrix} 0 & 0 & \cdots & 0 \\ a_{21} & \ddots & \ddots & \vdots \\ \vdots & \ddots & \ddots & \vdots \\ a_{n1} & \cdots & a_{n,n-1} & 0 \end{bmatrix} + \begin{bmatrix} 0 & a_{12} & \cdots & a_{1n} \\ \vdots & \ddots & \ddots & \vdots \\ \vdots & \ddots & \ddots & a_{n-1,n} \\ 0 & \cdots & \cdots & 0 \end{bmatrix} \tag{3.3.4}$$

$$= D + L + U.$$

The equation $Ax = b$, or $(D + L + U)x = b$, is then transformed into

$$Dx = -(L + U)x + b,$$

and, if D^{-1} exists, that is, if $a_{ii} \neq 0$ for each i ,then

$$x = -D^{-1}(L + U)x + D^{-1}b.$$

This results in the matrix form of the Jacobi iterative technique:

$$x^{(k)} = -D^{-1}(L + U)x^{(k-1)} + D^{-1}b, \quad k = 1,2,\cdots$$

Introducing the notation $M_J = -D^{-1}(L + U)$ and $d_J = D^{-1}b$, the Jacobi technique has the form

$$x^{(k)} = M_J x^{(k-1)} + d_J, \quad k = 1,2,3,\cdots \tag{3.3.5}$$

In practice, the iterative formulae defined in Eq. (3.3.3) are used in computation and Eq. (3.3.5) for theoretical purposes.

3.3.2 Gauss-Seidel iterative method (Gauss-Seidel 迭代法)

Definition 3.10 For the linear system

$$\begin{cases} E_1 : a_{11}x_1 + a_{12}x_2 + \cdots + a_{1n}x_n = b_1 \\ E_2 : a_{21}x_1 + a_{22}x_2 + \cdots + a_{2n}x_n = b_2 \\ \vdots \qquad \vdots \\ E_n : a_{n1}x_1 + a_{n2}x_2 + \cdots + a_{nn}x_n = b_n \end{cases}$$

if $a_{ii} \neq 0$, we convert it to the form

$$x_i = \frac{1}{a_{ii}}\left[-\sum_{j=1}^{i-1} a_{ij}x_j - \sum_{j=i+1}^{n} a_{ij}x_j + b_i\right] \tag{3.3.6}$$

for $i = 1, 2, \cdots, n$, and generate each $x^{(k)}$ for $k \geqslant 1$ by

$$x_i^{(k)} = \frac{1}{a_{ii}}\left[-\sum_{j=1}^{i-1} a_{ij}x_j^{(k)} - \sum_{j=i+1}^{n} a_{ij}x_j^{(k-1)} + b_i\right] \tag{3.3.7}$$

for $i = 1, 2, \cdots, n$ .The iterative method is called Gauss-Seidel iterative method (Gauss-Seidel 迭代法).

Comparing the Jacobi iterative formulae (Jacobi 迭代公式)

$$x_i^{(k)} = \frac{1}{a_{ii}}\left[\sum_{\substack{j=1 \\ j\neq i}}^{n}\left(-a_{ij}x_j^{(k-1)}\right) + b_i\right], \quad i = 1, 2, \cdots, n$$

with Gauss-Seidel iterative formulae (Gauss-Seidel 迭代公式)

$$x_i^{(k)} = \frac{1}{a_{ii}}\left[-\sum_{j=1}^{i-1} a_{ij}x_j^{(k)} - \sum_{j=i+1}^{n} a_{ij}x_j^{(k-1)} + b_i\right], \quad i = 1, 2, \cdots, n$$

for $i = 1, 2, \cdots, n$, we can draw the following conclusion:

In Jacobi iteration method (3.3.3), the components of $x^{(k-1)}$ are used to compute $x_i^{(k)}$. Since, for $i > 1$, $x_1^{(k)}, x_2^{(k)}, \cdots, x_{i-1}^{(k)}$ have already been computed and are probably better approximations to the actual solution $x_1, x_2, \cdots, x_{i-1}$ than $x_1^{(k-1)}, x_2^{(k-1)}, \cdots, x_{i-1}^{(k-1)}$, it seems more reasonable to compute $x_i^{(k)}$ using these most recently calculated values. This is the ideal of the Gauss-Seidel iterative method.

We can also write the Gauss-Seidel method in matrix form. With the definitions of $\boldsymbol{D}$, $\boldsymbol{L}$ and $\boldsymbol{U}$ given in (3.3.4):

$$\boldsymbol{A} = \boldsymbol{D} + \boldsymbol{L} + \boldsymbol{U},$$

we have the Gauss-Seidel iterative method represented by

$$\boldsymbol{D}\boldsymbol{x}^{(k)} = -\boldsymbol{L}\boldsymbol{x}^{(k)} - \boldsymbol{U}\boldsymbol{x}^{(k-1)} + \boldsymbol{b}$$

or

$$\boldsymbol{x}^{(k)} = -(\boldsymbol{D}+\boldsymbol{L})^{-1}\boldsymbol{U}\boldsymbol{x}^{(k-1)} + (\boldsymbol{D}+\boldsymbol{L})^{-1}\boldsymbol{b} \quad \text{for} \quad k = 1, 2, \cdots \tag{3.3.8}$$

Let $\boldsymbol{M}_G = -(\boldsymbol{D}+\boldsymbol{L})^{-1}\boldsymbol{U}$ and $\boldsymbol{d}_G = (\boldsymbol{D}+\boldsymbol{L})^{-1}\boldsymbol{b}$, then the Gauss-Seidel technique has the form

$$\boldsymbol{x}^{(k)} = \boldsymbol{M}_G\boldsymbol{x}^{(k-1)} + \boldsymbol{d}_G \quad \text{for} \quad k = 1, 2, \cdots \tag{3.3.9}$$

For the lower-triangular matrix (下三角矩阵) $\boldsymbol{D} + \boldsymbol{L}$ to be nonsingular (非奇异的), it is necessary and sufficient that $a_{ii} \neq 0$ for each $a_{ii} \neq 0$.

3.3.3 General iteration method (一般迭代法)

To study the convergence of general iteration techniques, we need to analyze the formula

$$\boldsymbol{x}^{(k)} = \boldsymbol{M}\boldsymbol{x}^{(k-1)} + \boldsymbol{d} \quad \text{for each} \quad k = 1, 2, \cdots$$

where $\boldsymbol{x}^{(0)}$ is arbitrary.

Theorem 3.9 If the spectral radius satisfies $\rho(\boldsymbol{T})<1$, then $(\boldsymbol{I}-\boldsymbol{T})^{-1}$ exists, and

$$(\boldsymbol{I}-\boldsymbol{T})^{-1}=\boldsymbol{I}+\boldsymbol{T}+\boldsymbol{T}^2+\cdots=\sum_{j=0}^{\infty}\boldsymbol{T}^j.$$

Theorem 3.10 For any $\boldsymbol{x}^{(0)}\in\mathbf{R}^n$, the sequence $\left\{\boldsymbol{x}^{(k)}\right\}_{k=0}^{\infty}$ defined by

$$\boldsymbol{x}^{(k)}=\boldsymbol{M}\boldsymbol{x}^{(k-1)}+\boldsymbol{d}, \text{ for each } k\geqslant 1$$

converges to the unique solution of $\boldsymbol{x}=\boldsymbol{M}\boldsymbol{x}+\boldsymbol{d}$ if and only if $\rho(\boldsymbol{M})<1$.

Theorem 3.11 If A is strictly diagonally dominant, then for any choice of $\boldsymbol{x}^{(0)}$, both the Jacobi and Gauss-Seidel methods give sequences $\left\{\boldsymbol{x}^{(k)}\right\}_{k=0}^{\infty}$ that converge to the unique solution of $\boldsymbol{A}\boldsymbol{x}=\boldsymbol{b}$.

3.3.4 Examples (例题)

Example 3.8 The linear system $\boldsymbol{A}\boldsymbol{x}=\boldsymbol{b}$ given by

$$\begin{cases} E_1: 10x_1 - x_2 + 2x_3 = 6 \\ E_2: -x_1 + 11x_2 - x_3 + 3x_4 = 25 \\ E_3: 2x_1 - x_2 + 10x_3 - x_4 = -11 \\ E_4: 3x_2 - x_3 + 8x_4 = 15 \end{cases}$$

has the unique solution $\boldsymbol{x}=(1,2,-1,1)^{\mathrm{T}}$. Use Jacobi iterative method (Jacobi 迭代方法) to compute the approximation solution of the given system.

Solution. To use Jacobi iterative method, solve equation E_i for x_i, for each $i=1,2,3,4$, to obtain

$$\begin{cases} x_1 = \frac{1}{10}x_2 - \frac{1}{5}x_3 + \frac{3}{5} \\ x_2 = \frac{1}{11}x_1 + \frac{1}{11}x_3 - \frac{3}{11}x_4 + \frac{25}{11} \\ x_3 = \frac{-1}{5}x_1 + \frac{1}{10}x_2 + \frac{1}{10}x_4 - \frac{11}{10} \\ x_4 = -\frac{3}{8}x_2 + \frac{1}{8}x_3 + \frac{15}{8} \end{cases}.$$

For an initial approximation, we let $\boldsymbol{x}^{(0)}=(0,0,0,0)^{\mathrm{T}}$. Then $\boldsymbol{x}^{(k)}$ is given by

$$\begin{cases} x_1^{(k)} = \frac{1}{10}x_2^{(k-1)} - \frac{1}{5}x_3^{(k-1)} + \frac{3}{5} \\ x_2^{(k)} = \frac{1}{11}x_1^{(k-1)} + \frac{1}{11}x_3^{(k-1)} - \frac{3}{11}x_4^{(k-1)} + \frac{25}{11} \\ x_3^{(k)} = \frac{-1}{5}x_1^{(k-1)} + \frac{1}{10}x_2^{(k-1)} + \frac{1}{10}x_4^{(k-1)} - \frac{11}{10} \\ x_4^{(k)} = -\frac{3}{8}x_2^{(k-1)} + \frac{1}{8}x_3^{(k-1)} + \frac{15}{8} \end{cases}.$$

and are presented in following Table 3.1.

Table 3.1 Jacobi iterative method

k	x_1^k	x_2^k	x_3^k	x_4^k
0	0.0000	0.0000	0.0000	0.0000
1	0.6000	2.2727	−1.1000	1.8750
2	1.0473	1.7159	−0.8052	0.8852
3	0.9326	2.0530	−1.0493	1.1309
4	1.0152	1.9537	−0.9681	0.9736
5	0.9890	2.0114	−1.0103	1.0214
6	1.0032	1.9922	−0.9945	0.9944
7	0.9981	2.0023	−1.0020	1.0036
8	1.0006	1.9987	−0.9990	0.9989
9	0.9997	2.0004	−1.0004	1.0006
10	1.0001	1.9998	−0.9998	0.9998

The unique solution is $\boldsymbol{x}=(1,2,-1,1)^{\mathrm{T}}$.

Example 3.9 Try to use the Gauss-Seidel iterative method (Gauss-Seidel 迭代方法) to solve the linear system

$$\begin{cases} E_1: 10x_1 - x_2 + 2x_3 = 6 \\ E_2: -x_1 + 11x_2 - x_3 + 3x_4 = 25 \\ E_3: 2x_1 - x_2 + 10x_3 - x_4 = -11 \\ E_4: 3x_2 - x_3 + 8x_4 = 15 \end{cases}.$$

Solution. For using Gauss-Seidel iterative method, we covert the linear system to

$$\begin{cases} x_1 = \frac{1}{10}x_2 - \frac{1}{5}x_3 + \frac{3}{5} \\ x_2 = \frac{1}{11}x_1 + \frac{1}{11}x_3 - \frac{3}{11}x_4 + \frac{25}{11} \\ x_3 = \frac{-1}{5}x_1 + \frac{1}{10}x_2 + \frac{1}{10}x_4 - \frac{11}{10} \\ x_4 = -\frac{3}{8}x_2 + \frac{1}{8}x_3 + \frac{15}{8} \end{cases}.$$

For an initial approximation, we let $\boldsymbol{x}^{(0)}=(0,0,0,0)^{\mathrm{T}}$. Then $\boldsymbol{x}^{(k)}$ is given by

$$\begin{cases} x_1^{(k)} = \frac{1}{10}x_2^{(k-1)} - \frac{1}{5}x_3^{(k-1)} + \frac{3}{5} \\ x_2^{(k)} = \frac{1}{11}x_1^{(k)} + \frac{1}{11}x_3^{(k-1)} - \frac{3}{11}x_4^{(k-1)} + \frac{25}{11} \\ x_3^{(k)} = \frac{-1}{5}x_1^{(k)} + \frac{1}{10}x_2^{(k)} + \frac{1}{10}x_4^{(k-1)} - \frac{11}{10} \\ x_4^{(k)} = -\frac{3}{8}x_2^{(k)} + \frac{1}{8}x_3^{(k)} + \frac{15}{8} \end{cases},$$

and are presented in following Table 3.2.

Table 3.2 Gauss-Seidel iterative method

k	$x_1^{(k)}$	$x_2^{(k)}$	$x_3^{(k)}$	$x_4^{(k)}$
0	0.0000	0.0000	0.0000	0.0000
1	0.6000	2.3272	−0.9873	0.8789
2	1.0300	2.0370	−1.0140	0.9844
3	1.0065	2.0036	−1.0025	0.9983
4	1.0009	2.0003	−1.0003	0.9999
5	1.0001	2.0000	−1.0000	1.0000

The unique solution is $\boldsymbol{x} = (1,2,-1,1)^{\mathrm{T}}$.

3.4 Convergence analysis and SOR iterative method (收敛性分析与 SOR 迭代法)

- Convergence analysis (收敛性分析)
- SOR iterative method (SOR 迭代法)
- SOR iterative method in matrix form (矩阵形式的 SOR 迭代法)
- Examples (例题)

3.4.1 Convergence analysis (收敛性分析)

To study the convergence of general iterative techniques, we consider the formula

$$\boldsymbol{x}^{(k)} = \boldsymbol{M}\boldsymbol{x}^{(k-1)} + \boldsymbol{d}, \quad \text{for } k = 1,2,3,\cdots \tag{3.4.1}$$

where $\boldsymbol{x}^{(0)}$ is arbitrary.

Lemma 3.12 If the spectral radius $\rho(\boldsymbol{M})$ satisfies $\rho(\boldsymbol{M})<1$, then $(\boldsymbol{I}-\boldsymbol{M})^{-1}$ exists, and

$$(\boldsymbol{I}-\boldsymbol{M})^{-1} = \boldsymbol{I} + \boldsymbol{M} + \boldsymbol{M}^2 + \cdots = \sum_{k=0}^{\infty} \boldsymbol{M}^k.$$

Theorem 3.13 For any $\boldsymbol{x}^{(0)} \in \mathbf{R}^n$, the sequence $\left\{\boldsymbol{x}^{(k)}\right\}_{k=0}^{\infty}$ defined by

$$\boldsymbol{x}^{(k)} = \boldsymbol{M}\boldsymbol{x}^{(k-1)} + \boldsymbol{d}, \quad \text{for } k = 1,2,3,\cdots$$

converges to the unique solution of $\boldsymbol{x} = \boldsymbol{M}\boldsymbol{x} + \boldsymbol{d}$, if and only if $\rho(\boldsymbol{M})<1$.

Corollary 3.14 If $\|\boldsymbol{M}\| = q < 1$ for any natural matrix norm and $\boldsymbol{d}$ is a given vector, then the sequence $\left\{\boldsymbol{x}^{(k)}\right\}_{k=0}^{\infty}$ defined by $\boldsymbol{x}^{(k)} = \boldsymbol{M}\boldsymbol{x}^{(k-1)} + \boldsymbol{d}$, converges, for any $\boldsymbol{x}^{(0)} \in \mathbf{R}^n$, to a vector $\boldsymbol{x} \in \mathbf{R}^n$, and for $k = 1,2,3,\cdots$, the following error bounds hold:

(a) $\left\|\boldsymbol{x} - \boldsymbol{x}^{(k)}\right\| \leqslant q^k \left\|\boldsymbol{x} - \boldsymbol{x}^{(0)}\right\|$; (3.4.2)

(b) $\left\|\boldsymbol{x}-\boldsymbol{x}^{(k)}\right\| \leqslant \dfrac{q^k}{1-q}\left\|\boldsymbol{x}^{(1)}-\boldsymbol{x}^{(0)}\right\|$. (3.4.3)

The relationship of the rapidity of convergence to the spectral radius of the iteration matrix M can be seen from Corollary 3.14. Since the inequalities hold for any natural matrix norm, it follows from the statement after Theorem 3.8 that

$$\left\|\boldsymbol{x}-\boldsymbol{x}^{(k)}\right\| \approx \rho(\boldsymbol{M})^k\left\|\boldsymbol{x}-\boldsymbol{x}^{(0)}\right\|.$$

Thus, it is desirable to select the iterative technique with minimal $\rho(\boldsymbol{M})<1$ for a particular system $\boldsymbol{Ax}=\boldsymbol{b}$.

Theorem 3.15 If A is strictly diagonally dominant (严格对角占优的), then for any choice of $\boldsymbol{x}^{(0)}$, both the Jacobi and Gauss-Seidel methods give sequences $\left\{\boldsymbol{x}^{(k)}\right\}_{k=0}^{\infty}$ that converge to the unique solution of $\boldsymbol{Ax}=\boldsymbol{b}$.

No general results exist to tell which of the two techniques, Jacobi or Gauss-Seidel, will be most successful for an arbitrary linear system. In special cases, however, the answer is known, as is demonstrated in the following theorem.

Theorem 3.16 (Stein-Rosenberg) If $a_{ij} \leqslant 0$, for each $i \neq j$ and $a_{ii}>0$, for each $i=1,2,\cdots,n$, then one and only one of the following statements holds:

(a) $0 \leqslant \rho(\boldsymbol{M}_G)<\rho(\boldsymbol{M}_J)<1$;

(b) $1<\rho(\boldsymbol{M}_J)<\rho(\boldsymbol{M}_G)$;

(c) $\rho(\boldsymbol{M}_J)=\rho(\boldsymbol{M}_G)=0$;

(d) $\rho(\boldsymbol{M}_J)=\rho(\boldsymbol{M}_G)=1$.

For the special case described in Theorem 3.16, we see from part (a) that when one method gives convergence, then both give convergence, and the Gauss-Seidel method converges faster than the Jacobi method. Part (b) indicates that when one method diverges, then both diverge, and the divergence is more pronounced for the Gauss-Seidel method.

3.4.2 SOR iterative method (SOR 迭代法)

Given the linear system

$$\begin{cases} a_{11}x_1+a_{12}x_2+\cdots+a_{1n}x_n=b_1 \\ a_{21}x_1+a_{22}x_2+\cdots+a_{2n}x_n=b_2 \\ \qquad\vdots \\ a_{n1}x_1+a_{n2}x_2+\cdots+a_{nn}x_n=b_n \end{cases}, \tag{3.4.4}$$

we establish the iterative formulae:

$$x_i^{(k)}=(1-\omega)x_i^{(k-1)}+\frac{\omega}{a_{ii}}\left[b_i-\sum_{j=1}^{i-1}a_{ij}x_j^{(k)}-\sum_{j=i+1}^{n}a_{ij}x_j^{(k-1)}\right] \tag{3.4.5}$$

for $i=1,2,\cdots,n$ and $k=1,2,\cdots$, and call this iterative method the relaxation method (松

弛法).

For choices of ω with $0 < \omega < 1$, the procedures are called under-relaxation methods (低松弛法) and can be used to obtain convergence of some systems that are not convergent by the Gauss-Seidel method.

For choices of ω with $\omega > 1$, the procedures are called over-relaxation methods (超松弛法), which are used to accelerate the convergence for systems that are convergent by the Gauss-Seidel technique. These methods are abbreviated SOR, for Successive Over-Relaxation (逐次超松弛法), and are particularly useful for solving the linear systems that occur in the numerical solution of certain partial-differential equations.

In particular, the relaxation method defined in (3.4.5) is Gauss-Seidel iterative method with $\omega = 1$:

$$x_i^{(k)} = \frac{1}{a_{ii}}\left[b_i - \sum_{j=1}^{i-1} a_{ij}x_j^{(k)} - \sum_{j=i+1}^{n} a_{ij}x_j^{(k-1)}\right]$$

for $i = 1, 2, \cdots, n$ and $k = 1, 2, \cdots$.

3.4.3 SOR iterative method in matrix form (矩阵形式的 SOR 迭代法)

To determine the matrix of the SOR method:

$$x_i^{(k)} = (1-\omega)x_i^{(k-1)} + \frac{\omega}{a_{ii}}\left[b_i - \sum_{j=1}^{i-1} a_{ij}x_j^{(k)} - \sum_{j=i+1}^{n} a_{ij}x_j^{(k-1)}\right],$$

we rewrite this as

$$a_{ii}x_i^{(k)} + \omega\sum_{j=1}^{i-1} a_{ij}x_j^{(k)} = (1-\omega)a_{ii}x_i^{(k-1)} - \omega\sum_{j=i+1}^{n} a_{ij}x_j^{(k-1)} + \omega b_i,$$

so that in vector form, we have

$$(\boldsymbol{D} + \omega\boldsymbol{L})\boldsymbol{x}^{(k)} = [(1-\omega)\boldsymbol{D} - \omega\boldsymbol{U}]\boldsymbol{x}^{(k-1)} + \omega\boldsymbol{b}$$

or

$$\boldsymbol{x}^{(k)} = (\boldsymbol{D} + \omega\boldsymbol{L})^{-1}[(1-\omega)\boldsymbol{D} - \omega\boldsymbol{U}]\boldsymbol{x}^{(k-1)} + \omega(\boldsymbol{D} + \omega\boldsymbol{L})^{-1}\boldsymbol{b}. \tag{3.4.6}$$

If we let $\boldsymbol{M}_\omega = (\boldsymbol{D} + \omega\boldsymbol{L})^{-1}[(1-\omega)\boldsymbol{D} - \omega\boldsymbol{U}]$ and $\boldsymbol{d}_\omega = \omega(\boldsymbol{D} + \omega\boldsymbol{L})^{-1}\boldsymbol{b}$, the SOR technique has the form

$$\boldsymbol{x}^{(k)} = \boldsymbol{M}_\omega\boldsymbol{x}^{(k-1)} + \boldsymbol{d}_\omega. \tag{3.4.7}$$

SOR iterative method is

$$x_i^{(k)} = (1-\omega)x_i^{(k-1)} + \frac{\omega}{a_{ii}}\left[b_i - \sum_{j=1}^{i-1} a_{ij}x_j^{(k)} - \sum_{j=i+1}^{n} a_{ij}x_j^{(k-1)}\right]$$

for $i = 1, 2, \cdots, n$ and $k = 1, 2, \cdots$, and its matrix form is

$$\boldsymbol{x}^{(k)} = \boldsymbol{M}_\omega\boldsymbol{x}^{(k-1)} + \boldsymbol{d}_\omega, \text{ for } k = 1, 2, \cdots$$

with the iterative matrix $\boldsymbol{M}_\omega = (\boldsymbol{D} + \omega\boldsymbol{L})^{-1}[(1-\omega)\boldsymbol{D} - \omega\boldsymbol{U}]$ and $\boldsymbol{d}_\omega = \omega(\boldsymbol{D} + \omega\boldsymbol{L})^{-1}\boldsymbol{b}$.

The obvious question to ask is how the appropriate value of ω is chosen. Although no complete answer to this question is known for the general $n\times n$ linear system, the following results can be used in certain situations.

Theorem 3.17 (Kahan) If $a_{ii}\neq 0$ for each $i=1,2,\cdots,n$, then $\rho(\boldsymbol{M}_\omega)\geqslant|1-\omega|$. This implies that the SOR method can converge only if $0<\omega<2$.

Theorem 3.18 (Ostrowski-Reich) If A is a positive definite matrix and $0<\omega<2$, then the SOR method converges for any choice of initial approximate vector $\boldsymbol{x}^{(0)}$.

Theorem 3.19 If A is a positive definite and tridiagonal, then $\rho(\boldsymbol{M}_G)=\left|\rho(\boldsymbol{M}_J)\right|^2<1$, and the optimal choice of ω for the SOR method is

$$\omega=\frac{2}{1+\sqrt{1-\left|\rho(M_J)\right|^2}}.$$

With this choice of ω we have $\rho(\boldsymbol{M}_G)=\omega-1$.

3.4.4 Examples (例题)

Example 3.10 The linear system $\boldsymbol{Ax}=\boldsymbol{b}$ given by

$$\begin{cases}4x_1-3x_2 \qquad = 24\\ 3x_1+4x_2-x_3=30\\ \qquad -x_2+4x_3=-24\end{cases}\tag{3.4.8}$$

has the solution $\boldsymbol{x}=(3,\ 4,-5)^{\mathrm{T}}$. Use the Gauss-Seidel iterative method and the SOR iterative method with $\omega=1.25$ to solve the system (3.4.10), using $\boldsymbol{x}^{(0)}=(1,\ 1,\ 1)^{\mathrm{T}}$ for both methods.

Solution. For each $k=1,2,3,\cdots,$ the equations for the Gauss-Seidel method are

$$\begin{cases}x_1^{(k)}= \qquad +0.75x_2^{(k-1)} \qquad +6\\ x_2^{(k)}=-0.75x_1^{(k)} \qquad +0.25x_3^{(k-1)}+7.5\\ x_3^{(k)}= \qquad 0.25x_2^{(k)} \qquad -6\end{cases}$$

and the equations for the SOR method with $\omega=1.25$ are

$$\begin{cases}x_1^{(k)}=(1-\omega)x_1^{(k-1)}+\omega\left(\qquad 0.75x_2^{(k-1)} \qquad +6\right)\\ x_2^{(k)}=(1-\omega)x_2^{(k-1)}+\omega\left(-0.75x_1^{(k)} \qquad +0.25x_3^{(k-1)}+7.5\right)\\ x_3^{(k)}=(1-\omega)x_3^{(k-1)}+\omega\left(\qquad 0.25x_2^{(k)} \qquad -6\right)\end{cases}$$

The first seven iterates for each method are listed in Table 3.3 and Table 3.4. For the iterates to be accurate to seven decimal places, the Gauss-Seidel method requires 34 iterations, as opposed to 14 iterations for the SOR method with $\omega=1.25$. The actual solution $\boldsymbol{x}=(3,\ 4,-5)^{\mathrm{T}}$.

Table 3.3 (Gauss-Seidel)

k	x_1^k	x_2^k	x_3^k
0	1.000000	1.000000	1.000000
1	5.250000	3.812500	−5.046875
2	3.1406250	3.8828125	−5.0292969
3	3.0878906	3.9267578	−5.0183105
4	3.0549316	3.9542236	−5.0114441
5	3.0343323	3.9713898	−5.0071525
6	3.0214577	3.9821186	−5.0044703
7	3.0134110	3.9888241	−5.0027940

Table 3.4 (SOR with $\omega=1.25$)

k	x_1^k	x_2^k	x_3^k
0	1.000000	1.000000	1.000000
1	6.312500	3.5195313	−6.6501465
2	2.6223145	3.9585266	−4.6004238
3	3.1333027	4.0102646	−5.0966863
4	2.9570512	4.0074838	−4.9734897
5	3.0037211	4.0029250	−5.0057135
6	2.9963276	4.0009262	−4.9982822
7	3.0000498	4.0002586	−5.0003486

Example 3.11 The matrix

$$\boldsymbol{A}=\begin{bmatrix}4 & 3 & 0\\ 3 & 4 & -1\\ 0 & -1 & 4\end{bmatrix}$$

given in Example 3.9, is positive definite and tridiagonal, so Theorem 3.19 applies. Since

$$\boldsymbol{M}_J=-\boldsymbol{D}^{-1}(\boldsymbol{L}+\boldsymbol{U})=\begin{bmatrix}1/4 & 0 & 0\\ 0 & 1/4 & 0\\ 0 & 0 & 1/4\end{bmatrix}\begin{bmatrix}0 & -3 & 0\\ -3 & 0 & 1\\ 0 & 1 & 0\end{bmatrix}=\begin{bmatrix}0 & -0.75 & 0\\ -0.75 & 0 & 0.25\\ 0 & 0.25 & 0\end{bmatrix},$$

we have

$$\boldsymbol{M}_J-\lambda\boldsymbol{I}=\begin{bmatrix}-\lambda & -0.75 & 0\\ -0.75 & -\lambda & 0.25\\ 0 & 0.25 & -\lambda\end{bmatrix}$$

so $\det\left(\boldsymbol{M}_J-\lambda\boldsymbol{I}\right)=-\lambda\left(\lambda^2-0.625\right)$.

Thus

$$\rho\left(\boldsymbol{M}_J\right)=\sqrt{0.625}$$

and

$$\omega=\frac{2}{1+\sqrt{1-\left|\rho\left(\boldsymbol{M}_J\right)\right|^2}}=\frac{2}{1+\sqrt{1-0.625}}\approx 1.24.$$

This explains the rapid convergence obtained in Example 3.9 when using $\omega=1.25$.

3.5 Condition number and iterative refinement (条件数和迭代优化)

- Condition number (条件数)
- Iterative refinement (迭代优化)
- Examples (例题)

3.5.1 Condition number (条件数)

Definition 3.11 Suppose $\tilde{\boldsymbol{x}} \in \mathbf{R}^n$ is an approximation to the solution of the linear system defined by $\boldsymbol{Ax} = \boldsymbol{b}$. The residual vector (残差向量) for $\tilde{\boldsymbol{x}}$ with respect to this system is

$$\boldsymbol{r} = \boldsymbol{b} - \boldsymbol{A}\tilde{\boldsymbol{x}}. \tag{3.5.1}$$

It seems intuitively reasonable that if $\tilde{\boldsymbol{x}}$ is an approximation to the solution $\boldsymbol{x}$ of $\boldsymbol{Ax} = \boldsymbol{b}$ and the residual vector $\boldsymbol{r} = \boldsymbol{b} - \boldsymbol{A}\tilde{\boldsymbol{x}}$ has the property that if $\|\boldsymbol{r}\| = \|\boldsymbol{b} - \boldsymbol{A}\tilde{\boldsymbol{x}}\|$ is small, then $\|\boldsymbol{x} - \tilde{\boldsymbol{x}}\|$ would be small as well. This is often the case, but certain system, which occurs frequently in practice, fails to have this property.

Theorem 3.20 Suppose that $\tilde{\boldsymbol{x}}$ is an approximation to the solution of $\boldsymbol{Ax} = \boldsymbol{b}$, $\boldsymbol{A}$ is a nonsingular matrix, and $\boldsymbol{r}$ is the residual vector for $\tilde{\boldsymbol{x}}$, $\boldsymbol{r} = \boldsymbol{b} - \boldsymbol{A}\tilde{\boldsymbol{x}}$. Then for any natural norm

$$\|\boldsymbol{x} - \tilde{\boldsymbol{x}}\| = \|\boldsymbol{A}^{-1}\|\|\boldsymbol{r}\| \tag{3.5.2}$$

and if $\tilde{\boldsymbol{x}} \neq \boldsymbol{0}$ and $\boldsymbol{b} \neq \boldsymbol{0}$

$$\frac{\|\boldsymbol{x} - \tilde{\boldsymbol{x}}\|}{\|\boldsymbol{x}\|} \leqslant \|\boldsymbol{A}\|\|\boldsymbol{A}^{-1}\|\frac{\|\boldsymbol{r}\|}{\|\boldsymbol{b}\|}. \tag{3.5.3}$$

The inequalities (3.5.2) and (3.5.3) imply that $\|\boldsymbol{A}^{-1}\|$ and $\|\boldsymbol{A}\|\|\boldsymbol{A}^{-1}\|$ provide an indication of the connection between the residual vector and the accuracy of the approximation. In general, the relative error

$$\frac{\|\boldsymbol{x} - \tilde{\boldsymbol{x}}\|}{\|\boldsymbol{x}\|}$$

is of most interest, and, by inequality (3.5.3), this error is bounded by the product of $\|\boldsymbol{A}\|\|\boldsymbol{A}^{-1}\|$ and the relative residual $\dfrac{\|\boldsymbol{r}\|}{\|\boldsymbol{b}\|}$ for this approximation. Any convenient norm can be used for this approximation; the only requirement is that it (should) be used consistently throughout.

Definition 3.12 The condition number of the nonsingular matrix $\boldsymbol{A}$ relative to a norm $\|\cdot\|$ is

$$K(\boldsymbol{A}) = \|\boldsymbol{A}\|\|\boldsymbol{A}^{-1}\|. \tag{3.5.4}$$

A matrix $\boldsymbol{A}$ is well-conditioned (良态的) if $K(\boldsymbol{A})$ is close to 1, and is ill conditioned (病态的) when $K(\boldsymbol{A})$ is significantly greater than 1.

The condition number defined in (3.5.4) can be also denoted

$$\text{Cond}\ (\boldsymbol{A}) = \|\boldsymbol{A}\|\|\boldsymbol{A}^{-1}\|$$

and conditioning in this context refers to the relative security that a small residual vector implies a correspondingly accurate approximate solution.

With this notation, the inequalities (3.5.2) and (3.5.3) become

$$\|\boldsymbol{x} - \tilde{\boldsymbol{x}}\| \leqslant K(\boldsymbol{A})\frac{\|\boldsymbol{r}\|}{\|\boldsymbol{A}\|} \quad \text{and} \quad \frac{\|\boldsymbol{x} - \tilde{\boldsymbol{x}}\|}{\|\boldsymbol{x}\|} \leqslant K(\boldsymbol{A})\frac{\|\boldsymbol{r}\|}{\|\boldsymbol{b}\|}$$

and $K(A) \geqslant 1$. In fact, $K(A) = \| A \| \left\| A^{-1} \right\| \geqslant \left\| AA^{-1} \right\| = \| \boldsymbol{I} \| = 1$.

Lemma 3.21 If the approximation solution to the linear system $\boldsymbol{Ax} = \boldsymbol{b}$ is being determined using *t-digit* arithmetic and Gauss elimination, then the residual vector for the approximation $\tilde{\boldsymbol{x}}$, $\boldsymbol{r} = \boldsymbol{b} - \boldsymbol{A}\tilde{\boldsymbol{x}}$ has

$$\|\boldsymbol{r}\| \approx 10^{-t} \|A\| \|\tilde{x}\|. \tag{3.5.5}$$

From this approximation, an estimate for the effective condition number in *t-digit* arithmetic can be obtained without having to invert the matrix A: In actuality, this approximation assumes that all the arithmetic operations in the Gauss elimination technique are performed using *t-digit* arithmetic but that the operations needed to determine the residual are done in double-precision (that is, *t-digit*) arithmetic. This technique does not add significantly to the computational effort and eliminates much of the loss of accuracy involved with the subtraction of the nearly equal numbers that occur in the calculation of the residual.

The approximation for the *t-digit* condition number $K(A)$ comes from consideration of the linear system

$$\boldsymbol{Ay} = \boldsymbol{r}.$$

The solution to this system can be readily approximated since the multipliers for the Gauss elimination method have already been calculated. In fact, $\tilde{\boldsymbol{y}}$, the approximate solution of $\boldsymbol{Ay} = \boldsymbol{r}$, satisfies

$$\tilde{\boldsymbol{y}} \approx \boldsymbol{A}^{-1}\boldsymbol{r} = \boldsymbol{A}^{-1}\left(\boldsymbol{b} - \boldsymbol{A}\tilde{\boldsymbol{x}}\right) = \boldsymbol{A}^{-1}\boldsymbol{b} - \boldsymbol{A}^{-1}\boldsymbol{A}\tilde{\boldsymbol{x}} = \boldsymbol{x} - \tilde{\boldsymbol{x}}, \tag{3.5.6}$$

and

$$\boldsymbol{x} \approx \tilde{\boldsymbol{x}} + \tilde{\boldsymbol{y}}.$$

So $\tilde{\boldsymbol{y}}$ is an estimate of the error produced when $\tilde{\boldsymbol{x}}$ approximates the solution $\boldsymbol{x}$ to the original system. Equations (3.5.5) and (3.5.6) imply that

$$\tilde{\boldsymbol{y}} \approx \| \boldsymbol{x} - \tilde{\boldsymbol{x}} \| = \left\| A^{-1}\boldsymbol{r} \right\| \leqslant \left\| A^{-1} \right\| \| \boldsymbol{r} \| \approx \left\| A^{-1} \right\| \left(10^{-t} \| A \| \ \| \tilde{\boldsymbol{x}} \| \right) = 10^{-t} \| \tilde{\boldsymbol{x}} \| K(A).$$

This gives an approximation for the condition number involved with solving the system $\boldsymbol{Ax} = \boldsymbol{b}$ using Gauss elimination and the *t-digit* type of arithmetic just described:

$$K(A) \approx 10^{t} \frac{\| \tilde{\boldsymbol{y}} \|}{\| \tilde{\boldsymbol{x}} \|}.$$

Theorem 3.22 If the approximation solution $\tilde{\boldsymbol{x}}$ to the linear system $\boldsymbol{Ax} = \boldsymbol{b}$ is being determined using *t-digit* arithmetic and Gauss elimination, then

$$K(A) \approx 10^{t} \frac{\| \tilde{\boldsymbol{y}} \|}{\| \boldsymbol{x} \|} \tag{3.5.7}$$

where $\tilde{\boldsymbol{y}}$ is the approximation solution to the linear system

$$\boldsymbol{Ay} = \boldsymbol{r}.$$

3.5.2 Iterative refinement (迭代优化)

In the subsection 3.5, we have that $\tilde{\boldsymbol{y}}$, the approximate solution of $\boldsymbol{A}\boldsymbol{y}=\boldsymbol{r}$, satisfies

$$\tilde{\boldsymbol{y}} \approx \boldsymbol{A}^{-1}\boldsymbol{r} = \boldsymbol{A}^{-1}(\boldsymbol{b}-\boldsymbol{A}\tilde{\boldsymbol{x}}) = \boldsymbol{A}^{-1}\boldsymbol{b} - \boldsymbol{A}^{-1}\boldsymbol{A}\tilde{\boldsymbol{x}} = \boldsymbol{x}-\tilde{\boldsymbol{x}} \tag{3.5.8}$$

and

$$\boldsymbol{x} \approx \tilde{\boldsymbol{x}} + \tilde{\boldsymbol{y}}.$$

In general, $\tilde{\boldsymbol{x}}+\tilde{\boldsymbol{y}}$ is a more accurate approximation to the solution of the linear system $\boldsymbol{A}\boldsymbol{x}=\boldsymbol{b}$ than the original approximation $\tilde{\boldsymbol{x}}$.

The method using the above assumption is called iterative refinement (迭代优化), or iterative improvement (迭代改进), and consists of performing iterations on the system whose right -hand side is the residual vector for successive approximations until satisfactory accuracy results.

Iterative Refinement. Suppose $\tilde{\boldsymbol{x}}^{(1)}=\tilde{\boldsymbol{x}}$ is the approximate solution to the system $\boldsymbol{A}\boldsymbol{x}=\boldsymbol{b}$ and $\tilde{\boldsymbol{y}}^{(1)}=\tilde{\boldsymbol{y}}$ is the approximate solution to the system $\boldsymbol{A}\boldsymbol{y}=\boldsymbol{r}^{(1)}$, where $\boldsymbol{r}^{(1)}=\boldsymbol{r}=\boldsymbol{b}-\boldsymbol{A}\tilde{\boldsymbol{x}}^{(1)}$. Then

$$\tilde{\boldsymbol{x}}^{(2)} = \tilde{\boldsymbol{x}}^{(1)} + \tilde{\boldsymbol{y}}^{(1)}.$$

Compute $\boldsymbol{r}^{(2)}=\boldsymbol{b}-\boldsymbol{A}\tilde{\boldsymbol{x}}^{(2)}$ and solve the system $\boldsymbol{A}\tilde{\boldsymbol{y}}^{(2)}=\boldsymbol{r}^{(2)}$, which gives $\tilde{\boldsymbol{y}}^{(2)}$ and

$$\tilde{\boldsymbol{x}}^{(3)} = \tilde{\boldsymbol{x}}^{(2)} + \tilde{\boldsymbol{y}}^{(2)}.$$

Continuing this process until $\left\|\tilde{\boldsymbol{y}}^{(k)}\right\|_\infty \leqslant 10^{-t}$, and we get a sequence of approximation vectors $\{\tilde{\boldsymbol{x}}^{(k)}\}$.

If the process is applied using *t-digit* arithmetic and if $K(\boldsymbol{A})\approx 10^q$, then after k iterations of iterative refinement the solution has approximately the smaller of t and $k(t-q)$ correct digits, that is, $\min\{t, k(t-q)\}$ correct digits.

If the system is well-conditioned, one or two iterations will indicate that the solution is accurate. There is the possibility of significant improvement on ill-conditioned systems unless the matrix $\boldsymbol{A}$ is so ill-conditioned that $K_\infty(\boldsymbol{A})\approx 10^t$. In that situation, increased precision should be used for the calculations.

Throughout this section it has been assumed that in the linear system $\boldsymbol{A}\boldsymbol{x}=\boldsymbol{b}$, $\boldsymbol{A}$ and $\boldsymbol{b}$ can be represented exactly. Realistically, the entries a_{ij} and b_j will be altered or perturbed by an amount $\delta_{a_{ij}}$ and δ_{b_j}, causing the linear system

$$(\boldsymbol{A}+\delta\boldsymbol{A})\boldsymbol{x}=\boldsymbol{b}+\delta\boldsymbol{b}$$

to be solved in place of $\boldsymbol{A}\boldsymbol{x}=\boldsymbol{b}$. Normally, if $\|\delta\boldsymbol{A}\|$ and $\|\delta\boldsymbol{b}\|$ are small (on the order of 10^{-t}), the *t-digit* arithmetic should yield a solution $\tilde{\boldsymbol{x}}$ for which $\|\boldsymbol{x}-\tilde{\boldsymbol{x}}\|$ is correspondingly small.

However, in the case of ill-conditioned systems, we have seen that even if A and b are represented exactly, rounding errors can cause $\|x-\tilde{x}\|$ to be large. The following theorem relates the perturbations of linear systems to the condition number of a matrix.

Theorem 3.23 Suppose A is nonsingular and

$$\|\delta A\| < \frac{1}{\|A^{-1}\|}.$$

The solution $\tilde{x}$ to $(A+\delta A)x = b+\delta b$ approximates the solution x of $Ax=b$ with the error estimate

$$\frac{\| x-\tilde{x} \|}{\| x \|} \leqslant \frac{K(A)\| A \|}{\| A \| - K(A) \| \delta A \|}\left(\frac{\| \delta b \|}{\| b \|} + \frac{\| \delta A \|}{\| A \|}\right). \tag{3.5.9}$$

The estimate in inequality (3.5.9) states that if the matrix A is well conditioned (that is, $K(A)$ is not too large), then small changes in A and b produce correspondingly small changes in the solution x. If, on the other hand, A is ill-conditioned, then small changes in A and b may produce large changes in x.

3.5.3 Examples (例题)

Example 3.12 The linear system $Ax=b$ given by

$$\begin{bmatrix} 1 & 2 \\ 1.0001 & 2 \end{bmatrix}\begin{bmatrix} x_1 \\ x_2 \end{bmatrix} = \begin{bmatrix} 3 \\ 3.0001 \end{bmatrix}$$

has the unique solution $x=(1,1)^{\mathrm{T}}$. Determine the residual vector for the poor approximation $\tilde{x}=(3,-0.0001)^{\mathrm{T}}$.

Solution. We have

$$r = b - A\tilde{x} = \begin{bmatrix} 3 \\ 3.0001 \end{bmatrix} - \begin{bmatrix} 1 & 2 \\ 1.0001 & 2 \end{bmatrix}\begin{bmatrix} 3 \\ 3.0001 \end{bmatrix} = \begin{bmatrix} 0.0002 \\ 0 \end{bmatrix},$$

so $\|r\|_\infty = 0.0002$. Although the norm of the residual vector is small, the approximation $\tilde{x}=(3,-0.0001)^{\mathrm{T}}$ is obviously quite poor; in fact, $\|x-\tilde{x}\|_\infty = 2$.

Example 3.13 Determine the condition number for the matrix

$$A = \begin{bmatrix} 1 & 2 \\ 1.0001 & 2 \end{bmatrix}.$$

Solution. We saw in Example 3.12 that the very poor approximation $(3,-0.0001)^{\mathrm{T}}$ to the exact solution $(1,1)^{\mathrm{T}}$ had a residual vector with small norm, so we should expect the condition number of A to be large. In fact, we have $\|A\|_\infty = \max\{|1|+|2|, |1.0001|+|2|\} = 3.0001$, which would not be considered large. However

$$A^{-1}=\begin{bmatrix}-10000 & 10000\\ 5000.5 & -5000\end{bmatrix},\quad \text{so}\quad \left\|A^{-1}\right\|_{\infty}=20000.$$

And for the infinity norm, $K(A)=20000\times 3.0001=60002$. The size of the condition number for this example should certainly keep us from making hasty accuracy decisions based on the residual of an approximation.

Example 3.14 Suppose the linear system is given by

$$\begin{bmatrix}3.3330 & 15920 & -10.333\\ 2.2220 & 16.710 & 9.6120\\ 1.5611 & 5.1791 & 1.6852\end{bmatrix}\begin{bmatrix}x_1\\ x_2\\ x_2\end{bmatrix}=\begin{bmatrix}15913\\ 28.544\\ 8.4254\end{bmatrix}$$

which has the exact solution $\boldsymbol{x}=(1,1,1)^{\mathrm{T}}$. Compute $K_{\infty}(A)$ using (3.5.7) and compare with the exact value of $K_{\infty}(A)$ obtained by

$$K_{\infty}(A)=\|A\|_{\infty}\left\|A^{-1}\right\|_{\infty}$$

where

$$A=\begin{bmatrix}3.3330 & 15920 & -10.333\\ 2.2220 & 16.710 & 9.6120\\ 1.5611 & 5.1791 & 1.6852\end{bmatrix}.$$

Solution. The augmented matrix of the original linear system is

$$\tilde{A}=\left[\begin{array}{ccc:c}3.3330 & 15920 & -10.333 & 15913\\ 2.2220 & 16.710 & 9.6120 & 28.544\\ 1.5611 & 5.1791 & 1.6852 & 8.4254\end{array}\right].$$

Using Gauss elimination and five-digit rounding arithmetic leads successively to the augmented matrices

$$\left[\begin{array}{ccc:c}3.3330 & 15920 & -10.333 & 15913\\ 0 & -10596 & 16.501 & 10580\\ 0 & -7451.4 & 6.5250 & -7444.9\end{array}\right]$$

and

$$\left[\begin{array}{ccc:c}3.3330 & 15920 & -10.333 & 15913\\ 0 & -10596 & 16.501 & 10580\\ 0 & 0 & -5.0790 & -4.7000\end{array}\right].$$

The approximate solution to this system is

$$\tilde{\boldsymbol{x}}=(1.2001,0.99991,0.92538)^{\mathrm{T}}.$$

The residual vector corresponding to $\tilde{\boldsymbol{x}}$ is computed in double precision to be

$$r = b - A\tilde{x} = \begin{bmatrix} 15913 \\ 28.544 \\ 8.4254 \end{bmatrix} - \begin{bmatrix} 3.3330 & 15920 & -10.333 \\ 2.2220 & 16.710 & 9.6120 \\ 1.5611 & 5.1791 & 1.6852 \end{bmatrix} \begin{bmatrix} 1.2001 \\ 0.99991 \\ 0.92538 \end{bmatrix}$$

$$= \begin{bmatrix} 15913 \\ 28.544 \\ 8.4254 \end{bmatrix} - \begin{bmatrix} 15913.00518 \\ 28.26987086 \\ 8.611560367 \end{bmatrix} = \begin{bmatrix} -0.00518 \\ 0.27412914 \\ -0.186160367 \end{bmatrix},$$

So

$$\|r\|_\infty = 0.27413.$$

The estimate for the condition number given in the preceding discussion is obtained by first solving the system $Ay = r$ for $\tilde{y}$:

$$\begin{bmatrix} 3.3330 & 15920 & -10.333 \\ 2.2220 & 16.710 & 9.6120 \\ 1.5611 & 5.1791 & 1.6852 \end{bmatrix} \begin{bmatrix} y_1 \\ y_2 \\ y_3 \end{bmatrix} = \begin{bmatrix} -0.00518 \\ 0.27412914 \\ -0.186160367 \end{bmatrix}.$$

This implies that $\tilde{y} = \left(-2.0008, 8.9987 \times 10^{-5}, 0.074607\right)^{\mathrm{T}}$, so

$$\| \tilde{y} \|_\infty = 2.0008.$$

3.6 本章算法程序及实例 (Algorithms and examples)

3.6.1 雅可比迭代法 (Jacobi iterative method)

```
%%雅可比迭代法函数
function x=jacobi(A,b,x0,errors,M)
D=diag(diag(A));     %求A的对角矩阵
L=-tril(A,-1);       %求A的下三角阵
U=-triu(A,1);        %求A的上三角阵
B=D\(L+U);
F=D\b;
x=B*x0+F;
n=1;                 %迭代次数
while norm(x-x0)>=errors
    x0=x;
    x=B*x0+F;
    n=n+1;
    if(n>=M)
        disp('Warning: 迭代次数太多，可能不收敛！');
        return
    end
```

```
end
```

例 3.15 采用雅可比迭代法求解方程组

$$\begin{pmatrix} 6 & 2 & 1 & -2 \\ 2 & 5 & 0 & -2 \\ -2 & 0 & 8 & 5 \\ 1 & 3 & 2 & 7 \end{pmatrix}\begin{pmatrix} x_1 \\ x_2 \\ x_3 \\ x_4 \end{pmatrix}=\begin{pmatrix} 4 \\ 7 \\ -1 \\ 0 \end{pmatrix}.$$

```
%高斯消元法的MATLAB主程序
clc; clear all;
A=[6 2 1 -2;2 5 0 -2;-2 0 8 5;1 3 2 7];      %输入矩阵系数
b=[4;7;-1;0];
x0=[2;2;2;2];          %初始值
M=100;    %迭代次数
errors=1e-3;
x=jacobi(A,b,x0,errors,M);
disp(x);
结果 0.0520; 1.1508; 0.2447; -0.5706
```

3.6.2 高斯-赛德尔迭代法 (Gauss-Seidel iterative method)

```
%%高斯-赛德尔迭代法函数
function x=gauss_seidel(A,b,x0,errors,M)
D=diag(diag(A));        %求A的对角矩阵
L=-tril(A,-1);          %求A的下三角阵
U=-triu(A,1);           %求A的上三角阵
G=(D-L)\U;
F=(D-L)\b;
x=G*x0+F;
n=1;                    %迭代次数
while norm(x-x0)>=errors
   x0=x;
   x=G*x0+F;
   n=n+1;
   if(n>=M)
      disp('Warning: 迭代次数太多，可能不收敛！');
      return
   end
end
```

例 3.16 采用高斯-赛德尔迭代法求解方程组

$$\begin{pmatrix} 6 & 2 & 1 & -2 \\ 2 & 5 & 0 & -2 \\ -2 & 0 & 8 & 5 \\ 1 & 3 & 2 & 7 \end{pmatrix}\begin{pmatrix} x_1 \\ x_2 \\ x_3 \\ x_4 \end{pmatrix}=\begin{pmatrix} 4 \\ 7 \\ -1 \\ 0 \end{pmatrix}.$$

```
%高斯-赛德尔迭代法的MATLAB主程序
clc; clear all;
A=[6 2 1 -2;2 5 0 -2;-2 0 8 5;1 3 2 7];      %输入矩阵系数
b=[4;7;-1;0];
x0=[2;2;2;2];          %初始值
M=100;   %迭代次数
errors=1e-3;
x=gauss_seidel(A,b,x0,errors,M);
disp(x)
结果 0.0520; 1.1508; 0.2447; -0.5706
```

3.6.3 SOR 迭代法 (SOR iterative method)

```
%%SOR迭代法，加权
function x=SOR(A,b,x0,w,errors,M)
D=diag(diag(A));       %求A的对角矩阵
L=-tril(A,-1);         %求A的下三角阵
U=-triu(A,1);          %求A的上三角阵
B=inv(D-L*w)*((1-w)*D+w*U);
F=w*inv((D-L*w))*b;
x=B*x0+F;
n=1;                   %迭代次数
while norm(x-x0)>=errors
   x0=x;
   x=B*x0+F;
   n=n+1;
   if(n>=M)
      disp('Warning: 迭代次数太多，可能不收敛！');
      return
   end
end
```

例 3.17 采用 SOR 迭代法求解方程组

$$\begin{pmatrix} 6 & 2 & 1 & -2 \\ 2 & 5 & 0 & -2 \\ -2 & 0 & 8 & 5 \\ 1 & 3 & 2 & 7 \end{pmatrix}\begin{pmatrix} x_1 \\ x_2 \\ x_3 \\ x_4 \end{pmatrix}=\begin{pmatrix} 4 \\ 7 \\ -1 \\ 0 \end{pmatrix}.$$

```
%SOR消元法的MATLAB程序
clc; clear all;
A=[6 2 1 -2;2 5 0 -2;-2 0 8 5;1 3 2 7];      %输入矩阵系数
b=[4;7;-1;0];
x0=[2;2;2;2];         %初始值
w=1.2;         %权系数w满足0<=w<=2
M=100;         %迭代次数
errors=1e-3;
x=SOR(A,b,x0,w,errors,M);
disp(x)
结果 0.0520; 1.1508; 0.2447; -0.5706
```

3.7 本章要点 (Highlights)

1. 范数定义在赋范线性空间中，常常被用来度量某个向量空间 (或矩阵) 中的每个向量的长度或大小.
2. 范数满足一定的条件，即：非负性、齐次性、三角不等式.
3. 特征值，是线性代数中的一个重要概念. 设 $\boldsymbol{A}$ 是 n 阶方阵，如果存在数 m 和非零的 n 维列向量 $\boldsymbol{x}$ ，使得 $\boldsymbol{Ax}=m\boldsymbol{x}$ 成立，则称 m 是 $\boldsymbol{A}$ 的一个特征值.
4. 一个线性变换通常可以由其特征值和特征向量完全描述.
5. 迭代是重复反馈过程的活动，其目的通常是为了逼近所需目标或结果. 每一次对过程的重复称为一次“迭代”，而每一次迭代得到的结果会作为下一次迭代的初始值.
6. 雅可比迭代格式为

$$x_i^{(k)}=\frac{1}{a_{ii}}\left[-\sum_{j=1}^{i-1}a_{ij}x_j^{(k-1)}-\sum_{j=i+1}^{n}a_{ij}x_j^{(k-1)}+b_i\right],\quad i=1,2,\cdots,n;\quad k=1,2,3,\cdots$$

矩阵形式是

$$\boldsymbol{x}^{(k)}=\boldsymbol{M}_J\boldsymbol{x}^{(k-1)}+\boldsymbol{d}_J,\quad k=1,2,3,\cdots$$

其中迭代矩阵为 $\boldsymbol{M}_J=-\boldsymbol{D}^{-1}(\boldsymbol{L}+\boldsymbol{U})$ 以及 $\boldsymbol{d}_J=\boldsymbol{D}^{-1}\boldsymbol{b}$.

7. 高斯-赛德尔迭代方法：

$$x_i^{(k)}=\frac{1}{a_{ii}}\left[-\sum_{j=1}^{i-1}a_{ij}x_j^{(k)}-\sum_{j=i+1}^{n}a_{ij}x_j^{(k-1)}+b_i\right],\quad i=1,2,\cdots,n;\quad k=1,2,3,\cdots$$

矩阵形式是

$$\boldsymbol{x}^{(k)}=\boldsymbol{M}_G\boldsymbol{x}^{(k-1)}+\boldsymbol{d}_G,\quad k=1,2,3,\cdots$$

其中迭代矩阵为 $\boldsymbol{M}_G=-(\boldsymbol{D}+\boldsymbol{L})^{-1}\boldsymbol{U}$ 以及 $\boldsymbol{d}_G=(\boldsymbol{D}+\boldsymbol{L})^{-1}\boldsymbol{b}$.

8. SOR 迭代法:

$$x_i^{(k)}=(1-\omega)x_i^{(k-1)}+\frac{\omega}{a_{ii}}\left[b_i-\sum_{j=1}^{i-1}a_{ij}x_j^{(k)}-\sum_{j=i+1}^{n}a_{ij}x_j^{(k-1)}\right],i=1,2,\cdots,n;\ k=1,2,\cdots$$

矩阵形式是

$$\boldsymbol{x}^{(k)}=\boldsymbol{M}_\omega\boldsymbol{x}^{(k-1)}+\boldsymbol{d}_\omega,\quad k=1,2,\cdots$$

其中迭代矩阵为 $\boldsymbol{M}_\omega=(\boldsymbol{D}+\omega\boldsymbol{L})^{-1}[(1-\omega)\boldsymbol{D}-\omega\boldsymbol{U}]$ 以及 $\boldsymbol{d}_\omega=\omega(\boldsymbol{D}+\omega\boldsymbol{L})^{-1}\boldsymbol{b}$.

9. 条件数定义为：矩阵的范数，乘以矩阵的逆矩阵的范数.
10. 通过迭代优化可以提高迭代结果的精度.

3.8 问题讨论 (Questions for discussion)

1. 什么是向量范数？给出三种常见的向量范数，并说明它们的区别.
2. 什么是矩阵范数？给出三种常见的矩阵范数，并说明它们的区别.
3. 什么是矩阵 $\boldsymbol{A}$ 的特征值和特征向量？
4. 写出求解线性方程组 $\boldsymbol{A}\boldsymbol{x}=\boldsymbol{b}$ 的迭代法的一般形式，并给出它收敛的充要条件.
5. 给出迭代法收敛的充分条件、误差估计和收敛速度.
6. 写出求解线性方程组 $\boldsymbol{A}\boldsymbol{x}=\boldsymbol{b}$ 的雅可比迭代法与高斯-赛德尔迭代法的计算公式，它们的基本区别是什么？
7. 什么叫做矩阵 $\boldsymbol{A}$ 严格对角占优，什么叫做矩阵 A 不可约？
8. 给出求解线性方程组 $\boldsymbol{A}\boldsymbol{x}=\boldsymbol{b}$ 的 SOR 迭代法的计算公式.
9. 如何判断迭代法的收敛性？
10. 将雅可比迭代法、高斯-赛德尔迭代法和具有最低松弛参数 SOR 迭代法，按照收敛速度的快慢进行排序.
11. 什么是残差向量？怎么计算残差向量？
12. 什么情形下矩阵 $\boldsymbol{A}$ 是良态的？什么情形下矩阵 $\boldsymbol{A}$ 是病态的？举例说明.
13. 什么是条件数？常用的条件数有哪些？
14. 什么是迭代优化，迭代优化的基本思想是什么？

3.9 关键术语 (Key terms)

vector norm (向量范数)
Euclidean norm (欧几里得范数)
distances (距离)
converge to (收敛至)

matrix norm (矩阵范数)
characteristic polynomial (特征多项式)
eigenvalues(characteristic values) (特征值)
eigenvector (characteristic vector) (特征向量)
spectral radius (谱半径)
iterative matrix (迭代矩阵)
Jacobi iterative method (雅可比迭代法)
Gauss-Seidel iterative method (高斯-赛德尔迭代法)
strictly diagonally dominant (严格对角占优的)
relaxation method (松弛法)
under-relaxation methods (低松弛法)
over-relaxation methods (超松弛法)
Successive over-relaxation (SOR) (逐次超松弛)
residual vector (残差向量)
condition number (条件数)
well-conditioned (良态的)
ill-conditioned (病态的)
iterative refinement (迭代优化)

3.10 延伸阅读 (Extending reading)

3.10.1 背景知识

1. 标量是一个实数 (或复数)，通常用斜体或希腊字母表示. 向量用黑体字母表示.

2. 柯西-施瓦茨不等式 (Cauchy-Schwartz inequality) 有很多种形式，因此有很多发现者. 奥古斯丁 •路易斯 •柯西 (Augustine Louis Cauchy) (1789～1857) 在 1821 年的《分析课程》中描述了这个不等式，《分析课程》是第一本严谨的微积分著作. 这个不等式的积分形式出现在 1859 年维克托 • 雅科夫列维奇 • 本尼亚科夫斯基 (Victor Yakovlev Benyakovsky) (1804～1889) 的著作中，而赫尔曼 • 阿曼德斯 • 施瓦茨 (Herman Armands Schwartz) (1843～1921) 在 1885 年使用了这个不等式的二重积分形式.

3. 每个向量范数产生一个相关的自然矩阵范数.

4. 卡尔 • 古斯塔夫 • 雅各布 • 雅可比 (Karl Gustav Jacob Jacobi) (1804～1851) 最初因其在数论和椭圆函数领域的工作而得到认可，但他的数学兴趣和能力非常广泛. 他有着强烈的个性，在建立以研究为导向的态度方面发挥了重要作用，这种态度成为 19 世纪德国大学数学复兴的核心.

5. 菲利普 • 路德维希 • 赛德尔 (Philipp Ludwig Seidel) (1821～1896) 作为雅可比的助手，解决了高斯关于最小二乘工作所产生的线性方程组的问题. 这些方程通常有非对角线元素，这些元素比对角线上的元素要小得多，所以迭代法特别有效.

6. 詹姆斯·哈代·威尔金森 (James Hardy Wilkinson) (1919～1986) 以其在求解线性方程和特征值问题的数值方法方面的广泛工作而闻名. 他还发展了逆向误差分析技术.

3.10.2 数学家传记：高斯 (Gauss)

约翰·卡尔·弗里德里希·高斯 (Johann Carl Friedrich Gauss) (1777 年 4 月 30 日～1855 年 2 月 23 日)，德国著名数学家、物理学家、天文学家、几何学家、大地测量学家，毕业于 Carolinum 学院 (现布伦瑞克工业大学). 1796 年，高斯证明了可以尺规作正十七边形. 1807 年高斯成为哥廷根大学教授和哥廷根天文台台长. 1818～1826 年间，汉诺威公国的大地测量工作由高斯主导. 1840 年高斯与韦伯一同画出世界上第一张地球磁场图. 高斯被认为是世界上最伟大的数学家之一，享有“数学王子”的美誉.

当高斯 12 岁时，已经开始怀疑元素几何学中的基础证明. 当他 16 岁时，预测在欧氏几何之外必然会产生一门完全不同的几何学，即非欧几里得几何学. 他导出了二项式定理的一般形式，将其成功地运用在无穷级数，并发展了数学分析的理论. 高斯的老师布吕特纳与他的助手马丁·巴特尔斯很早就认识到了高斯在数学上异乎寻常的天赋，同时卡尔·威廉·费迪南德布伦瑞克公爵也对这个天才儿童留下了深刻印象，于是他们从高斯 14 岁起便资助其学习与生活. 这也使高斯能够在公元 1792～1795 年在 Carolinum 学院 (布伦瑞克工业大学的前身) 学习. 18 岁时，高斯转入哥廷根大学学习，在他 19 岁时，第一个成功地证明了正十七边形可以用尺规作图完成.

17 岁的高斯发现了质数分布定理和最小二乘法. 通过对足够多的测量数据的处理，可以得到一个新的、概率性质的测量结果. 在这些基础之上，高斯随后专注于曲面与曲线的计算，并成功得到高斯钟形曲线(正态分布曲线)，其函数被命名为标准正态分布 (或高斯分布)，并在概率计算中大量使用.

高斯总结了复数的应用，并且严格证明了每一个 n 阶的代数方程必有 n 个实数或者复数解. 在他的第一本著名的著作《算术研究》中，做出了二次互反律的证明，成为数论继续发展的重要基础. 在这部著作的第一章，他导出了三角形全等定理的概念. 高斯在最小二乘法基础上创立了测量平差理论，测算出了小行星谷神星的运行轨迹.

3.10.3 数学家传记：雅可比 (Jacobi)

卡尔・古斯塔夫・雅各布・雅可比 (Carl Gustav Jacob Jacobi，1804～1851 年)，德国数学家. 1804 年 12 月 10 日生于普鲁士的波茨坦，1851 年 2 月 18 日卒于柏林. 雅可比是数学史上最勤奋的学者之一，与欧拉一样也是一位在数学上多产的数学家，是被广泛认可的历史上最伟大的数学家之一. 雅可比善于处理各种繁复的代数问题，在纯粹数学和应用数学上都有非凡的贡献，他所理解的数学有一种强烈的柏拉图式的格调，其数学成就对后人影响颇为深远. 在他逝世后，狄利克雷称他为拉格朗日以来德国科学院成员中最卓越的数学家.

雅可比 1816 年 11 月进入波茨坦大学预科学习，1821 年春毕业. 当时他的希腊语、拉丁语和历史成绩都很优异. 尤其在数学方面，他掌握的知识远远超过学校所教授的内容. 他还自学了 L.欧拉的《无穷小分析引论》，并且试图求解五次代数方程.

1821 年 4 月，雅可比在柏林大学开始了两年的学习生活，他对哲学、古典文学和数学都颇有兴趣. 该校的校长评价说，从一开始雅可比就显示出他是一个“全才”. 像高斯一样，要不是数学强烈吸引着他，他很可能在语言上取得很高成就. 雅可比最后还是决定全力投身数学. 1825 年，他获得柏林大学理学博士学位之后，留校任教. 1825 年到 1826 年，他主讲关于三维空间曲线和曲面的解析理论课程. 年仅 21 岁的雅可比善于将自己的观点贯穿在教学之中，启发学习，引导独立思考，是当时最吸引人的数学教师，他的成功引起普鲁士教育部的注意.

1826 年 5 月，雅可比到柯尼斯堡大学任教. 在柯尼斯堡大学的 18 年间，雅可比不知疲倦地工作着，在科学研究和教学上都做出惊人的成绩. 他在数学方面最主要的成就是和挪威数学家尼尔斯・亨利克・阿贝尔 (Niels Henrik Abel，1802.8.5～1829.4.6) 相互独立地奠定了椭圆函数论的基础，引入并研究了 θ 函数和其他一些超越函数. 这些工作使法国数学家阿德利昂 •玛利埃 •勒让德 (Adrien Marie Legendre, 1752.9.18～1833.1.10) 在这一领域的工作黯然失色，但无私的勒让德赞扬和支持他和阿贝尔的工作. 他对阿贝尔函数也做了研究，还发现了超椭圆函数. 他对椭圆函数理论的透彻研究在数学界引起轰动，从而与阿贝尔齐名. 雅可比在椭圆函数理论、数学分析、数论、几何学、力学方面的主要论文都发表在克雷勒的《纯粹和应用数学》杂志上，平均每期有三篇雅可比的文章. 这使得他很快获得国际声誉. 当时，他同数学家贝塞尔、物理学家冯·诺伊曼三人成为德国数学复兴的核心. 他不仅是伦敦皇家学会会员，还是彼得堡、维也纳、巴黎、马德里等科学院院士.

现代数学许多定理、公式和函数恒等式、方程、积分、曲线、矩阵、根式、行列式以及许多数学符号都冠以雅可比的名字，可见雅可比的成就对后人影响之深. 雅可比几乎与阿贝尔同时各自独立地发现了椭圆函数，是椭圆函数理论的奠基人. 雅可比在函数行列式方面有一篇著名的论文：《论行列式的形成与性质》(1841). 文中给出了函数行列式的导数公式，还利用函数行列式证明了函数之间相关或无关的条件是：雅可比行列式等于零或不等于零. 他还给出了雅可比行列式的乘积定理. 雅可比是将椭圆函数理论应用于数论研究的第一人.

3.11　习题 (Exercises)

1. Find l_∞ and l_2 norms of the vectors.

   a. $\boldsymbol{x}=\left(3,-4,0,\dfrac{3}{2}\right)^{\mathrm{T}}$;

   b. $\boldsymbol{x}=(2,1,-3,4)^{\mathrm{T}}$;

   c. $\boldsymbol{x}=\left(\sin k,\cos k,2^k\right)^{\mathrm{T}}$ for a fixed positive integer k;

   d. $\boldsymbol{x}=\left(4/(k+1),2/k^2,k^2\mathrm{e}^{-k}\right)^{\mathrm{T}}$ for a fixed positive integer k.

2. a. Verify that the function $\|\cdot\|_1$, defined on $\mathbf{R}^n$ by

$$\|\boldsymbol{x}\|_1=\sum_{i=1}^{n}|x_i|$$

   is a norm on $\mathbf{R}^n$;

   b. Find $\|\boldsymbol{x}\|_1$ for the vectors given in Exercise 1;

   c. Prove that for all $\boldsymbol{x}\in\mathbf{R}^n$, $\|\boldsymbol{x}\|_1\geqslant\|\boldsymbol{x}\|_2$.

3. Prove that the following sequences are convergent, and find their limits.

   a. $\boldsymbol{x}^{(k)}=\left(1/k,\mathrm{e}^{1-k},-2/k^2\right)^{\mathrm{T}}$;

   b. $\boldsymbol{x}^{(k)}=\left(\mathrm{e}^{-k}\cos k,k\sin(1/k),3+k^{-2}\right)^{\mathrm{T}}$;

   c. $\boldsymbol{x}^{(k)}=\left(k\mathrm{e}^{-k^2},(\cos k)/k,\sqrt{k^2+k}-k\right)^{\mathrm{T}}$;

d. $\boldsymbol{x}^{(k)}=\left(e^{1-k},\left(k^2+1\right)/\left(1-k^2\right),\left(1/k^2\right)\left(1+3+5+\cdots+(2k-1)\right)\right)^{\mathrm{T}}$.

4. Find the l_∞ norm of the matrices.

a. $\begin{bmatrix}10 & 15\\ 0 & 1\end{bmatrix}$; b. $\begin{bmatrix}10 & 0\\ 15 & 1\end{bmatrix}$;

c. $\begin{bmatrix}2 & -1 & 0\\ -1 & 2 & -1\\ 0 & -1 & 2\end{bmatrix}$; d. $\begin{bmatrix}4 & -1 & 7\\ -1 & 4 & 0\\ -7 & 0 & 4\end{bmatrix}$.

5. The following linear systems $\boldsymbol{Ax}=\boldsymbol{b}$ have $\boldsymbol{x}$ as the actual solution and $\tilde{\boldsymbol{x}}$ as an approximate solution. Compute $\|\boldsymbol{x}-\tilde{\boldsymbol{x}}\|_\infty$ and $\|\boldsymbol{A}\tilde{\boldsymbol{x}}-\boldsymbol{b}\|_\infty$.

a. $\begin{cases}\frac{1}{2}x_1+\frac{1}{3}x_2=\frac{1}{63}\\ \frac{1}{3}x_1+\frac{1}{4}x_2=\frac{1}{168}\\ \boldsymbol{x}=\left(\frac{1}{7},-\frac{1}{6}\right)^{\mathrm{T}}\\ \tilde{\boldsymbol{x}}=(0.142,-0.166)^{\mathrm{T}}\end{cases}$; b. $\begin{cases}x_1+2x_2+3x_3=1\\ 2x_1+3x_2+4x_3=-1\\ 3x_1+4x_2+6x_3=2\\ \boldsymbol{x}=(0,-7,5)^{\mathrm{T}}\\ \tilde{\boldsymbol{x}}=(-0.33,-7.9,5.8)^{\mathrm{T}}\end{cases}$;

c. $\begin{cases}x_1+2x_2+3x_3=1\\ 2x_1+3x_2+4x_3=-1\\ 3x_1+4x_2+6x_3=2\\ \boldsymbol{x}=(0,-7,5)^{\mathrm{T}}\\ \tilde{\boldsymbol{x}}=(-0.2,-7.5,5.4)^{\mathrm{T}}\end{cases}$; d. $\begin{cases}0.04x_1+0.01x_2-0.01x_3=0.06\\ 0.2x_1+0.5x_2-0.2x_3=0.3\\ x_1+2x_2+4x_3=11\\ \boldsymbol{x}=(1.827586,0.6551724,1.965517)^{\mathrm{T}}\\ \tilde{\boldsymbol{x}}=(1.8,0.64,1.9)^{\mathrm{T}}\end{cases}$.

6. The matrix norm $\|\cdot\|_1$, defined by $\|\boldsymbol{A}\|_1=\max\limits_{\|\boldsymbol{x}\|_1=1}\|\boldsymbol{Ax}\|_1$, can be computed using the formula

$$\|\boldsymbol{A}\|_1=\max_{1\leqslant j\leqslant n}\sum_{i=1}^{n}\left|a_{ij}\right|$$

where the vector norm $\|\cdot\|_1$ is defined in Exercise 2. Find $\|\cdot\|_1$ for the matrices in Exercise 4.

7. Prove that if $\|\cdot\|$ is a vector norm on $\mathbf{R}^n$, then $\|\boldsymbol{A}\|=\max\limits_{\|\boldsymbol{x}\|_1=1}\|\boldsymbol{Ax}\|$ is a matrix norm.

8. Show that the Cauchy-Buniakowsky-Schwarz Inequality can be strengthened to

$$\sum_{i=1}^{n}x_iy_i\leqslant\sum_{i=1}^{n}\left|x_iy_i\right|\leqslant\left(\sum_{i=1}^{n}x_i^2\right)^{1/2}\left(\sum_{i=1}^{n}y_i^2\right)^{1/2}.$$

9. Compute the eigenvalues and associated eigenvectors of the following matrices.

a. $\begin{bmatrix}2 & -1\\ -1 & 2\end{bmatrix}$; b. $\begin{bmatrix}0 & 1\\ 1 & 1\end{bmatrix}$; c. $\begin{bmatrix}0 & 1/2\\ 1/2 & 0\end{bmatrix}$;

d. $\begin{bmatrix} 2 & 1 & 0 \\ 1 & 2 & 0 \\ 0 & 0 & 3 \end{bmatrix}$; e. $\begin{bmatrix} -1 & 2 & 0 \\ 0 & 3 & 4 \\ 0 & 0 & 7 \end{bmatrix}$; f. $\begin{bmatrix} 2 & 1 & 1 \\ 2 & 3 & 2 \\ 1 & 1 & 2 \end{bmatrix}$.

10. Compute the eigenvalues and associated eigenvectors of the following matrices.

a. $\begin{bmatrix} 1 & 1 \\ -2 & -2 \end{bmatrix}$; b. $\begin{bmatrix} -1 & -1 \\ 1/3 & 1/6 \end{bmatrix}$; c. $\begin{bmatrix} 3 & 4 \\ 1 & 0 \end{bmatrix}$;

d. $\begin{bmatrix} 3 & 2 & -1 \\ 1 & -2 & 3 \\ 2 & 0 & 4 \end{bmatrix}$; e. $\begin{bmatrix} 1/2 & 0 & 0 \\ -1 & 1/2 & 0 \\ 2 & 2 & -1/3 \end{bmatrix}$; f. $\begin{bmatrix} 2 & -1 & 0 \\ 0 & 2 & 4 \\ 0 & 0 & 2 \end{bmatrix}$.

11. Find the complex eigenvalues and associated eigenvectors for the following matrices.

a. $\begin{bmatrix} 2 & 2 \\ -1 & 2 \end{bmatrix}$; b. $\begin{bmatrix} 1 & 2 \\ -1 & 2 \end{bmatrix}$.

12. Find the complex eigenvalues and associated eigenvectors for the following matrices.

a. $\begin{bmatrix} 1 & 0 & 2 \\ 0 & 1 & -1 \\ -1 & 1 & 1 \end{bmatrix}$; b. $\begin{bmatrix} 0 & 1 & -2 \\ 1 & 0 & 0 \\ 1 & 1 & 1 \end{bmatrix}$.

13. Let $\boldsymbol{A}_1 = \begin{bmatrix} 1 & 0 \\ -1/4 & 1/2 \end{bmatrix}$ and $\boldsymbol{A}_2 = \begin{bmatrix} 1/2 & 0 \\ 16 & 1/2 \end{bmatrix}$. Show that $\boldsymbol{A}_1$ is not convergent, but $\boldsymbol{A}_2$ is convergent.

14. Show that if $\boldsymbol{A}$ is symmetric, then $\|\boldsymbol{A}\|_2 = \rho(\boldsymbol{A})$.

15. Show that if $\|\cdot\|$ is any natural norm, then $\left(\|\boldsymbol{A}^{-1}\|\right)^{-1} \leqslant |\lambda| \leqslant \|\boldsymbol{A}\|$ for any eigenvalue λ of the nonsingular matrix $\boldsymbol{A}$.

16. Find the first two iterations of the Jacobi method for the following linear systems, using $\boldsymbol{x}^{(0)} = 0$:

a. $\begin{cases} 3x_1 - x_2 + x_3 = 1 \\ 3x_1 + 6x_2 + 2x_3 = 0 \\ 3x_1 + 3x_2 + 7x_3 = 4 \end{cases}$; b. $\begin{cases} 10x_1 - x_2 = 9 \\ -x_1 + 10x_2 - 2x_3 = 7 \\ -2x_2 + 10x_3 = 6 \end{cases}$;

c. $\begin{cases} 10x_1 + 5x_2 = 6 \\ 5x_1 + 10x_2 - 4x_3 = 25 \\ -4x_2 + 8x_3 - x_4 = -11 \\ -x_3 + 5x_4 = -11 \end{cases}$; d. $\begin{cases} 4x_1 + x_2 + x_3 + x_5 = 6 \\ -x_1 - 3x_2 + x_3 + x_4 = 6 \\ 2x_1 + x_2 + 5x_3 - x_4 - x_5 = 6 \\ -x_1 - x_2 - x_3 + 4x_4 = 6 \\ 2x_2 - x_3 + x_4 + 4x_5 = 6 \end{cases}$.

17. Find the first two iterations of the Jacobi method for the following linear systems, using $\boldsymbol{x}^{(0)} = 0$:

a. $\begin{cases} 4x_1 + x_2 - x_3 = 5 \\ -x_1 + 3x_2 + x_3 = -4 \\ 2x_1 + 2x_2 + 5x_3 = 1 \end{cases}$; b. $\begin{cases} -2x_1 + x_2 + \frac{1}{2}x_3 = 4 \\ x_1 - 2x_2 - \frac{1}{2}x_3 = -4 \\ x_2 + 2x_3 = 0 \end{cases}$;

c. $\begin{cases} 4x_1 + x_2 - x_3 + x_4 = -2 \\ x_1 + 4x_2 - x_3 - x_4 = -1 \\ -x_1 - x_2 + 5x_3 + x_4 = 0 \\ x_1 - x_2 + x_3 + 3x_4 = 1 \end{cases}$; d. $\begin{cases} 4x_1 - x_2 - x_4 = 0 \\ -x_1 + 4x_2 - x_3 - x_5 = 5 \\ -x_2 + 4x_3 - x_6 = 0 \\ -x_1 + 4x_4 - x_5 = 6 \\ -x_2 - x_4 + 4x_5 - x_6 = -2 \\ -x_3 - x_5 + 4x_6 = 6 \end{cases}$.

18. Repeat Exercise 16 using the Gauss-Seidel method.

19. Repeat Exercise 17 using the Gauss-Seidel method.

20. Use the Jacobi method to solve the linear systems in Exercise16, with $TOL = 10^{-3}$ in the l_∞ norm.

21. Use the Jacobi method to solve the linear systems in Exercise17, with $TOL = 10^{-3}$ in the l_∞ norm.

22. Use the Gauss-Seidel method to solve the linear systems in Exercise16, with $TOL = 10^{-3}$ in the l_∞ norm.

23. Use the Gauss-Seidel method to solve the linear systems in Exercise17, with $TOL = 10^{-3}$ in the l_∞ norm.

24. The linear system

$$\begin{cases} x_1 - x_2 + x_3 = -12 \\ x_1 + 2x_2 + 2x_3 = 4 \\ -x_1 - x_2 + 2x_3 = -5 \end{cases}$$

has the solution $(1, 2, -1)^{\mathrm{T}}$.

a. Show that $\rho(\boldsymbol{M}_J) = \frac{\sqrt{5}}{2} > 1$;

b. Show that the Jacobi method with $\boldsymbol{x}^{(0)} = 0$ fails to give a good approximation after 25 iterations;

c. Show that $\rho(\boldsymbol{M}_G) = \frac{1}{2}$;

d. Use the Gauss-Seidel method with $\boldsymbol{x}^{(0)} = 0$ to approximate the solution to the linear system to within 10^{-5} in the l_∞ norm.

25. The linear system

$$\begin{cases} x_1 - x_3 = 0.2 \\ \dfrac{-1}{2}x_1 + x_2 - \dfrac{1}{4}x_3 = -1.425 \\ x_1 - \dfrac{1}{2}x_2 + x_3 = 2 \end{cases}$$

has the solution $(0.9,-0.8,0.7)^{\mathrm{T}}$.

a. Is the coefficient matrix

$$\boldsymbol{A} = \begin{bmatrix} 1 & 0 & -1 \\ \dfrac{-1}{2} & 1 & \dfrac{-1}{4} \\ 1 & \dfrac{-1}{2} & 1 \end{bmatrix}$$

strictly diagonally dominant?

b. Compute the spectral radius of the Gauss-Seidel matrix $\boldsymbol{M}_G$.

c. Use the Gauss-Seidel iterative method to approximate the solution to the linear system with a tolerance of 10^{-2} and a maximum of 300 iterations.

d. What happens in part (c) when the system is changed to

$$\begin{cases} x_1 - 2x_3 = 0.2 \\ \dfrac{-1}{2}x_1 + x_2 - \dfrac{1}{4}x_3 = -1.425 \\ x_1 - \dfrac{1}{2}x_2 + x_3 = 2 \end{cases}.$$

26. Repeat Exercise 25 using the Jacobi method.

27. Find the first two iterations of the SOR method with $\omega = 1.1$ for the following linear systems, using $\boldsymbol{x}^{(0)} = 0$:

a. $\begin{cases} 3x_1 - x_2 + x_3 = 1 \\ 3x_1 + 6x_2 + 2x_3 = 0 \\ 3x_1 + 3x_2 + 7x_3 = 4 \end{cases}$; b. $\begin{cases} 10x_1 - x_2 = 9 \\ -x_1 + 10x_2 - 2x_3 = 7 \\ -2x_2 + 10x_3 = 6 \end{cases}$;

c. $\begin{cases} 10x_1 + 5x_2 = 6 \\ 5x_1 + 10x_2 - 4x_3 = 25 \\ 2x_1 + x_2 + 5x_3 - x_4 - x_5 = 6 \end{cases}$; d. $\begin{cases} 4x_1 + x_2 + x_3 + x_5 = 6 \\ -x_1 - 3x_2 + x_3 + x_4 = 6 \\ 2x_1 + x_2 + 5x_3 - x_4 - x_5 = 6 \\ -x_1 - x_2 - x_3 + 4x_4 = 6 \\ 2x_2 - x_3 + x_4 + 4x_5 = 6 \end{cases}$.

28. Find the first two iterations of the SOR method with $\omega = 1.1$ for the following linear systems, using $\boldsymbol{x}^{(0)} = 0$:

a. $\begin{cases} 4x_1 + x_2 - x_3 = 5 \\ -x_1 + 3x_2 + x_3 = -4 \\ 2x_1 + 2x_2 + 5x_3 = 1 \end{cases}$; b. $\begin{cases} -2x_1 + x_2 + \frac{1}{2}x_3 = 4 \\ x_1 - 2x_2 - \frac{1}{2}x_3 = -4 \\ x_2 + 2x_3 = 0 \end{cases}$;

c. $\begin{cases} 4x_1 + x_2 - x_3 + x_4 = -2 \\ x_1 + 4x_2 - x_3 - x_4 = -1 \\ -x_1 - x_2 + 5x_3 + x_4 = 0 \\ x_1 - x_2 + x_3 + 3x_4 = 1 \end{cases}$; d. $\begin{cases} 4x_1 - x_2 = 0 \\ -x_1 + 4x_2 - x_3 = 5 \\ -x_2 + 4x_3 = 0 \\ 4x_4 - x_5 = 6 \\ -x_4 + 4x_5 - x_6 = -2 \\ -x_5 + 4x_6 = 6 \end{cases}$.

29. Repeat Exercise 27 using $\omega = 1.3$.

30. Repeat Exercise 28 using $\omega = 1.3$.

31. Use the SOR method with $\omega = 1.2$ to solve the linear systems in Exercise 27 with a tolerance $TOL = 10^{-3}$ in the l_∞ norm.

32. Use the SOR method with $\omega = 1.2$ to solve the linear systems in Exercise 28 with a tolerance $TOL = 10^{-3}$ in the l_∞ norm.

33. Determine which matrices in Exercise 27 are tridiagonal and positive definite. Repeat Exercise 27 for these matrices using the optimal choice of ω.

34. Determine which matrices in Exercise 28 are tridiagonal and positive definite. Repeat Exercise 28 for these matrices using the optimal choice of ω.

35. Compute the condition numbers of the following matrices relative to $\|\cdot\|_\infty$

a. $\begin{bmatrix} \frac{1}{2} & \frac{1}{3} \\ \frac{1}{3} & \frac{1}{4} \end{bmatrix}$; b. $\begin{bmatrix} 3.9 & 1.6 \\ 6.8 & 2.9 \end{bmatrix}$;

c. $\begin{bmatrix} 1 & 2 \\ 1.00001 & 2 \end{bmatrix}$; d. $\begin{bmatrix} 1.003 & 58.09 \\ 5.550 & 321.8 \end{bmatrix}$.

36. Compute the condition numbers of the following matrices relative to $\|\cdot\|_\infty$

a. $\begin{bmatrix} 0.03 & 58.9 \\ 5.31 & -6.10 \end{bmatrix}$; b. $\begin{bmatrix} 58.9 & 0.03 \\ -6.10 & 5.31 \end{bmatrix}$;

c. $\begin{bmatrix} 1 & -1 & -1 \\ 0 & 1 & -1 \\ 0 & 0 & -1 \end{bmatrix}$; d. $\begin{bmatrix} 0.04 & 0.01 & -0.01 \\ 0.2 & 0.5 & -0.2 \\ 1 & 2 & 4 \end{bmatrix}$.

37. The following linear systems $\boldsymbol{Ax} = \boldsymbol{b}$ have $\boldsymbol{x}$ as the actual solution and $\tilde{\boldsymbol{x}}$ as an approximate solution. Using the results of Exercise 35, compute

$$\|\boldsymbol{x} - \tilde{\boldsymbol{x}}\|_\infty \quad \text{and} \quad K_\infty(\boldsymbol{A})\frac{\|\boldsymbol{b} - \boldsymbol{A}\tilde{\boldsymbol{x}}\|_\infty}{\|\boldsymbol{A}\|_\infty}.$$

a. $\begin{cases} \dfrac{1}{2}x_1 + \dfrac{1}{3}x_2 = \dfrac{1}{63} \\ \dfrac{1}{3}x_1 + \dfrac{1}{4}x_2 = \dfrac{1}{168} \\ \boldsymbol{x} = \left(\dfrac{1}{7}, -\dfrac{1}{6}\right)^{\mathrm{T}} \\ \tilde{\boldsymbol{x}} = (0.142, -0.166)^{\mathrm{T}} \end{cases}$;　b. $\begin{cases} 3.9x_1 + 1.6x_2 = 5.5 \\ 6.8x_1 + 2.9x_2 = 9.7 \\ \boldsymbol{x} = (1,1)^{\mathrm{T}} \\ \tilde{\boldsymbol{x}} = (0.98, 1.1)^{\mathrm{T}} \end{cases}$;

c. $\begin{cases} x_1 + 2x_2 = 3 \\ 1.0001x_1 + 2x_2 = 3.0001 \\ \boldsymbol{x} = (1,1)^{\mathrm{T}} \\ \tilde{\boldsymbol{x}} = (0.96, 1.02)^{\mathrm{T}} \end{cases}$;　d. $\begin{cases} 1.003x_1 + 58.09x_2 = 68.12 \\ 5.550x_1 + 321.8x_2 = 377.3 \\ \boldsymbol{x} = (10,1)^{\mathrm{T}} \\ \tilde{\boldsymbol{x}} = (-10,1)^{\mathrm{T}} \end{cases}$.

38. The following linear systems $\boldsymbol{Ax} = \boldsymbol{b}$ have $\boldsymbol{x}$ as the actual solution and $\tilde{\boldsymbol{x}}$ as an approximate solution. Using the results of Exercise 36, compute

$$\|\boldsymbol{x} - \tilde{\boldsymbol{x}}\|_\infty \quad \text{and} \quad K_\infty(\boldsymbol{A})\frac{\|\boldsymbol{b} - \boldsymbol{A}\tilde{\boldsymbol{x}}\|_\infty}{\|\boldsymbol{A}\|_\infty}.$$

a. $\begin{cases} 0.03x_1 + 58.9x_2 = 59.2 \\ 5.31x_1 - 6.10x_2 = 47.0 \\ \boldsymbol{x} = (10,1)^{\mathrm{T}} \\ \tilde{\boldsymbol{x}} = (30.0, 0.990)^{\mathrm{T}} \end{cases}$;　b. $\begin{cases} 58.9x_1 + 0.03x_2 = 59.2 \\ -6.10x_1 + 5.31x_2 = 47.0 \\ \boldsymbol{x} = (1,10)^{\mathrm{T}} \\ \tilde{\boldsymbol{x}} = (1.02, 9.98)^{\mathrm{T}} \end{cases}$;

c. $\begin{cases} x_1 - x_2 - x_3 = 2\pi \\ x_2 - x_3 = 0 \\ -x_3 = \pi \\ \boldsymbol{x} = (0, -\pi, -\pi)^{\mathrm{T}} \\ \tilde{\boldsymbol{x}} = (-0.1, -3.15, -3.14)^{\mathrm{T}} \end{cases}$;　d. $\begin{cases} 0.04x_1 + 0.01x_2 - 0.01x_3 = 0.06, \\ 0.2x_1 + 0.5x_2 - 0.2x_3 = 0.3 \\ x_1 + 2x_2 + 4x_3 = 1 \\ \boldsymbol{x} = (1.827586, 0.6551724,\ 1.965517)^{\mathrm{T}} \\ \tilde{\boldsymbol{x}} = (1.8,\ 0.64,\ 1.9)^{\mathrm{T}} \end{cases}$.

39. The linear system

$$\begin{bmatrix} 1 & 2 \\ 1.0001 & 2 \end{bmatrix}\begin{bmatrix} x_1 \\ x_2 \end{bmatrix} = \begin{bmatrix} 3 \\ 3.0001 \end{bmatrix}$$

has solution $(1,1)^{\mathrm{T}}$. Change $\boldsymbol{A}$ slightly to

$$\begin{bmatrix} 1 & 2 \\ 0.9999 & 2 \end{bmatrix}$$

and consider the linear system

$$\begin{bmatrix} 1 & 2 \\ 0.9999 & 2 \end{bmatrix}\begin{bmatrix} x_1 \\ x_2 \end{bmatrix} = \begin{bmatrix} 3 \\ 3.0001 \end{bmatrix}.$$

Compute the new solution using five-digit rounding arithmetic, and compare the actual error to the estimate. Is $\boldsymbol{A}$ ill-conditioned?

40. The linear system

$$\begin{bmatrix} 1 & 2 \\ 1.00001 & 2 \end{bmatrix} \begin{bmatrix} x_1 \\ x_2 \end{bmatrix} = \begin{bmatrix} 3 \\ 3.00001 \end{bmatrix}$$

has solution $(1,1)^{\mathrm{T}}$. Use seven-digit rounding arithmetic to find the solution of the perturbed system

$$\begin{bmatrix} 1 & 2 \\ 1.000011 & 2 \end{bmatrix} \begin{bmatrix} x_1 \\ x_2 \end{bmatrix} = \begin{bmatrix} 3.00001 \\ 3.00003 \end{bmatrix}$$

and compare the actual error to the estimate. Is $\boldsymbol{A}$ ill-conditioned?

Chapter 4　Solutions of Equations in One Variable (一元方程求根)

本章内容 (Contents)

4.1　Bisection method (二分法)
4.2　Fixed-point iteration and error analysis (不动点迭代及误差分析)
4.3　Newton's method (牛顿法)
4.4　The secant method (弦截法)

学习目标 (Learning Objectives)

- 了解二分法的基本思想及其局限性
- 会用二分法求方程的根
- 了解不动点迭代法的基本思想
- 掌握方程求根迭代法的收敛性分析
- 了解方程求根迭代法收敛阶的概念
- 理解 Newton 迭代法的基本思想
- 掌握 Newton 迭代法的收敛性
- 熟练应用 Newton 迭代法求方程的根
- 理解弦截法的基本思想
- 熟练应用弦截法求方程的根
- 了解牛顿法处理重根的基本思想和方法

This chapter deals with one of the most basic and important problems of numerical approximation: finding a root, or solution, of an equation of the form $f(x)=0$, for a given function f. A root of this equation is also called a zero of the function f. Some equations are easy to solve if their relations are linear. So in this chapter we turn our attention on the root-finding problem of nonlinear equations.

4.1 Bisection method (二分法)

For stating the first technique of finding a root of an equation, we recall some theorems studied in calculus.

Theorem 4.1 (Intermediate Value Theorem 介值定理) If $f \in C[a,b]$ and c is any number between $f(a)$ and $f(b)$, then there exists at least one number ξ in (a,b) for which $f(\xi)=c$.

Corollary 4.2 (Zero-Point Theorem 零点定理) If $f \in C[a,b]$ with $f(a)$ and $f(b)$ of opposite sign, then there exists at least one number x^* in (a,b) for which $f(x^*)=0$.

The first technique, based on **Zero-Point Theorem**, is called the **bisection method** (二分法). Suppose there exists a root x^* of the equation $f(x)=0$, the method calls for a repeated halving of subintervals of $[a,b]$ and at each step, locating the half containing x^*.

Note

Although there may be more than one root of $f(x)=0$ in the interval (a,b), we assume for simplicity that the root in this interval is unique.

Bisection Method Suppose $f \in C[a,b]$ with $f(a)$ and $f(b)$ of opposite sign, set $a=a_1$, and $b=b_1$, and let x_1 be the midpoint (中点) of $[a,b]$; that is,

$$x_1 = \frac{a+b}{2}.$$

If $f(x_1)=0$, then $x^*=x_1$, and the process is completed.

If $f(x_1)\neq 0$, then $f(x_1)$ has the same sign as either $f(a_1)$ or $f(b_1)$.

When $f(x_1)$ and $f(a_1)$ have the same sign, $x^* \in (x_1,b_1)$, and we set $a_2=x_1$, and $b_2=b_1$.

When $f(x_1)$ and $f(a_1)$ have opposite signs, $x^* \in (a_1,x_1)$, and we set $a_2=a_1$, and $b_2=x_1$.

Then we apply the process to the interval $[a_2,b_2]$.

Then the method successively divides the interval in half and replaces one endpoint (端点) with the midpoint so that again the root is bracketed. (See Figure 4.1)

Theorem 4.3 Suppose that $f \in C[a,b]$ and $f(a)f(b)<0$. The bisection method generates a sequence $\{x_n\}_{n=1}^{\infty}$ approximating a zero x^* of f with

$$\left|x^* - x_n\right| \leqslant \frac{b-a}{2^n}, \text{ for each } n=1,2,\cdots.$$

Proof. For each $n=1,2,\cdots$, we have

$$b_n - a_n = \frac{b-a}{2^{n-1}} \quad \text{and} \quad x^* \in (a_n,b_n).$$

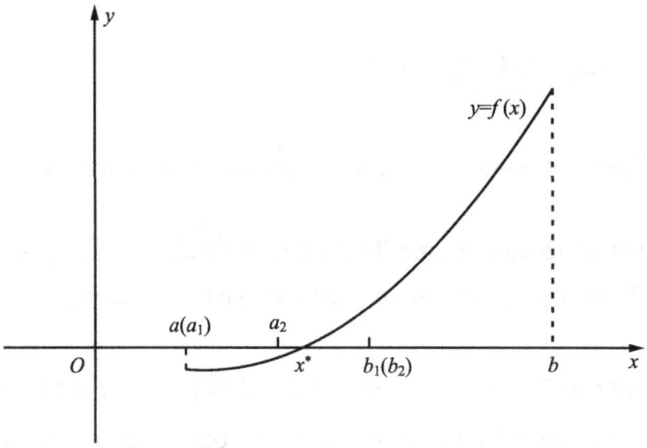

Figure 4.1　Bisection method

Since

$$x_n = \frac{a_n + b_n}{2} \quad \text{for all} \quad n = 1, 2, \cdots,$$

it follows that

$$\left| x^* - x_n \right| \leqslant \frac{b_n - a_n}{2} = \frac{b - a}{2^n}.$$

Example 4.1　The equation $f(x) = x^3 + 3x^2 - 7 = 0$ has a root in $[1, 1.5]$ since $f(1) = -3$ and $f(1.5) = 3.125$. The bisection method gives the values in Table 4.1.

Table 4.1　The approximations by the bisection method

n	a_n	b_n	$x_n = (a_n + b_n)/2$	$f(x_n)$
1	1.0	1.5	1.25	−0.35937
2	1.25	1.5	1.375	1.27148
3	1.25	1.375	1.3125	−0.42895
4	1.25	1.3125	1.28125	0.02811
5	1.25	1.28125	1.26563	−0.16729
6	1.26563	1.28125	1.27344	−0.07000
7	1.27344	1.28125	1.27734	−0.02105

After 7 iterations, $x_7 = 1.27734$ approximates the root x^* with an error

$$\left| x^* - x_7 \right| \leqslant \frac{1.5 - 1}{2^7} = 0.00390625.$$

Note

(1) The bisection method is slow to converge and cannot be used for finding roots of even multiplicity, since $f(a)$ and $f(b)$ will have the same sign.

(2) The bisection method will not be able to distinguish between a singularity and a root.

(3) The bisection method cannot be used for finding a complex root.

(4) The method always converges to a solution if the solution is real and simple or of odd multiplicity, and for that reason it is often used as a starter for the more efficient methods we will present later in this chapter.

4.2 Fixed-point iteration and error analysis (不动点迭代及误差分析)

- Fixed-point iteration (不动点迭代法)
- Convergence analysis and error estimation (收敛性分析和误差估计)
- The order of convergence (收敛阶)

4.2.1 Fixed-point iteration (不动点迭代法)

Fixed-point iteration is an iterative technique for finding a fixed point of a function. In this section, we use this iterative technique to convert the root-finding problem of the equation $f(x)=0$ into the fixed-point problem of the function φ.

Definition 4.1 A number x^* is called a **fixed point** (不动点) for a given function φ if $\varphi(x^*)=x^*$.

An iterative technique to solve the nonlinear equation

$$f(x)=0 \tag{4.2.1}$$

starts with an initial approximation x_0 to the solution x^* and generates a sequence $\{x_k\}_{k=0}^{\infty}$ that converges to x^*.

Fixed-point iterative technique involves a process that converts the system $f(x)=0$ into an equivalent system of the form

$$x=\varphi(x) \tag{4.2.2}$$

for an appropriate function φ.

After the initial value (初值) x_0 is selected, the sequence of approximation solutions is generated by computing

$$x_k=\varphi(x_{k-1}), \tag{4.2.3}$$

for each $k=1,2,3,\cdots$, and the function φ is called the **iterative function** (迭代函数) (See Figure 4.2).

4.2.2 Convergence analysis and error estimation (收敛性分析和误差估计)

Theorem 4.4 If the iterative function φ satisfies

(1) $\varphi\in C[a,b]$ and $\varphi(x)\in[a,b]$ for all $x\in[a,b]$, and

(2) $\varphi'(x)$ exists on (a,b) and a positive constant $L<1$ exists with

$$|\varphi'(x)|\leqslant L,\ \text{for all}\ x\in(a,b),$$

then, for any number x_0 in $[a,b]$, the sequence $\{x_k\}$, defined by

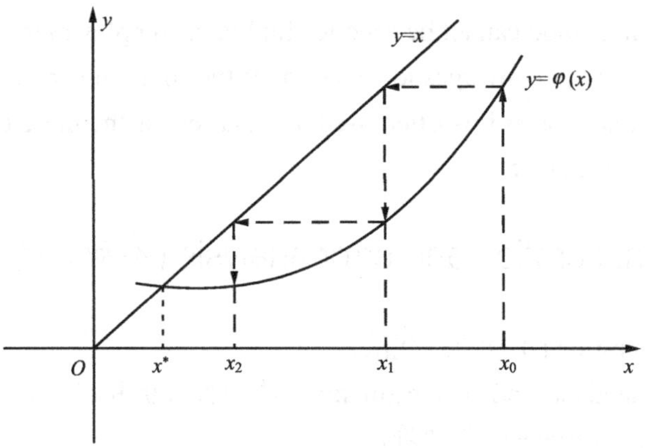

Figure 4.2 Fixed-point iterative process

$$x_k = \varphi(x_{k-1}), \text{ for } k = 1,2,3,\cdots,$$

converges to the unique root x^* of $x=\varphi(x)$ in $[a,b]$. Moreover the following error bounds hold.

(a) $\left|x^* - x_k\right| \leqslant \dfrac{L}{1-L}\left|x_k - x_{k-1}\right|$, (4.2.4)

(b) $\left|x^* - x_k\right| \leqslant \dfrac{L^k}{1-L}\left|x_1 - x_0\right|$. (4.2.5)

Proof. Since φ maps $[a,b]$ into itself, the sequence $\{x_k\}_{k=0}^{\infty}$ is defined, and $x_k \in [a,b]$ for all k. The fact that $|\varphi'(x)| \leqslant L$ and the Mean Value Theorem implies that, for each k,

$$\left|x^* - x_k\right| = \left|\varphi(x^*) - \varphi(x_{k-1})\right| = \left|\varphi'(\xi_k)(x^* - x_{k-1})\right| \leqslant L\left|x^* - x_{k-1}\right|, \tag{4.2.6}$$

where ξ_k lies between x^* and x_{k-1}. Applying this inequality inductively gives

$$\left|x^* - x_k\right| \leqslant L\left|x^* - x_{k-1}\right| \leqslant L^2\left|x^* - x_{k-2}\right| \leqslant \cdots \leqslant L^k\left|x^* - x_0\right|.$$

Since $0 < L < 1$, we have $\lim\limits_{k\to\infty} L^k = 0$ and therefore

$$\lim_{k\to\infty}\left|x^* - x_k\right| = \lim_{k\to\infty} L^k\left|x^* - x_0\right| = 0.$$

Hence, $\{x_k\}_{k=0}^{\infty}$ converges to x^*.

(a) Eq. (4.2.6) implies that

$$\left|x^* - x_k\right| \leqslant L\left|x^* - x_{k-1}\right| = L\left|(x^* - x_k) + (x_k - x_{k-1})\right| \leqslant L\left|x^* - x_k\right| + L\left|x_k - x_{k-1}\right|.$$

Thus,

$$\left|x^* - x_k\right| \leqslant \frac{L}{1-L}\left|x_k - x_{k-1}\right|. \tag{4.2.7}$$

(b) Since

$$\left|x_k - x_{k-1}\right| = \left|\varphi(x_{k-1}) - \varphi(x_{k-2})\right| = \left|\varphi'(\eta_k)\right|\left|x_{k-1} - x_{k-2}\right| \leqslant L\left|x_{k-1} - x_{k-2}\right|,$$

where η_k lies between x_k and x_{k-1}. Applying this inequality inductively gives

$$|x_k - x_{k-1}| \leqslant L|x_{k-1} - x_{k-2}| \leqslant L^2 |x_{k-2} - x_{k-3}| \leqslant \cdots \leqslant L^{k-1}|x_1 - x_0|. \tag{4.2.8}$$

Combining Eq.(4.2.7) and Eq.(4.2.8) implies

$$|x^* - x_k| \leqslant \frac{L^k}{1-L}|x_1 - x_0|.$$

Note

(1) If the sequence $\{x_k\}_{k=0}^{\infty}$, defined by $x_k = \varphi(x_{k-1})$, converges, then for a given tolerance $\varepsilon > 0$, the stopping-technique for the iterative method is

$$|x_k - x_{k-1}| < \frac{1-L}{L}\varepsilon. \tag{4.2.9}$$

(2) For a given tolerance $\varepsilon > 0$, the number of the iterative steps can be estimated by

$$k > \frac{1}{\ln L}\ln\frac{\varepsilon(1-L)}{|x_1 - x_0|}. \tag{4.2.10}$$

Example 4.2 Use a fixed-point iteration to find a solution accurate to within 10^{-8} for the equation $f(x) = x^3 + 4x^2 - 10 = 0$.

Solution. Since

$$f(1) = 1^3 + 4\times 1^2 - 10 = -5 < 0, \quad f(1.5) = 1.5^3 + 4\times 1.5^2 - 10 = 2.375 > 0,$$

there exists at least a root of $f(x) = 0$ in $(1, 1.5)$ from **Zero-Point Theorem**.

Since again $f'(x) = 3x^2 + 8x > 0$ for all $x \in (1, 1.5)$, which means the function f is monotone (单调的) on $(1, 1.5)$, there exists only one root x^* of $f(x) = 0$ in $(1, 1.5)$.

We manipulate the equation $x^3 + 4x - 10 = 0$ as follows:

$$x = \frac{1}{2}\sqrt{10 - x^3}.$$

For the function $\varphi(x) = \frac{1}{2}\sqrt{10 - x^3}$,

$$|\varphi'(x)| = \left|-\frac{3x^2}{4\sqrt{10 - x^3}}\right| \leqslant |\varphi'(1.5)| \approx 0.66 < 1 \quad \text{on} \quad [1, 1.5].$$

Since $\varphi'(x) < 0$ on $[1, 1.5]$,

$$1 < 1.28 \approx \varphi(1.5) \leqslant \varphi(x) \leqslant \varphi(1) = 1.5,$$

for all $x \in [1, 1.5]$. This shows that φ maps the interval $[1, 1.5]$ into itself.

Theorem 4.4 confirms the convergence of the sequence $\{x_k\}_{k=0}^{\infty}$ defined by

$$x_k = \varphi(x_{k-1}) = \frac{1}{2}\sqrt{10 - x_{k-1}^3}, \quad k \geqslant 1.$$

With $x_0 = 1.25$, the approximations in Table 4.2 are generated and compared with the true root $x^* = 1.3652300134140969\cdots$.

Table 4.2 The approximations by the fixed-point iteration

k	x_k	$\lvert x_k - x_{k-1}\rvert$
0	1.25	
1	1.4183507147	0.16835
2	1.3366635022	0.08169
3	1.3794767424	0.04281
4	1.3578398600	0.02164
⋮	⋮	⋮
25	1.3652300192	0.1702×10^{-7}
26	1.3652300105	0.8714×10^{-8}
27	1.3652300149	0.4461×10^{-8}

Theorem 4.5 Let x^* be a root of the equation $x=\varphi(x)$. If a number $\delta>0$ exists such that the function φ' is continuous in $(x^*-\delta,\ x^*+\delta)$, and

$$\left|\varphi'(x^*)\right|<1, \tag{4.2.11}$$

then the sequence $\{x_k\}_{k=0}^{\infty}$, defined by $x_k=\varphi(x_{k-1})$, converges to x^* for any $x_0\in(x^*-\delta,\ x^*+\delta)$.

4.2.3 The order of convergence (收敛阶)

Definition 4.2 Suppose $\{x_k\}_{k=0}^{\infty}$ is a sequence that converges to x^*, with $x_k\neq x^*$ for all k, and denote $e_k=x^*-x_k$. If constants $c(\neq 0)$ and $p>0$ exist with

$$\lim_{k\to\infty}\frac{e_k}{e_{k-1}^p}=c, \tag{4.2.12}$$

then $\{x_k\}_{k=0}^{\infty}$ is said to converge to x^* of order p (p 阶收敛于 x^*), with asymptotic error constant (渐近误差常数) c.

An iterative technique of the form $x_k=\varphi(x_{k-1})$ is said to be of order p if the sequence $\{x_k\}_{k=0}^{\infty}$ converges to the solution $x^*=\varphi(x^*)$ of order p.

Note

(1) The sequence $\{x_k\}_{k=0}^{\infty}$ is linearly convergent if $p=1$.

(2) The sequence $\{x_k\}_{k=0}^{\infty}$ is quadratically convergent (二次收敛) if $p=2$.

Theorem 4.6 Suppose the sequence $\{x_k\}_{k=0}^{\infty}$ is defined by $x_k=\varphi(x_{k-1})$, and x^* is the root of the equation $x=\varphi(x)$. If a number $\delta>0$ exists such that the function φ is p times differentiable in $(x^*-\delta,\ x^*+\delta)$, and

$$\varphi'(x^*)=\varphi''(x^*)=\cdots=\varphi^{(p-1)}(x^*)=0,\ \text{but}\ \ \varphi^{(p)}(x^*)\neq 0, \tag{4.2.13}$$

then the sequence $\{x_k\}_{k=0}^{\infty}$ converges to x^* of order p, where $p\geqslant 1$ is an integer.

Note

(1) The sequence $\{x_k\}_{k=0}^{\infty}$ is linearly convergent if $\left|\varphi'(x^*)\right|<1$ but $\varphi'(x^*)\neq 0$.

(2) The sequence $\{x_k\}_{k=0}^{\infty}$ is quadratically convergent if $\varphi'(x^*)=0$ but $\varphi''(x^*)\neq 0$.

4.3 Newton's method (牛顿法)

- Newton's method and convergence analysis (牛顿法及其收敛性分析)
- How to handle multiple roots using Newton's method (如何采用牛顿法处理重根问题)

4.3.1 Newton's method and convergence analysis (牛顿法及其收敛性分析)

Definition 4.3 A number x^* is called a simple root (单根) of the equation $f(x)=0$, if for $x\neq x^*$, we can write

$$f(x)=(x-x^*)g(x)\text{, where }\lim_{x\to x^*}g(x)\neq 0. \tag{4.3.1}$$

Theorem 4.7 The function $f\in C[a,b]$ has a simple zero at x^* in (a,b) if $f(x^*)=0$, but $f'(x^*)\neq 0$.

Newton's (or the Newton-Raphson) method is one of the most widely used numerical methods for solving nonlinear equations. Newton's method can be derived by using the Taylor's series expansion of the function $f(x)$ about a number x_0 as

$$f(x)=f(x_0)+(x-x_0)f'(x_0)+\frac{1}{2!}(x-x_0)^2f''(x_0)+\cdots, \tag{4.3.2}$$

where the function f and its derivatives $f', f'',\cdots$ are evaluated at x_0. By considering only the first two terms in the expression (4.3.2), we have

$$f(x)\approx f(x_0)+(x-x_0)f'(x_0). \tag{4.3.3}$$

In order to find the root of $f(x)=0$, we set

$$f(x_0)+(x-x_0)f'(x_0)=0, \tag{4.3.4}$$

which obtains the approximation to the root of $f(x)=0$ as

$$x=x_1=x_0-\frac{f(x_0)}{f'(x_0)}, \tag{4.3.5}$$

where x_1 denotes an improved approximation to the root.

To further improve the approximation to the root, we use x_1 in place of x_0 on the right-hand side of Eq.(4.3.5) to obtain x_2. This iterative procedure can be generalized as

$$x_k=x_{k-1}-\frac{f(x_{k-1})}{f'(x_{k-1})},\quad k=1,2,\cdots. \tag{4.3.6}$$

The procedure is shown graphically in Figure. 4.3.

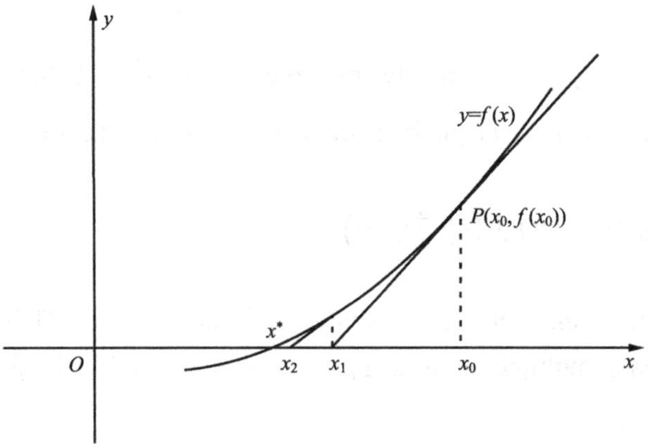

Figure 4.3　Newton's method

Definition 4.4　Suppose x^* is a simple root of the equation $f(x)=0$, and there exists a $\delta>0$ such that the function f has the continuous derivative and $f'(x)\neq 0$ on $(x^*-\delta,\ x^*+\delta)$. Newton's method is an iterative technique of the form $x_k=\varphi(x_{k-1})$, for which

$$x_k = x_{k-1} - \frac{f(x_{k-1})}{f'(x_{k-1})},\ \text{for}\ \ k=1,2,3,\cdots, \tag{4.3.7}$$

where

$$\varphi(x) = x - \frac{f(x)}{f'(x)} \tag{4.3.8}$$

is the iterative function. The iterative form defined in Eq. (4.3.7) is called **Newton's iterative formula** (Newton 迭代公式). Newton's method is also called the **tangent method** (切线法) .

Theorem 4.8　Let $f\in C^2[a,b]$. If x^* is a simple root of the equation $f(x)=0$, then a number $\delta>0$ exists such that the sequence $\{x_k\}_{k=1}^{\infty}$ generated by Newton's method converges to x^* at least quadratically for any initial approximation $x_0\in(x^*-\delta,\ x^*+\delta)\subset[a,b]$.

Proof.　Since the iterative function of Newton's method is $\varphi(x)=x-\dfrac{f(x)}{f'(x)}$,

$$\varphi'(x) = 1 - \frac{[f'(x)]^2 - f(x)f''(x)}{[f'(x)]^2} = \frac{f(x)f''(x)}{[f'(x)]^2},$$

and

$$\varphi'(x^*) = \frac{f(x^*)f''(x^*)}{[f'(x^*)]^2} = 0.$$

The fact shows the conclusion of Theorem 4.8 holds from Theorem 4.6.

Example 4.3　Let $f(x)=x^3+4x-10$ and $x_0=1.25$. Use Newton's method to find the solution accurate to within 10^{-8} for the equation $f(x)=0$.

Solution.　Since $f'(x)=3x^2+8x$, from Eq. (4.3.7), the sequence $\{x_k\}_{k=1}^{\infty}$ is generated by

$$x_k = x_{k-1} - \frac{f(x_{k-1})}{f'(x_{k-1})} = x_{k-1} - \frac{x_{k-1}^3 + 4x_{k-1}^2 - 10}{3x_{k-1}^2 + 8x_{k-1}}, \quad \text{for } k = 1, 2, \cdots$$

With $x_0 = 1.25$, the approximations in Table 4.3 are generated. An excellent approximation is obtained with $n = 4$ and compared with the true root $x^* = 1.36523001341409 69\cdots$.

Table 4.3 The approximations by Newton's method

k	x_k	$\lvert x_k - x_{k-1}\rvert$
0	1.25	
1	1.3723404255	0.122340
2	1.3652546708	0.007086
3	1.3652300137	0.000023
4	1.3652300134	0.29806×10^{-9}

Note

(1) Newton's method converges very fast in most cases. However, it may not converge, if the initial value x_0 is very far from the exact root.

(2) The following convergence criteria can be used for stopping the iterative procedure of Eq. (4.3.7) for a given tolerance $\varepsilon > 0$:

$$\lvert x_k - x_{k-1}\rvert \leqslant \varepsilon, \tag{4.3.9}$$

and

$$\lvert f(x_k)\rvert \leqslant \varepsilon. \tag{4.3.10}$$

(3) Newton's method can also be used for finding complex roots. In this case, we need to perform the computations using complex arithmetic with a complex number as the initial value for the root.

4.3.2 How to handle multiple roots using Newton's method (如何采用牛顿法处理重根问题)

Definition 4.5 A solution x^* of $f(x) = 0$ is a **zero of multiplicity *m*** (m 重零点) ($m \geqslant 2$ is an integer) of f if for $x \neq x^*$,

$$f(x) = (x - x^*)^m q(x), \text{ where } \lim_{x \to x^*} q(x) \neq 0. \tag{4.3.11}$$

Theorem 4.9 The function $f \in C^m[a,b]$ has a zero of multiplicity m at x^* in (a, b) if

$$f(x^*) = f'(x^*) = \cdots = f^{(m-1)}(x^*) = 0, \text{ but } f^{(m)}(x^*) \neq 0. \tag{4.3.12}$$

Since the function value $f(x)$ and its derivative $f'(x)$ are zero at a multiple root x^*, Newton's method, which contains derivatives in the denominator, cannot be applied conveniently to find the root x^*.

The following theorem shows that Newton's method converges only linearly for multiple roots. However, a slight modification for Newton's method can lead to quadratic convergence.

Theorem 4.10 Let $f \in C^2[a,b]$ and $x^* \in (a,b)$ be a root of multiplicity m $(m \geqslant 2)$ of the equation $f(x)=0$. Then a number $\delta > 0$ exists such that for $x_0 \in (x^* - \delta,\ x^* + \delta) \subset [a,b]$,

(a) the sequence $\{x_k\}$ defined by Newton's method

$$x_k = x_{k-1} - \frac{f(x_{k-1})}{f'(x_{k-1})}, \quad \text{for} \quad k = 1, 2, \cdots,$$

converges linearly to x^*;

(b) the sequence $\{x_k\}$ defined by the **modified Newton's method** (改进的牛顿迭代法)

$$x_k = x_{k-1} - m\frac{f(x_{k-1})}{f'(x_{k-1})}, \quad \text{for } k = 1, 2, \cdots, \tag{4.3.13}$$

converges at least quadratically to x^*.

Proof. Since $x_0 \in (a,b)$ is a root of multiplicity m $(m \geqslant 2)$ of the equation $f(x)=0$, from Definition 4.5, $f(x)$ can be written as

$$f(x) = (x - x^*)^m q(x), \text{ where } \lim_{x \to x^*} q(x) \neq 0.$$

(a) Since the iterative function of Newton's method is

$$\varphi(x) = x - \frac{f(x)}{f'(x)} = x - \frac{(x - x^*)q(x)}{mq(x) + (x - x^*)q'(x)},$$

we have

$$\varphi'(x) = 1 - \frac{q(x)}{mq(x) + (x - x^*)q'(x)} - (x - x^*)\left[\frac{q(x)}{mq(x) + (x - x^*)q'(x)}\right]'.$$

Then

$$0 < \varphi'(x^*) = 1 - \frac{1}{m} < 1$$

since $m \geqslant 2$. This shows the sequence $\{x_k\}$ defined by Newton's method converges linearly to x^* from Theorem 4.5 and Theorem 4.6.

(b) Since the iterative function of the modified Newton's method is

$$\psi(x) = x - m\frac{f(x)}{f'(x)} = x - m\frac{(x - x^*)q(x)}{mq(x) + (x - x^*)q'(x)},$$

$$\psi'(x) = 1 - \frac{mq(x)}{mq(x) + (x - x^*)q'(x)} x - m(x - x^*)\left[\frac{q(x)}{mq(x) + (x - x^*)q'(x)}\right]'.$$

Then

$$\psi'(x^*) = 1 - m \times \frac{1}{m} = 0.$$

This shows the sequence $\{x_k\}$ defined by the modified Newton's method converges at least quadratically to x^* from Theorem 4.6.

Example 4.4 Use Newton's method to find the approximation to the multiple root $x^*=1$ of $f(x)=x^3-2x^2+x=0$.

Solution. Since

$$f(1)=1^3-2\times1^2+1=0,\quad f'(1)=3\times1^2-4\times1+1=0,$$

but $f''(1)=6\times1-4\neq0$, $f(x)$ has a zero of multiplicity two at $x^*=1$.

The sequence $\{x_k\}$ generated by Newton's method applied to $f(x)$ with $x_0=2$ are shown in Table 4.4.

Table 4.4 The approximations to the multiple root by Newton's method

k	x_k	$\|x_k-x_{k-1}\|$	$f(x_k)$
0	2		
1	1.6	0.4	0.576000
2	1.347368	0.252632	0.162580
3	1.193617	0.153852	0.044696
⋮	⋮	⋮	⋮
16	1.000028	0.000028	0.781×10^{-9}
17	1.000014	0.000014	0.195×10^{-9}
18	1.000007	0.699×10^{-5}	0.488×10^{-9}

The sequence $\{x_k\}$ is clearly converging to $x^*=1$, but not quadratically.

Example 4.5 Use the modified Newton's method to find the approximation to the multiple root $x^*=1$ of $f(x)=x^3-2x^2+x=0$, which was solved by Newton's method in Example 4.4.

Solution. From Example 4.4, we know the multiplicity of the root $x^*=1$ is $m=2$, then the modified Newton's method applied to $f(x)=0$ yields

$$x_k=x_{k-1}-m\frac{f(x_{k-1})}{f'(x_{k-1})}=x_{k-1}-2\frac{x_{k-1}^3-2x_{k-1}^2+x_{k-1}}{3x_{k-1}^2-4x_{k-1}+1},\quad \text{for}\quad k=1,2,\cdots. \tag{4.3.14}$$

The sequence $\{x_k\}$ generated by Eq.(4.3.14) with $x_0=2$ are shown in Table 4.5 and then compared with the results listed in Table 4.5.

Table 4.5 The approximations to the multiple root by the modified Newton's method

k	x_k	$\|x_k-x_{k-1}\|$	$f(x_k)$
0	2		
1	1.2	0.8	0.048000

续表

k	x_k	$\lvert x_k - x_{k-1}\rvert$	$f(x_k)$
2	1.015385	0.184615	0.000240
3	1.000116	0.015269	0.134×10^{-8}
4	1.000000	0.000116	0

Eq. (4.3.13) is not very practical since its use requires a prior knowledge of the multiplicity of the root.

The following theorem gives another approach with quadratic convergence for finding multiple roots without the knowledge of the multiplicity of these roots .

Theorem 4.11 Suppose $f\in C^2[a,b]$ and $x^*\in(a,b)$ is a root of multiplicity $m\ (m\geqslant 2)$ of the equation $f(x)=0$. Then x^* is a simple root of the equation $\mu(x)=0$, where

$$\mu(x)=\frac{f(x)}{f'(x)}, \tag{4.3.15}$$

and a number $\delta>0$ exists such that for $x_0\in(x^*-\delta,\ x^*+\delta)\subset(a,b)$, the sequence $\{x_k\}$ defined by the following iterative method

$$x_k=x_{k-1}-\frac{\mu(x_{k-1})}{\mu'(x_{k-1})},\quad \text{for}\quad k=1,2,\cdots \tag{4.3.16}$$

converges quadratically at least to x^*.

Note

The iterative function of the method defined in Eq. (4.3.16) is

$$\varphi(x)=x-\frac{\mu(x)}{\mu'(x)}=x-\frac{f(x)/f'(x)}{\{[f'(x)]^2-f(x)f''(x)\}/[f'(x)]^2}$$

or

$$\varphi(x)=x-\frac{f(x)f'(x)}{[f'(x)]^2-f(x)f''(x)}. \tag{4.3.17}$$

Example 4.6 Use the method defined in Eq. (4.3.16) to find the approximation to the multiple root $x^*=1$ of $f(x)=x^3-2x^2+x=0$, which was considered in Example 4.4 and Example 4.5.

Solution. Since $f(x)=x^3-2x^2+x$, $f'(x)=3x^2-4x+1$, and

$$\mu(x)=\frac{f(x)}{f'(x)}=\frac{x^2-x}{3x-1}.$$

Applying Newton's method to μ gives

$$x_k=x_{k-1}-\frac{\mu(x_{k-1})}{\mu'(x_{k-1})}=x_{k-1}-\frac{(x_{k-1}^2-x_{k-1})(3x_{k-1}-1)}{3x_{k-1}^2-2x_{k-1}+1},\quad \text{for}\quad k=1,2,\cdots. \tag{4.3.18}$$

The terms generated by Eq. (4.3.18) with $x_0 = 2$ are shown in Table 4.6.

Table 4.6 Approximations with quadratic convergence for finding multiple roots

k	x_k	$\lvert x_k - x_{k-1}\rvert$	$f(x_k)$
0	2		
1	0.888889	1.111111	0.059259
2	0.992248	0.103359	–0.003891
3	0.999969	0.007721	–0.000015
4	1.000000	0.000031	0

4.4 The secant method (弦截法)

Newton's method requires the derivative of the function, $f'(x)$. In many cases, such as polynomial equations, the derivative of $f(x)$ can be obtained easily. However, in some problems, such as those involving transcendental functions, the differentiation of the function $f(x)$ may be quite complicated. In some cases, the function $f(x)$ may not be available in explicit form. In those cases above, the secant method can be used instead of Newton's method.

The secant method is similar to Newton's method, but is different in that the derivative $f'(x)$ is approximated by using two consecutive iterative values of $f(x)$.

Definition 4.6 The secant method is an iterative technique as

$$x_k = x_{k-1} - \frac{x_{k-1} - x_{k-2}}{f(x_{k-1}) - f(x_{k-2})} f(x_{k-1}), \quad \text{for each} \quad k = 2,3,\cdots. \tag{4.4.1}$$

Note

(1) Eq. (4.4.1) is obtained from Eq. (4.3.2) by using an approximation for the derivative $f'(x_{k-1})$ as

$$f'(x_{k-1}) \approx \frac{f(x_{k-1}) - f(x_{k-2})}{x_{k-1} - x_{k-2}}. \tag{4.4.2}$$

(2) Two initial values x_0 and x_1 are required to start the iterative process according to Eq. (4.4.1).

Geometrical Interpretation Starting with two initial approximations x_0 and x_1, the approximation x_2 is the x-intercept of the line joining $(x_0, f(x_0))$ and $(x_1, f(x_1))$. The approximation x_3 is the x-intercept of the line joining $(x_1, f(x_1))$ and $(x_2, f(x_2))$, and so on (See Figure. 4.4).

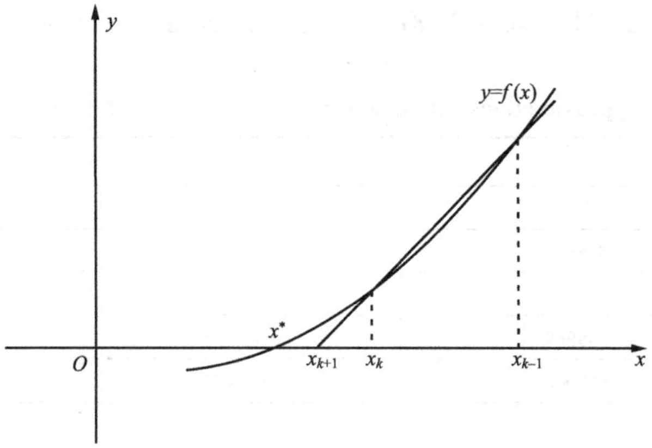

Figure 4.4 The secant method

Theorem 4.12 If x^* is a simple root of the equation $f(x)=0$, and there exists a small number $\delta>0$ such that the function $f(x)$ has two continuous derivatives and $f'(x)\neq 0$ in $(x^*-\delta,\ x^*+\delta)$, then the sequence $\{x_k\}_{k=1}^{\infty}$, generated by the secant method, converges to x^* of order

$$p=\frac{1+\sqrt{5}}{2}\approx 1.618,$$

for any two initial approximations $x_0, x_1 \in (x^*-\delta,\ x^*+\delta)$.

Example 4.7 Use the secant method to find the solution to $f(x)=x^3+4x-10=0$ accurate to within 10^{-8}.

Solution. With two initial approximations $x_0=1.25$, $x_1=1.3$, the approximations listed in Table 4.7 are generated by the iterative process from Eq. (4.4.1)

$$\begin{aligned} x_k &= x_{k-1}-\frac{x_{k-1}-x_{k-2}}{f(x_{k-1})-f(x_{k-2})}f(x_{k-1}) \\ &= x_{k-1}-\frac{(x_{k-1}-x_{k-2})(x_{k-1}^3+4x_{k-1}^2-10)}{(x_{k-1}^3+4x_{k-1}^2-10)-(x_{k-2}^3+4x_{k-2}^2-10)}, \quad \text{for } k=2,3,\cdots. \end{aligned}$$

An approximation we want is obtained with $k=6$.

Table 4.7 The approximations by the secant method

k	x_k	$\lvert x_k - x_{k-1}\rvert$
0	1.25	
1	1.3	
2	1.3691759244	0.069176
3	1.3651009356	0.004075
4	1.3652297641	0.000129
5	1.3652300134	0.249×10^{-6}
6	1.3652300134	0.158×10^{-10}

4.5 本章算法程序及实例 (Algorithms and examples)

4.5.1 二分法求方程的根 (Root finding by bisection method)

```
%% 二分法求方程的根，[a1,b1],[a2,b2],…,[an,bn]，函数，误差，初始区间
function x_root=Bisection_Method(fun,e,a,b)
p=-1;
while (fun(a)*fun(b)<=0) && (abs(a-b)>e)
    c=(a+b)/2;
    if fun(c)*fun(b)<=0
        a=c;
        p=p+1;
    else
        p=p+1;
        b=c;
    end
end
x_root=c;
end
```

例 4.8 用二分法求方程 $e^x+10x-2=0$ 在 $[0,1]$ 内的根，使误差小于 $\left|x_p-x^*\right|<10^{-3}$.

MATLAB 主程序：

```
clc; clear all;
fun=@(x) exp(x)+10*x-2;
x_root=Bisection_Method(fun,1e-3,0,1);
disp(x_root);
结果: 0.0908
```

4.5.2 不动点迭代法求方程的根 (Root finding by fix point iteration)

```
%%x_(k+1)=phi(x_k)，函数自取，初始点，误差，并计算迭代次数
function x_root=Fixed_Point_Iter(fun,x0,errors,m) %m表示迭代次数
p=1;
for i=1:m
x1=fun(x0);
    if abs(x1-x0)<errors
       disp('近似解为：');
       disp(x1);
       disp('迭代次数：');
```

```
        disp(p);
        break
    end
    x0=x1;
    p=p+1;
end
x_root=x1;
if p==m
disp('超出迭代次数');
end
```

例 4.9 用不动点方法求方程 $e^x+10x-2=0$ 在 $x_0=1$ 附近的根，使误差小于 $|x_{p+1}-x_p|<10^{-3}$.

解 令 $x=\dfrac{2-e^x}{10}$,得到迭代函数 $\varphi(x)=\dfrac{2-e^x}{10}$，运行 MATLAB 主程序：

```
clc; clear all;
fun=@(x)(2-exp(x))/10;
x0=1;
error=1e-3;
m=20;
x_root=Fixed_Point_Iter(fun,x0,errors,m);
disp(x_root);
```

结果迭代次数5次，x_root=0.0905.

4.5.3 牛顿法求方程的根 (Root finding by Newton's method)

```
%% 牛顿迭代法：x_(k+1)=x_k-f(x_k)/f'(x_k)，函数，导函数，初值，误差，次数
function x_root=Newton_Iterative_root (fun,fun_d,x0,errors,m)
%x0为初值，m为迭代次数；
p=0;
for i=1:m
    x1=x0-fun(x0)/fun_d(x0);
    if abs(x0-x1)<errors
        disp('迭代次数：');
        disp(p);
        break
    end
    x0=x1;
    p=p+1;
end
```

```
x_root=x1;
if p==m
disp('超出迭代次数');
end
```

例 4.10 用牛顿迭代法求方程 $e^x+10x-2=0$ 在 $x_0=1$ 附近的根，使误差小于 $|x_{p+1}-x_p|<10^{-3}$.

直接运行 MATLAB 主程序：

```
clc; clear all;
fun=@(x) exp(x)+10*x-2;%%直接代入函数
fun_d=@(x) exp(x)+10;%%输入函数的导函数
x0=1;
errors=1e-3;
m=20;
x_root=Newton_Iterative_root (fun,fun_d,x0,errors, m);
disp(x_root);
```

结果迭代次数2次，x_root=0.0908.

4.5.4 牛顿法求一元方程重根(未知重数) (Multiple root finding by Newton's method)

```
%%迭代格式：x_(k+1)=x_k-f(x_k)*f'(x_k)/(f'(x_k)^2-f(x_k)*f''(x_k)),
function x_root=Newton_Iterative_multiple_root(fun,fun_d,fun_dd,x0,
errors,m)
%x0为初值，m为迭代次数；
p=0;
for i=1:m
    x1=x0-(fun(x0)*fun_d(x0))/(fun_d(x0)^2-fun(x0)*fun_dd(x0));
    if abs(x0-x1)<errors
       disp('迭代次数：');
       disp(p);
       break
    end
    x0=x1;
    p=p+1;
end
x_root=x1;
if p==m
disp('超出迭代次数');
end
```

例 4.11　用重根格式求方程 $x^3-3x^2+4=0$ 在 $x_0=1$ 附近的根，使误差小于 $|x_{p+1}-x_p|<10^{-3}$.

运行 MATLAB 主程序：

```
clc; clear all;
fun=@(x) x^3-3*x^2+4;  %原函数
fun_d=@(x) 3*x^2-6*x;  %一阶导函数
fun_dd=@(x) 6*x-6;     %二阶导函数
x0=1;
errors=1e-3;
m=20;
x_root=Newton_Iterative_multiple_root(fun,fun_d,fun_dd,x0,errors,m);
disp(x_root);
```

结果迭代次数3次，x_root= 2.0000.

4.5.5 割线法求方程的根 (Root finding by secant method)

```
%%迭代格式：用差商代替导数，用牛顿迭代法先求x1,或自定义x1
function x_root=Secant_root(fun,fun_d,x0,errors,m)%x0为初值，m为迭代次数;
x1=x0-fun(x0)/fun_d(x0);  %利用牛顿法找出另一个初值，也可以自定义x1;
p=0;
for i=1:m
    x2=x1-fun(x0)*(x1-x0)/(fun(x1)-fun(x0));
    if abs(x2-x1)<errors
       disp('迭代次数：');
       disp(p);
       break
    end
    x0=x1;
    x1=x2;
    p=p+1;
end
x_root=x0;
if p==m
     disp('超出迭代次数');
end
```

例 4.12　用弦截法求方程 $e^x+10x-2=0$ 在 $x_0=1$ 附近的根，使误差小于 $|x_{p+1}-x_p|<10^{-4}$.

运行 MATLAB 主程序：

```
clc; clear all;
fun=@(x) exp(x)+10*x-2;
fun_d=@(x) exp(x)+10;
x0=1;
errors=1e-4;
m=10000;
x_root=Secant_root(fun,fun_d,x0,errors, m);
disp(x_root);
结果迭代次数4282次, x_root=0.0904.
```

4.6 本章要点 (Highlights)

1. 二分法的理论基础是零点定理，因此它只能求单实根或奇数重实根，不能求偶数重实根或复根.
2. 二分法运算简单，方法可靠，对 $f(x)$ 要求不高，但收敛速度慢，仅有线性收敛，因此常常用来确定其他迭代法的初始值.
3. 不动点迭代法是将方程求根的问题转化为求函数的不动点问题，从而得到求方程根的迭代法.
4. 不动点迭代法是一种逐次逼近的方法，具有原理简单，计算方便，在计算机上容易实现等优点. 迭代法使用的关键是构造收敛的迭代公式，主要是迭代函数 $\varphi(x)$ 和初值 x_0 的选取：迭代函数 $\varphi(x)$ 必须满足定理 4.4 的条件，才能保证迭代格式产生的序列 $\{x_k\}_{k=0}^{\infty}$ 收敛到方程的根 x^*；初值 x_0 就在满足定理 4.4 的条件的区间内选取.
5. 收敛阶是衡量收敛快慢的标准，本章遇到的收敛阶主要是 1 阶和 2 阶，1 阶收敛也称为线性收敛；2 阶收敛也称为平方收敛.
6. 在方程求根的数值方法中，牛顿法是一种最常用的方法；它不但可以求单实根，还可以求重根和复根，并且构造方便、收敛速度快 (至少平方收敛)；但它对初值要求较高，初值的选取必须在根的附近，而且需要函数的导数信息.
7. 牛顿法虽然能求重根，但收敛速度变慢，只有线性收敛. 改进的牛顿法(4.3.13)求重根，可达到平方收敛，但必须先知道重根的重数，这是它的不足.
8. 用迭代格式(4.3.16)求重根，相当于将求重根转化为求单根，而求单根的牛顿法至少是平方收敛的，因此用迭代格式(4.3.16)求重根可以达到平方收敛.
9. 为了避免求函数的导数值，可采用弦截法. 但弦截法需要两个初值，可用二分法确定它的初值. 弦截法对初值要求较高，初值的选取必须在根的附近.
10. 弦截法是超线性收敛的，虽然收敛阶比牛顿法低，但是不需要计算导数，总体计算耗时并不比牛顿法长，因此在工程计算中，也是常用的方法之一.

4.7 问题讨论 (Questions for discussion)

1. 当求方程根的迭代法收敛速度比较慢时，如何对其加速？
2. 定理 4.4 中迭代函数 $\varphi(x)$ 必须满足的条件能否放宽？
3. 牛顿法和弦截法都对初值要求较高，初值的选取必须在根的附近；如何用二分法确定它们的初值？选取几个方程，用二分法分别确定它们的初值.
4. 为什么收敛阶能作为衡量收敛快慢的标准？
5. 取 Taylor 公式的前 2 项构造的 Newton 迭代法至少是平方收敛的；若取 Taylor 公式的更多项，能否构造出更高收敛阶的迭代法？
6. 不知道重根的重数时，如何估计重根的重数？
7. 如何用牛顿法求方程的复根？
8. 弦截法是超线性收敛的，虽然收敛阶比牛顿法低，但是不需要计算导数，总体计算耗时未必比牛顿法长. 在什么情况下，弦截法耗时比牛顿法长？在什么情况下，弦截法耗时比牛顿法短？选取几个方程，分别用牛顿法和弦截法进行验证.

4.8 关键术语 (Key terms)

root (根)
nonlinear equation (非线性方程)
bisection (二分法)
fixed-point (不动点)
root-finding problem (求根问题)
fixed-point problem (求不动点问题)
interval (区间)
endpoint (端点)
midpoint (中点)
initial value (初值)
iterative function (迭代函数)
neighbourhood (邻域)
asymptotic (渐近的)
order of convergence (收敛阶)
linearly convergent (线性收敛的)
quadratically convergent (平方收敛的)
tangent (切线，正切)
simple root (单根)
monotone (单调的)
multiple root (重根)

multiplicity (重数)
complex root (复根)
modified (改进的)
secant method (弦截法)
numerator (分子)
denominator (分母)

4.9 延伸阅读 (Extending reading)

非线性方程组的不动点迭代法、牛顿迭代法及拟牛顿迭代法

设非线性方程组 (system of nonlinear equations)

$$\begin{cases} f_1(x_1,x_2,\cdots,x_n)=0, \\ f_2(x_1,x_2,\cdots,x_n)=0, \\ \quad\cdots\cdots \\ f_n(x_1,x_2,\cdots,x_n)=0, \end{cases} \tag{4.9.1}$$

其中 $f_i\ (i=1,2,\cdots,n)$ 是变量 $x_1,x_2,\cdots,x_n$ 的 n 元函数，且至少有一个是自变量 $x_i\ (i=1,2,\cdots,n)$ 的非线性函数. 若记

$$\boldsymbol{x}=(x_1,x_2,\cdots,x_n)^{\mathrm{T}},\quad \boldsymbol{F}(\boldsymbol{x})=\left(f_1(\boldsymbol{x}),f_2(\boldsymbol{x}),\cdots,f_n(\boldsymbol{x})\right)^{\mathrm{T}},$$

则(4.9.1)式可写成

$$\boldsymbol{F}(\boldsymbol{x})=\boldsymbol{0}. \tag{4.9.2}$$

求解非线性方程组，形式上，只要把单变量函数 $f(x)$ 看成向量函数 $\boldsymbol{F}(\boldsymbol{x})$，就可将单个非线性方程的求根方法用于求非线性方程组. 下面主要介绍不动点迭代法、牛顿迭代法及拟牛顿迭代法. 因为非线性方程组(4.9.1)可能无解，有唯一解或无穷多解，所以有关它的求解问题一般要比单个非线性方程困难.

1. 解非线性方程组的不动点迭代法

首先将(4.9.2)式改写为

$$\boldsymbol{x}=\boldsymbol{\Phi}(\boldsymbol{x}), \tag{4.9.3}$$

式中，向量函数 $\boldsymbol{\Phi}(\boldsymbol{x})$ 为连续函数. 若向量 $\boldsymbol{x}^*$ 满足 $\boldsymbol{x}^*=\boldsymbol{\Phi}(\boldsymbol{x}^*)$，则称 $\boldsymbol{x}^*$ 为函数 $\boldsymbol{\Phi}(\boldsymbol{x})$ 的不动点， $\boldsymbol{x}^*$ 也就是方程组(4.9.1)的一个解.

由(4.9.3)式可构造不动点迭代格式

$$\boldsymbol{x}^{(k+1)}=\boldsymbol{\Phi}(\boldsymbol{x}^{(k)}),\quad k=0,1,2,\cdots \tag{4.9.4}$$

若由(4.9.4)式产生的向量序列 $\{\boldsymbol{x}^{(k)}\}$ 满足 $\lim\limits_{k\to\infty}\boldsymbol{x}^{(k)}=\boldsymbol{x}^*$，则 $\boldsymbol{x}^*$ 为向量函数 $\boldsymbol{\Phi}(\boldsymbol{x})$ 的不动点.

例 4.13 用不动点迭代法解下列非线性方程组

$$\begin{cases} x_1^2 - 10x_1 + x_2^2 + 8 = 0, \\ x_1x_2^2 + x_1 - 10x_2 + 8 = 0. \end{cases}$$

解　将原方程组改写成等价形式

$$\begin{cases} x_1 = \dfrac{1}{10}(x_1^2 + x_2^2 + 8), \\ x_2 = \dfrac{1}{10}(x_1x_2^2 + x_1 + 8). \end{cases} \tag{4.9.5}$$

记

$$\boldsymbol{x} = \begin{bmatrix} x_1 \\ x_2 \end{bmatrix}, \qquad \boldsymbol{\Phi}(\boldsymbol{x}) = \begin{bmatrix} \varphi_1(\boldsymbol{x}) \\ \varphi_2(\boldsymbol{x}) \end{bmatrix} = \begin{bmatrix} \dfrac{1}{10}(x_1^2 + x_2^2 + 8) \\ \dfrac{1}{10}(x_1x_2^2 + x_1 + 8) \end{bmatrix},$$

则(4.9.5)式可写为

$$\boldsymbol{x} = \boldsymbol{\Phi}(\boldsymbol{x}).$$

由此构造不动点迭代格式

$$\begin{cases} x_1^{(k+1)} = \dfrac{1}{10}((x_1^{(k)})^2 + (x_2^{(k)})^2 + 8), \\ x_2^{(k+1)} = \dfrac{1}{10}(x_1^{(k)}(x_2^{(k)})^2 + x_1^{(k)} + 8), \end{cases} \quad k=0,1,2,\cdots \tag{4.9.6}$$

即

$$\boldsymbol{x}^{(k+1)} = \boldsymbol{\Phi}(\boldsymbol{x}^{(k)}), \quad k=0,1,2,\cdots$$

取初始近似 $\boldsymbol{x}^{(0)} = (0, 0)^{\mathrm{T}}$，按迭代格式(4.9.6)计算得 $\boldsymbol{x}^*$ 近似值，见表 4.8.

表 4.8　不动点迭代法求解非线性方程组

k	0	1	2	3	4	…	10
$x_1^{(k)}$	0	0.8	0.9280000	0.9728317	0.9893656	…	0.9999571
$x_2^{(k)}$	0	0.8	0.9312000	0.9732700	0.9894351	…	0.9999571

其收敛性类似于单个方程的情况.

定理 4.13　设函数 $\boldsymbol{\Phi}(\boldsymbol{x})$ 的定义域 $D \subset \mathbf{R}^n$，且

(1) 存在闭集 $D_0 \subset D$，实数 $L \in (0,1)$，对 $\forall \boldsymbol{x}, \boldsymbol{y} \in D_0$，有

$$\|\boldsymbol{\Phi}(\boldsymbol{x}) - \boldsymbol{\Phi}(\boldsymbol{y})\| \leqslant L\|\boldsymbol{x} - \boldsymbol{y}\|,$$

(2) 对 $\forall \boldsymbol{x} \in D_0$，有 $\boldsymbol{\Phi}(\boldsymbol{x}) \in D_0$，

则 $\boldsymbol{\Phi}(\boldsymbol{x})$ 在 D_0 中有唯一的不动点 $\boldsymbol{x}^*$，且对任意 $\boldsymbol{x}^{(0)} \in D_0$，由迭代格式(4.9.4)产生的向量序列 $\{\boldsymbol{x}^{(k)}\}$ 收敛到 $\boldsymbol{x}^*$，并有误差估计

$$\| \boldsymbol{x}^* - \boldsymbol{x}^{(k)} \| \leqslant \frac{1}{1-L} \| \boldsymbol{x}^{(k+1)} - \boldsymbol{x}^{(k)} \| \leqslant \frac{L^k}{1-L} \| \boldsymbol{x}^{(1)} - \boldsymbol{x}^{(0)} \|.$$

例 4.14　对于例 4.13 中的方程组，设 $D_0 = \{\boldsymbol{x} = (x_1, x_2)^{\mathrm{T}} \mid 0 \leqslant x_1, x_2 \leqslant 1.5\}$，证明：对任意初始点 $\boldsymbol{x}^{(0)} \in D_0$，由迭代格式(4.9.6)生成的序列 $\{\boldsymbol{x}^{(k)}\}$ 均收敛到 $\boldsymbol{x}^*$.

证明　首先，对任意 $\boldsymbol{x} = (x_1, x_2)^{\mathrm{T}} \in D_0$，可以验证

$$0 < \varphi_1(\boldsymbol{x}) \leqslant 1, \quad 0 < \varphi_2(\boldsymbol{x}) \leqslant 1.2875.$$

因此当 $\boldsymbol{x} \in D_0$，有 $\boldsymbol{\Phi}(\boldsymbol{x}) \in D_0$.

其次，对一切 $\boldsymbol{x} = (x_1, x_2)^{\mathrm{T}} \in D_0$，$\boldsymbol{y} = (y_1, y_2)^{\mathrm{T}} \in D_0$，都有

$$\begin{aligned}
\left|\varphi_1(\boldsymbol{x}) - \varphi_1(\boldsymbol{y})\right| &= \frac{1}{10} | (x_1 + y_1)(x_1 - y_1) + (x_2 + y_2)(x_2 - y_2) | \\
&\leqslant \frac{3}{10}(| x_1 - y_1 | + | x_2 - y_2 |) = 0.3 \| \boldsymbol{x} - \boldsymbol{y} \|_1, \\
\left|\varphi_2(\boldsymbol{x}) - \varphi_2(\boldsymbol{y})\right| &= \frac{1}{10} | x_1 - y_1 + x_1 x_2^2 - y_1 y_2^2 | \\
&= \frac{1}{10} | x_1 - y_1 + x_1 x_2^2 - y_1 x_2^2 + y_1 x_2^2 - y_1 y_2^2) | \\
&= \frac{1}{10} | (1 + x_2^2)(x_1 - y_1) + y_1 (x_2 + y_2)(x_2 - y_2) | \\
&= \frac{1}{10}(3.25 | x_1 - y_1 | + 4.5 | x_2 - y_2 |) \\
&\leqslant \frac{4.5}{10}(| x_1 - y_1 | + | x_2 - y_2 |) = 0.45 \| \boldsymbol{x} - \boldsymbol{y} \|_1.
\end{aligned}$$

从而有

$$\|\boldsymbol{\Phi}(\boldsymbol{x}) - \boldsymbol{\Phi}(\boldsymbol{y})\| = \left|\varphi_1(\boldsymbol{x}) - \varphi_1(\boldsymbol{y})\right| + \left|\varphi_2(\boldsymbol{x}) - \varphi_2(\boldsymbol{y})\right| \leqslant 0.75 \|\boldsymbol{x} - \boldsymbol{y}\|_1.$$

由定理 4.13 知，$\boldsymbol{\Phi}(\boldsymbol{x})$ 在 D_0 上有唯一不动点 $\boldsymbol{x}^*$，因此对任意的初始点 $\boldsymbol{x}^{(0)} \in D_0$，由迭代格式(4.9.6)生成的序列 $\left\{\boldsymbol{x}^{(k)}\right\}$ 均收敛到 $\boldsymbol{x}^*$.

定理 4.14　设函数 $\boldsymbol{\Phi}(\boldsymbol{x})$ 的定义域 $D \subset \mathbf{R}^n$ 内有不动点 $\boldsymbol{x}^*$，$\boldsymbol{\Phi}(\boldsymbol{x})$ 的分量函数 $\varphi_i\left(i = 1, 2, \cdots, n\right)$ 有连续的偏导数，且谱半径

$$\rho\left(\boldsymbol{\Phi}'(\boldsymbol{x}^*)\right) < 1, \tag{4.9.7}$$

则存在 $\boldsymbol{x}^*$ 的一个邻域 U，对任意 $\boldsymbol{x}_0 \in U$，迭代格式(4.9.4)产生的序列 $\left\{\boldsymbol{x}^{(k)}\right\}$ 收敛到 $\boldsymbol{x}^*$. 其中 $\boldsymbol{\Phi}(\boldsymbol{x})$ 的导数为 Jacobi 矩阵(Jacobi matrix)

$$\boldsymbol{\Phi}'(\boldsymbol{x})=\begin{bmatrix}\dfrac{\partial\varphi_1(\boldsymbol{x})}{\partial x_1} & \dfrac{\partial\varphi_1(\boldsymbol{x})}{\partial x_2} & \cdots & \dfrac{\partial\varphi_1(\boldsymbol{x})}{\partial x_n}\\ \dfrac{\partial\varphi_2(\boldsymbol{x})}{\partial x_1} & \dfrac{\partial\varphi_2(\boldsymbol{x})}{\partial x_2} & \cdots & \dfrac{\partial\varphi_2(\boldsymbol{x})}{\partial x_n}\\ \vdots & \vdots & & \vdots\\ \dfrac{\partial\varphi_n(\boldsymbol{x})}{\partial x_1} & \dfrac{\partial\varphi_n(\boldsymbol{x})}{\partial x_2} & \cdots & \dfrac{\partial\varphi_n(\boldsymbol{x})}{\partial x_n}\end{bmatrix}.$$

根据矩阵谱半径与 1-范数的关系，可得到如下结论：

对于定义在 D 上的函数

$$\boldsymbol{\Phi}(\boldsymbol{x})=\left(\varphi_1(x_1,x_2,\cdots,x_n),\ \varphi_2(x_1,x_2,\cdots,x_n),\ \cdots,\ \varphi_n(x_1,x_2,\cdots,x_n)\right)^{\mathrm{T}},$$

若存在常数 $0<K<1$ 使得，当 $\boldsymbol{x}\in D$ 时，

$$\left|\frac{\partial\varphi_i(\boldsymbol{x})}{\partial x_j}\right|\leqslant\frac{K}{n},\quad i,j=1,2,\cdots,n,$$

则矩阵 $\boldsymbol{\Phi}'(\boldsymbol{x})$ 的谱半径 $\rho\left(\boldsymbol{\Phi}'(\boldsymbol{x}^*)\right)<1$.

例如，在例 4.13 中，由

$$\boldsymbol{\Phi}'(\boldsymbol{x})=\frac{1}{10}\begin{bmatrix}2x_1 & 2x_2\\ x_2^2+1 & 2x_1x_2\end{bmatrix}$$

可知，对一切 $\boldsymbol{x}\in D_0$，都有 $\left|\dfrac{\partial\varphi_i(\boldsymbol{x})}{\partial x_j}\right|\leqslant\dfrac{K}{2}(i,j=1,2)$，其中 $K=0.9<1$，因此在 D_0 内有 $\rho\left(\boldsymbol{\Phi}'(\boldsymbol{x})\right)<1$，满足定理 4.14 的条件，故迭代序列 $\left\{\boldsymbol{x}^{(k)}\right\}$ 局部收敛.

2. 解非线性方程组的 Newton 迭代法

设向量 $\boldsymbol{x}^*$ 是方程组(4.9.1)的解，向量 $\boldsymbol{x}^{(0)}$ 是方程组的初始近似解，假定 $\boldsymbol{F}(\boldsymbol{x})$ 在 $\boldsymbol{x}^*$ 可微，将 $\boldsymbol{F}(\boldsymbol{x})$ 在 $\boldsymbol{x}^{(0)}$ 处作多元函数的 Taylor 展开，并取其线性部分：

$$\begin{cases}0=f_1(\boldsymbol{x}^*)\approx f_1(\boldsymbol{x}^{(0)})+\dfrac{\partial f_1(\boldsymbol{x}^{(0)})}{\partial x_1}(x_1^*-x_1^{(0)})+\dfrac{\partial f_1(\boldsymbol{x}^{(0)})}{\partial x_2}(x_2^*-x_2^{(0)})+\cdots+\dfrac{\partial f_1(\boldsymbol{x}^{(0)})}{\partial x_n}(x_n^*-x_n^{(0)}),\\ 0=f_2(\boldsymbol{x}^*)\approx f_2(\boldsymbol{x}^{(0)})+\dfrac{\partial f_2(\boldsymbol{x}^{(0)})}{\partial x_1}(x_1^*-x_1^{(0)})+\dfrac{\partial f_2(\boldsymbol{x}^{(0)})}{\partial x_2}(x_2^*-x_2^{(0)})+\cdots+\dfrac{\partial f_2(\boldsymbol{x}^{(0)})}{\partial x_n}(x_n^*-x_n^{(0)}),\\ \cdots\\ 0=f_n(\boldsymbol{x}^*)\approx f_n(\boldsymbol{x}^{(0)})+\dfrac{\partial f_n(\boldsymbol{x}^{(0)})}{\partial x_1}(x_1^*-x_1^{(0)})+\dfrac{\partial f_n(\boldsymbol{x}^{(0)})}{\partial x_2}(x_2^*-x_2^{(0)})+\cdots+\dfrac{\partial f_n(\boldsymbol{x}^{(0)})}{\partial x_n}(x_n^*-x_n^{(0)}),\end{cases}$$

即

$$0=\boldsymbol{F}(\boldsymbol{x}^*)\approx\boldsymbol{F}(\boldsymbol{x}^{(0)})+\boldsymbol{F}'(\boldsymbol{x}^{(0)})(\boldsymbol{x}^*-\boldsymbol{x}^{(0)}),$$

其中，$\boldsymbol{F}'(\boldsymbol{x}^{(0)})$ 为 Jacobi 矩阵 $\boldsymbol{F}'(\boldsymbol{x})$ 在 $\boldsymbol{x}^{(0)}$ 处的值

$$\boldsymbol{F}'(\boldsymbol{x}^{(0)})=\begin{bmatrix}\dfrac{\partial f_1(\boldsymbol{x}^{(0)})}{\partial x_1} & \dfrac{\partial f_1(\boldsymbol{x}^{(0)})}{\partial x_2} & \cdots & \dfrac{\partial f_1(\boldsymbol{x}^{(0)})}{\partial x_n}\\ \dfrac{\partial f_2(\boldsymbol{x}^{(0)})}{\partial x_1} & \dfrac{\partial f_2(\boldsymbol{x}^{(0)})}{\partial x_2} & \cdots & \dfrac{\partial f_2(\boldsymbol{x}^{(0)})}{\partial x_n}\\ \vdots & \vdots & & \vdots\\ \dfrac{\partial f_n(\boldsymbol{x}^{(0)})}{\partial x_1} & \dfrac{\partial f_n(\boldsymbol{x}^{(0)})}{\partial x_2} & \cdots & \dfrac{\partial f_n(\boldsymbol{x}^{(0)})}{\partial x_n}\end{bmatrix}.$$

若 Jacobi 矩阵 $\boldsymbol{F}'(\boldsymbol{x}^{(0)})$ 非奇异，则方程组

$$\boldsymbol{F}(\boldsymbol{x}^{(0)})+\boldsymbol{F}'(\boldsymbol{x}^{(0)})(\boldsymbol{x}-\boldsymbol{x}^{(0)})=\boldsymbol{0}$$

有唯一解

$$\boldsymbol{x}^{(1)}=\boldsymbol{x}^{(0)}-\left[\boldsymbol{F}'(\boldsymbol{x}^{(0)})\right]^{-1}\boldsymbol{F}(\boldsymbol{x}^{(0)}).$$

一般地，当 $\boldsymbol{F}'(\boldsymbol{x}^{(k)})$ 非奇异时，可得 Newton 迭代法

$$\boldsymbol{x}^{(k+1)}=\boldsymbol{x}^{(k)}-\left[\boldsymbol{F}'(\boldsymbol{x}^{(k)})\right]^{-1}\boldsymbol{F}(\boldsymbol{x}^{(k)}),\quad k=0,1,2,\cdots. \tag{4.9.8}$$

在第 k 步计算过程中，不仅要算出雅可比矩阵 $\boldsymbol{F}'(\boldsymbol{x}^{(k)})$，还要求逆矩阵，计算量较大. 因此，通常把第 k 步迭代分成两步：

记 $\boldsymbol{x}^{(k+1)}-\boldsymbol{x}^{(k)}=\Delta\boldsymbol{x}^{(k)}$，则(4.9.8)式变为

$$\boldsymbol{F}'(\boldsymbol{x}^{(k)})\Delta\boldsymbol{x}^{(k)}=-\boldsymbol{F}(\boldsymbol{x}^{(k)}),\quad k=0,1,2,\cdots. \tag{4.9.9}$$

解此线性方程组，求出解向量 $\Delta\boldsymbol{x}^{(k)}$ 后，再令 $\boldsymbol{x}^{(k+1)}=\boldsymbol{x}^{(k)}+\Delta\boldsymbol{x}^{(k)}$，避免了求矩阵的逆.

定理 4.15　设 $\boldsymbol{F}(\boldsymbol{x})$ 的定义域 $D\subset\mathbf{R}^n$，$\boldsymbol{F}(\boldsymbol{x}^*)=\boldsymbol{0}$. 若有 $\boldsymbol{x}^*$ 的开邻域 $U_0\subset D$，$\boldsymbol{F}'(\boldsymbol{x})$ 在其上连续，$\boldsymbol{F}'(\boldsymbol{x}^*)$ 可逆，则

(1) Newton 迭代法产生的序列 $\{\boldsymbol{x}^{(k)}\}$ 在 $\boldsymbol{x}^*$ 的某个邻域 U 上超线性收敛于 $\boldsymbol{x}^*$；

(2) 若再加上条件：存在常数 $K>0$，使得

$$\left\|\boldsymbol{F}'(\boldsymbol{x})-\boldsymbol{F}'(\boldsymbol{x}^*)\right\|\leqslant K\left\|\boldsymbol{x}-\boldsymbol{x}^*\right\|,\quad \forall\boldsymbol{x}\in U,$$

则 $\{\boldsymbol{x}^{(k)}\}$ 至少平方收敛.

例 4.15　用 Newton 迭代法解下列非线性方程组

$$\begin{cases}x_1^2-10x_1+x_2^2+8=0,\\ x_1x_2^2+x_1-10x_2+8=0.\end{cases}$$

解　记 $\boldsymbol{F}(\boldsymbol{x})=\begin{bmatrix}x_1^2-10x_1+x_2^2+8\\ x_1x_2^2+x_1-10x_2+8\end{bmatrix}$，则 $\boldsymbol{F}'(\boldsymbol{x})=\begin{bmatrix}2x_1-10 & 2x_2\\ x_2^2+1 & 2x_1x_2-10\end{bmatrix}$.

取初值 $\boldsymbol{x}^{(0)}=(0,0)^{\mathrm{T}}$，解方程组

$$\boldsymbol{F}'(\boldsymbol{x}^{(0)})\Delta\boldsymbol{x}^{(0)}=-\boldsymbol{F}(\boldsymbol{x}^{(0)}),$$

即

$$\begin{bmatrix} -10 & 0 \\ 1 & -10 \end{bmatrix} \Delta \boldsymbol{x}^{(0)} = \begin{bmatrix} -8 \\ 8 \end{bmatrix}.$$

求得 $\Delta \boldsymbol{x}^{(0)} = \begin{bmatrix} \Delta x_1^{(0)} \\ \Delta x_2^{(0)} \end{bmatrix} = \begin{bmatrix} 0.8 \\ 0.88 \end{bmatrix}$，然后计算得

$$\boldsymbol{x}^{(1)} = \boldsymbol{x}^{(0)} + \Delta \boldsymbol{x}^{(0)} = (0.8,\ 0.88)^{\mathrm{T}}.$$

类似地，可得 $\boldsymbol{x}^{(2)}, \boldsymbol{x}^{(3)}, \cdots$，具体结果见表 4.9.

表 4.9　Newton 迭代法求解非线性方程组

k	0	1	2	3	4
$x_1^{(k)}$	0	0.8	0.9917872	0.9999752	1.0000000
$x_2^{(k)}$	0	0.8	0.9917117	0.9999685	1.0000000

可见，Newton 迭代法的收敛速度比不动点迭代法的收敛速度要快.

3. 解非线性方程组的拟 Newton 法

在求解非线性方程组(4.9.2)的 Newton 法

$$\boldsymbol{x}^{(k+1)} = \boldsymbol{x}^{(k)} - \left[\boldsymbol{F}'(\boldsymbol{x}^{(k)})\right]^{-1} \boldsymbol{F}(\boldsymbol{x}^{(k)}), \quad k = 0,1,2,\cdots$$

中，

$$\boldsymbol{F}'(\boldsymbol{x}^{(k)}) = \begin{bmatrix} \dfrac{\partial f_1(\boldsymbol{x}^{(k)})}{\partial x_1} & \dfrac{\partial f_1(\boldsymbol{x}^{(k)})}{\partial x_2} & \cdots & \dfrac{\partial f_1(\boldsymbol{x}^{(k)})}{\partial x_n} \\ \dfrac{\partial f_2(\boldsymbol{x}^{(k)})}{\partial x_1} & \dfrac{\partial f_2(\boldsymbol{x}^{(k)})}{\partial x_2} & \cdots & \dfrac{\partial f_2(\boldsymbol{x}^{(k)})}{\partial x_n} \\ \vdots & \vdots & & \vdots \\ \dfrac{\partial f_n(\boldsymbol{x}^{(k)})}{\partial x_1} & \dfrac{\partial f_n(\boldsymbol{x}^{(k)})}{\partial x_2} & \cdots & \dfrac{\partial f_n(\boldsymbol{x}^{(k)})}{\partial x_n} \end{bmatrix} \in \mathbf{R}^{n \times n}$$

是 $\boldsymbol{F}(\boldsymbol{x})$ 的 Jacobi 矩阵在 $\boldsymbol{x}^{(k)}$ 处的值. 当 $\boldsymbol{F}(\boldsymbol{x})$ 复杂时，$\boldsymbol{F}'(\boldsymbol{x}^{(k)})$ 的计算量较大，且求解困难. 在实际计算中，为减少计算量，避免每步都重新计算 $\boldsymbol{F}'(\boldsymbol{x}^{(k)})$，类似于割线法的思路，只要求新的 $\boldsymbol{F}'(\boldsymbol{x}^{(k+1)})$ 满足方程

$$\boldsymbol{F}'(\boldsymbol{x}^{(k+1)})\left(\boldsymbol{x}^{(k+1)} - \boldsymbol{x}^{(k)}\right) = \boldsymbol{F}(\boldsymbol{x}^{(k+1)}) - \boldsymbol{F}(\boldsymbol{x}^{(k)}), \quad k = 0,1,\cdots. \tag{4.9.10}$$

这表明矩阵 $\boldsymbol{F}'(\boldsymbol{x}^{(k+1)})$ 关于点 $\boldsymbol{x}^{(k)}$ 及 $\boldsymbol{x}^{(k+1)}$ 具有差商性质，即当 $n=1$ 时，$\boldsymbol{F}'(\boldsymbol{x}^{(k+1)})$ 即为 $\boldsymbol{F}(\boldsymbol{x})$ 关于 $\boldsymbol{x}^{(k)}$ 及 $\boldsymbol{x}^{(k+1)}$ 的差商. 但当 $n>1$ 时，$\boldsymbol{F}'(\boldsymbol{x}^{(k+1)})$ 并不确定，为此限制 $\boldsymbol{F}'(\boldsymbol{x}^{(k+1)})$ 是由 $\boldsymbol{F}'(\boldsymbol{x}^{(k)})$ 的一个低秩修正矩阵得到的，即

$$\boldsymbol{F}'(\boldsymbol{x}^{(k+1)}) = \boldsymbol{F}'(\boldsymbol{x}^{(k)}) + \Delta \boldsymbol{F}'(\boldsymbol{x}^{(k)}), \quad \operatorname{rank}\left(\Delta \boldsymbol{F}'(\boldsymbol{x}^{(k)})\right) = m \geqslant 1 \tag{4.9.11}$$

式中，$\Delta \boldsymbol{F}'(\boldsymbol{x}^{(k)})$ 是秩为 m 的修正矩阵，称由(4.9.8)式、(4.9.10)式、(4.9.11)式组成的迭代

算法为**拟 Newton 法**(quasi Newton method). 此方法只需对给出的初始近似 $\boldsymbol{x}^{(0)}$ 及矩阵 $\boldsymbol{F}'(\boldsymbol{x}^{(0)})$，用这三式逐次计算得到 $\{\boldsymbol{x}^{(k)}\}$ 及 $\{\boldsymbol{F}'(\boldsymbol{x}^{(k)})\}$，从而避免了每步都要计算 $\boldsymbol{F}(\boldsymbol{x})$ 的 Jacobi 阵，减少了计算量. 根据 $\Delta\boldsymbol{F}'(\boldsymbol{x}^{(k)})$ 的不同取法，可得不同的拟 Newton 法.

在拟 Newton 法中，若矩阵 $\boldsymbol{F}'(\boldsymbol{x}^{(k)})$ 非奇异，可令 $\boldsymbol{H}^{(k)}=\left(\boldsymbol{F}'(\boldsymbol{x}^{(k)})\right)^{-1}$，于是得到与(4.9.8)式、(4.9.10)式、(4.9.11)式互逆的迭代格式

$$\begin{cases}\boldsymbol{x}^{(k+1)}=\boldsymbol{x}^{(k)}-\boldsymbol{H}^{(k)}\boldsymbol{F}(\boldsymbol{x}^{(k)}),\\ \boldsymbol{H}^{(k+1)}\left(\boldsymbol{F}(\boldsymbol{x}^{(k+1)})-\boldsymbol{F}(\boldsymbol{x}^{(k)})\right)=\boldsymbol{x}^{(k+1)}-\boldsymbol{x}^{(k)},\\ \boldsymbol{H}^{(k+1)}=\boldsymbol{H}^{(k)}+\Delta\boldsymbol{H}^{(k)},\quad k=0,1,2\cdots.\end{cases} \tag{4.9.12}$$

式中，修正矩阵 $\Delta\boldsymbol{H}^{(k)}$ 的秩 $\operatorname{rank}\left(\Delta\boldsymbol{H}^{(k)}\right)=m\geqslant 1$.

可以看出，迭代格式(4.9.12)不用求逆就能逐次递推算出 $\left\{\boldsymbol{H}^{(k)}\right\}$，而迭代格式(4.9.8)需要求逆才能算出 $\boldsymbol{x}^{(k+1)}$，因此在实际计算中可根据具体情况选用其中一种迭代格式.

使用拟 Newton 法时，需要确定修正矩阵 $\Delta\boldsymbol{F}'(\boldsymbol{x}^{(k)})$ 和 $\Delta\boldsymbol{H}^{(k)}$，在此只介绍取 $\operatorname{rank}\left(\Delta\boldsymbol{F}'(\boldsymbol{x}^{(k)})\right)=\operatorname{rank}\left(\Delta\boldsymbol{H}^{(k)}\right)=1$ 的方法，称为**秩 1 拟 Newton 法**.

设

$$\Delta\boldsymbol{F}'(\boldsymbol{x}^{(k)})=\boldsymbol{u}^{(k)}\left(\boldsymbol{v}^{(k)}\right)^{\mathrm{T}},\quad \boldsymbol{u}^{(k)},\boldsymbol{v}^{(k)}\in\mathbf{R}^n, \tag{4.9.13}$$

$\boldsymbol{u}^{(k)},\boldsymbol{v}^{(k)}$ 待定. 记 $\boldsymbol{r}^{(k)}=\boldsymbol{x}^{(k+1)}-\boldsymbol{x}^{(k)}$，$\boldsymbol{y}^{(k)}=\boldsymbol{F}(\boldsymbol{x}^{(k+1)})-\boldsymbol{F}(\boldsymbol{x}^{(k)})$，则(4.9.10)式变为

$$\boldsymbol{F}'(\boldsymbol{x}^{(k+1)})\boldsymbol{r}^{(k)}=\boldsymbol{y}^{(k)}. \tag{4.9.14}$$

将(4.9.13)式代入(4.9.11)式，得

$$\boldsymbol{F}'(\boldsymbol{x}^{(k+1)})=\boldsymbol{F}'(\boldsymbol{x}^{(k)})+\boldsymbol{u}^{(k)}\left(\boldsymbol{v}^{(k)}\right)^{\mathrm{T}}.$$

代入(4.9.14)式，得

$$\left(\boldsymbol{F}'(\boldsymbol{x}^{(k)})+\boldsymbol{u}^{(k)}\left(\boldsymbol{v}^{(k)}\right)^{\mathrm{T}}\right)\boldsymbol{r}^{(k)}=\boldsymbol{y}^{(k)}$$

或

$$\boldsymbol{u}^{(k)}\left(\boldsymbol{v}^{(k)}\right)^{\mathrm{T}}\boldsymbol{r}^{(k)}=\boldsymbol{y}^{(k)}-\boldsymbol{F}'(\boldsymbol{x}^{(k)})\boldsymbol{r}^{(k)}.$$

若 $\left(\boldsymbol{v}^{(k)}\right)^{\mathrm{T}}\boldsymbol{r}^{(k)}\neq 0$，则有

$$\boldsymbol{u}^{(k)}=\frac{\boldsymbol{y}^{(k)}-\boldsymbol{F}'(\boldsymbol{x}^{(k)})\boldsymbol{r}^{(k)}}{\left(\boldsymbol{v}^{(k)}\right)^{\mathrm{T}}\boldsymbol{r}^{(k)}},$$

将它代入(4.9.13)式，得

$$\Delta\boldsymbol{F}'(\boldsymbol{x}^{(k)})=\frac{\left[\boldsymbol{y}^{(k)}-\boldsymbol{F}'(\boldsymbol{x}^{(k)})\boldsymbol{r}^{(k)}\right]\left(\boldsymbol{v}^{(k)}\right)^{\mathrm{T}}}{\left(\boldsymbol{v}^{(k)}\right)^{\mathrm{T}}\boldsymbol{r}^{(k)}}. \tag{4.9.15}$$

若取 $\boldsymbol{v}^{(k)}=\boldsymbol{r}^{(k)}\neq\boldsymbol{0}$，即 $\left(\boldsymbol{r}^{(k)}\right)^{-1}\boldsymbol{r}^{(k)}\neq\boldsymbol{0}$，由(4.9.15)式得

$$\Delta\boldsymbol{F}'(\boldsymbol{x}^{(k)})=\frac{\left[\boldsymbol{y}^{(k)}-\boldsymbol{F}'(\boldsymbol{x}^{(k)})\boldsymbol{r}^{(k)}\right]\left(\boldsymbol{r}^{(k)}\right)^{\mathrm{T}}}{\left(\boldsymbol{r}^{(k)}\right)^{\mathrm{T}}\boldsymbol{r}^{(k)}}. \tag{4.9.16}$$

于是得到一个**秩 1 拟 Newton 法**

$$\begin{cases}\boldsymbol{x}^{(k+1)}=\boldsymbol{x}^{(k)}-\left(\boldsymbol{F}'(\boldsymbol{x}^{(k)})\right)^{-1}\boldsymbol{F}(\boldsymbol{x}^{(k)}),\\ \boldsymbol{F}'(\boldsymbol{x}^{(k+1)})=\boldsymbol{F}'(\boldsymbol{x}^{(k)})+\dfrac{\left[\boldsymbol{y}^{(k)}-\boldsymbol{F}'(\boldsymbol{x}^{(k)})\boldsymbol{r}^{(k)}\right]\left(\boldsymbol{r}^{(k)}\right)^{\mathrm{T}}}{\left(\boldsymbol{r}^{(k)}\right)^{\mathrm{T}}\boldsymbol{r}^{(k)}},\\ k=0,1,2,\cdots.\end{cases} \tag{4.9.17}$$

(4.9.17)式称为 **Broyden 秩 1 方法**.

与(4.9.17)式互逆的 Broyden 秩 1 方法为

$$\begin{cases}\boldsymbol{x}^{(k+1)}=\boldsymbol{x}^{(k)}-\boldsymbol{H}^{(k)}\boldsymbol{F}(\boldsymbol{x}^{(k)}),\\ \boldsymbol{H}(\boldsymbol{x}^{(k+1)})=\boldsymbol{H}(\boldsymbol{x}^{(k)})+\dfrac{\left[\boldsymbol{r}^{(k)}-\boldsymbol{H}(\boldsymbol{x}^{(k)})\boldsymbol{y}^{(k)}\right]\left(\boldsymbol{r}^{(k)}\right)^{\mathrm{T}}\boldsymbol{H}(\boldsymbol{x}^{(k)})}{\left(\boldsymbol{r}^{(k)}\right)^{\mathrm{T}}\boldsymbol{H}(\boldsymbol{x}^{(k)})\boldsymbol{y}^{(k)}},\\ k=0,1,2,\cdots.\end{cases} \tag{4.9.18}$$

其中，$\left(\boldsymbol{r}^{(k)}\right)^{\mathrm{T}}\boldsymbol{H}(\boldsymbol{x}^{(k)})\boldsymbol{y}^{(k)}\neq 0$.

实际应用(4.9.17)式或(4.9.18)式求方程组(4.9.1)的近似解时，只要选择较好的初始近似 $\boldsymbol{x}^{(0)}$ 和初始矩阵 $\boldsymbol{A}^{(0)}$ 和 $\boldsymbol{H}^{(0)}$，一般可得到较好的近似解，迭代序列 $\left\{\boldsymbol{x}^{(k)}\right\}$ 具有超线性收敛速度.

例 4.16　用(4.9.18)式求下列方程组

$$\boldsymbol{F}(x)=\begin{bmatrix}x_1^{\ 2}-x_2-1\\ x_1^{\ 2}-4x_1+x_2^{\ 2}-x_2+3.25\end{bmatrix}=0$$

的解，取 $\boldsymbol{x}^{(0)}=(0,0)^{\mathrm{T}}$.

解　$\boldsymbol{F}(\boldsymbol{x}^{(0)})=(-1,\ 3.25)^{\mathrm{T}}$，因为

$$\boldsymbol{F}'(\boldsymbol{x})=\begin{bmatrix}2x_1 & -1\\ 2x_1-4 & 2x_2-1\end{bmatrix},$$

所以

$$\boldsymbol{F}'(\boldsymbol{x}^{(0)})=\begin{bmatrix}0 & -1\\ -4 & -1\end{bmatrix}.$$

取

$$H(x^{(0)}) = \left(F'(x^{(0)})\right)^{-1} = \begin{bmatrix} 0.25 & -0.25 \\ -1 & 0 \end{bmatrix},$$

用(4.9.18)式迭代，可求得

$$x^{(1)} = (1.0625,\ -1)^{\mathrm{T}}, \qquad r^{(0)} = (1.0625,\ -1)^{\mathrm{T}},$$

$$F(x^{(1)}) = (1.12890625,\ 2.12890625)^{\mathrm{T}}, \quad y^{(0)} = (2.12890625,\ -1.121937)^{\mathrm{T}},$$

$$H^{(1)} = \begin{bmatrix} 0.3557441 & -0.2721932 \\ -0.5224991 & -0.1002162 \end{bmatrix}.$$

重复以上步骤，共迭代 11 次，得解 $x^{11} = (1.54634088332,\ 1.39117631279)^{\mathrm{T}}$.若用 Newton 法(4.9.8)，取相同的初始近似 $x^{(0)}$，达到同一精度只需迭代 7 次，但它每步计算量比 Broyden 大得多.

4.10 习题 (Exercises)

1. Use the bisection method to find the solution accurate to within 0.5×10^{-2} for the equation $x^3+4x^2-10=0$ on the interval [1, 2].
2. Prove that the equation $x^3-x^2-1=0$ has one root $x^*\in(1,2)$, and use the bisection method to find an approximation with accuracy 10^{-2} to x^*. Determine the numbers of iterations necessary for the bisection method to solve the equation with accuracy 10^{-3} and 10^{-4}.
3. The following three methods are proposed to compute the root of $x^3-x^2-1=0$ near 1.5. Try to test the local convergence of them in the near neighborhood of 1.5. Use the method with the highest order of convergence to solve the equation with accuracy 10^{-2}, assuming $x_0=1.5$.

   a. $x_k = 1+1/x_{k-1}^2$; b. $x_k=\sqrt[3]{1+x_{k-1}^2}$; c. $x_k=\sqrt{1/(x_{k-1}-1)}$.
4. Use a fixed-point iteration method to determine a root of the equation $x^3-x^2-1=0$ on the interval [1, 2] accurate to within 0.5×10^{-4}.
5. Prove Theorem 4.5.
6. Prove Theorem 4.6.
7. Use Newton's method and the secant method respectively to find solutions accurate to within 10^{-5} for the following problems:

   a. $x^3-x^2-1=0$, [1, 2]; b. $x^3-3x^2-x+9=0$, [–2, –1];

   c. $x(x+1)^2-1=0$, in the near neighborhood of 0.4.
8. Use Newton's method to compute $\sqrt{115}$ with accuracy 10^{-6}.
9. Construct Newton's iteration algorithm for computing $\sqrt[3]{a}$, and discuss the convergence of the algorithm.
10. Let $x^4-4x^2+4=0$ which has a root of multiplicity two at $x^*=\sqrt{2}$. Use Newton's

method and the modified Newton's method respectively to find x_3 using $x_0 = 1.5$.

11. Prove Theorem 4.12.
12. Suppose the function f has the property: there exist two positive numbers m, M with $m \leqslant f'(x) \leqslant M$, for any x. Try to prove that for any fixed number λ with $0 < \lambda < 2/M$, the iteration scheme, $x_k = x_{k-1} - \lambda f(x_{k-1})$, converges to the root x^* of $f(x) = 0$.

Chapter 5 Interpolation by Polynomials

(多项式插值)

本章内容 (Contents)

学习目标 (Learning Objectives)

- 掌握插值的基本思想
- 掌握 Lagrange 基函数的基本性质
- 熟练应用 Lagrange 基函数构造插值多项式
- 能够应用 Lagrange 误差公式对插值余项进行估计
- 了解 Neville 插值的思想和方法
- 熟悉差商的基本性质并能够熟练计算差商
- 熟练应用差商表构造 Newton 插值多项式
- 了解 Hermite 插值概念
- 了解三次样条的基本概念以及构造三次样条的 m-方法和 M-方法

The problem of polynomial interpolation is to find a polynomial that agrees exactly with some given information of a real-valued function $f(x)$ of a single real variable x. The information may be a finite set of values, say, $\{f(x_1), f(x_2), \cdots, f(x_n)\}$ of the function $f(x)$ at $n+1$ numbers $\{x_1, x_2, \cdots, x_n\}$, or it may involve the derivatives of the function $f(x)$ at some of these numbers. In this Chapter we will introduce different ways, including the Lagrange interpolation formula, the Neville's interpolation method, the Newton's interpolatory divided difference formula and the Hermite interpolation method, to get the interpolating polynomials. The polynomial interpolation at least serves two purposes. One purpose is to predict or to estimate or even to reconstruct the unknown (or lost) information by the polynomial interpolating the information known in advance, and another purpose is to approximate the function $f(x)$ by its interpolation polynomial when $f(x)$ is so complex that it cannot be directly used to work out the result as will be encountered in the numerical quadrature.

5.1 Lagrange interpolation (Lagrange 插值)

- Linear interpolation (线性插值)
- Quadratic interpolation (二次插值)
- *n*th-order polynomial interpolation (*n* 次多项式插值)
- Uniqueness of interpolation (插值的唯一性)
- Lagrange error formula (Lagrange 误差公式)
- Examples (例题)

5.1.1 Linear interpolation (线性插值)

Let us begin with two simple interpolation problems: linear interpolation and quadratic interpolation before we describe the problem of general polynomial interpolation.

Problem Determine a polynomial $P_1(x)$ of degree one that passes through two distinct points (x_0, y_0) and (x_1, y_1), where

$$y_i = f(x_i), i = 0,1.$$

Solution. Define

$$L_0(x) = \frac{x - x_1}{x_0 - x_1}, L_1(x) = \frac{x - x_0}{x_1 - x_0}.$$

Then

$$L_0(x_0) = 1, L_0(x_1) = 0; L_1(x_0) = 0, L_1(x_1) = 1.$$

Let

$$P_1(x) = a_0 L_0(x) + a_1 L_1(x),$$

where a_0, a_1 are two unknowns to be determined. Then $P_1(x)$ is a polynomial of degree one. Since $y = P_1(x)$ passes through two distinct points (x_0, y_0) and (x_1, y_1), we have

$$P_1(x_0) = y_0 = f(x_0), P_1(x_1) = y_1 = f(x_1),$$

which implies

$$a_0 = f(x_0), a_1 = f(x_1).$$

As a result, we have

$$P_1(x) = f(x_0)L_0(x) + f(x_1)L_1(x),$$

and this linear polynomial is clearly unique.

5.1.2 Quadratic interpolation (二次插值)

Problem Determine a polynomial $P_2(x)$ of degree two such that

$$P_2(x_i)=f(x_i), i=0,1,2,$$

where x_i, $i=0,1,2$, are distinct from one another.

Solution. Define

$$L_0(x)=\frac{(x-x_1)(x-x_2)}{(x_0-x_1)(x_0-x_2)},$$

$$L_1(x)=\frac{(x-x_0)(x-x_2)}{(x_1-x_0)(x_1-x_2)},$$

$$L_2(x)=\frac{(x-x_0)(x-x_1)}{(x_2-x_0)(x_2-x_1)}.$$

Then $L_0(x), L_1(x)$ and $L_2(x)$ are all quadratic polynomials which have the following properties

$$L_0(x_0)=1, L_0(x_1)=0, L_0(x_2)=0;$$

$$L_1(x_0)=0, L_1(x_1)=1, L_1(x_2)=0;$$

$$L_2(x_0)=0, L_2(x_1)=0, L_2(x_2)=1.$$

Let

$$P_2(x)=a_0L_0(x)+a_1L_1(x)+a_2L_2(x)$$

where a_0, a_1, a_2 are the unknowns to be determined. Then $P_2(x)$ is a polynomial of degree two. From

$$P_2(x_i)=f(x_i), i=0,1,2,$$

it follows

$$a_0=P_2(x_0)=f(x_0),\ a_1=P_2(x_1)=f(x_1), a_2=P_2(x_2)=f(x_2).$$

Therefore

$$P_2(x)=f(x_0)L_0(x)+f(x_1)L_1(x)+f(x_2)L_2(x).$$

Remark One may also solve the above quadratic interpolation problem in a way as follows. Suppose

$$P_2(x)=b_0+b_1x+b_2x^2.$$

Then using the interpolation conditions

$$P_2(x_i)=f(x_i), i=0,1,2$$

gives the following system of linear equations:

$$b_0+b_1x_0+b_2x_0^2=f(x_0),$$

$$b_0+b_1x_1+b_2x_1^2=f(x_1),$$

$$b_0+b_1x_2+b_2x_2^2=f(x_2).$$

By Gramer's rule, we get

$$b_0 = \frac{\begin{vmatrix} f(x_0) & x_0 & x_0^2 \\ f(x_1) & x_1 & x_1^2 \\ f(x_2) & x_2 & x_2^2 \end{vmatrix}}{\begin{vmatrix} 1 & x_0 & x_0^2 \\ 1 & x_1 & x_1^2 \\ 1 & x_2 & x_2^2 \end{vmatrix}}, \quad b_1 = \frac{\begin{vmatrix} 1 & f(x_0) & x_0^2 \\ 1 & f(x_1) & x_1^2 \\ 1 & f(x_2) & x_2^2 \end{vmatrix}}{\begin{vmatrix} 1 & x_0 & x_0^2 \\ 1 & x_1 & x_1^2 \\ 1 & x_2 & x_2^2 \end{vmatrix}}, \quad b_2 = \frac{\begin{vmatrix} 1 & x_0 & f(x_0) \\ 1 & x_1 & f(x_1) \\ 1 & x_2 & f(x_2) \end{vmatrix}}{\begin{vmatrix} 1 & x_0 & x_0^2 \\ 1 & x_1 & x_1^2 \\ 1 & x_2 & x_2^2 \end{vmatrix}}.$$

Consequently

$$P_2(x) = \frac{\begin{vmatrix} f(x_0) & x_0 & x_0^2 \\ f(x_1) & x_1 & x_1^2 \\ f(x_2) & x_2 & x_2^2 \end{vmatrix} + x\begin{vmatrix} 1 & f(x_0) & x_0^2 \\ 1 & f(x_1) & x_1^2 \\ 1 & f(x_2) & x_2^2 \end{vmatrix} + x^2\begin{vmatrix} 1 & x_0 & f(x_0) \\ 1 & x_1 & f(x_1) \\ 1 & x_2 & f(x_2) \end{vmatrix}}{\begin{vmatrix} 1 & x_0 & x_0^2 \\ 1 & x_1 & x_1^2 \\ 1 & x_2 & x_2^2 \end{vmatrix}}.$$

Apparently it is not a good practice to find out the quadratic interpolating polynomial by solving the system of linear equations via Gramer's rule, compared with the expression by means of $L_0(x), L_1(x), L_2(x)$. And this shows how important to choose a good basis to express the polynomial.

5.1.3 *n*th-order polynomial interpolation (*n* 次多项式插值)

Problem Determine a polynomial of degree n that passes through $n+1$ distinct points

$$(x_i, y_i) = (x_i, f(x_i)), i = 0, 1, \cdots, n.$$

Solution. The skills used for solving the linear and quadratic interpolation problems inspire us to define the following $n+1$ polynomials:

$$L_k(x) = \frac{\prod_{j=0, j\neq k}^{n} (x - x_j)}{\prod_{j=0, j\neq k}^{n} (x_k - x_j)}, k = 0, 1, \cdots, n. \tag{5.1.1}$$

Then all $L_k(x)$ are polynomials of degree n which satisfy

$$L_k(x_i) = \delta_{k,i} = \begin{cases} 1, & i = k; \\ 0, & i \neq k, \end{cases} \quad k, i \in \{0, 1, \cdots, n\}$$

where $\delta_{k,i}$ is the Kronecker's symbol (Kronecker 记号).

And we call $\{L_k(x)\}_{k=0}^{n}$ the Lagrange basis functions (拉格朗日基函数).

Let

$$P_n(x) = \sum_{k=0}^{n} c_k L_k(x)$$

be the polynomial of degree n that passes through $n+1$ distinct points

$$(x_i, y_i) = (x_i, f(x_i)), i = 0,1,\cdots,n.$$

Then from

$$P_n(x_i) = f(x_i), i = 0,1,\cdots,n$$

it follows

$$\sum_{k=0}^{n} c_k L_k(x_i) = f(x_i), i = 0,1,\cdots,n,$$

namely,

$$c_i = f(x_i), i = 0,1,\cdots,n.$$

Therefore the required polynomial is given by

$$P_n(x) = \sum_{k=0}^{n} f(x_k) L_k(x). \tag{5.1.2}$$

The polynomial $P_n(x)$ given by (5.1.2) is called the nth Lagrange interpolating polynomial (n 次 Lagrange 插值多项式) and $x_i, i = 0,1,\cdots,n,$ are called the interpolation nodes or interpolation knots.

5.1.4 Uniqueness of interpolation (插值的唯一性)

Theorem 5.1 (Uniqueness Theorem) Given $n+1$ support points (支撑点) $\{(x_i, y_i)\}_{i=0}^{n}$, where $x_i \neq x_j$ if $i \neq j$. If $P_n(x)$ and $Q_n(x)$ are both polynomials of degree n interpolating the given data, i.e.,

$$P_n(x_i) = Q_n(x_i) = y_i, \quad i = 0,1,\cdots,n,$$

then

$$P_n(x) \equiv Q_n(x).$$

Proof. Let

$$R(x) = P_n(x) - Q_n(x).$$

Then $R(x)$ is a polynomial of degree at most n which satisfies

$$R(x_i) = 0, \quad i = 0,1,\cdots,n.$$

Suppose

$$R(x) = a_0 + a_1 x + a_2 x^2 + \cdots + a_n x^n.$$

Then we get the following system of linear equations

$$\begin{pmatrix} 1 & x_0 & \cdots & x_0^n \\ 1 & x_1 & \cdots & x_1^n \\ \vdots & \vdots & & \vdots \\ 1 & x_n & \cdots & x_n^n \end{pmatrix}\begin{pmatrix} a_0 \\ a_1 \\ \vdots \\ a_n \end{pmatrix} = \begin{pmatrix} 0 \\ 0 \\ \vdots \\ 0 \end{pmatrix}.$$

Since

$$\begin{vmatrix} 1 & x_0 & \cdots & x_0^n \\ 1 & x_1 & \cdots & x_1^n \\ \vdots & \vdots & & \vdots \\ 1 & x_n & \cdots & x_n^n \end{vmatrix} = \prod_{0 \leqslant j < i \leqslant n} (x_i - x_j) \neq 0,$$

we have

$$a_0 = a_1 = \cdots = a_n = 0.$$

Therefore $R(x) \equiv 0$, which implies

$$P_n(x) \equiv Q_n(x).$$

The proof of Theorem 5.1 is completed.

Remark Theorem 5.1 can also be proved as follows. Let

$$R(x) = P_n(x) - Q_n(x).$$

Then $R(x)$ is a polynomial of degree at most n and it satisfies

$$R(x_i) = 0, i = 0, 1, \cdots, n$$

which implies that $R(x)$ can be divided by $x-x_i$, $i=0,1,\cdots,n$, and hence by $(x-x_0)(x-x_1)\cdots(x-x_n)$ since $x_0, x_1, \cdots, x_n$ are distinct from one another. This is possible only when $R(x) \equiv 0$, and thus $P_n(x) \equiv Q_n(x)$.

5.1.5 Lagrange error formula (Lagrange 误差公式)

Theorem 5.2 (Error Theorem) Suppose $x_0, x_1, \cdots, x_n$ are distinct numbers in the interval $[a,b]$ and $f \in C^{n+1}[a,b]$. Then for each x in $[a,b]$, there exists $\xi(x)$ in (a,b) such that

$$\begin{aligned} & f(x) - P_n(x) \\ = & \frac{f^{(n+1)}(\xi(x))}{(n+1)!}(x-x_0)(x-x_1)\cdots(x-x_n) \end{aligned} \tag{5.1.3}$$

where $P_n(x)$ is the nth Lagrange interpolating polynomial.

Proof. In the case when $x = x_k, k \in \{0,1,\cdots,n\}$, Theorem 5.2 is clearly true since $P_n(x_k) = f(x_k)$. If $x \neq x_k, \forall k \in \{0,1,\cdots,n\}$, let

$$g(t) = f(t) - P_n(t) - [f(x) - P_n(x)]\prod_{i=0}^{n}\frac{t-x_i}{x-x_i}.$$

Then for $k \in \{0,1,\cdots,n\}$,

$$g(x_k) = f(x_k) - P_n(x_k) - [f(x) - P_n(x)]\prod_{i=0}^{n}\frac{x_k - x_i}{x - x_i}$$
$$= 0$$

and

$$g(x) = f(x) - P_n(x) - [f(x) - P_n(x)]\prod_{i=0}^{n}\frac{x - x_i}{x - x_i}$$
$$= 0.$$

Since $f \in C^{n+1}[a,b]$, clearly $g \in C^{n+1}[a,b]$. By the Generalized Rolle's Theorem, there exists a number $\xi(x)$ in (a,b) such that $g^{(n+1)}(\xi(x)) = 0$, namely

$$0 = f^{(n+1)}(\xi(x)) - P_n^{(n+)}(\xi(x)) - [f(x) - P_n(x)]\frac{d^{n+1}}{dt^{n+1}}\left(\prod_{i=0}^{n}\frac{t - x_i}{x - x_i}\right)\Bigg|_{t=\xi(x)}$$
$$= f^{(n+1)}(\xi(x)) - [f(x) - P_n(x)]\frac{(n+1)!}{\prod_{i=0}^{n}(x - x_i)}.$$

Therefore

$$f(x) - P_n(x)$$
$$= \frac{f^{(n+1)}(\xi(x))}{(n+1)!}(x - x_0)(x - x_1)\cdots(x - x_n).$$

The proof of Theorem 5.2 is completed.

5.1.6 Examples (例题)

Example 5.1 Suppose $x_0, x_1, \cdots, x_n$ are $n+1$ distinct numbers and $L_i(x), i = 0,1,\cdots,n$ are the Lagrange basis functions. Show

(1) $\sum_{i=0}^{n} L_i(x) \equiv 1$;

(2) $\sum_{i=0}^{n} x_i^k L_i(x) \equiv x^k,\ k = 0,1,\cdots,n$;

(3) $\sum_{i=0}^{n} (x - x_i)^k L_i(x) \equiv 0,\ k = 0, 1, \cdots, n$;

(4) $\sum_{i=0}^{n} x_i^k L_i(0) \equiv \begin{cases} 1, & k = 0; \\ 0, & k = 1,2,\cdots,n; \\ a_n, & k = n+1 \end{cases}$

where $a_n = (-1)^n x_0 x_1 \cdots x_n$.

Proof. For $f \in C^{n+1}[a,b]$, let

$$P_n(f;x) = \sum_{k=0}^{n} f(x_k) L_k(x)$$

and

$$R_n(f;x)=\frac{f^{(n+1)}(\xi(x))}{(n+1)!}(x-x_0)(x-x_1)\cdots(x-x_n).$$

Then by Theorem 5.2, we can write

$$f(x)=P_n(f;x)+R_n(f;x). \tag{5.1.4}$$

(1) Let $f(x)\equiv 1$. Then

$$P_n(1;x)=\sum_{k=0}^{n}L_k(x),\ R_n(1;x)=0,$$

which implies $1\equiv P_n(1;x)+R_n(1;x)=\sum_{k=0}^{n}L_k(x)$.

(2) Let $f(x)\equiv x^k,\ k=0,1,\cdots,n$. Then

$$P_n(x^k;x)=\sum_{i=0}^{n}x_i^k L_i(x),\ R_n(x^k;x)=0$$

and using (5.1.4) gives

$$\sum_{i=0}^{n}x_i^k L_i(x)\equiv x^k,\ k=0,1,\cdots,n.$$

(3) Since

$$(x-x_i)^k=(-1)^k(x_i-x)^k=(-1)^k\sum_{j=0}^{k}\binom{k}{j}x_i^j(-x)^{k-j},$$

by (2) we have

$$\begin{aligned}
&\sum_{i=0}^{n}(x-x_i)^k L_i(x)\\
&=\sum_{i=0}^{n}(-1)^k\sum_{j=0}^{k}\binom{k}{j}x_i^j(-x)^{k-j}L_i(x)\\
&=(-1)^k\sum_{j=0}^{k}\binom{k}{j}(-x)^{k-j}\sum_{i=0}^{n}x_i^j L_i(x)\\
&=(-1)^k\sum_{j=0}^{k}\binom{k}{j}(-x)^{k-j}x^j\\
&=(-1)^k(x+(-x))^k\\
&=0.
\end{aligned}$$

(4) By (2) we have

$$\sum_{i=0}^{n}x_i^k L_i(x)\equiv x^k,\ k=0,1,\cdots,n.$$

Therefore

$$\sum_{i=0}^{n} L_i(0)x_i^k \equiv \begin{cases} 1, k=0; \\ 0, k=1,2,\cdots,n. \end{cases}$$

Let $f(x) \equiv x^{n+1}$. Then using Eq. (5.1.4) gives

$$x^{n+1} = \sum_{i=0}^{n} x_i^{n+1} L_i(x) + (x-x_0)(x-x_1)\cdots(x-x_n).$$

Setting $x=0$ in the above expression results in

$$0 = \sum_{i=0}^{n} x_i^{n+1} L_i(0) + (-x_0)(-x_1)\cdots(-x_n).$$

Hence

$$\sum_{i=0}^{n} x_i^{n+1} L_i(0) = (-1)^n x_0 x_1 \cdots x_n.$$

Example 5.2 Suppose $f(x) \in C^2[a,b]$ and $f(a)=f(b)=0$. Show

$$\max_{a\leqslant x\leqslant b} |f(x)| \leqslant \frac{1}{8}(b-a)^2 \max_{a\leqslant x\leqslant b} |f''(x)|.$$

Proof. Constructing a linear interpolating polynomial $P_1(x)$ passing through the two points $(a, f(a))$ and $(a, f(b))$ yields

$$\begin{aligned} f(x) &= P_1(f;x) + R_1(f;x) \\ &= f(a)L_0(x) + f(b)L_1(x) + \frac{f''(\xi)}{2}(x-a)(x-b) \\ &= \frac{f''(\xi)}{2}(x-a)(x-b). \end{aligned}$$

So for $x \in [a,b]$

$$\begin{aligned} |f(x)| &= \frac{|f''(\xi)|}{2} \left|\frac{x-a}{b-a}\right| \cdot \left|\frac{b-x}{b-a}\right| (b-a)^2 \\ &\leqslant \max_{a\leqslant x\leqslant b} \frac{|f''(x)|}{2} (b-a)^2 \cdot \max_{0\leqslant t\leqslant 1} t(1-t) \\ &= \frac{(b-a)^2}{8} \max_{a\leqslant x\leqslant b} |f''(x)|. \end{aligned}$$

Example 5.3 Given the data in the table below

i	0	1	2
x_i	0	1	3
y_i	1	3	2

Find the quadratic polynomial $P_2(x)$ interpolating the prescribed values (预先给定的值).

Solution. Using $x_0=0, x_1=1, x_2=3$, one can work out the corresponding Lagrange basis functions as follows

$$L_0(x)=\frac{(x-x_1)(x-x_2)}{(x_0-x_1)(x_0-x_2)}=\frac{1}{3}(x-1)(x-3),$$

$$L_1(x)=\frac{(x-x_0)(x-x_2)}{(x_1-x_0)(x_1-x_2)}=-\frac{1}{2}x(x-3),$$

$$L_2(x)=\frac{(x-x_0)(x-x_1)}{(x_2-x_0)(x_2-x_1)}=\frac{1}{6}x(x-1).$$

As a result,

$$\begin{aligned}P_2(x)&=y_0L_0(x)+y_1L_1(x)+y_2L_2(x)\\&=\frac{1}{3}(x-1)(x-3)-\frac{3}{2}x(x-3)+\frac{1}{3}x(x-1)\\&=-\frac{5}{6}x^2+\frac{17}{6}x+1.\end{aligned}$$

Example 5.4 Suppose a table is to be prepared for the function

$$f(x)=\sin x$$

for $x\in[0,\pi/2]$. Assume that the step size (步长) , i.e., the difference between adjacent x -values, is h. What should h be for quadratic interpolation (二次插值) to give an absolute error of at most 10^{-6}?

Solution. Suppose x_{k-1},x_k,x_{k+1} are three distinct numbers in $[0,\pi/2]$ used for the quadratic interpolation. For

$$x\in[x_{k-1},x_{k+1}],$$

let

$$\frac{x-x_k}{h}=t,\ t\in[-1,1].$$

Then from $x_{k-1}=x_k-h$ and $x_{k+1}=x_k+h$ we have

$$\frac{x-x_{k-1}}{h}=\frac{x-x_k}{h}+1=t+1,$$

$$\frac{x-x_{k+1}}{h}=\frac{x-x_k}{h}-1=t-1.$$

Using the Lagrange error formula gives

$$|f(x)-P_2(x)|$$

$$
\begin{aligned}
&= \frac{1}{3!}\left|f^{(3)}(\xi)\right| \cdot \left|(x-x_{k-1})(x-x_k)(x-x_{k+1})\right| \\
&= \frac{|\cos\xi|}{6} \cdot \left|(x-x_{k-1})(x-x_k)(x-x_{k+1})\right| \\
&\leqslant \frac{1}{6} \max_{x_{k-1}\leqslant x\leqslant x_{k+1}} \cdot \left|(x-x_{k-1})(x-x_k)(x-x_{k+1})\right| \\
&\leqslant \frac{1}{6} h^3 \max_{-1\leqslant t\leqslant 1} \left|t(t^2-1)\right|.
\end{aligned}
$$

Let

$$g(t) = t - t^3.$$

Solving

$$g'(t) = 1 - 3t^2 = 0$$

gives the stationary points (驻点) $t = \pm\sqrt{3}/3$. Therefore

$$
\begin{aligned}
&\max_{-1\leqslant t\leqslant 1} |g(t)| \\
&= \max_{-1\leqslant t\leqslant 1} \left\{|g(-1)|, |g(1)|, |g(\sqrt{3}/3)|, |g(-\sqrt{3}/3)|\right\} \\
&= \frac{2\sqrt{3}}{9}.
\end{aligned}
$$

Consequently

$$|f(x) - P_2(x)| \leqslant \frac{1}{6} h^3 \cdot \frac{2\sqrt{3}}{9} = \frac{\sqrt{3}}{27} h^3.$$

To make

$$|f(x) - P_2(x)| \leqslant 10^{-6},$$

it suffices to make

$$\frac{\sqrt{3}}{27} h^3 \leqslant 10^{-6},$$

and thus one should choose the step size (步长) h such that

$$h \leqslant \left(\frac{27}{\sqrt{3}} \times 10^{-6}\right)^{1/3} \approx 2.498 \times 10^{-2}.$$

5.2 Neville interpolation (Neville 插值)

As we know, the nth Lagrange interpolating polynomial is closely related to the $n+1$ Lagrange basis functions. Once the support data are given, the corresponding Lagrange basis functions can be determined and the Lagrange interpolating polynomial is done. In other words, the Lagrange method provides a one-step solution to the interpolation problem. However if one more support point is added, then all the Lagrange basis functions have to be reconstructed, which is one of the disadvantages of the Lagrange method. Instead of solving the interpolation problem all at once, one might consider first solving the problem for smaller sets of support points and then updating these solutions to the full interpolation problem.

Let $f(x)$ be a function defined at distinct numbers $x_0, x_1, \cdots, x_n$. Denote by $P_{m_1,m_2,\cdots,m_k}(x)$ the $(k-1)$th Lagrange polynomial interpolating $f(x)$ at the k points $x_{m_1}, x_{m_2}, \cdots, x_{m_k}$, where $m_1, m_2, \cdots, m_k$ are k distinct integers with $0 \leqslant m_i \leqslant n$ for each i.

Let

$$S_{i,j} = \{(x_k, f(x_k)) \mid k = i, i+1, \cdots, j\}.$$

Then the set of support points $S_{0,n}$ can be decomposed into various subsets. In particular, $S_{0,n}$ can be decomposed into two maximal subsets $S_{0,n-1}$ and $S_{1,n}$. Similarly $S_{0,n-1}$ can be decomposed into two maximal subsets $S_{0,n-2}$ and $S_{1,n-1}$, and $S_{1,n}$ can be decomposed into two maximal subsets $S_{1,n-1}$ and $S_{2,n}$. This process can be repeated until each subset only contains two support points, namely

$$\begin{aligned} S_{0,n} &= S_{0,n-1} \cup S_{1,n} \\ &= S_{0,n-2} \cup S_{1,n-1} \cup S_{2,n} \\ &= \cdots = S_{0,1} \cup S_{1,2} \cup S_{2,3} \cup \cdots \cup S_{n-1,n}. \end{aligned}$$

Now first we construct linear interpolation $P_{i,i+1}(x)$ on each subset $S_{i,i+1}$ for $i = 0,1,\cdots,n-1$. The second step is to obtain quadratic interpolation polynomials $P_{i,i+1,i+2}(x)$ based on $P_{i,i+1}(x)$ and $P_{i+1,i+2}(x)$ on each subset $S_{i,i+1,i+2}$ for $i = 0,1,\cdots,n-2$. The nth step (last step) is to obtain the interpolation polynomial $P_{0,1,\cdots,n}(x)$ of degree n based on $P_{0,1,\cdots,n-1}(x)$ and $P_{1,2,\cdots,n}(x)$ on the set of support points $S_{0,n}$.

For instance, to construct a cubic Lagrange polynomial $P_3(x)$ interpolating four given support points

$$(x_i, y_i) = (x_i, f(x_i)), i = 0,1,2,3$$

one can proceed as follows (refer to the tableau) .

	$k=0$	$k=1$	$k=2$	$k=3$	$k=4$
x_0	y_0				
x_1	y_1	$P_{01}(x)$			
x_2	y_2	$P_{12}(x)$	$P_{012}(x)$		
x_3	y_3	$P_{23}(x)$	$P_{123}(x)$	$P_{0123}(x)$	
x_4	y_4	$P_{34}(x)$	$P_{234}(x)$	$P_{1234}(x)$	$P_{01234}(x)$

Step 1. Based on the Lagrange interpolation formula, four linear interpolating polynomials $P_{01}(x), P_{12}(x)$, $P_{23}(x)$ and $P_{34}(x)$ are obtained below (for $i = 0,1,2,3$)

$$P_{i,i+1}(x)=y_i\frac{x-x_{i+1}}{x_i-x_{i+1}}+y_{i+1}\frac{x-x_i}{x_{i+1}-x_i}$$
$$=y_i\frac{x_{i+1}-x}{x_{i+1}-x_i}+y_{i+1}\frac{x-x_i}{x_{i+1}-x_i}.$$

Step 2. Use two neighboring linear interpolation $P_{i,i+1}, P_{i+1,i+2}$ to construct the quadratic interpolating polynomial $P_{i,i+1,i+2}, i=0,1,2$

$$P_{i,i+1,i+2}(x)$$
$$=\frac{x_{i+2}-x}{x_{i+2}-x_i}P_{i,i+1}(x)+\frac{x-x_i}{x_{i+2}-x_i}P_{i+1,i+2}(x).$$

Clearly, $P_{i,i+1,i+2}$ is a quadratic polynomial satisfying

$$P_{i,i+1,i+2}(x_i)=\frac{x_{i+2}-x_i}{x_{i+2}-x_i}P_{i,i+1}(x_i)$$
$$=P_{i,i+1}(x_i)=y_i,$$

$$P_{i,i+1,i+2}(x_{i+1})=\frac{x_{i+2}-x_{i+1}}{x_{i+2}-x_i}P_{i,i+1}(x_{i+1})+\frac{x_{i+1}-x_i}{x_{i+2}-x_i}P_{i+1,i+2}(x_{i+1})$$
$$=y_{i+1},$$

$$P_{i,i+1,i+2}(x_{i+2})=\frac{x_{i+2}-x_i}{x_{i+2}-x_i}P_{i+1,i+2}(x_{i+2})$$
$$=P_{i+1,i+2}(x_{i+2})=y_{i+2},$$

which means that $P_{i,i+1,i+2}(x)$ is the quadratic interpolating polynomial passing through the support points $(x_i, y_i),(x_{i+1}, y_{i+1})$ and (x_{i+2}, y_{i+2}).

Step 3. Use two neighboring quadratic interpolating polynomial $P_{i,i+1,i+2}(x)$ and $P_{i+1,i+2,i+3}(x)$ to construct the cubic interpolating polynomial $P_{i,i+1,i+2,i+3}(x), i=0,1$

$$P_{i,i+1,i+2,i+3}(x)=\frac{x_{i+3}-x}{x_{i+3}-x_i}P_{i,i+1,i+2}(x)+\frac{x-x_i}{x_{i+3}-x_i}P_{i+1,i+2,i+3}(x).$$

It is easy to verify that

$$P_{i,i+1,i+2,i+3}(x_j)=y_j, j=i,i+1,i+2,i+3$$

The above technique to construct interpolating polynomial is called Neville's method. As illustrated in the tableau, the first column of the tableau contains the prescribed support ordinates (纵坐标) y_i, subsequent columns are filled by calculating each entry recursively from its two "neighboring" in the previous column, according to the following recursion:

$$P_{i_0i_1\cdots i_k}(x)=\frac{x_{i_k}-x}{x_{i_k}-x_{i_0}}P_{i_0i_1\cdots i_{k-1}}(x)+\frac{x-x_{i_0}}{x_{i_k}-x_{i_0}}P_{i_1i_2\cdots i_k}(x).$$

Using the data given in Example 5.3, we can construct the quadratic interpolating polynomial by Neville's method as follows

	$k=0$	$k=1$	$k=2$
$x_0=0$	$y_0=1$		
$x_1=1$	$y_1=3$	$P_{01}(x)$	
$x_2=3$	$y_2=2$	$P_{12}(x)$	$P_{012}(x)$

$$\begin{aligned}P_{01}(x)&=y_0\frac{x-x_1}{x_0-x_1}+y_1\frac{x-x_0}{x_1-x_0}\\&=-(x-1)+3x=2x+1,\\P_{12}(x)&=y_1\frac{x-x_2}{x_1-x_2}+y_2\frac{x-x_1}{x_2-x_1}\\&=-\frac{3}{2}(x-3)+(x-1)=-\frac{x}{2}+\frac{7}{2},\\P_{012}(x)&=\frac{x_2-x}{x_2-x_0}P_{01}(x)+\frac{x-x_0}{x_2-x_0}P_{12}(x)\\&=\frac{3-x}{3}(2x+1)+\frac{x}{3}\left(-\frac{x}{2}+\frac{7}{2}\right)\\&=-\frac{5}{6}x^2+\frac{17}{6}x+1.\end{aligned}$$

Example 5.5　Neville's algorithm is used to approximate (逼近) $f(0)$ using $f(-2), f(-1), f(1)$ and $f(2)$. Suppose $f(-1)$ was overstated (高估) by 2 and $f(1)$ was understated (低估) by 3. Determine the error in the original calculation of the value of the interpolating polynomial to approximate $f(0)$.

Solution.　Let

$$\begin{aligned}&x_0=-2, x_1=-1, x_2=1, x_3=2,\\&y_0=f(x_0)=f(-2), y_1=f(-1)+2,\\&y_2=f(1)-3, y_3=f(2).\end{aligned}$$

Using $\{(x_i,y_i)\}_{i=0}^3$ as support points, we have by Neville's algorithm

$$\begin{aligned}P_{01}(0)&=y_0\frac{x_1-0}{x_1-x_0}+y_1\frac{0-x_0}{x_1-x_0}\\&=-y_0+2y_1=2y_1-y_0,\\P_{12}(0)&=y_1\frac{x_2-0}{x_2-x_1}+y_2\frac{0-x_1}{x_2-x_1}\\&=\frac{1}{2}(y_1+y_2),\\P_{23}(0)&=y_2\frac{x_3-0}{x_3-x_2}+y_3\frac{0-x_2}{x_3-x_2}\\&=2y_2-y_3,\end{aligned}$$

$$P_{012}(0)=\frac{x_2-0}{x_2-x_0}P_{01}(0)+\frac{0-x_0}{x_2-x_0}P_{12}(0)$$
$$=\frac{1}{3}(3y_1+y_2-y_0),$$
$$P_{123}(0)=\frac{x_3-0}{x_3-x_1}P_{12}(0)+\frac{0-x_1}{x_3-x_1}P_{23}(0)$$
$$=\frac{1}{3}(y_1+3y_2-y_3),$$
$$P_{0123}(0)=\frac{x_3-0}{x_3-x_0}P_{012}(0)+\frac{0-x_0}{x_3-x_0}P_{123}(0)$$
$$=\frac{2}{3}y_1+\frac{2}{3}y_2-\frac{1}{6}y_0-\frac{1}{6}y_3$$
$$=\frac{2}{3}f(-1)+\frac{2}{3}f(1)-\frac{1}{6}f(-2)-\frac{1}{6}f(2)-\frac{2}{3}$$
$$=\overline{P}_{0123}(0)-\frac{2}{3}$$

where $\overline{P}_{0123}(x)$ is the cubic interpolant obtained by using the support points $\{(x_i,f(x_i))\}_{i=0}^{3}$.

Therefore the error in the original calculation of the value of the interpolating polynomial to approximate $f(0)$ is given by

$$\left|P_{0123}(0)-\overline{P}_{0123}(0)\right|=\frac{2}{3}.$$

5.3 Newton interpolation (Newton 插值)

- Definition of divided differences (差商的定义)
- Newton's expansion of a function (函数的 Newton 展开)
- Properties of divided differences (差商的性质)
- Computation of Newton's interpolant (Newton 插值的计算)
- The relationship between divided differences and derivatives (差商与导数的关系)
- Relations between Newton's expansion and Taylor's expansion (Newton 展开与 Taylor 展开之间的关系)
- Comparisons among Lagrange, Neville and Newton interpolations (Lagrange 插值, Neville 插值与 Newton 插值之间的比较)
- Newton forward divided-difference formula (Newton 向前差商公式)
- Newton forward difference formula (Newton 向前差分公式)
- Newton backward divided-difference formula (Newton 向后差商公式)
- Newton backward-difference formula (Newton 向后差分公式)

Suppose that $P_n(x)$ is the interpolation polynomial of degree n that agrees with the function $f(x)$ at the distinct numbers $x_0, x_1, \cdots, x_n$. Let

$$N_0(x)=1;\ N_i(x)=(x-x_0)(x-x_1)\cdots(x-x_{i-1}),\ i=1,2,\cdots,n.$$

Then $N_i(x)$ is a polynomial of degree i which satisfies

$$N_i(x_j)=0,\ j=0,1,\cdots,i-1;\ i=1,2,\cdots,n.$$

Let $P_n(x)$ be expressed in the form

$$P_n(x)=a_0N_0(x)+a_1N_1(x)+\cdots+a_nN_n(x). \tag{5.3.1}$$

Evaluating $P_n(x)$ at x_0 leaves only the constant term a_0, that is,

$$a_0=P_n(x_0)=f(x_0).$$

When $P_n(x)$ is evaluated at x_1, the only nonzero terms in the evaluation of $P_n(x_1)$ are the constant and linear terms,

$$f(x_0)+a_1(x_1-x_0)=P_n(x_1)=f(x_1).$$

So

$$a_1=\frac{f(x_1)-f(x_0)}{x_1-x_0}.$$

From $P_n(x_2)=f(x_2)$ it follows

$$a_0+a_1N_1(x_2)+a_2N_2(x_2)=f(x_2),$$

namely,

$$f(x_0)+a_1(x_2-x_0)+a_2(x_2-x_0)(x_2-x_1)=f(x_2).$$

Solving for a_2 gives

$$a_2=\frac{f(x_2)-f(x_0)-a_1(x_2-x_0)}{(x_2-x_0)(x_2-x_1)}=\frac{\dfrac{f(x_2)-f(x_0)}{x_2-x_0}-\dfrac{f(x_1)-f(x_0)}{x_1-x_0}}{x_2-x_1}.$$

5.3.1 Definition of divided differences (差商的定义)

$f[x_i]=f(x_i)$ is called the zeroth divided difference of the function $f(x)$ with respect to x_i;

$$f[x_i,x_{i+1}]=\frac{f[x_{i+1}]-f[x_i]}{x_{i+1}-x_i}=\frac{f(x_{i+1})-f(x_i)}{x_{i+1}-x_i} \tag{5.3.2}$$

is called the first divided difference (一阶差商) of the function $f(x)$ with respect to x_i and x_{i+1};

$$f[x_i,x_{i+1},x_{i+2}]=\frac{f[x_i,x_{i+2}]-f[x_i,x_{i+1}]}{x_{i+2}-x_{i+1}} \tag{5.3.3}$$

is called the second divided difference (二阶差商) of the function $f(x)$ with respect to

x_i, x_{i+1} and x_{i+2}.

Generally,

$$f\left[x_i, x_{i+1}, \cdots, x_{i+k-1}, x_{i+k}\right] = \frac{f\left[x_i, x_{i+1}, \cdots, x_{i+k-2}, x_{i+k}\right] - f\left[x_i, x_{i+1}, \cdots, x_{i+k-1}\right]}{x_{i+k} - x_{i+k-1}} \tag{5.3.4}$$

is called the kth divided difference (k 阶差商) of the function $f(x)$ with respect to $x_i, x_{i+1}, \cdots, x_{i+k}$.

5.3.2 Newton's expansion of a function (函数的 Newton 展开)

Proposition Let $X = \{x_0, x_1, \cdots\}$ and $f(x)$ be defined on X. If all the divided differences $f[x_0, x_1, \cdots, x_i], i = 0, 1, \cdots$, exist, then we have

$$\begin{aligned} f(x) &= f[x_0] + \sum_{k=1}^{n} f[x_0, x_1, \cdots, x_k] N_k(x) \\ &\quad + f[x, x_0, x_1, \cdots, x_n] N_{n+1}(x) \\ &= P_n(x) + R_n(x) \end{aligned} \tag{5.3.5}$$

where

$$N_k(x) = (x - x_0)(x - x_1)\cdots(x - x_{k-1}),\ k = 1, 2, \cdots, n,$$

$$P_n(x) = f[x_0] + \sum_{k=1}^{n} f[x_0, x_1, \cdots, x_k] N_k(x) \tag{5.3.6}$$

is the polynomial of degree n which agrees with $f(x)$ at $n+1$ numbers $x_0, x_1, \cdots, x_n$, and is called the nth Newton's interpolating polynomial (n 次 Newton 插值多项式) for $f(x)$.

$$R_n(x) = f[x, x_0, x_1, \cdots, x_n] N_{n+1}(x) \tag{5.3.7}$$

is the remainder term (余项) or error term (误差项) . And Eq.(5.3.5) is called the **Newton's interpolatory divided difference formula** (Newton 插值差商公式) .

Proof. By definition of divided difference

$$f[x_0, x] = \frac{f(x) - f(x_0)}{x - x_0}$$

which implies

$$f(x) - f(x_0) = (x - x_0) f[x_0, x]$$

i.e.

$$f(x) = f(x_0) + (x - x_0) f[x_0, x]. \tag{5.3.8}$$

From

$$f[x_0, x_1, x] = \frac{f[x_0, x] - f[x_0, x_1]}{x - x_1}$$

it follows

$$f[x_0,x]=f[x_0,x_1]+(x-x_1)f[x_0,x_1,x]. \tag{5.3.9}$$

Substituting Eq. (5.3.9) into Eq. (5.3.8) yields

$$\begin{aligned} f(x)&=f[x_0]+(x-x_0)\big(f[x_0,x_1]+(x-x_1)f[x_0,x_1,x]\big)\\ &=f[x_0]+(x-x_0)f[x_0,x_1]+(x-x_0)(x-x_1)f[x_0,x_1,x]. \end{aligned} \tag{5.3.10}$$

Similarly

$$f[x_0,x_1,x]=f[x_0,x_1,x_2]+(x-x_2)f[x_0,x_1,x_2,x]. \tag{5.3.11}$$

Substituting the above expression into (5.3.10) results in

$$\begin{aligned} f(x)&=f[x_0]+(x-x_0)f[x_0,x_1]\\ &\quad+(x-x_0)(x-x_1)f[x_0,x_1,x_2]\\ &\quad+(x-x_0)(x-x_1)(x-x_2)f[x_0,x_1,x_2,x]. \end{aligned} \tag{5.3.12}$$

This process can be repeated as many times as we want, and after n times we obtain

$$\begin{aligned} f(x)&=f[x_0]+\sum_{i=1}^{n}(x-x_0)\cdots(x-x_{i-1})\cdot f[x_0,x_1,\cdots,x_i]\\ &\quad+(x-x_0)(x-x_1)\cdots(x-x_n)\cdot f[x_0,x_1,\cdots,x_n,x]\\ &=P_n(x)+R_n(x). \end{aligned} \tag{5.3.13}$$

Since

$$R_n(x)=(x-x_0)(x-x_1)\cdots(x-x_n)\cdot f[x_0,x_1,\cdots,x_n,x],$$

we have

$$R_n(x_i)=0, i=0,1,\cdots,n$$

which implies

$$P_n(x_i)=f(x_i), i=0,1,\cdots,n.$$

Therefore $P_n(x)$ is a polynomial of degree n that agrees with $f(x)$ at $n+1$ numbers $x_i, i=0,1,\cdots,n$.

The proof is completed.

5.3.3 Properties of divided differences (差商的性质)

(1) If $f(x)=cg(x)$, where c is a constant, then

$$f[x_0,x_1,\cdots,x_n]=cg[x_0,x_1,\cdots,x_n].$$

(2) If $f(x)=g(x)\pm h(x)$, then

$$f[x_0,x_1,\cdots,x_n]=cg[x_0,x_1,\cdots,x_n]\pm h[x_0,x_1,\cdots,x_n].$$

(3) $f[x_0,x_1,\cdots,x_n]=\sum_{i=0}^{n}\dfrac{f(x_i)}{N'_{n+1}(x_i)}$

where $N_{n+1}(x)=\prod_{i=0}^{n}(x-x_i)$.

(4) $f[x_0, x_1, \cdots, x_n]$ is independent of the order of $x_0, x_1, \cdots, x_n$.

(5) If $f(x) = g(x)h(x)$, then

$$f[x_0, x_1, \cdots, x_n] = \sum_{i=0}^{n} g[x_0, x_1, \cdots, x_i] h[x_i, x_{i+1}, \cdots, x_n].$$

(6) If the $(k+1)st$ divided difference $f[x, x_0, x_1, \cdots, x_k]$ is a polynomial in x of degree m, then the $(k+2)nd$ divided difference $f[x, x_0, x_1, \cdots, x_k, x_{k+1}]$ is a polynomial in x of degree $m-1$.

Proof. Properties (1) and (2) are obvious. Now we prove Properties (3) and (4).

As we know, if $P_n(x)$ is the polynomial of degree n that agrees with $f(x)$ at $n+1$ number $x_0, x_1, \cdots, x_n$, then $P_n(x)$ has three representations as follows:

(a) Using Lagrange interpolation formula gives

$$\begin{aligned} P_n(x) &= \sum_{i=0}^{n} f(x_i) L_i(x) \\ &= \sum_{i=0}^{n} f(x_i) \frac{N_{n+1}(x)}{(x - x_i) N'_{n+1}(x_i)}. \end{aligned}$$

(b) Using Newton's interpolatory (divided difference) formula gives

$$P_n(x) = f[x_0] + \sum_{i=1}^{n} f[x_0, x_1, \cdots, x_i] N_i(x)$$

where

$$N_i(x) = (x - x_0)(x - x_1) \cdots (x - x_{i-1}).$$

(c) Using the expression in terms of power basis gives

$$\begin{aligned} P_n(x) &= a_0 + a_1 x + a_2 x^2 + \cdots + a_n x^n \\ &= (-1)^n \frac{\begin{vmatrix} 1 & x & \cdots & x^n & 0 \\ 1 & x_0 & \cdots & x_0^n & f(x_0) \\ 1 & x_1 & \cdots & x_1^n & f(x_1) \\ \vdots & \vdots & & \vdots & \vdots \\ 1 & x_n & \cdots & x_n^n & f(x_n) \end{vmatrix}}{\begin{vmatrix} 1 & x_0 & \cdots & x_0^{n-1} & x_0^n \\ 1 & x_1 & \cdots & x_1^{n-1} & x_1^n \\ \vdots & \vdots & & \vdots & \vdots \\ 1 & x_n & \cdots & x_n^{n-1} & x_n^n \end{vmatrix}}. \end{aligned} \tag{5.3.14}$$

Comparing the leading coefficients (首项系数) in (a) ,(b) and (c) yields

$$
\begin{aligned}
&f\left[x_0,x_1,\cdots,x_n\right]\\
&=\sum_{i=0}^{n}\frac{f\left(x_i\right)}{N'_{n+1}\left(x_i\right)}\\
&=\frac{\begin{vmatrix}1 & x_0 & \cdots & x_0^{n-1} & f\left(x_0\right)\\ 1 & x_1 & \cdots & x_1^{n-1} & f\left(x_1\right)\\ \vdots & \vdots & & \vdots & \vdots\\ 1 & x_n & \cdots & x_n^{n-1} & f\left(x_n\right)\end{vmatrix}}{\begin{vmatrix}1 & x_0 & \cdots & x_0^{n-1} & x_0^n\\ 1 & x_1 & \cdots & x_1^{n-1} & x_1^n\\ \vdots & \vdots & & \vdots & \vdots\\ 1 & x_n & \cdots & x_n^{n-1} & x_n^n\end{vmatrix}}.
\end{aligned}
\tag{5.3.15}
$$

Since

$$
f\left[x_0,x_1,\ldots,x_n\right]=\frac{\begin{vmatrix}1 & x_0 & \cdots & x_0^{n-1} & f\left(x_0\right)\\ 1 & x_1 & \cdots & x_1^{n-1} & f\left(x_1\right)\\ \vdots & \vdots & & \vdots & \vdots\\ 1 & x_n & \cdots & x_n^{n-1} & f\left(x_n\right)\end{vmatrix}}{\begin{vmatrix}1 & x_0 & \cdots & x_0^{n-1} & x_0^n\\ 1 & x_1 & \cdots & x_1^{n-1} & x_1^n\\ \vdots & \vdots & & \vdots & \vdots\\ 1 & x_n & \cdots & x_n^{n-1} & x_n^n\end{vmatrix}},
$$

we draw the conclusion that $f\left[x_0,x_1,\cdots,x_n\right]$ is independent of the order of $x_0,x_1,\cdots,x_n$. The Property (3) and Property (4) are proved.

Remark Property (3) can also be proved by induction.

First of all, Property (3) is clearly true for $k=1$ since

$$
\begin{aligned}
f\left[x_0,x_1\right]&=\frac{f\left(x_1\right)-f\left(x_0\right)}{x_1-x_0}\\
&=\frac{f\left(x_0\right)}{x_0-x_1}+\frac{f\left(x_1\right)}{x_1-x_0}\\
&=\frac{f\left(x_0\right)}{N'_2\left(x_0\right)}+\frac{f\left(x_1\right)}{N'_2\left(x_1\right)}.
\end{aligned}
$$

Now assume that Property (3) holds for $k=m$, which means

$$
\begin{aligned}
&f\left[x_0,x_1,\cdots,x_m\right]\\
&=\sum_{i=0}^{m}\frac{f\left(x_i\right)}{\left(x_i-x_0\right)\cdots\left(x_i-x_{i-1}\right)\left(x_i-x_{i+1}\right)\cdots\left(x_i-x_m\right)},
\end{aligned}
$$

$$f[x_1,x_2,\cdots,x_{m+1}]$$
$$=\sum_{i=1}^{m+1}\frac{f(x_i)}{(x_i-x_1)\cdots(x_i-x_{i-1})(x_i-x_{i+1})\cdots(x_i-x_{m+1})}.$$

Then we have

$$f[x_0,x_1,\cdots,x_{m+1}]=\frac{f[x_1,x_2,\cdots,x_{m+1}]-f[x_0,x_1,\cdots,x_m]}{x_{m+1}-x_0}$$
$$=\frac{1}{x_{m+1}-x_0}\cdot\left[\sum_{i=1}^{m}\frac{f(x_i)}{(x_i-x_1)\cdots(x_i-x_{i-1})(x_i-x_{i+1})\cdots(x_i-x_{m+1})}\right.$$
$$+\frac{f(x_{m+1})}{(x_{m+1}-x_1)(x_{m+1}-x_2)\cdots(x_{m+1}-x_m)}-\frac{f(x_0)}{(x_0-x_1)(x_0-x_2)\cdots(x_0-x_m)}$$
$$\left.-\sum_{i=1}^{m}\frac{f(x_i)}{(x_i-x_0)\cdots(x_i-x_{i-1})(x_i-x_{i+1})\cdots(x_i-x_m)}\right]$$
$$=\frac{f(x_0)}{(x_0-x_1)(x_0-x_2)\cdots(x_0-x_m)(x_0-x_{m+1})}$$
$$+\frac{f(x_{m+1})}{(x_{m+1}-x_0)(x_{m+1}-x_1)\cdots(x_{m+1}-x_m)}$$
$$+\frac{1}{x_{m+1}-x_0}\sum_{i=1}^{m}\frac{(x_i-x_0)f(x_i)-(x_i-x_{m+1})f(x_i)}{(x_i-x_0)\cdots(x_i-x_{i-1})(x_i-x_{i+1})\cdots(x_i-x_{m+1})}$$
$$=\sum_{i=0}^{m+1}\frac{f(x_i)}{(x_i-x_0)\cdots(x_i-x_{i-1})(x_i-x_{i+1})\cdots(x_i-x_{m+1})}$$
$$=\sum_{i=0}^{m+1}\frac{f(x_i)}{N'_{m+2}(x_i)}.$$

So Property (3) is proved by induction.

Proof of Property (5).

We use the induction to complete the proof.

Since

$$f[x_0,x_1]=(gh)[x_0,x_1]$$
$$=\frac{(gh)(x_1)-(gh)(x_0)}{x_1-x_0}$$
$$=\frac{g(x_1)h(x_1)-g(x_0)h(x_1)+g(x_0)h(x_1)-g(x_0)h(x_0)}{x_1-x_0}$$
$$=\frac{[g(x_1)-g(x_0)]h(x_1)+g(x_0)[h(x_1)-h(x_0)]}{x_1-x_0}$$
$$=g(x_0)h[x_0,x_1]+g[x_0,x_1]h(x_1),$$

Property (5) holds for $n=1$. Assume Property (5) holds for $n=k$, namely,

$$
\begin{aligned}
& f[x_0,x_1,\cdots,x_{k-1},x_k] \\
&= \sum_{i=0}^{k} g[x_0,x_1,\cdots,x_i]h[x_i,x_{i+1},\cdots,x_k] \\
&= \sum_{i=0}^{k-1} g[x_0,x_1,\cdots,x_i]h[x_i,x_{i+1},\cdots,x_k] \\
&\quad + g[x_0,x_1,\cdots,x_k]h[x_k], \\
&\quad f[x_0,x_1,\cdots,x_{k-1},x_{k+1}] \\
&= \sum_{i=0}^{k-1} g[x_0,x_1,\cdots,x_i]h[x_i,x_{i+1},\cdots,x_{k-1},x_{k+1}] \\
&\quad + g[x_0,x_1,\cdots,x_{k-1},x_{k+1}]h[x_{k+1}].
\end{aligned}
$$

Then for $n=k+1$, we have

$$
\begin{aligned}
& f[x_0,x_1,\cdots,x_k,x_{k+1}] \\
&= \frac{1}{x_{k+1}-x_k}\left(f[x_0,x_1,\cdots,x_{k-1},x_{k+1}] - f[x_0,x_1,\cdots,x_{k-1},x_k]\right) \\
&= \frac{1}{x_{k+1}-x_k}\left(\begin{aligned}
&\sum_{i=0}^{k-1} g[x_0,x_1,\cdots,x_i]h[x_i,x_{i+1},\cdots,x_{k-1},x_{k+1}] \\
&+ g[x_0,x_1,\cdots,x_{k-1},x_{k+1}]h[x_{k+1}] \\
&- \sum_{i=0}^{k-1} g[x_0,x_1,\cdots,x_i]h[x_i,x_{i+1},\cdots,x_k] \\
&- g[x_0,x_1,\cdots,x_k]h[x_k],
\end{aligned}\right) \\
&= \frac{1}{x_{k+1}-x_k}\left(\begin{aligned}
&\sum_{i=0}^{k-1} g[x_0,x_1,\cdots,x_i]\left(h[x_i,x_{i+1},\cdots,x_{k-1},x_{k+1}] - h[x_i,x_{i+1},\cdots,x_k]\right) \\
&+ g[x_0,x_1,\cdots,x_{k-1},x_{k+1}]h[x_{k+1}] - g[x_0,x_1,\cdots,x_k]h[x_k]
\end{aligned}\right) \\
&= \frac{1}{x_{k+1}-x_k}\left(\begin{aligned}
&\sum_{i=0}^{k-1} g[x_0,x_1,\cdots,x_i]\left(h[x_i,x_{i+1},\cdots,x_{k-1},x_{k+1}] - h[x_i,x_{i+1},\cdots,x_k]\right) \\
&+ g[x_0,x_1,\cdots,x_{k-1},x_{k+1}]h[x_{k+1}] - g[x_0,x_1,\cdots,x_{k-1},x_k]h[x_{k+1}] \\
&+ g[x_0,x_1,\cdots,x_{k-1},x_k]h[x_{k+1}] - g[x_0,x_1,\cdots,x_k]h[x_k]
\end{aligned}\right) \\
&= \sum_{i=0}^{k-1} g[x_0,x_1,\cdots,x_i]\left(\frac{h[x_i,x_{i+1},\cdots,x_{k-1},x_{k+1}] - h[x_i,x_{i+1},\cdots,x_k]}{x_{k+1}-x_k}\right) \\
&\quad + \frac{g[x_0,x_1,\cdots,x_{k-1},x_{k+1}] - g[x_0,x_1,\cdots,x_{k-1},x_k]}{x_{k+1}-x_k}h[x_{k+1}] \\
&\quad + g[x_0,x_1,\cdots,x_{k-1},x_k]\frac{h[x_{k+1}]-h[x_k]}{x_{k+1}-x_k} \\
&= \sum_{i=0}^{k-1} g[x_0,x_1,\cdots,x_i]h[x_i,x_{i+1},\cdots,x_k,x_{k+1}] + g[x_0,x_1,\cdots,x_k,x_{k+1}]h[x_{k+1}]
\end{aligned}
$$

$$+g\left[x_0,x_1,\cdots,x_{k-1},x_k\right]h\left[x_k,x_{k+1}\right]$$
$$=\sum_{i=0}^{k+1}g\left[x_0,x_1,\cdots,x_i\right]h\left[x_i,x_{i+1},\cdots,x_k,x_{k+1}\right].$$

Therefore Property (5) holds for $n=k+1$. And Property (5) is proved by induction.

Proof of Property (6).

Suppose $q_m(x)=f\left[x,x_0,x_1,\cdots,x_k\right]$ is a polynomial in x of degree m, then

$$f\left[x,x_0,x_1,\cdots,x_k,x_{k+1}\right]$$
$$=\frac{f\left[x,x_0,x_1,\cdots,x_k\right]-f\left[x_0,x_1,\cdots,x_{k+1}\right]}{x-x_{k+1}}$$
$$=\frac{q_m(x)-q_m(x_{k+1})}{x-x_{k+1}}.$$

Since $q_m(x)=f\left[x,x_0,x_1,\cdots,x_k\right]$ is a polynomial in x of degree m, we have

$$q_m(x)$$
$$=q_m(x_{k+1})+\sum_{j=1}^{m}\frac{q_m^{(j)}(x_{k+1})}{j!}(x-x_{k+1})^j.$$

Therefore

$$\frac{q_m(x)-q_m(x_{k+1})}{x-x_{k+1}}=\sum_{j=0}^{m-1}\frac{q_m^{(j+1)}(x_{k+1})}{(j+1)!}(x-x_{k+1})^j$$

is a polynomial in x of degree $m-1$, i.e., $f\left[x,x_0,x_1,\cdots,x_k,x_{k+1}\right]$ is a polynomial in x of degree $m-1$.

5.3.4 Computation of Newton's interpolant (Newton 插值的计算)

We can obtain the Newton's interpolant by computing all divided differences as in the following tableau

x_0	$f(x_0)$				
x_1	$f(x_1)$	$f[x_0,x_1]$			
x_2	$f(x_2)$	$f[x_1,x_2]$	$f[x_0,x_1,x_2]$		
x_3	$f(x_3)$	$f[x_2,x_3]$	$f[x_1,x_2,x_3]$	$f[x_0,x_1,x_2,x_3]$	
x_4	$f(x_4)$	$f[x_3,x_4]$	$f[x_2,x_3,x_4]$	$f[x_1,x_2,x_3,x_4]$	$f[x_0,x_1,x_2,x_3,x_4]$

Algorithm (算法) Set

$$\begin{array}{lllll}
x_0 & f(x_0)=F_{0,0} & & & \\
x_1 & f(x_1)=F_{1,0} & f\left[x_0,x_1\right]=F_{1,1} & & \\
x_2 & f(x_2)=F_{2,0} & f\left[x_1,x_2\right]=F_{2,1} & F_{2,2} & \\
x_3 & f(x_3)=F_{3,0} & f\left[x_2,x_3\right]=F_{3,1} & F_{3,2} & \\
\vdots & \vdots & \vdots & \vdots & \ddots \\
x_n & f(x_n)=F_{n,0} & f\left[x_{n-1},x_n\right]=F_{n,1} & F_{n,2}\cdots & F_{n,n}
\end{array}$$

where

$$F_{i,j}=f\left[x_{i-j},x_{i-j+1},\cdots,x_i\right].$$

Step 1. For $j=1,2,\cdots,n$

For $i=j,j+1,\cdots,n$

Set $F_{i,j}=\dfrac{F_{i,j-1}-F_{i-1,j-1}}{x_i-x_{i-j}}$

Step 2. Output$\left(F_{0,0},F_{1,1},\cdots,F_{n,n}\right)$

Step 3. $p(x)=\sum_{i=0}^{n}F_{i,i}\prod_{j=0}^{i-1}\left(x-x_j\right)$.

Example 5.6 Use the data given in the following table to construct a polynomial of degree three

x_i	0	1	2	3
$f(x_i)$	0	1	8	27

Solution. Let us first list all the necessary divided differences to be determined in the tableau below

x_0	$f(x_0)$			
x_1	$f(x_1)$	$f[x_0,x_1]$		
x_2	$f(x_2)$	$f[x_1,x_2]$	$f[x_0,x_1,x_2]$	
x_3	$f(x_3)$	$f[x_2,x_3]$	$f[x_1,x_2,x_3]$	$f[x_0,x_1,x_2,x_3]$

By definition we have

$$f\left[x_0,x_1\right]=\frac{f\left(x_1\right)-f\left(x_0\right)}{x_1-x_0}=1,$$

$$f\left[x_1,x_2\right]=\frac{f\left(x_2\right)-f\left(x_1\right)}{x_2-x_1}=7,$$

$$f\left[x_2,x_3\right]=\frac{f\left(x_3\right)-f\left(x_2\right)}{x_3-x_2}=19,$$

$$f\left[x_0,x_1,x_2\right]=\frac{f\left[x_1,x_2\right]-f\left[x_0,x_1\right]}{x_2-x_0}=3,$$

$$f\left[x_1,x_2,x_3\right]=\frac{f\left[x_2,x_3\right]-f\left[x_1,x_2\right]}{x_3-x_1}=6,$$

$$f\left[x_0,x_1,x_2,x_3\right]=\frac{f\left[x_1,x_2,x_3\right]-f\left[x_0,x_1,x_2\right]}{x_3-x_0}=1.$$

Therefore the polynomial of degree three is

$$
\begin{aligned}
&P_3(x)\\
&=f[x_0]+(x-x_0)f[x_0,x_1]\\
&\quad+(x-x_0)(x-x_1)f[x_0,x_1,x_2]\\
&\quad+(x-x_0)(x-x_1)(x-x_2)f[x_0,x_1,x_2,x_3]\\
&=0+x\cdot 1+3\cdot x(x-1)+1\cdot x(x-1)(x-2)\\
&=x^3.
\end{aligned}
$$

5.3.5 The relationship between divided differences and derivatives (差商与导数的关系)

Theorem 5.3 Suppose that $f\in C^n[a,b]$ and $x_0,x_1,\cdots,x_n$ are $n+1$ numbers in $[a,b]$. Then a number ξ exists in (a,b) with

$$
f[x_0,x_1,\cdots,x_n]=\frac{f^{(n)}(\xi)}{n!}. \tag{5.3.16}
$$

Proof. Suppose

$$
P_n(x)=f[x_0]+\sum_{k=1}^{n}f[x_0,x_1,\cdots,x_k]\cdot(x-x_0)(x-x_1)\cdots(x-x_{k-1}).
$$

Then

$$
P_n(x_i)=f(x_i),\quad i=0,1,\cdots,n
$$

which means $f(x)-P_n(x)$ has $n+1$ zeros in $[a,b]$. The Generalized Rolle's Theorem implies that there exists a number ξ among $x_0,x_1,\cdots,x_n$ such that

$$
(f-P_n)^{(n)}(\xi)=0.
$$

Since $P_n(x)$ is a polynomial of degree n with leading coefficient $f[x_0,x_1,\cdots,x_n]$, we have

$$
P_n^{(n)}(\xi)=P_n^{(n)}(x)=n!f[x_0,x_1,\cdots,x_n].
$$

As a result,

$$
f^{(n)}(\xi)-n!f[x_0,x_1,\cdots,x_n]=0,
$$

i.e.

$$
f[x_0,x_1,\cdots,x_n]=\frac{f^{(n)}(\xi)}{n!}.
$$

Especially

$$
\begin{aligned}
&f\Bigg[\overbrace{x_0,x_0,\cdots,x_0}^{n+1}\Bigg]\\
&=\lim_{\substack{x_i\to x_0\\ i=1,2,\cdots,n}} f[x_0,x_1,\cdots,x_n]\\
&=\frac{f^{(n)}(x_0)}{n!}.
\end{aligned}
$$

5.3.6 Relations between Newton's expansion and Taylor's expansion (Newton 展开与 Taylor 展开之间的关系)

As we know, the Newton's expansion with remainder term (带余项的 Newton 展开) reads as follows

$$f(x)=f[x_0]+\sum_{i=1}^{n}(x-x_0)\cdots(x-x_{i-1})f[x_0,x_1,\cdots,x_i]$$
$$+(x-x_0)(x-x_1)\cdots(x-x_n)f[x_0,x_1,\cdots,x_n,x],$$

and the Taylor's expansion about $x=x_0$ with remainder term ($x=x_0$ 处带余项的 Taylor 展开) reads as follows

$$f(x)=f(x_0)+\sum_{i=1}^{n}\frac{f^{(i)}(x_0)}{i!}(x-x_0)^i+\frac{f^{(n+1)}(\xi)}{(n+1)!}(x-x_0)^{n+1}.$$

Note that

$$\lim_{\substack{x_j\to x_0\\ j=1,2,\cdots,i}} f[x_0,x_1,\cdots,x_i]=\frac{f^{(i)}(x_0)}{i!},\quad i=1,2,\cdots,n$$

$$\lim_{\substack{x_j\to x_0\\ j=1,2,\cdots,i-1}} (x-x_0)\cdots(x-x_{i-1})=(x-x_0)^i$$

and

$$f[x_0,x_1,\cdots,x_n,x]=\frac{f^{(n+1)}(\xi)}{(n+1)!}.$$

Therefore Taylor's polynomial can be understood as the limit of Newton's interpolatory polynomial (Taylor 多项式可以视为 Newton 插值多项式的极限).

5.3.7 Comparisons among Lagrange, Neville and Newton interpolations (Lagrange 插值, Neville 插值与 Newton 插值之间的比较)

As we know, given distinct interpolation nodes, one can use Lagrange interpolation, Neville interpolation or Newton interpolation to get the interpolating polynomial. However, if multiple nodes (重结点) are involved, only Newton's interpolatory divided difference formula is still available to be used for constructing the required polynomial while neither Lagrange interpolation nor Neville interpolation, works in this case when the denominator is zero.

Example 5.7 Use the data given in the following table to construct a quadratic interpolating polynomial $p_2(x)$.

x_i	0	1	2
$f(x_i)$	1	2	5

Solution 1 (**Lagrange interpolation**).

Since $x_0 = 0, x_1 = 1, x_2 = 2$, we have

$$L_0(x) = \frac{(x-x_1)(x-x_2)}{(x_0-x_1)(x_0-x_2)} = \frac{1}{2}(x-1)(x-2),$$

$$L_1(x) = \frac{(x-x_0)(x-x_2)}{(x_1-x_0)(x_1-x_2)} = -x(x-2),$$

$$L_2(x) = \frac{(x-x_0)(x-x_1)}{(x_2-x_0)(x_2-x_1)} = \frac{1}{2}x(x-1).$$

Using the Lagrange interpolation formula gives

$$\begin{aligned} p_2(x) &= f(x_0)L_0(x) + f(x_1)L_1(x) + f(x_2)L_2(x) \\ &= \frac{1}{2}(x-1)(x-2) - 2x(x-2) + \frac{5}{2}x(x-1) \\ &= 1 + x^2. \end{aligned}$$

Solution 2 (**Neville interpolation**).

With the given data, we have

$$p_{01}(x) = f(x_0)\frac{x_1 - x}{x_1 - x_0} + f(x_1)\frac{x - x_0}{x_1 - x_0} = 1 + x,$$

$$p_{12}(x) = f(x_1)\frac{x_2 - x}{x_2 - x_1} + f(x_2)\frac{x - x_1}{x_2 - x_1} = 3x - 1,$$

$$p_2(x) = p_{012}(x) = p_{01}(x)\frac{x_2 - x}{x_2 - x_0} + p_{12}(x)\frac{x - x_0}{x_2 - x_0} = 1 + x^2.$$

Remark One can alternatively obtain the quadratic interpolating polynomial in the following way

$$p_{01}(x) = f(x_0)\frac{x_1 - x}{x_1 - x_0} + f(x_1)\frac{x - x_0}{x_1 - x_0} = 1 + x,$$

$$p_{02}(x) = f(x_0)\frac{x_2 - x}{x_2 - x_0} + f(x_2)\frac{x - x_0}{x_2 - x_0} = 1 + 2x,$$

$$p_2(x) = p_{012}(x) = p_{01}(x)\frac{x_2 - x}{x_2 - x_1} + p_{02}(x)\frac{x - x_1}{x_2 - x_1} = 1 + x^2.$$

Solution 3 (**Newton interpolation**).

First of all, we establish the following table of divided differences.

$$\begin{array}{llll} x_0=0 & f(x_0)=1 & & \\ x_1=1 & f(x_1)=2 & f[x_0,x_1]=1 & \\ x_2=2 & f(x_2)=5 & f[x_1,x_2]=3 & f[x_0,x_1,x_2]=1 \end{array}$$

Using the Newton's interpolatory divided difference formula gives

$$\begin{aligned} p_2(x) &= f(x_0)+(x-x_0)f[x_0,x_1]+(x-x_0)(x-x_1)f[x_0,x_1,x_2] \\ &= 1+x+x(x-1) \\ &= 1+x^2. \end{aligned}$$

Example 5.8 Use the data given in the following table to construct a quadratic interpolating polynomial $p_2(x)$ such that $p_2(x_0)=f(x_0)$, $p_2(x_1)=f(x_1)$, and $p_2'(x_1)=f'(x_1)$.

x_i	0	1
$f(x_i)$	1	2
$f'(x_i)$		2

Solution. In this case, the node $x_1=1$ is a multiple node where two pieces of information $f(x_1)=2$ and $f'(x_1)=2$ are involved, and neither the Lagrange interpolation nor the Neville interpolation, can be used to get the quadratic interpolating polynomial $p_2(x)$. Fortunately the Newton interpolation still works. One can establish the following table of divided differences.

$$\begin{array}{llll} x_0=0 & f(x_0)=1 & & \\ x_1=1 & f(x_1)=2 & f[x_0,x_1]=1 & \\ x_1=1 & f(x_1)=2 & f[x_1,x_1]=2 & f[x_0,x_1,x_1]=1 \end{array}$$

Using the Newton's interpolatory divided difference formula gives

$$\begin{aligned} p_2(x) &= f(x_0)+(x-x_0)f[x_0,x_1]+(x-x_0)(x-x_1)f[x_0,x_1,x_1] \\ &= 1+x+x(x-1) \\ &= 1+x^2. \end{aligned}$$

To sum up, the Lagrange interpolation, the Neville interpolation and the Newton interpolation can be compared with one another in terms of the requirements for nodes, recursion and intermediate values as in the following Table 5.1.

Table 5.1 Comparison among three methods

Methods	requirements for nodes	recursion	intermediate values
Lagrange	distinct	No	No
Neville	distinct	Yes	Yes
Newton	arbitrary	Yes	Yes

5.3.8 Newton forward divided-difference formula (Newton 向前差商公式)

If $x_0,x_1,\cdots,x_n$ are of equal spacing, say,

$$x_i-x_{i-1}=h,\quad i=1,2,\cdots,n,$$

so $x_i=x_0+ih$. Suppose $x=x_0+sh$, then $x-x_i=(s-i)h$

and

$$\begin{aligned}&(x-x_0)(x-x_1)\cdots(x-x_{i-1})\\&=sh\cdot(s-1)h\cdots(s-i+1)h\\&=h^i i!\binom{s}{i}\end{aligned}$$

In this case, the Newton's interpolatory polynomial $P_n(x)$ can be expressed as

$$\begin{aligned}P_n(x)&=f[x_0]+\sum_{i=1}^{n}f[x_0,x_1,\cdots,x_i]\cdot(x-x_0)(x-x_1)\cdots(x-x_{i-1})\\&=f(x_0)+\sum_{i=1}^{n}\binom{s}{i}h^i i!f[x_0,x_1,\cdots,x_i].\end{aligned}$$

This formula is called the **Newton forward divided-difference formula** (Newton 向前差商公式).

5.3.9 Newton forward difference formula (Newton 向前差分公式)

We introduce the following notation for forward difference

$$\Delta f(x_i)=f(x_i+h)-f(x_i), \tag{5.3.17}$$

then

$$\begin{aligned}&\Delta^2 f(x_i)\\&=\Delta(\Delta f(x_i))\\&=\Delta(f(x_i+h)-f(x_i))\\&=\Delta f(x_i+h)-\Delta f(x_i)\\&=f(x_i+2h)-f(x_i+h)-f(x_i+h)+f(x_i)\\&=f(x_i+2h)-2f(x_i+h)+f(x_i).\end{aligned} \tag{5.3.18}$$

And in general,

$$\begin{aligned}\Delta^k f(x_i)&=\Delta(\Delta^{k-1}f(x_i))\\&=\Delta^{k-1}f(x_i+h)-\Delta^{k-1}f(x_i).\end{aligned} \tag{5.3.19}$$

Now

$$f[x_0,x_1]=\frac{f(x_1)-f(x_0)}{x_1-x_0}=\frac{1}{h}\Delta f(x_0),$$

$$f[x_i,x_{i+1}]=\frac{f(x_{i+1})-f(x_i)}{x_{i+1}-x_i}=\frac{1}{h}\Delta f(x_i),$$

$$
\begin{aligned}
f[x_0,x_1,x_2] &= \frac{f[x_1,x_2]-f[x_0,x_1]}{x_2-x_0} \\
&= \frac{1}{2h}\left(\frac{1}{h}\Delta f(x_1)-\frac{1}{h}\Delta f(x_0)\right) \\
&= \frac{1}{2h^2}\Delta^2 f(x_0),
\end{aligned}
$$

$$
f[x_1,x_2,x_3] = \frac{1}{2h^2}\Delta^2 f(x_1),
$$

$$
\begin{aligned}
f[x_0,x_1,x_2,x_3] &= \frac{f[x_1,x_2,x_3]-f[x_0,x_1,x_2]}{x_3-x_0} \\
&= \frac{1}{3h}\left(\frac{1}{2h^2}\Delta^2 f(x_1)-\frac{1}{2h^2}\Delta^2 f(x_0)\right) \\
&= \frac{1}{3!h^3}\Delta^3 f(x_0).
\end{aligned}
$$

We can prove by induction

$$
f[x_0,x_1,\cdots,x_k] = \frac{1}{k!h^k}\Delta^k f(x_0) \tag{5.3.20}
$$

and from

$$
\Delta^i = (E-I)^i = \sum_{j=0}^{i}\binom{i}{j}E^j(-1)^{i-j}I^{i-j}
$$

it follows

$$
\Delta^i f(x_0) = \sum_{j=0}^{i}\binom{i}{j}(-1)^{i-j}f(x_j), \tag{5.3.21}
$$

where E is the shift operator satisfying $Ef(x_i)=f(x_i+h)$ and I is the identity operator satisfying $If(x_i)=f(x_i)$.

Hence the Newton's interpolatory polynomial $P_n(x)$ can be expressed via forward difference as follows

$$
\begin{aligned}
P_n(x) &= P_n(x_0+sh) \\
&= f(x_0)+\sum_{k=1}^{n}\binom{s}{k}h^k k! f[x_0,x_1,\cdots,x_k] \\
&= f(x_0)+\sum_{k=1}^{n}\binom{s}{k}h^k k!\frac{1}{k!h^k}\Delta^k f(x_0) \\
&= f(x_0)+\sum_{k=1}^{n}\binom{s}{k}\Delta^k f(x_0) \\
&= f(x_0)+\sum_{k=1}^{n}\binom{s}{k}\sum_{j=0}^{k}\binom{k}{j}(-1)^{k-j}f(x_j).
\end{aligned} \tag{5.3.22}
$$

The above formula is called **Newton forward-difference formula** (Newton 向前差分公式).

5.3.10 Newton backward divided-difference formula (Newton 向后差商公式)

If the interpolating nodes (插值结点) $x_0, x_1, \cdots, x_n$ are reordered as $x_n, x_{n-1}, \cdots, x_0$, it is not difficult to verify that the nth Newton's interpolatory polynomial $P_n(x)$ is of the following form

$$\begin{aligned}
&P_n(x)\\
&=f[x_n]+(x-x_n)f[x_n,x_{n-1}]\\
&\quad+(x-x_n)(x-x_{n-1})f[x_n,x_{n-1},x_{n-2}]\\
&\quad+\cdots\\
&\quad+(x-x_n)(x-x_{n-1})\cdots(x-x_1)\cdot f[x_n,x_{n-1},\cdots,x_1,x_0]\\
&=f[x_n]+\sum_{k=1}^{n}(x-x_n)(x-x_{n-1})\cdots(x-x_{n-k+1})\cdot f[x_n,x_{n-1},\cdots x_{n-k}].
\end{aligned}\tag{5.3.23}$$

In fact, for $i\in\{0,1,\cdots,n\}$,

$$\begin{aligned}
P_n(x_i)&=f[x_n]+(x_i-x_n)f[x_n,x_{n-1}]+\cdots\\
&\quad+(x_i-x_n)(x_i-x_{n-1})\cdots(x_i-x_{i+1})\cdot f[x_n,x_{n-1},\cdots,x_{i+1},x_i]\\
&=f[x_n]+(x_i-x_n)f[x_n,x_{n-1}]+\cdots\\
&\quad+(x_i-x_n)(x_i-x_{n-1})\cdots(x_i-x_{i+1})\\
&\quad\cdot\frac{f[x_n,x_{n-1},\cdots,x_{i+2},x_i]-f[x_n,x_{n-1},\cdots,x_{i+2},x_{i+1}]}{x_i-x_{i+1}}\\
&=f[x_n]+(x_i-x_n)f[x_n,x_{n-1}]+\cdots\\
&\quad+(x_i-x_n)(x_i-x_{n-1})\cdots(x_i-x_{i+2})\\
&\quad\cdot f[x_n,x_{n-1},\cdots,x_{i+2},x_i]\\
&=f[x_n]+(x_i-x_n)f[x_n,x_{n-1}]+\cdots\\
&\quad+(x_i-x_n)(x_i-x_{n-1})\cdots(x_i-x_{i+2})\\
&\quad\cdot\frac{f[x_n,x_{n-1},\cdots,x_{i+3},x_i]-f[x_n,x_{n-1},\cdots,x_{i+2}]}{x_i-x_{i+2}}\\
&=f[x_n]+(x_i-x_n)f[x_n,x_{n-1}]+\cdots\\
&\quad+(x_i-x_n)(x_i-x_{n-1})\cdots(x_i-x_{i+3})\\
&\quad\cdot f[x_n,x_{n-1},\cdots,x_{i+3},x_i]\\
&=\cdots\\
&=f[x_n]+(x_i-x_n)f[x_n,x_i]\\
&=f[x_n]+(x_i-x_n)\frac{f(x_i)-f(x_n)}{x_i-x_n}\\
&=f(x_i).
\end{aligned}$$

If the nodes are equally spaced with

$$x = x_n + sh = x_i + (n-i)h + sh = x_i + (s+n-i)h,$$

then we have

$$x - x_i = (s+n-i)h$$

and

$$\begin{aligned}&(x-x_n)(x-x_{n-1})\cdots(x-x_{n-k+1})\\&= sh\cdot(s+1)h\cdots(s+k-1)h\\&= h^k s(s+1)\cdots(s+k-1).\end{aligned}$$

Therefore

$$\begin{aligned}P_n(x) &= P_n(x_n+sh)\\&= f[x_n] + \sum_{k=1}^{n} s(s+1)\cdots(s+k-1)h^k f[x_n, x_{n-1}, \cdots, x_{n-k}]\\&= f(x_n) + \sum_{k=1}^{n}(-1)^k(-s)(-s-1)\cdots(-s-k+1)h^k f[x_n, x_{n-1}, \cdots, x_{n-k}](t=-s)\\&= f(x_n) + \sum_{k=1}^{n}(-1)^k t(t-1)\cdots(t-k+1)h^k f[x_n, x_{n-1}, \cdots, x_{n-k}]\\&= f(x_n) + \sum_{k=1}^{n}(-1)^k \frac{t!}{(t-k)!k!}k!h^k f[x_n, x_{n-1}, \cdots, x_{n-k}]\\&= f(x_n) + \sum_{k=1}^{n}(-1)^k \binom{t}{k}k!h^k f[x_n, x_{n-1}, \cdots, x_{n-k}]\\&= f(x_n) + \sum_{k=1}^{n}(-1)^k \binom{-s}{k}k!h^k f[x_n, x_{n-1}, \cdots, x_{n-k}].\end{aligned} \tag{5.3.24}$$

This form is called the **Newton backward divided-difference formula** (Newton 向后差商公式).

5.3.11 Newton backward-difference formula (Newton 向后差分公式)

Let us introduce the backward-difference (向后差分) as below

$$\nabla f(x_i) = f(x_i) - f(x_i - h). \tag{5.3.25}$$

Then

$$\begin{aligned}&\nabla^2 f(x_i)\\&= \nabla f(x_i) - \nabla f(x_i - h)\\&= f(x_i) - f(x_i - h) - f(x_i - h) + f(x_i - 2h)\\&= f(x_i) - 2f(x_i - h) + f(x_i - 2h),\end{aligned} \tag{5.3.26}$$

$$\begin{aligned}
&\nabla^k f(x_i) \\
&= \nabla\left(\nabla^{k-1} f(x_i)\right) \\
&= \nabla^{k-1} f(x_i) - \nabla^{k-1} f(x_i - h).
\end{aligned} \tag{5.3.27}$$

As we know

$$Ef(x_i) = f(x_i + h), \quad E^{-1} f(x_i) = f(x_i - h),$$

then we have

$$\begin{aligned}
\nabla f(x_i) &= \left(I - E^{-1}\right) f(x_i), \\
\nabla^2 f(x_i) &= \left(I - E^{-1}\right)^2 f(x_i), \\
\nabla^k f(x_i) &= \left(I - E^{-1}\right)^k f(x_i) \\
&= \sum_{j=0}^{k} (-1)^j \binom{k}{j} \left(E^{-1}\right)^j I^{k-j} f(x_i) \\
&= \sum_{j=0}^{k} (-1)^j \binom{k}{j} f(x_{i-j}).
\end{aligned}$$

In this case, we have

$$f[x_n, x_{n-1}] = \frac{f(x_n) - f(x_{n-1})}{x_n - x_{n-1}} = \frac{1}{h} \nabla f(x_n),$$

$$\begin{aligned}
f[x_n, x_{n-1}, x_{n-2}] &= \frac{f[x_n, x_{n-1}] - f[x_{n-1}, x_{n-2}]}{x_n - x_{n-2}} \\
&= \frac{1}{2h}\left(\frac{1}{h} \nabla f(x_n) - \frac{1}{h} \nabla f(x_{n-1})\right) \\
&= \frac{1}{2h^2} \nabla^2 f(x_n).
\end{aligned}$$

In general,

$$f[x_n, x_{n-1}, \cdots, x_{n-k}] = \frac{1}{k! h^k} \nabla^k f(x_n). \tag{5.3.28}$$

Note that

$$\begin{aligned}
\binom{-s}{k} &= \frac{(-s)(-s-1)\cdots(-s-k+1)}{k!} \\
&= \frac{(-1)^k s(s+1)\cdots(s+k-1)}{k!}
\end{aligned}$$

i.e.

$$(-1)^k \binom{-s}{k} = \frac{s(s+1)\cdots(s+k-1)}{k!}.$$

As a consequence, the nth Newton's interpolatory polynomial can now be rewritten as

$$\begin{aligned}
P_n(x) &= P(x_n + sh) \\
&= f[x_n] + \sum_{k=1}^{n} s(s+1)\cdots(s+k-1)h^k f[x_n, x_{n-1}, \cdots, x_{n-k}] \\
&= f(x_n) + \sum_{k=1}^{n} (-1)^k (-s)(-s-1)\cdots(-s-k+1)h^k \frac{1}{k!h^k}\nabla^k f(x_n), t=-s \\
&= f(x_n) + \sum_{k=1}^{n} (-1)^k \frac{t(t-1)\cdots(t-k+1)}{k!}\nabla^k f(x_n) \\
&= f(x_n) + \sum_{k=1}^{n} (-1)^k \frac{t!}{(t-k)!k!}k!\frac{1}{k!}\nabla^k f(x_n) \\
&= f(x_n) + \sum_{k=1}^{n} (-1)^k \binom{t}{k}\nabla^k f(x_n) \\
&= f(x_n) + \sum_{k=1}^{n} (-1)^k \binom{-s}{k}\nabla^k f(x_n).
\end{aligned}$$

Therefore

$$P_n(x) = f(x_n) + \sum_{k=1}^{n} (-1)^k \binom{-s}{k}\nabla^k f(x_n). \tag{5.3.29}$$

This form is called **Newton backward difference formula** (Newton 向后差分公式).

To sum up, we have

$$\begin{aligned}
P_n(x) &= P_n(x_0 + sh) \\
&= f(x_0) + \sum_{i=0}^{n} \binom{s}{i} h^i i! f[x_0, x_1, \cdots, x_i] \quad \text{(Newton 向前差商公式)} \\
&= f(x_0) + \sum_{k=1}^{n} \binom{s}{k} \Delta^k f(x_0) \quad \text{(Newton 向前差分公式)} \\
&= f(x_0) + \sum_{k=1}^{n} \binom{s}{k} \sum_{j=0}^{k} \binom{k}{j} (-1)^{k-j} f(x_j),
\end{aligned}$$

$$\begin{aligned}
P_n(x) &= P_n(x_n + sh) \\
&= f(x_n) + \sum_{k=1}^{n} (-1)^k \binom{-s}{k} k!h^k \\
&\quad \cdot f[x_n, x_{n-1}, \cdots, x_{n-k}] \quad \text{(Newton 向后差商公式)} \\
&= f(x_n) + \sum_{k=1}^{n} (-1)^k \binom{-s}{k} \nabla^k f(x_n) \quad \text{(Newton 向后差分公式)} \\
&= f(x_n) + \sum_{k=1}^{n} (-1)^k \binom{-s}{k} \sum_{j=0}^{k} (-1)^j \binom{k}{j} f(x_{n-j}).
\end{aligned}$$

Example 5.9 A fourth-degree polynomial $P(x)$ satisfies

$$\Delta^4 P(0)=24,\ \Delta^3 P(0)=6,\ \Delta^2 P(0)=0,$$

where

$$\Delta P(x)=P(x+1)-P(x).$$

Compute $\Delta^2 P(10)$.

Solution. Let

$$x_i=i, i=0,1,2,3,4.$$

Then $h=1$. By Newton forward-difference formula,

$$\begin{aligned}
P(x)&=P(x_0+sh)\\
&=P(s)\\
&=P(x_0)+\sum_{k=1}^{4}\binom{s}{k}\Delta^k P(x_0)\\
&=P(0)+\sum_{k=1}^{4}\binom{s}{k}\Delta^k P(0),\\
\Delta P(s)&=P(s+1)-P(s)\\
&=P(0)+\sum_{k=1}^{4}\binom{s+1}{k}\Delta^k P(0)-P(0)\\
&\quad-\sum_{k=1}^{4}\binom{s}{k}\Delta^k P(0)\\
&=\sum_{k=1}^{4}\left(\binom{s+1}{k}-\binom{s}{k}\right)\Delta^k P(0)\\
&=\sum_{k=1}^{4}\binom{s}{k-1}\Delta^k P(0)
\end{aligned}$$

and

$$\begin{aligned}
\Delta^2 P(s)&=\Delta P(s+1)-\Delta P(s)\\
&=\sum_{k=1}^{4}\left(\binom{s+1}{k-1}-\binom{s}{k-1}\right)\Delta^k P(0)\\
&=\sum_{k=2}^{4}\binom{s}{k-2}\Delta^k P(0)\\
&=\Delta^2 P(0)+s\cdot\Delta^3 P(0)+\frac{s(s-1)}{2}\Delta^4 P(0).
\end{aligned}$$

Therefore

$$\Delta^2 P(10)=10\times 6+\frac{10\times 9}{2}\times 24=1140.$$

Example 5.10 The following data are given for a polynomial $P(x)$ of unknown degree

x	0	1	2
$P(x)$	2	−1	4

Determine the coefficient of x^2 in $P(x)$ if all third-order forward differences are 1.

Solution. Let $x_i = i, i = 0,1,\cdots$. All third-order forward differences are 1 implies

$$P[x_j, x_{j+1}, x_{j+2}, x_{j+3}] = \frac{1}{3!h^3}\Delta^3 P(x_j) = \frac{1}{3!} = \frac{1}{6}, \quad \text{for all } x_j,$$

$$P[x_0, x_1, x_2, \cdots, x_n] = 0 \text{ for } n \geqslant 4.$$

As a consequence,

$$\begin{aligned} P(x) &= P(x_0) + (x - x_0)P[x_0, x_1] \\ &\quad + (x - x_0)(x - x_1)P[x_0, x_1, x_2] \\ &\quad + (x - x_0)(x - x_1)(x - x_2)P[x_0, x_1, x_2, x] \\ &= 2 + P[x_0, x_1]x + P[x_0, x_1, x_2]x(x-1) \\ &\quad + \frac{1}{6}x(x-1)(x-2) \end{aligned}$$

where

$$P[x_0, x_1] = \frac{P(x_1) - P(x_0)}{x_1 - x_0} = -3,$$

$$P[x_1, x_2] = \frac{P(x_2) - P(x_1)}{x_2 - x_1} = 5,$$

$$P[x_0, x_1, x_2] = \frac{P[x_1, x_2] - P[x_0, x_1]}{x_2 - x_0} = 4.$$

Therefore the coefficient of x^2 in $P(x)$ is

$$P[x_0, x_1, x_2] - 3 \times \frac{1}{6} = 4 - 0.5 = 3.5.$$

5.4 Hermite interpolation (Hermite 插值)

- Two-point Hermite interpolation (两点 Hermite 插值)
- General Hermite interpolation (一般 Hermite 插值)

5.4.1 Two-point Hermite interpolation (两点 Hermite 插值)

Problem Find a cubic polynomial $H_3(x)$, such that

$$H_3(x_0) = y_0, \quad H_3(x_1) = y_1;$$
$$H_3'(x_0) = y_0', \quad H_3'(x_1) = y_1'.$$

Solution 1. Construct cubic basis functions $\varphi_i(x), \psi_i(x), i = 0, 1,$ for which

$$\begin{cases}\varphi_i(x_j)=\delta_{i,j}, & \varphi_i'(x_j)=0, \quad i,j=0,1;\\ \psi_i'(x_j)=\delta_{i,j}, & \psi_i(x_j)=0, \quad i,j=0,1.\end{cases}$$

Then

$$H_3(x)=\sum_{i=0}^{1}(y_i\varphi_i(x)+y_i'\psi_i(x)).$$

Since $\varphi_0(x_0)=1$, $\varphi_0(x_1)=0$, $\varphi_0'(x_0)=0$, $\varphi_0'(x_1)=0$, we can suppose

$$\varphi_0(x)=(x-x_1)^2[a(x-x_0)+b].$$

From $\varphi_0(x_0)=1$, it follows

$$b=\frac{1}{(x_1-x_0)^2}$$

and $\varphi_0'(x_0)=0$ leads to

$$2b(x_0-x_1)+a(x_1-x_0)^2=0.$$

Solving for a yields

$$a=\frac{2b}{x_1-x_0}=\frac{2}{(x_1-x_0)^3}.$$

As a consequence

$$\varphi_0(x)=(1+2\frac{x-x_0}{x_1-x_0})(\frac{x-x_1}{x_0-x_1})^2.$$

Similarly we can get

$$\varphi_1(x)=(1+2\frac{x-x_1}{x_0-x_1})(\frac{x-x_0}{x_1-x_0})^2,$$

$$\psi_0(x)=(x-x_0)(\frac{x-x_1}{x_0-x_1})^2,$$

$$\psi_1(x)=(x-x_1)(\frac{x-x_0}{x_1-x_0})^2.$$

Solution 2. We can make use of Newton's interpolatory divided-difference formula to construct the Hermite interpolating polynomial. For convenience, set

$$y_i=f(x_i), \quad y_i'=f'(x_i), \quad i=0,1.$$

Let

$$t_0=x_0, \quad t_1=x_0, \quad t_2=x_1, \quad t_3=x_1.$$

Then using Newton's formula gives

$$\begin{aligned}H_3(x)&=f(t_0)+(x-t_0)f[t_0,t_1]+(x-t_0)(x-t_1)f[t_0,t_1,t_2]+(x-t_0)(x-t_1)(x-t_2)f[t_0,t_1,t_2,t_3]\\ &=f(x_0)+(x-x_0)f[x_0,x_0]+(x-x_0)^2f[x_0,x_0,x_1]+(x-x_0)^2(x-x_1)f[x_0,x_0,x_1,x_1]\end{aligned}$$

where

$$f[x_0,x_0]=f'(x_0),$$

$$f[x_0,x_0,x_1]=\frac{f[x_0,x_1]-f[x_0,x_0]}{x_1-x_0}$$
$$=\frac{f(x_1)-f(x_0)-(x_1-x_0)f'(x_0)}{(x_1-x_0)^2},$$

$$f[x_0,x_0,x_1,x_1]=\frac{f[x_0,x_1,x_1]-f[x_0,x_0,x_1]}{x_1-x_0}$$
$$=\frac{1}{(x_1-x_0)^3}\{(x_1-x_0)[f'(x_0)+f'(x_1)]-2[f(x_1)-f(x_0)]\}.$$

Since

$$f(x)=H_3(x)+(x-x_0)^2(x-x_1)^2 f[x_0,x_0,x_1,x_1,x],$$

we see

$$f(x)-H_3(x)=(x-x_0)^2(x-x_1)^2\frac{f^{(4)}(\xi)}{4!},$$

and herewith

$$f'(x_i)-H_3'(x_i)=0,\quad i=0,1;$$
$$f(x_i)-H_3(x_i)=0,\quad i=0,1.$$

which implies that $H_3(x)$ is the unique cubic polynomial satisfying

$$H_3'(x_i)=f'(x_i),\quad i=0,1,$$
$$H_3(x_i)=f(x_i),\quad i=0,1.$$

5.4.2 General Hermite interpolation (一般 Hermite 插值)

Problem Given two groups of discrete data $\{(x_i,f(x_i))\}_{i=0}^{n}$ and $\{(x_i,f'(x_i))\}_{i=0}^{n}$, find out the polynomial $H_{2n+1}(x)$ of degree $2n+1$, so that

$$H_{2n+1}(x_i)=f(x_i),\quad i=0,1,\cdots,n$$

and

$$H_{2n+1}'(x_i)=f'(x_i),\quad i=0,1,\cdots,n.$$

Solution. Let

$$t_{2k}=x_k,\quad t_{2k+1}=x_k$$

for $k=0,1,\cdots,n$.

Using Newton's interpolatory divided-difference formula gives,

$$\begin{aligned}H_{2n+1}(x)=&f(t_0)+(x-t_0)f[t_0,t_1]\\&+(x-t_0)(x-t_1)f[t_0,t_1,t_2]\\&+(x-t_0)(x-t_1)(x-t_2)f[t_0,t_1,t_2,t_3]\\&+(x-t_0)\cdots(x-t_3)f[t_0,t_1,t_2,t_3,t_4]\end{aligned}$$

$$
\begin{aligned}
&+\cdots \\
&+(x-t_0)\cdots(x-t_{2n})f[t_0,\cdots,t_{2n+1}] \\
=\ &f(x_0)+(x-x_0)f[x_0,x_0] \\
&+(x-x_0)^2 f[x_0,x_0,x_1] \\
&+(x-x_0)^2(x-x_1)f[x_0,x_0,x_1,x_1] \\
&+(x-x_0)^2(x-x_1)^2 f[x_0,x_0,x_1,x_1,x_2] \\
&+\cdots \\
&+(x-x_0)^2(x-x_1)^2\cdots(x-x_{n-1})^2(x-x_n)f[x_0,x_0,x_1,x_1,\cdots,x_{n-1},x_{n-1},x_n,x_n]
\end{aligned}
$$

which is a polynomial of degree $2n+1$ satisfying

$$H_{2n+1}(x_i)=f(x_i),\ H'_{2n+1}(x_i)=f'(x_i),\ i=0,1,\cdots,n$$

since

$$f(x)=H_{2n+1}(x)+(x-x_0)^2(x-x_1)^2\cdots(x-x_n)^2 f[x_0,x_0,x_1,x_1,\cdots,x_{n-1},x_{n-1},x_n,x_n,x].$$

The divided differences occurring in $H_{2n+1}(x)$ can be computed by means of the table below

$$
\begin{array}{llllll}
x_0 & f(x_0) & & & & \\
x_0 & f(x_0) & f[x_0,x_0] & & & \\
x_1 & f(x_1) & f[x_0,x_1] & f[x_0,x_0,x_1] & & \\
x_1 & f(x_1) & f[x_1,x_1] & f[x_0,x_1,x_1] & f[x_0,x_0,x_1,x_1] & \\
x_2 & f(x_2) & f[x_1,x_2] & f[x_1,x_1,x_2] & f[x_0,x_1,x_1,x_2] & f[x_0,x_0,x_1,x_1,x_2] \\
x_2 & f(x_2) & f[x_2,x_2] & f[x_1,x_2,x_2] & f[x_1,x_1,x_2,x_2] & f[x_0,x_1,x_1,x_2,x_2] \quad f[x_0,x_0,x_1,x_1,x_2,x_2]
\end{array}
$$

Similarly to the Solution 1 to the previous two-point Hermite interpolation problem, one can also obtain the general Hermit interpolation polynomial via basis functions, as stated in the following theorem.

Theorem 5.4 If $f\in C^1[a,b]$ and $x_0,x_1,\cdots,x_n\in[a,b]$ are distinct, then the unique polynomial of least degree agreeing with f and f' at $x_0,x_1,\cdots,x_n$ is the Hermite polynomial of degree at most $2n+1$ given by

$$H_{2n+1}(x)=\sum_{j=0}^{n}f(x_j)\varphi_j(x)+\sum_{j=0}^{n}f'(x_j)\psi_j(x)$$

where

$$\varphi_j(x)=[1-2(x-x_j)L'_j(x_j)]L_j^2(x)$$

and

$$\psi_j(x)=(x-x_j)L_j^2(x),$$

$$L_j(x)=\frac{\omega_n(x)}{(x-x_j)\omega'_n(x)}$$

with $\omega_n(x)=\prod_{k=0}^{n}(x-x_k)$.

Proof. We try to construct two groups of functions $\{\varphi_j(x)\}_{j=0}^n$ and $\{\psi_j(x)\}_{j=0}^n$, which are all polynomials of degree $2n+1$ and satisfy

$$\varphi_i(x_j)=\delta_{i,j},\quad \varphi_i'(x_j)=0,\quad i,j=0,1,\cdots,n;$$
$$\psi_i'(x_j)=\delta_{i,j},\quad \psi_i(x_j)=0,\quad i,j=0,1,\cdots,n.$$

Then

$$H_{2n+1}(x)=\sum_{j=0}^{n}[f(x_j)\varphi_j(x)+f'(x_j)\psi_j(x)]$$

is the polynomial of degree $2n+1$ satisfying

$$H_{2n+1}(x_i)=f(x_i),\qquad H_{2n+1}'(x_i)=f'(x_i)$$

for $i=0,1,\cdots,n$.

Now for $j=0,1,\cdots,n$,

$$\varphi_j(x_i)=\varphi_j'(x_i)=0,\quad i=0,1,\cdots,j-1,j+1,\cdots,n$$

implies that $\varphi_j(x)$ contains the factor

$$\frac{\prod_{k=0}^{n}(x-x_k)^2}{(x-x_j)^2}=\left[\frac{\omega_n(x)}{x-x_j}\right]^2.$$

Since $\varphi_j(x)$ is a polynomial of degree $2n+1$, we can write

$$\varphi_j(x)=[a_j+b_j(x-x_j)]\frac{\prod_{k=0}^{n}(x-x_k)^2}{(x-x_j)^2}.$$

Using

$$\varphi_j(x_j)=1$$

results in

$$a_j[\omega_n'(x_j)]^2=1,\text{ or }\ a_j=\frac{1}{[\omega_n'(x_j)]^2},$$

and using $\varphi_j'(x_j)=0$ leads to

$$b_j[\omega_n'(x_j)]^2+a_j\cdot 2\frac{\omega_n(x)}{x-x_j}\Big|_{x=x_j}\cdot\left[\frac{\omega_n(x)}{x-x_j}\right]'_{x=x_j}=0.$$

Therefore

$$\begin{aligned}b_j&=-2a_j\cdot\frac{\omega_n(x)}{(x-x_j)\omega_n'(x_j)}\Big|_{x=x_j}\cdot\left[\frac{\omega_n(x)}{(x-x_j)\omega_n'(x_j)}\right]'_{x=x_j}\\&=-2a_jL_j'(x_j)\\&=-2L_j'(x_j)/[\omega_n'(x_j)]^2.\end{aligned}$$

So

$$\varphi_j(x) = [1-2(x-x_j)L_j'(x_j)]\left[\frac{\omega_n(x)}{(x-x_j)\omega_n'(x_j)}\right]^2$$

$$= [1-2(x-x_j)L_j'(x_j)]L_j^2(x).$$

Again for $j=0,1,\cdots,n$,

$$\psi_j(x_i)=0, \qquad i=0,1,\cdots,n;$$

$$\psi_j'(x_i)=0, \qquad i=0,1,\cdots,j-1,j+1,\cdots,n$$

implies

$$\psi_j(x) = c_j(x-x_j)[\frac{\omega_n(x)}{x-x_j}]^2$$

$$= c_j[\omega_n'(x_j)]^2(x-x_j)L_j^2(x).$$

By $\psi_j'(x_j)=1$, we have

$$c_j[\omega_n'(x_j)]^2L_j^2(x_j)=1.$$

Namely

$$c_j = \frac{1}{[\omega_n'(x_j)]^2},$$

which yields

$$\psi_j(x)=(x-x_j)L_j^2(x).$$

The proof is completed.

5.4.3 Example (例题)

Example 5.11 Use the following values to construct the cubic Hermite interpolating polynomial $(i=0,1)$.

x_i	$f(x_i)$	$f'(x_i)$
1	1	3
2	8	12

Solution 1 (**direct method**).

Suppose the cubic Hermite interpolating polynomial $H_3(x)$ is expressed as follows

$$H_3(x)=a_0+a_1x+a_2x^2+a_3x^3.$$

Then using

$$H_3(x_0)=f(x_0), \quad H_3(x_1)=f(x_1);$$

$$H_3'(x_0)=f'(x_0), \quad H_3'(x_1)=f'(x_1)$$

gives

$$a_0 + a_1x_0 + a_2x_0^2 + a_3x_0^3 = f(x_0),$$
$$a_0 + a_1x_1 + a_2x_1^2 + a_3x_1^3 = f(x_1),$$
$$a_1 + 2a_2x_0 + 3a_3x_0^2 = f'(x_0),$$
$$a_1 + 2a_2x_1 + 3a_3x_1^2 = f'(x_1).$$

Substituting $x_0 = 1, x_1 = 2; f(x_0) = 1, f(x_1) = 8; f'(x_0) = 3, f'(x_1) = 12$ into above equations yields the following system of linear equations

$$a_0 + a_1 + a_2 + a_3 = 1,$$
$$a_0 + 2a_1 + 4a_2 + 8a_3 = 8,$$
$$a_1 + 2a_2 + 3a_3 = 3,$$
$$a_1 + 4a_2 + 12a_3 = 12.$$

Solving the above system gives

$$a_0 = 0, a_1 = 0, a_2 = 0, a_3 = 1.$$

As a result, $H_3(x) = x^3$.

Solution 2 (**Newton interpolation method**).

We use Newton's interpolatory divided-difference formula to construct the Hermite polynomial

$$H_3(x) = f(x_0) + (x - x_0)f[x_0, x_0] + (x - x_0)^2 f[x_0, x_0, x_1] + (x - x_0)^2(x - x_1)f[x_0, x_0, x_1, x_1].$$

Let all the involved divided differences be displayed in the following table

1	1			
1	1	$f[x_0, x_0] = 3$		
2	8	$f[x_0, x_1]$	$f[x_0, x_0, x_1]$	
2	8	$f[x_1, x_1] = 12$	$f[x_0, x_1, x_1]$	$f[x_0, x_0, x_1, x_1]$

where

$$f[x_0, x_1] = \frac{f(x_1) - f(x_0)}{x_1 - x_0} = 7,$$

$$f[x_0, x_0, x_1] = \frac{f[x_0, x_1] - f[x_0, x_0]}{x_1 - x_0} = 4,$$

$$f[x_0, x_1, x_1] = \frac{f[x_1, x_1] - f[x_0, x_1]}{x_1 - x_0} = 5,$$

$$f[x_0, x_0, x_1, x_1] = \frac{f[x_0, x_1, x_1] - f[x_0, x_0, x_1]}{x_1 - x_0} = 1.$$

Therefore

$$\begin{aligned}H_3(x) &= f(x_0) + (x - x_0)f[x_0, x_0] + (x - x_0)^2 f[x_0, x_0, x_1] + (x - x_0)^2(x - x_1)f[x_0, x_0, x_1, x_1]\\ &= 1 + 3(x - 1) + 4(x - 1)^2 + (x - 1)^2(x - 2)\\ &= 1 + 3(x - 1) + (x - 1)^2(x+2)\\ &= x^3.\end{aligned}$$

Solution 3 (**basis function method**).

Inspired by the Lagrange interpolation formula, one can write $H_3(x)$ as follows

$$H_3(x)=f(x_0)\varphi_0(x)+f(x_1)\varphi_1(x)+f'(x_0)\psi_0(x)+f'(x_1)\psi_1(x).$$

Now the interpolation conditions

$$H_3(x_0)=f(x_0),\quad H_3(x_1)=f(x_1);$$
$$H_3'(x_0)=f'(x_0),\quad H_3'(x_1)=f'(x_1)$$

turn out to be

$$f(x_0)\varphi_0(x_0)+f(x_1)\varphi_1(x_0)+f'(x_0)\psi_0(x_0)+f'(x_1)\psi_1(x_0)=f(x_0),$$
$$f(x_0)\varphi_0(x_1)+f(x_1)\varphi_1(x_1)+f'(x_0)\psi_0(x_1)+f'(x_1)\psi_1(x_1)=f(x_1),$$
$$f(x_0)\varphi_0'(x_0)+f(x_1)\varphi_1'(x_0)+f'(x_0)\psi_0'(x_0)+f'(x_1)\psi_1'(x_0)=f'(x_0),$$
$$f(x_0)\varphi_0'(x_1)+f(x_1)\varphi_1'(x_1)+f'(x_0)\psi_0'(x_1)+f'(x_1)\psi_1'(x_1)=f'(x_1).$$

To ensure the above equations to be true, one may smartly choose the four cubic basis functions $\varphi_0(x),\varphi_1(x),\psi_0(x),\psi_1(x)$ such that they satisfy the following conditions:

$$\varphi_0(x_0)=1,\quad \varphi_0(x_1)=0,\quad \varphi_0'(x_0)=0,\quad \varphi_0'(x_1)=0;$$
$$\varphi_1(x_0)=0,\quad \varphi_1(x_1)=1,\quad \varphi_1'(x_0)=0,\quad \varphi_1'(x_1)=0;$$
$$\psi_0(x_0)=0,\quad \psi_0(x_1)=0,\quad \psi_0'(x_0)=1,\quad \psi_0'(x_1)=0;$$
$$\psi_1(x_0)=0,\quad \psi_1(x_1)=0,\quad \psi_1'(x_0)=0,\quad \psi_1'(x_1)=1.$$

Using $\varphi_0(x_1)=0$, $\varphi_0'(x_1)=0$, we can write $\varphi_0(x)$ as follows

$$\varphi_0(x)=(x-x_1)^2[a(x-x_0)+b].$$

Substituting $\varphi_0(x_0)=1$ and $\varphi_0'(x_0)=0$ into the above expression results in

$$a=\frac{2}{(x_1-x_0)^3},\quad b=\frac{1}{(x_1-x_0)^2}.$$

Therefore
$$\varphi_0(x)=\left[1+2\frac{x-x_0}{x_1-x_0}\right]\left(\frac{x-x_1}{x_0-x_1}\right)^2.$$

Similarly one can get

$$\varphi_1(x)=\left(1+2\frac{x-x_1}{x_0-x_1}\right)\left(\frac{x-x_0}{x_1-x_0}\right)^2,$$

$$\psi_0(x)=(x-x_0)\left(\frac{x-x_1}{x_0-x_1}\right)^2,$$

$$\psi_1(x)=(x-x_1)\left(\frac{x-x_0}{x_1-x_0}\right)^2.$$

Substituting $x_0=1, x_1=2$ into above expressions gives

$$\varphi_0(x)=(2x-1)(x-2)^2,\quad \varphi_1(x)=(5-2x)(x-1)^2;$$
$$\psi_0(x)=(x-1)(x-2)^2,\quad \psi_1(x)=(x-2)(x-1)^2.$$

Consequently we have

$$\begin{aligned} H_3(x) &= (2x-1)(x-2)^2 + 8(5-2x)(x-1)^2 + 3(x-1)(x-2)^2 + 12(x-2)(x-1)^2 \\ &= x^3. \end{aligned}$$

5.5 Cubic spline interpolation (三次样条插值)

- Runge phenomenon (Runge 现象)
- Piecewise linear interpolation (分段线性插值)
- Piecewise cubic interpolation (分段三次插值)
- Definition of cubic splines (三次样条的定义)
- Derivation of cubic splines (三次样条的推导)
- Examples (例题)

5.5.1 Runge phenomenon (Runge 现象)

Let the interval $[-1,1]$ be equally divided into n subintervals. Denote by $P_n(x)$ the interpolating polynomial of degree n which agrees with $f(x)=\dfrac{1}{1+25x^2}$ as shown in Figure 5.1 at nodes $x_0, x_1, \cdots, x_n$, where $x_i=-1+2i/n$.

Figure 5.2~Figure 5.4 show that $P_n(x)$ provides poorer and poorer approximation to $f(x)$ as n increases, which gives a negative answer to the question: does the interpolating polynomial of higher degree result in better approximation?

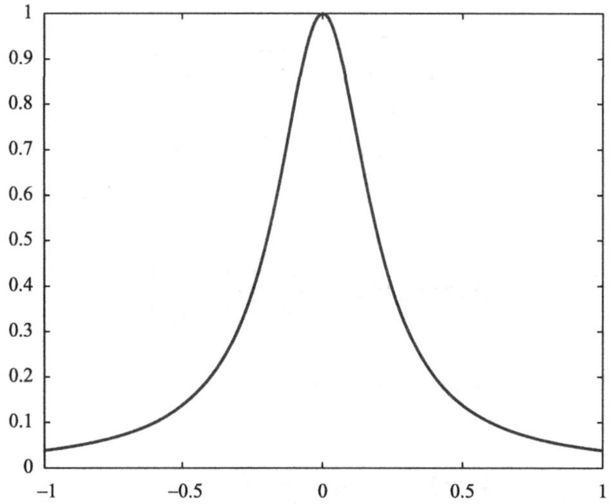

Figure 5.1 The function $f(x)=\dfrac{1}{1+25x^2}$

5.5.2 Piecewise linear interpolation (分段线性插值)

For a given partition of interval $[a,b]$, $\Delta: a=x_0<x_1<\cdots<x_n$, let $h_i=x_{i+1}-x_i$,

$i = 0,1,\cdots,n-1$. Then we can construct a piecewise linear function $S_1(x)$ whose restriction on the subinterval $[x_i, x_{i+1}]$ has the following form

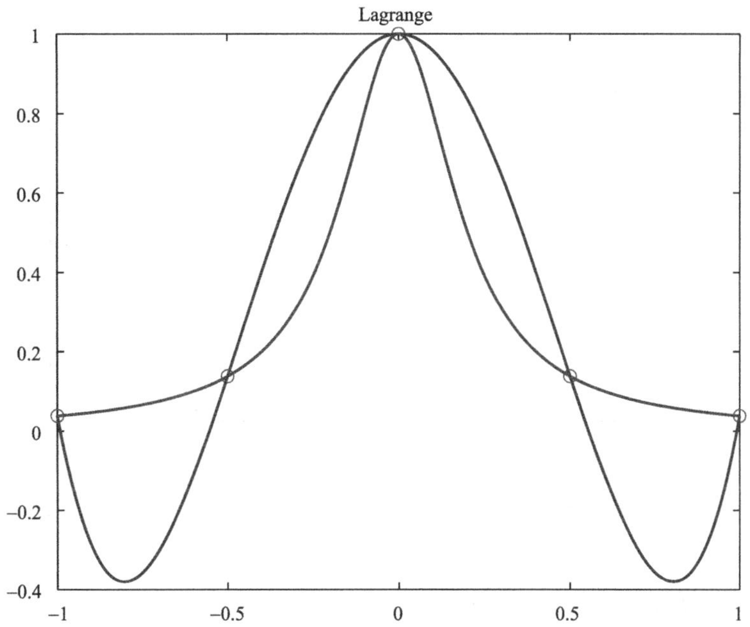

Figure 5.2 The quartic interpolating polynomial for $f(x)=\dfrac{1}{1+25x^2}$

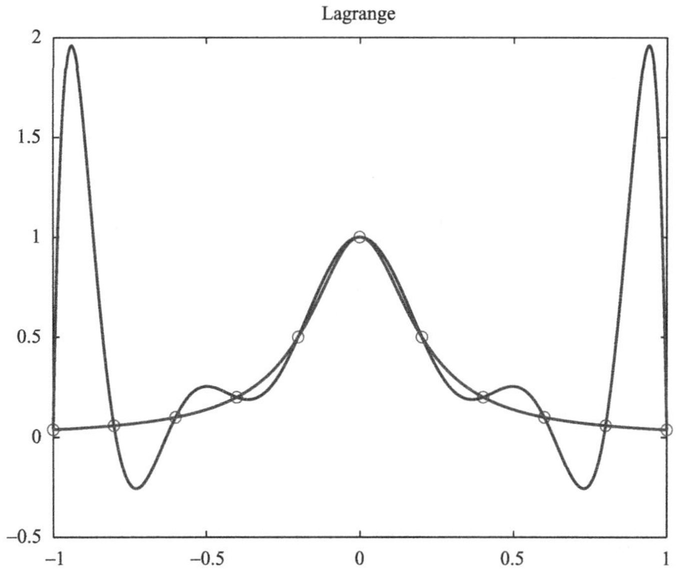

Figure 5.3 The tenth interpolating polynomial for $f(x)=\dfrac{1}{1+25x^2}$

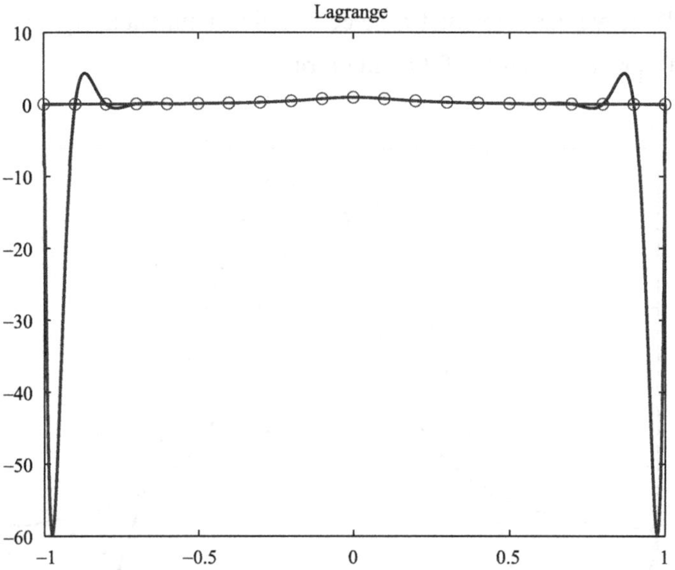

Figure 5.4 The twentieth interpolating polynomial for $f(x)=\dfrac{1}{1+25x^2}$

$$S_1(x)=f(x_i)\frac{x-x_{i+1}}{x_i-x_{i+1}}+f(x_{i+1})\frac{x-x_i}{x_{i+1}-x_i},\quad x_i\leqslant x\leqslant x_{i+1},\quad i=0,1,\cdots,n-1.$$

If $f''(x)$ exists on $[a,b]$, then $\forall x\in[a,b]$, there holds

$$|f(x)-S_1(x)|\leqslant\frac{h^2}{8}\max_{x\in[a,b]}|f''(x)|,\text{ where } h=\max_{0\leqslant i\leqslant n-1}h_i.$$

In fact, for $x\in[x_i,x_{i+1}]$ using the Lagrange interpolation with remainder term gives

$$f(x)-S_1(x)=\frac{f''(\xi_i)}{2!}(x-x_i)(x-x_{i+1}),\quad \xi_i\in(x_i,x_{i+1}).$$

Therefore setting $t=\dfrac{x-x_i}{h_i}$ gives

$$\begin{aligned}|f(x)-S_1(x)|&=\frac{1}{2}|f''(\xi_i)|\cdot|(x-x_i)(x-x_{i+1})|\\&=\frac{1}{2}|f''(\xi_i)|\cdot h_i^2t(1-t)\\&\leqslant\frac{1}{8}h_i^2|f''(\xi_i)|\\&\leqslant\frac{1}{8}h^2\max_{x\in[a,b]}|f''(x)|.\end{aligned}$$

5.5.3 Piecewise cubic interpolation (分段三次插值)

Let $[a,b]=\bigcup_{i=0}^{n-1}[x_i,x_{i+1}]$. Then one can construct a cubic Hermite interpolating polynomial on each $[x_i, x_{i+1}]$

$$S_3(x)=f(x_i)\varphi_i(x)+f(x_{i+1})\varphi_{i+1}(x)+f'(x_i)\psi_i(x)+f'(x_{i+1})\psi_{i+1}(x)$$

where

$$\varphi_i(x)=\left(1+2\frac{x-x_i}{x_{i+1}-x_i}\right)\left(\frac{x-x_{i+1}}{x_i-x_{i+1}}\right)^2=\frac{[h_i+2(x-x_i)](x-x_{i+1})^2}{h_i^3},$$

$$\varphi_{i+1}(x)=\left(1+2\frac{x-x_{i+1}}{x_i-x_{i+1}}\right)\left(\frac{x-x_i}{x_{i+1}-x_i}\right)^2=\frac{[h_i+2(x_{i+1}-x)](x-x_i)^2}{h_i^3},$$

$$\psi_i(x)=(x-x_i)\left(\frac{x-x_{i+1}}{x_i-x_{i+1}}\right)^2=\frac{1}{h_i^2}(x-x_i)(x-x_{i+1})^2,$$

$$\psi_{i+1}(x)=(x-x_{i+1})\left(\frac{x-x_i}{x_{i+1}-x_i}\right)^2=\frac{1}{h_i^2}(x-x_{i+1})(x-x_i)^2$$

or for $x\in[x_i,x_{i+1}]$

$$S_3(x)=f(x_i)+(x-x_i)f[x_i,x_i]+(x-x_i)^2 f[x_i,x_i,x_{i+1}]$$
$$+(x-x_i)^2(x-x_{i+1})f[x_i,x_i,x_{i+1},x_{i+1}].$$

If $f(x)\in C^4[a,b]$, then $\forall x\in[a,b]$, $|f(x)-S_3(x)|\leqslant\frac{h^4}{384}\max_{x\in[a,b]}\left|f^{(4)}(x)\right|$, where $h=\max_{0\leqslant i\leqslant n-1}h_i$.

As a matter of fact, using the Newton's interpolatory divided-difference formula with remainder term gives

$$f(x)-S_3(x)=(x-x_i)^2(x-x_{i+1})^2 f[x_i,x_i,x_{i+1},x_{i+1},x]$$
$$=(x-x_i)^2(x-x_{i+1})^2\frac{f^{(4)}(\xi_i)}{4!},\xi_i\in[x_i,x_{i+1}],$$

which leads to

$$|f(x)-S_3(x)|=(x-x_i)^2(x-x_{i+1})^2\left|\frac{f^{(4)}(\xi_i)}{4!}\right|$$
$$=h_i^4t^2(1-t)^2\cdot\frac{1}{4!}\left|f^{(4)}(\xi_i)\right|$$
$$\leqslant\frac{1}{16}h_i^4\cdot\frac{1}{24}\left|f^{(4)}(\xi_i)\right|$$
$$\leqslant\frac{h^4}{384}\max_{x\in[a,b]}\left|f^{(4)}(x)\right|.$$

Figure 5.5 and Figure 5.6 show that piecewise cubic interpolation can provide a good approximation to the given function.

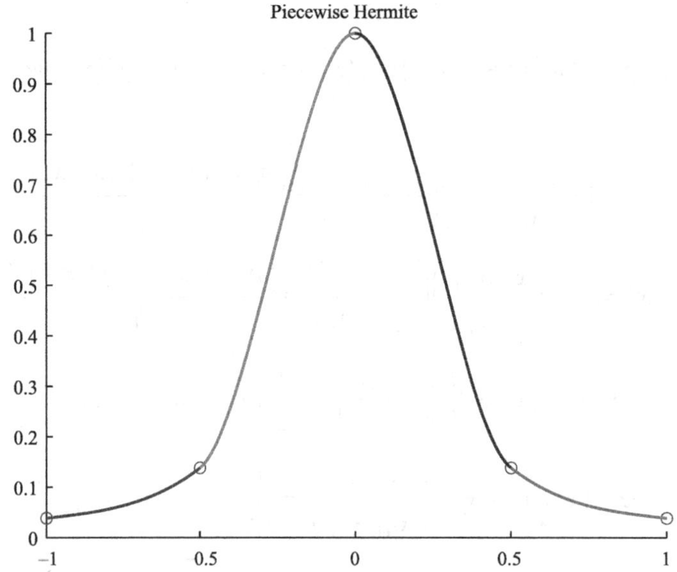

Figure 5.5 Piecewise cubic interpolating polynomials for $f(x)=\dfrac{1}{1+25x^2}$

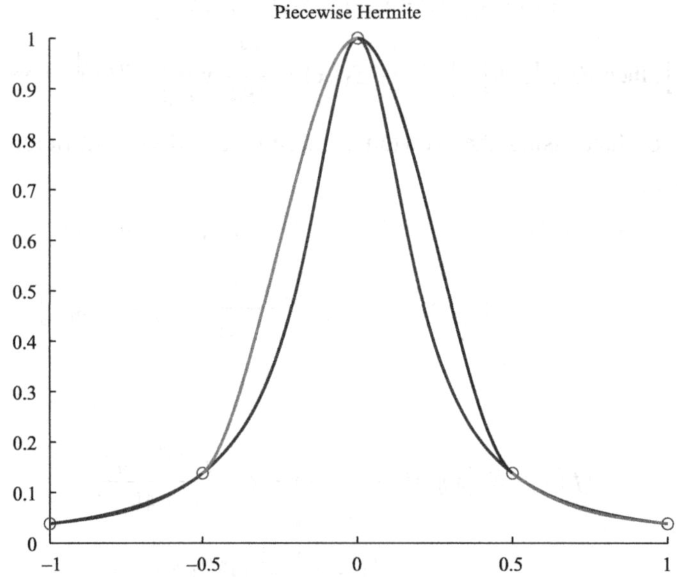

Figure 5.6 The comparison between the piecewise cubic polynomials and $f(x)=\dfrac{1}{1+25x^2}$

5.5.4 Definition of cubic splines (三次样条的定义)

For a partition $\Delta: a=x_0<x_1<\cdots<x_n=b$. If $S(x)$ is a cubic polynomial on each subinterval $[x_i,x_{i+1}]$ and $S(x)\in C^2[a,b]$, then $S(x)$ is called a cubic spline with respect to the knots (nodes)(结点) $x_0,x_1,\cdots,x_n$. If $S(x)$ agrees with a function $f(x)$ defined on $[a,b]$ at the

$n+1$ nodes $x_0, x_1, \cdots, x_n$, namely $S(x_i) = f(x_i), i = 0,1,\cdots,n$, then $S(x)$ is said to be a cubic spline interpolant to $f(x)$.

From the definition, we see that to solve for $S(x)$, one needs to determine $4n$ unknowns. A cubic spline interpolant $S(x)$ to $f(x)$ satisfies the following conditions

(1) $S(x_i) = f(x_i), \quad i = 0,1,\cdots,n;$

(2) $S(x_i - 0) = S(x_i + 0), \quad i = 1,2,\cdots,n-1;$

(3) $S'(x_i - 0) = S'(x_i + 0), \quad i = 1,2,\cdots,n-1;$

(4) $S''(x_i - 0) = S''(x_i + 0), \quad i = 1,2,\cdots,n-1.$

which offer $4n-2$ conditions. We need two other conditions. Usually we can use the following boundary conditions.

Clamped boundary (固支边界)

$$S'(x_0) = f'(x_0), \quad S'(x_n) = f'(x_n).$$

Free or natural boundary (自由边界或自然边界)

$$S''(x_0) = 0, \quad S''(x_n) = 0.$$

5.5.5 Derivation of cubic splines (三次样条的推导)

M-method: The first step for the M-method is based on the fact that the second derivative $S''(x)$ of the cubic spline $S(x)$ within each interval is a linear polynomial since each pair of knots is connected by a cubic polynomial. On this basis, the second derivatives can be represented by a first-order Lagrange interpolating polynomial as follows

$$S''(x) = M_i \frac{x - x_{i+1}}{x_i - x_{i+1}} + M_{i+1} \frac{x - x_i}{x_{i+1} - x_i} = \frac{1}{h_i}\left[M_{i+1}(x - x_i) - M_i(x - x_{i+1})\right] \tag{5.5.1}$$

where $M_i = S''(x_i), x \in [x_i, x_{i+1}]$.

Next, Eq. (5.5.1) can be integrated twice to yield an expression for $S(x)$:

$$S(x) = \frac{1}{6h_i}\left[M_{i+1}(x - x_i)^3 - M_i(x - x_{i+1})^3\right] + c_i(x - x_i) + d_i(x - x_{i+1}).$$

Two unknowns c_i and d_i can be determined by using the interpolation conditions:

$$S(x_i) = y_i = f(x_i), \quad S(x_{i+1}) = y_{i+1} = f(x_{i+1}).$$

Performing these evaluations results in the following cubic equation:

$$\begin{aligned} S(x) = \frac{1}{6h_i}\Big[&M_{i+1}(x - x_i)^3 - M_i(x - x_{i+1})^3 + \left(6y_{i+1} - h_i^2 M_{i+1}\right)(x - x_i) \\ &+ \left(h_i^2 M_i - 6y_i\right)(x - x_{i+1})\Big], \quad x \in [x_i, x_{i+1}], \quad i = 0,1,\cdots,n-1. \end{aligned} \tag{5.5.2}$$

To solve for $M_i, i = 0,1,\cdots,n$, we can use the continuity condition for first derivative of $S(x)$:

$$S'(x_i - 0) = S'(x_i + 0), \quad i = 1,2,\cdots,n-1.$$

Note that for $x \in [x_i, x_{i+1}]$

$$S'(x)=\frac{1}{2h_i}\left[M_{i+1}(x-x_i)^2-M_i(x-x_{i+1})^2\right]+f\left[x_i,x_{i+1}\right]+\frac{h_i}{6}(M_i-M_{i+1}) \tag{5.5.3}$$

and for $x\in[x_{i-1},x_i]$

$$S'(x)=\frac{1}{2h_{i-1}}\left[M_i(x-x_{i-1})^2-M_{i-1}(x-x_i)^2\right]+f\left[x_{i-1},x_i\right]+\frac{h_{i-1}}{6}(M_{i-1}-M_i). \tag{5.5.4}$$

We have

$$S'(x_i+0)=f\left[x_i,x_{i+1}\right]-\frac{1}{3}h_iM_i-\frac{1}{6}h_iM_{i+1}, \tag{5.5.5}$$

$$S'(x_i-0)=f\left[x_{i-1},x_i\right]+\frac{1}{6}h_{i-1}M_{i-1}+\frac{1}{3}h_{i-1}M_i. \tag{5.5.6}$$

Therefore $S'(x_i+0)=S'(x_i-0)$ implies

$$f\left[x_i,x_{i+1}\right]-\frac{1}{3}h_iM_i-\frac{1}{6}h_iM_{i+1}=f\left[x_{i-1},x_i\right]+\frac{1}{6}h_{i-1}M_{i-1}+\frac{1}{3}h_{i-1}M_i,$$

or

$$\frac{1}{6}h_{i-1}M_{i-1}+\frac{1}{3}h_{i-1}M_i+\frac{1}{3}h_iM_i+\frac{1}{6}h_iM_{i+1}=f\left[x_i,x_{i+1}\right]-f\left[x_{i-1},x_i\right],$$

or for $i=1,2,\cdots,n-1$

$$\frac{h_{i-1}}{h_{i-1}+h_i}M_{i-1}+2M_i+\frac{h_i}{h_{i-1}+h_i}M_{i+1}=\frac{6\left(f\left[x_i,x_{i+1}\right]-f\left[x_{i-1},x_i\right]\right)}{h_{i-1}+h_i}. \tag{5.5.7}$$

Let $\mu_i=\dfrac{h_{i-1}}{h_{i-1}+h_i},\lambda_i=\dfrac{h_i}{h_{i-1}+h_i}$. Then for $i=1,2,\cdots,n-1$, Eq. (5.5.7) can be rewritten as

$$\mu_iM_{i-1}+2M_i+\lambda_iM_{i+1}=6f\left[x_{i-1},x_i,x_{i+1}\right]. \tag{5.5.8}$$

Case 1　For the clamped boundary

$$S'(x_0)=f'(x_0),\quad S'(x_n)=f'(x_n)$$

It follows from (5.5.5) and (5.5.6)

$$f'(x_0)=S'(x_0)=S'(x_0+0)=f\left[x_0,x_1\right]-\frac{1}{3}h_0M_0-\frac{1}{6}h_0M_1,$$

$$f'(x_n)=S'(x_n)=S'(x_n-0)=f\left[x_{n-1},x_n\right]+\frac{1}{6}h_{n-1}M_{n-1}+\frac{1}{3}h_{n-1}M_n,$$

which are equivalent to

$$\begin{aligned}2M_0+M_1&=6\frac{f\left[x_0,x_1\right]-f'(x_0)}{h_0}\\&=6\frac{f[x_0,x_1]-f'(x_0)}{x_1-x_0}\\&=6f[x_0,x_0,x_1],\end{aligned} \tag{5.5.9}$$

$$\begin{aligned} M_{n-1}+2M_n &= 6\frac{f'(x_n)-f[x_{n-1},x_n]}{h_{n-1}} \\ &= 6\frac{f[x_n,x_n]-f[x_{n-1},x_n]}{x_n-x_{n-1}} \\ &= 6f[x_{n-1},x_n,x_n]. \end{aligned} \tag{5.5.10}$$

Now set

$$\begin{aligned} g_0 &= 6f[x_0,x_0,x_1], \\ g_n &= 6f[x_{n-1},x_n,x_n], \\ g_i &= 6f[x_{i-1},x_i,x_{i+1}], \quad i=1,2,\cdots,n-1. \end{aligned}$$

Combining Eq. (5.5.8) , (5.5.9) and (5.5.10) produces a linear system described by the vector equation $\boldsymbol{Ax}=\boldsymbol{b}$, where $\boldsymbol{A}$ is the $(n+1)\times(n+1)$ matrix as follows

$$\boldsymbol{A}=\begin{pmatrix} 2 & 1 & 0 & & & & \\ \mu_1 & 2 & \lambda_1 & & & & \\ 0 & \mu_2 & 2 & \lambda_2 & & & \\ & \ddots & \ddots & \ddots & & & \\ & & & \ddots & \ddots & \ddots & \\ & & & \mu_{n-2} & 2 & \lambda_{n-2} & 0 \\ & & & & \mu_{n-1} & 2 & \lambda_{n-1} \\ & & & & 0 & 1 & 2 \end{pmatrix}$$

$$\boldsymbol{x}=(M_0,M_1,\cdots,M_n)^{\mathrm{T}}, \quad \boldsymbol{b}=(g_0,g_1,\cdots,g_n)^{\mathrm{T}}.$$

The matrix $\boldsymbol{A}$ is strictly diagonally dominant and must be nonsingular.

Case 2 For the natural boundary

$s''=(x_0)=0, s''=(x_n)=0$, namely, $M_0=M_n=0$. Substituting $M_0=M_n=0$ into (5.5.8) produces the following system of linear equations

$$\begin{pmatrix} 2 & \lambda_1 & & & & & \\ \mu_2 & 2 & \lambda_2 & & & & \\ & \mu_3 & 2 & \lambda_3 & & & \\ & & \ddots & \ddots & \ddots & & \\ & & & \ddots & \ddots & \ddots & \\ & & & \mu_{n-3} & 2 & \lambda_{n-3} & \\ & & & & \mu_{n-2} & 2 & \lambda_{n-2} \\ & & & & 0 & \mu_{n-1} & 2 \end{pmatrix} \cdot \begin{pmatrix} M_1 \\ M_2 \\ \vdots \\ \vdots \\ \vdots \\ \vdots \\ \vdots \\ M_{n-1} \end{pmatrix} = \begin{pmatrix} g_1 \\ g_2 \\ \vdots \\ \vdots \\ \vdots \\ \vdots \\ \vdots \\ g_{n-1} \end{pmatrix}.$$

5.5.6 Examples (例题)

Example 5.12 Find natural cubic splines to the data given in the following table

	x_0	x_1	x_2	x_3
x	3.0	4.5	7.0	9.0
$f(x)$	2.5	1.0	2.5	0.5

Solution. From the table, one gets

$$h_0 = x_1 - x_0 = 1.5,\ h_1 = x_2 - x_1 = 2.5,\ h_2 = x_3 - x_2 = 2,\ M_0 = 0,\ M_3 = 0.$$

The first step is to employ Eq. (5.5.8) to generate the set of simultaneous equations that will be utilized to determine the second derivatives at the knots. In this example, we want to solve for M_i by the following equations $(i = 1,2)$:

$$\mu_i M_{i-1} + 2M_i + \lambda_i M_{i+1} = 6f[x_{i-1}, x_i, x_{i+1}],$$

that is

$$\begin{cases} \mu_1 M_0 + 2M_1 + \lambda_1 M_2 = 6f[x_0, x_1, x_2], \\ \mu_2 M_1 + 2M_2 + \lambda_2 M_3 = 6f[x_1, x_2, x_3], \end{cases} \tag{A}$$

where

$$\mu_1 = \frac{h_0}{h_0 + h_1} = \frac{3}{8},\ \mu_2 = \frac{h_1}{h_1 + h_2} = \frac{5}{9},\ \lambda_1 = 1 - \mu_1 = \frac{5}{8},\ \lambda_2 = 1 - \mu_2 = \frac{4}{9},$$

$$f[x_0, x_1, x_2] = \frac{f[x_1, x_2] - f[x_0, x_1]}{x_2 - x_1} = \frac{2}{5},\quad f[x_1, x_2, x_3] = \frac{f[x_2, x_3] - f[x_1, x_2]}{x_3 - x_1} = -\frac{16}{45}.$$

Substituting all the above together with $M_0 = M_3 = 0$ into (*A*) yields

$$\begin{cases} 2M_1 + \dfrac{5}{8}M_2 = \dfrac{12}{5}, \\ \dfrac{5}{9}M_1 + 2M_2 = -\dfrac{32}{15}. \end{cases} \tag{B}$$

Solving for M_1 and M_2 gives

$$M_1 = \frac{2208}{1315} = 1.67909,\ M_2 = -\frac{2016}{1315} = -1.53308.$$

By (5.5.2) we have

$$S(x) = \frac{1}{6h_i}\Big[(x - x_i)^3 M_{i+1} - (x - x_{i+1})^3 M_i + (6y_{i+1} - h_i^2 M_{i+1})(x - x_i)$$
$$+ (h_i^2 M_i - 6y_i)(x - x_{i+1})\Big],\ x \in (x_i, x_{i+1}),\ i = 0,1,2.$$

Let $S_i(x) = S(x),\ x \in (x_i, x_{i+1})$. Then we have

$$S_0(x) = \frac{1}{6h_0}\left\{(x - x_0)^3 M_1 - (x - x_1)^3 M_0 + \left[6y_1 - h_0^2 M_1\right](x - x_0) + \left[h_0^2 M_0 - 6y_0\right](x - x_1)\right\}$$
$$= \frac{1}{9}\left[\frac{2208}{1315}(x-3)^3 + \left(6 - \frac{9}{4}\cdot\frac{2208}{1315}\right)(x-3) - 15(x - 4.5)\right]$$
$$= 0.186565x^3 - 1.67909x^2 + 3.61749x + 1.72205,\quad x \in [3, 4.5];$$

$$S_1(x)=\frac{1}{6h_1}\left\{(x-x_1)^3M_2-(x-x_2)^3M_1+\left[6y_2-h_1^2M_2\right](x-x_1)+\left[h_1^2M_1-6y_1\right](x-x_2)\right\}$$
$$=-0.214144x^3+3.73049x^2-20.7256x+38.2367,\ x\in[4.5,7];$$
$$S_2(x)=\frac{1}{6h_2}\left\{(x-x_2)^3M_3-(x-x_3)^3M_2+\left[6y_3-h_2^2M_3\right](x-x_2)+\left[h_2^2M_2-6y_2\right](x-x_3)\right\}$$
$$=0.127757x^3-3.44943x^2+29.5338x-79.0354,\ x\in[7,9].$$

Example 5.13 Suppose

$$f(x)=\frac{1}{1+25x^2},$$

find out the cubic splines with natural boundary and clamped boundary respectively based on the following data.

x_i	-1	-0.5	0	0.5	1
$f(x_i)$	$\frac{1}{26}$	$\frac{4}{29}$	1	$\frac{4}{29}$	$\frac{1}{26}$

Solution. **Case 1 natural boundary**

First of all we establish the following tableau of divided differences

i	x_i	$f(x_i)$		
0	-1	$\frac{1}{26}$		
1	$-\frac{1}{2}$	$\frac{4}{29}$	$f[x_0,x_1]=\frac{75}{377}$	
2	0	1	$f[x_1,x_2]=\frac{50}{29}$	$f[x_0,x_1,x_2]=\frac{575}{377}$
3	$\frac{1}{2}$	$\frac{4}{29}$	$f[x_2,x_3]=-\frac{50}{29}$	$f[x_1,x_2,x_3]=-\frac{100}{29}$
4	1	$\frac{1}{26}$	$f[x_3,x_4]=-\frac{75}{377}$	$f[x_2,x_3,x_4]=\frac{575}{377}$

Since $h_i=\frac{1}{2}$, $i=0,1,2,3$, and $M_0=M_4=0$, we have

$$\mu_1=\mu_2=\mu_3=\lambda_1=\lambda_2=\lambda_3=\frac{1}{2}.$$

Therefore the system of equations

$$\begin{cases}\mu_1M_0+2M_1+\lambda_1M_2=6f[x_0,x_1,x_2]\\ \mu_2M_1+2M_2+\lambda_2M_3=6f[x_1,x_2,x_3]\\ \mu_3M_2+2M_3+\lambda_3M_4=6f[x_2,x_3,x_4]\end{cases}$$

turns out to be

$$\begin{cases} 2M_1 + \frac{1}{2}M_2 & = \frac{3450}{377}, \\ \frac{1}{2}M_1 + 2M_2 + \frac{1}{2}M_3 & = -\frac{600}{29}, \\ \frac{1}{2}M_2 + 2M_3 & = \frac{3450}{377}. \end{cases}$$

Solving for M_1, M_2 and M_3 gives

$$\begin{cases} M_1 = \frac{21600}{2639} = 8.18492, \\ M_2 = -\frac{38100}{2639} = -1.44372, \\ M_3 = \frac{21600}{2639} = 8.18492. \end{cases}$$

As a result,

$$S_0(x) = \frac{1}{3}\left[\frac{21600}{2639}(x+1)^3 + \left(\frac{24}{29} - \frac{1}{4}\cdot\frac{21600}{2639}\right)(x+1) - \frac{3}{13}\left(x+\frac{1}{2}\right)\right],\ x \in \left[-1, -\frac{1}{2}\right]$$

$$S_1(x) = \frac{1}{3}\left[-\frac{38100}{2639}\left(x+\frac{1}{2}\right)^3 - \frac{21600}{2639}x^3 + \left(6 + \frac{1}{4}\cdot\frac{38100}{2639}\right)\left(x+\frac{1}{2}\right) + \left(\frac{1}{4}\cdot\frac{38100}{2639} - \frac{24}{29}\right)x\right],$$

$$x \in \left[-\frac{1}{2}, 0\right]$$

$$S_2(x) = \frac{1}{3}\left[\frac{21600}{2639}x^3 + \frac{38100}{2639}\left(x-\frac{1}{2}\right)^3 + \left(\frac{24}{29} - \frac{1}{4}\cdot\frac{21600}{2639}\right)x - \left(\frac{1}{4}\cdot\frac{38100}{2639} + 6\right)\left(x-\frac{1}{2}\right)\right],$$

$$x \in \left[0, \frac{1}{2}\right]$$

$$S_3(x) = \frac{1}{3}\left[-\frac{21600}{2639}(x-1)^3 + \frac{3}{13}\left(x-\frac{1}{2}\right) + \left(\frac{1}{4}\cdot\frac{21600}{2639} - \frac{24}{29}\right)(x-1)\right],\ x \in \left[\frac{1}{2}, 1\right].$$

Case 2　clamped boundary (固支边界)

In this case, we get the following equations

$$\begin{cases} 2M_0 + M_1 & = \frac{3450}{377}, \\ M_0 + 4M_1 + M_2 & = \frac{6900}{377}, \\ M_1 + 4M_2 + M_3 & = -\frac{1200}{29}, \\ M_2 + 4M_3 + M_4 & = \frac{6900}{377}, \\ \frac{1}{2}M_2 + 2M_3 & = \frac{3450}{377}. \end{cases}$$

Solving for M_i, $i = 0,1,2,3,4$ gives

$$M_0 = \frac{425}{754}, M_1 = \frac{3025}{377}, M_2 = -\frac{10825}{75}, M_3 = \frac{3025}{377}, M_4 = \frac{425}{754}.$$

Example 5.14 (**m-method**, a method based on the Newton's interpolatory divided-difference formula)

Given the following data

x_i	0	1	2	3
$f(x_i)$	0	0	0	0

Find the cubic spline interpolant $S(x)$ with boundary conditions

$$S'(x_0) = f'(x_0) = 1, S'(x_3) = f'(x_3) = 0.$$

Solution. Let

$$S'(x_i) = m_i, i = 0,1,2,3.$$

Then $m_0 = 1, m_3 = 0$. Suppose $S_i(x)$ is the expression of $S(x)$ on $[x_i, x_{i+1}] = [i, i+1]$ for $i = 0,1,2$, then using Newton's interpolatory divided-difference formula yields

$$\begin{aligned} S_i(x) &= f(x_i) + (x - x_i)f[x_i, x_i] + (x - x_i)^2 f[x_i, x_i, x_{i+1}] + (x - x_i)^2 (x - x_{i+1}) f[x_i, x_i, x_{i+1}, x_{i+1}] \\ &= f(i) + (x - i)f[i,i] + (x - i)^2 f[i,i,i+1] + (x - i)^2 (x - i - 1) f[i,i,i+1,i+1], \end{aligned}$$

where

$$f[i,i] = f'(i) = m_i,$$

$$f[i,i,i+1] = \frac{f[i,i+1] - f[i,i]}{i+1-i} = f(i+1) - f(i) - m_i = -m_i,$$

$$f[i,i+1,i+1] = \frac{f[i+1,i+1] - f[i,i+1]}{i+1-i} = m_{i+1} - f(i+1) + f(i) = m_{i+1},$$

$$f[i,i,i+1,i+1] = \frac{f[i,i+1,i+1] - f[i,i,i+1]}{i+1-i} = m_{i+1} + m_i.$$

Therefore

$$\begin{aligned} S_i(x) &= m_i(x - i) - m_i(x - i)^2 + (m_{i+1} + m_i)(x - i)^2 (x - i - 1) \\ &= -m_i(x - i)(x - i - 1) + (m_{i+1} + m_i)(x - i)^2 (x - i - 1) \\ &= (x - i)(x - i - 1)[(m_{i+1} + m_i)(x - i) - m_i], i = 0,1,2. \end{aligned}$$

Differentiating $S_i(x)$ twice results in

$$S_i'(x) = m_i - 2m_i(x - i) + (m_i + m_{i+1})(x - i)^2 + 2(m_i + m_{i+1})(x - i)(x - i - 1),$$

$$S_i''(x) = -2m_i + 2(m_i + m_{i+1})(x - i) + 2(m_i + m_{i+1})(2x - 2i - 1).$$

$S(x) \in C^2$ implies

$$S_i''(x_{i+1}) = S_{i+1}''(x_{i+1}), \quad i = 0,1,$$

or

$$S_i''(i+1)=S_{i+1}''(i+1),\quad i=0,1,$$

$$-2m_i+2(m_i+m_{i+1})(i+1-i)+2(m_i+m_{i+1})(2i-2-2i-1)=-2m_{i+1}-2(m_{i+1}+m_{i+2}),i=0,1,$$

namely,

$$m_i+4m_{i+1}+m_{i+2}=0,\quad i=0,1,$$

or in details

$$\begin{cases}4m_1+m_2=-1,\\ m_1+4m_2=0.\end{cases}$$

Solving for m_1, m_2 gives

$$m_1=-\frac{4}{15},\quad m_2=\frac{1}{15}.$$

Consequently

$$S_0(x)=x(x-1)(\frac{11}{15}x-1)=\frac{1}{15}x(x-1)(11x-15),\quad x\in[0,1];$$

$$S_1(x)=\frac{1}{15}(x-1)(x-2)(7-3x),\quad x\in[1,2];$$

$$S_2(x)=\frac{1}{15}(x-3)^2(x-2),\quad x\in[2,3].$$

Example 5.15 A natural cubic spline S on $[0,2]$ is defined by

$$S_0(x)=1+2x-x^3,\qquad 0\leqslant x\leqslant 1;$$

$$S_1(x)=2+b(x-1)+c(x-1)^2+d(x-1)^3,\qquad 1\leqslant x\leqslant 2.$$

Find b, c and d.

Solution. From $S(x)\in C^2[0,2]$, it follows

$$S_0'(1)=S_1'(1),\quad S_0''(1)=S_1''(1).$$

The first equation is used to determine $b=-1$, and the second equation serves to solve for $c=-3$. $S(x)$ is a natural cubic spline on $[0,2]$ implies

$$S_0''(0)=0,\quad S_1''(2)=0.$$

The former is obvious and the latter means

$$2c+6d=0.$$

Therefore $d=-\frac{1}{3}c=1$.

Example 5.16 Determine the free cubic spline $s(x)$ that interpolates the data $f(0)=0$, $f(1)=1$, and $f(2)=2$.

Solution 1 (**direct method**). Suppose

$$s(x)=\begin{cases}s_0(x), & x\in[0,1];\\ s_1(x), & x\in[1,2].\end{cases}$$

Then $s_0(x)$ and $s_1(x)$ can be written as

$$s_0(x) = a_0 + a_1(x-1) + a_2(x-1)^2 + a_3(x-1)^3, \tag{5.5.11}$$

$$s_1(x) = b_0 + b_1(x-1) + b_2(x-1)^2 + b_3(x-1)^3. \tag{5.5.12}$$

Since the cubic spline $s(x)$ interpolates the data $f(0)=0, f(1)=1$, and $f(2)=2$, we have $s_0(0)=0, s_0(1)=s_1(1)=1$, and $s_1(2)=2$ which yields the following system of linear equations

$$\begin{cases} a_0 - a_1 + a_2 - a_3 = 0, \\ a_0 = 1, \\ b_0 = 1, \\ b_0 + b_1 + b_2 + b_3 = 2. \end{cases} \tag{5.5.13}$$

Since $s(x)$ is a free cubic spline, we have

$$s''(0) = 0, \quad s''(2) = 0.$$

Namely

$$s_0''(0) = 0, \quad s_1''(2) = 0.$$

From $s_0''(0) = 0$ it follows

$$a_2 = 3a_3. \tag{5.5.14}$$

From $s_1''(2) = 0$ it follows

$$b_2 = -3b_3. \tag{5.5.15}$$

The fact that $s(x)$ is a cubic spline implies

$$\begin{aligned} s_0(1-0) &= s_1(1+0), \\ s_0'(1-0) &= s_1'(1+0), \\ s_0''(1-0) &= s_1''(1+0), \end{aligned} \tag{5.5.16}$$

which gives

$$\begin{cases} a_0 = b_0, \\ a_1 = b_1, \\ a_2 = b_2. \end{cases} \tag{5.5.17}$$

Combining (5.5.13), (5.5.14), (5.5.15) and (5.5.17) gives

$$\begin{cases} a_0 = a_1 = 1, \\ a_2 = a_3 = 0, \\ b_0 = b_1 = 1, \\ b_2 = b_3 = 0. \end{cases} \tag{5.5.18}$$

Therefore

$$s_0(x) = 1 + x - 1 = x, \quad s_1(x) = 1 + x - 1 = x.$$

As a result,

$$s(x) = x, \quad x \in [0,2].$$

Solution 2 (**M-method**). Suppose

$$s(x)=\begin{cases} s_0(x), & x\in[0,1];\\ s_1(x), & x\in[1,2].\end{cases}$$

Let

$$M_0=s''(0),\quad M_1=s''(1),\quad M_2=s''(2).$$

Then

$$s_0''(x)=M_0\frac{x-1}{0-1}+M_1\frac{x}{1-0}=M_1x,$$

$$s_1''(x)=M_1\frac{x-2}{1-2}+M_2\frac{x-1}{2-1}=-M_1(x-2),$$

where the free boundary conditions $M_0=0$ and $M_2=0$ are used. Integrating $s_0''(x)$ and $s_1''(x)$ twice respectively results in

$$s_0(x)=\frac{1}{6}M_1x^3+ax+b(x-1),$$

$$s_1(x)=-\frac{1}{6}M_1(x-2)^3+c(x-1)+d(x-2).$$

Since $s(x)$ interpolates the data $f(0)=0, f(1)=1$, and $f(2)=2$,we have

$$s_0(0)=f(0)=0,\quad s_0(1)=f(1)=1;$$

$$s_1(1)=f(1)=1,\quad s_1(2)=f(2)=2.$$

Substituting the above interpolation conditions into the expressions of $s_0(x)$ and $s_1(x)$ respectively yields

$$a=1-\frac{1}{6}M_1, b=0;$$

$$c=2, d=\frac{1}{6}M_1-1.$$

As a result,

$$s_0(x)=\frac{1}{6}M_1x^3+(1-\frac{1}{6}M_1)x,$$

$$s_1(x)=-\frac{1}{6}M_1(x-2)^3+2(x-1)+(\frac{1}{6}M_1-1)(x-2).$$

Using $s'(1-0)=s'(1+0)$, namely, $s_0'(1)=s_1'(1)$, gives

$$1+\frac{1}{3}M_1=1-\frac{1}{3}M_1$$

which implies $M_1=0$. Therefore

$$s(x)=x,\quad x\in[0,2].$$

Solution 3 (**m-method**). Suppose

$$s(x)=\begin{cases} s_0(x), & x\in[0,1];\\ s_1(x), & x\in[1,2]. \end{cases}$$

Let

$$m_0=s'(0),\quad m_1=s'(1),\quad m_2=s'(2).$$

Then one can establish the tables of divided differences for $s_0(x)$ and $s_1(x)$ as follows

$0\ \ s_0(0)=0$

$0\ \ s_0(0)=0\ \ s_0[0,0]=m_0$

$1\ \ s_0(1)=1\ \ s_0[0,1]=1\quad s_0[0,0,1]=1-m_0$

$1\ \ s_0(1)=1\ \ s_0[1,1]=m_1\quad s_0[0,1,1]=m_1-1\quad s_0[0,0,1,1]=m_1+m_0-2$

$1\ \ s_1(1)=1$

$1\ \ s_1(1)=1\ \ s_1[1,1]=m_1$

$2\ \ s_1(2)=2\ \ s_1[1,2]=1\quad s_1[1,1,2]=1-m_1$

$2\ \ s_1(2)=2\ \ s_1[2,2]=m_2\ \ s_1[1,2,2]=m_2-1\quad s_1[1,1,2,2]=m_1+m_2-2$

By Newton's interpolation formula, we have

$$s_0(x)=m_0x+(1-m_0)x^2+(m_0+m_1-2)x^2(x-1),$$
$$s_1(x)=1+m_1(x-1)+(1-m_1)(x-1)^2+(m_1+m_2-2)(x-1)^2(x-2).$$

Hence

$$s_0''(x)=2(1-m_0)+(m_0+m_1-2)(6x-2),$$
$$s_1''(x)=2(1-m_1)+(m_1+m_2-2)(6x-8).$$

Using $s_0''(0)=0, s_1''(2)=0$ and $s_0''(1-0)=s_1''(1+0)$, we get the following equations

$$2m_0+m_1=3,$$
$$m_1+2m_2=3,$$
$$m_0+4m_1+m_2=6.$$

Solving the above equations for m_0, m_1, m_2 gives

$$m_0=m_1=m_2=1.$$

Consequently we have

$$s(x)=x,\quad x\in[0,2].$$

5.6 本章算法程序及实例 (Algorithms and examples)

5.6.1 拉格朗日插值 (Lagrange interpolation)

```
%% 计算Lagrange基函数
function s=lagrange(x0,y0,x);
n=length(x0);
m=length(x);
```

```
for i=1:m
    z=x(i);
    s=0.0;
    for k=1:n
        p=1.0;
        for j=1:n
            if j~=k
                p=p*(z-x0(j))/(x0(k)-x0(j));
            end
        end
        s=p*y0(k)+s;
    end
end
```

例 5.17 已知五点，如下

x	0	3	5	7	9
y	1	1.2	1.7	2.0	2.1

用 Lagrange 插值法求一个四次多项式，插值经过以上五点.

```
clc;clear all;
x0=[0   3   5   7   9];
y0=[1  1.2  1.7  2.0  2.1];
syms x;
s=simplify(lagrange(x0,y0,x));
n=length(x0);
s=vpa(s,6)
q=sym2poly(s);
x1=x0:0.01:x0*(n);
y1=polyval(q,x1);
plot(x0,y0,'o')
hold on
plot(x1,y1,'-'),title('Lagrange')
```

结果多项式函数为

```
s=0.000978836*x^4-0.0234921*x^3+0.17664*x^2-0.278254*x+1.0.
```

插值图形如图 5.7 所示.

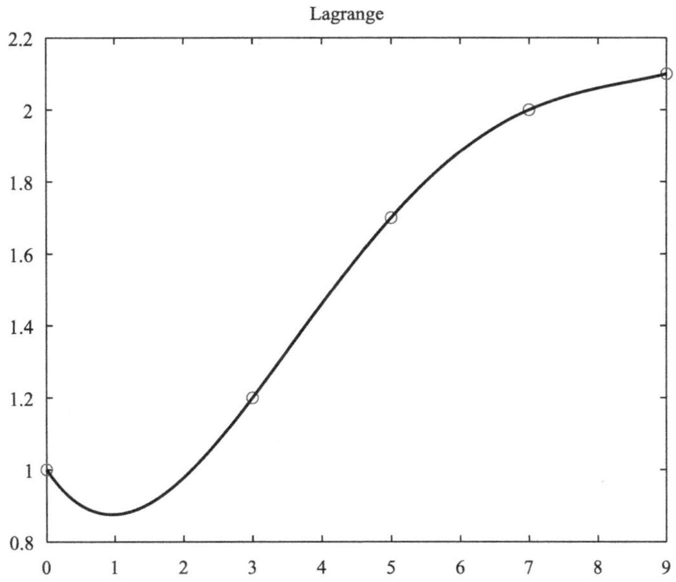

图 5.7 Lagrange 插值多项式

5.6.2 Neville 插值 (Neville interpolation)

```
function f=Neville(x,y,x0)
syms t;
n=length(x);
y1(1:n)=t;                                  %符号函数数组要赋初值
for i=1:n-1
   for j=i+1:n
      y1(j)=y(j)*(t-x(i))/(x(j)-x(i))+y(i)*(t-x(j))/(x(i)-x(j));
   end
   y=y1;
   simplify(y1);
end
if nargin==3
   f=subs(y1(n),'t',x0);                   %计算插值点的函数值
else
   simplify(y1(n));                        %化简
   f=collect(y1(n));                       %将插值多项式展开
end
```

例 5.18 已知五点，如下

x	0	3	5	7	9
y	1	1.2	1.7	2.0	2.1

用 Neville 插值法求一个四次多项式，插值经过以上五点.

```
clc;clear all;
x0=[0   3   5   7   9];
y0=[1  1.2  1.7  2.0   2.1];
syms x;
y=simplify(Neville(x0,y0,x));
n=length(x0);
s=vpa(y,6)
q=sym2poly(s);
x1=x0(1):0.01:x0(n);
y1=polyval(q,x1);
plot(x0,y0,'o')
hold on
plot(x1,y1,'-'),title('Neville')
```

结果多项式函数为

y=0.000978836*x^4-0.0234921*x^3+0.17664*x^2-0.278254*x+1.0.

插值图形如图 5.8 所示.

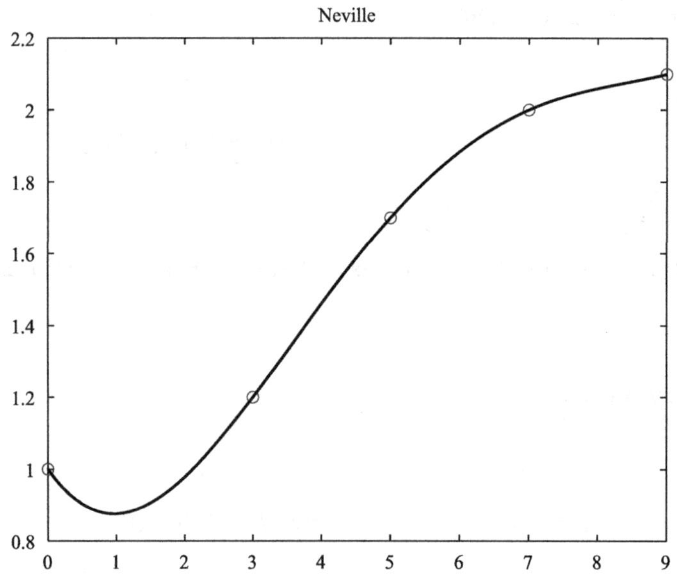

图 5.8　Neville 插值多项式

5.6.3　牛顿插值 (Newton interpolation)

```
function p=Newton_fun(x,xi,yi)
n=length(xi);
f=zeros(n,n);
```

```
for k=1:n
    f(k)=yi(k);
end
% 求差商表
for i=2:n         % 差商表从0阶开始；但是矩阵是从1维开始存储！
    for k=i:n
        f(k,i) =(f(k,i-1) -f(k-1,i-1) ) /(xi(k) -xi(k+1-i));
    end
end
disp('差商表如下：');
disp(f);
p=0;
for k=2:n
    t=1;
    for j=1:k-1
        t=t*(x-xi(j));
    end
    p=f(k,k)*t+p;
end
p=f(1,1)+p;
end
```

例 5.19 已知五点，如下

x	0	3	5	7	9
y	1	1.2	1.7	2.0	2.1

用 Newton 插值法求一个四次多项式，插值经过以上五点.

```
clc;clear all
x0=[0   3   5   7   9];
y0=[1  1.2  1.7  2.0  2.1];
x=sym('x');
p=simplify(Newton_fun(x,x0,y0));
n=length(x0);
s=vpa(p,6)
q=sym2poly(s);
x1=x0(1):0.01:x0(n);
y1=polyval(q,x1);
plot(x0,y0,'o')
```

```
hold on
plot(x1,y1,'-')
```

结果多项式函数为

```
p=0.000978836*x^4-0.0234921*x^3+0.17664*x^2-0.278254*x+1.0.
```

插值图形如图 5.9 所示.

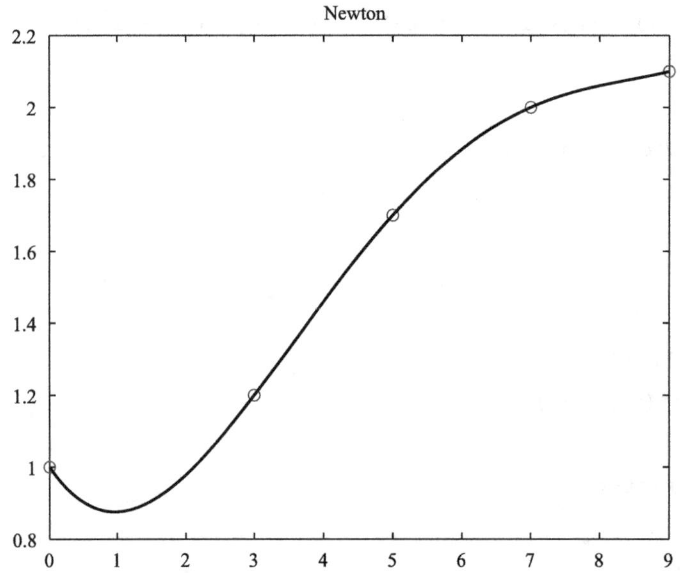

图 5.9　Newton 插值多项式

5.6.4　牛顿向前差商插值 (Interpolation by Newton forward divided differences)

```
function f=Newtonforward_fun(x,y,x0)
syms t;
n=length(x);
c(1:n)=0.0;
f=y(1);
y1=0;
for i=1:n-1
    for j=1:n-i
        y1(j)=y(j+1)-y(j);
    end
    c(i)=y1(1);
    l=t;
    for k=1:i-1
        l=l*(t-k);
```

```
        end
        f=f+c(i)*l/factorial(i);
        simplify(f);
        y=y1;
        if i==n-1
            if nargin==3
                f=subs(f,'t',(x0-x(1))/(x(2)-x(1)));
            else
                f=collect(f);
            end
        end
end
```

例 5.20 已知五点，如下

x	1	3	5	7	9
y	1	1.2	1.7	2.0	2.1

用 Newton 向前差商插值求一个四次多项式，插值经过以上五点.

```
clc;clear all
x0=[1   3   5   7   9];
y0=[1  1.2  1.7  2.0  2.1];
x=sym('x');
f=simplify(newtonforward_fun(x0,y0,x));
n=length(x0);
s=vpa(f,6)
q=sym2poly(s);
x1=x0(1):0.01:x0(n);
y1=polyval(q,x1);
plot(x0,y0,'o')
hold on
plot(x1,y1,'-'),title('Newtonforward')
```

结果函数为

```
f=0.00130208*x^4-0.03125*x^3+0.243229*x^2-0.51875*x+1.30547.
```

插值图形如图 5.10 所示.

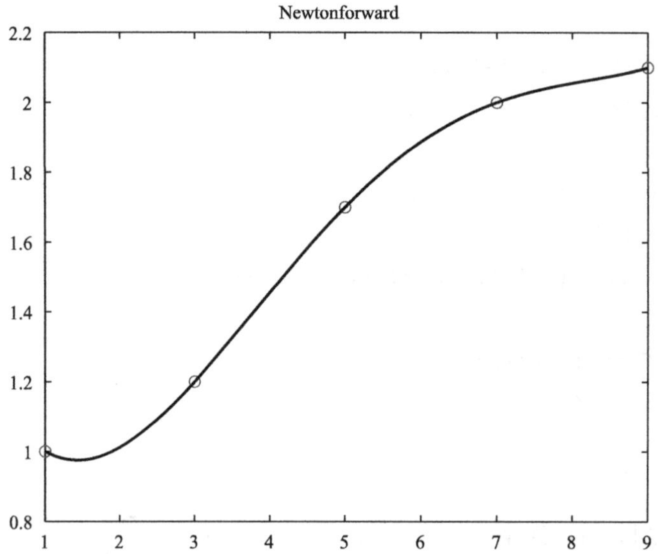

图 5.10　Newton 向前差商插值

5.6.5　牛顿向后差商插值 (Interpolation by Newton backward divided differences)

```
function f=Newtonback_fun(x,y,x0)   %数据必须是等距结点
syms t;
n=length(x);
c(1:n)=0.0;
f=y(n);
y1=0;
for i=1:n-1
    for j=i+1:n
        y1(j)=y(j)-y(j-1);
    end
    c(i)=y1(n);
    l=t;
    for k=1:i-1
        l=l*(t+k);
    end
    f=f+c(i)*l/factorial(i);
    simplify(f);
    y=y1;
    if i==n-1
    if nargin==3
            f=subs(f,'t',(x0-x(n))/(x(2)-x(1)));
        else
```

```
            f=collect(f) ;
        end
    end
end
```

例 5.21 已知五点，如下

x	1	3	5	7	9
y	1	1.2	1.7	2.0	2.1

用 Newton 向后差商插值求一个四次多项式，插值经过以上五点.

```
clc;clear all
x0=[1   3   5   7   9];    %数据必须为等距结点
y0=[1  1.2  1.7  2.0  2.1];
x=sym('x');
f=simplify(Newtonback_fun(x0,y0,x));
n=length(x0);
s=vpa(f,6)
q=sym2poly(s);
x1=x0(1):0.01:x0(n);
y1=polyval(q,x1);
plot(x0,y0,'o')
hold on
plot(x1,y1,'-'),title('Newtonbcak')
```

结果函数为

```
 f=0.00130208*x^4-0.03125*x^3+0.243229*x^2-0.51875*x+1.30547.
```

插值图像如图 5.11.

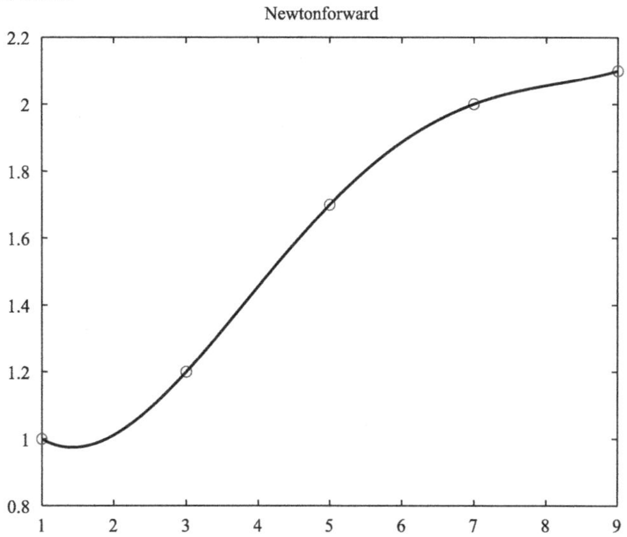

图 5.11 Newton 向后差商插值

5.6.6 埃尔米特插值 (Hermite interpolation)

```
function y=hermite(x0,y0,y1,x);
n=length(x0);m=length(x);
for k=1:m
yy=0.0;
    for i=1:n
        h=1.0;
        a=0.0;
        for j=1:n
            if j~=i
                h=h*((x(k)-x0(j))/(x0(i) -x0(j)))^2;
                a=1/(x0(i)-x0(j))+a;
            end
        end
        yy=yy+h*((x0(i)-x(k))*(2*a*y0(i)-y1(i))+y0(i));
    end
    y(k)=yy;
end
```

例 5.22 已知三点

x	0	1
y	0	1
y'	0	2

用 Hermite 插值法构造一个最高次数为三次的多项式，满足以上条件.

```
clc;clear all
x0=[0 1]; y0=[0 1]; y1=[0 2];
x=sym('x');
f=simplify(hermite(x0,y0,y1,x));
n=length(x0);
f=vpa(f,6)
q=sym2poly(s);
x2=x0(1):0.01:x0(n);
y2=polyval(q,x2);
plot(x0,y0,'o')
hold on
plot(x2,y2,'-'), title('Hermite')
```

结果函数为

```
f=x^2.
```

插值图像如 5.12 所示.

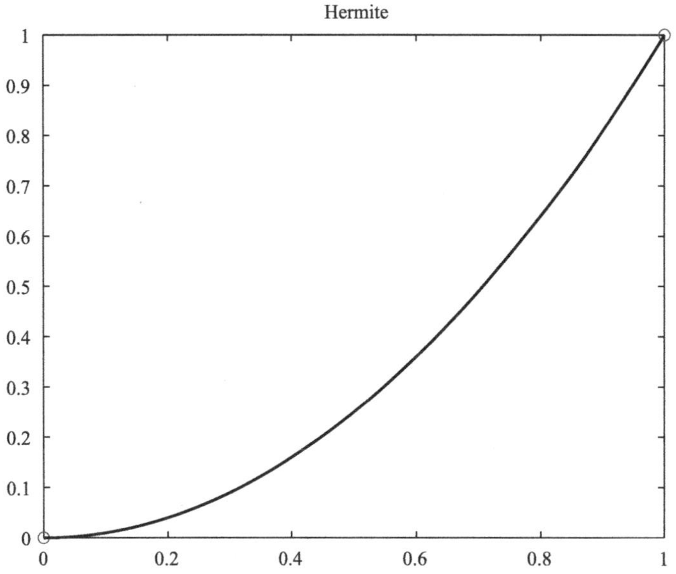

图 5.12　Hermite 插值

5.6.7　分段线性插值 (Piecewise linear interpolation)

例 5.23　已知五点，如下

x	0	3	5	7	9
y	1	1.2	1.7	2.0	2.1

用分段线性插值方法，构造分段函数插值通过以上五点.

直接运行程序如下

```
clc; clear all
x0=[0   3   5   7   9];
y0=[1  1.2  1.7  2.0  2.1];
syms x;
n=length(x0);
for i=1:n-1
  sx=y0(i)*(x-x0(i+1))/(x0(i)-x0(i+1))+y0(i+1)*(x-x0(i))/(x0(i+1)-x0(i));
  sx=vpa(sx,3);
  fprintf('sx=%s \t x∈(%d,%d) \n',char(sx),x0(i),x0(i+1));  %分段输出函数
   %画图
```

```
    coeff=vpa(sym2poly(sx),3); %提取多项式的系数，并数值化
    q=double(coeff);  %将系数数值化
     x1=x0(i):0.01:x0(i+1);
    y1=polyval(q,x1);
    hold on
    plot(x0,y0,'o',x1,y1,'-'),title('Piecewise Linear Interpolation')
end
```

结果函数如下，插值图形如图5.13.

```
sx=0.0667*x+1.0  x∈(0,3)
sx=0.25*x+0.45   x∈(3,5)
sx=0.15*x+0.95   x∈(5,7)
sx=0.05*x+1.65   x∈(7,9)
```

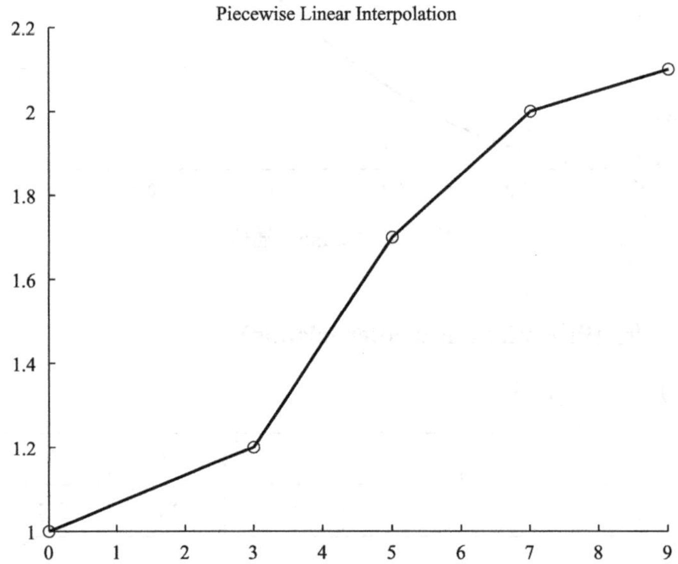

图 5.13　分段线性插值

5.6.8　分段三次插值 (Piecewise cubic interpolation)

例 5.24　已知五点，如下

x	0	3	5	7	9
y	1	1.2	1.7	2.0	2.1
y'	0	0.3	0.5	0.9	1.0

用 Hermite 插值法构造一个最高次数为三次的多项式，满足以上条件.

直接运行程序：

```
clc;clear all
```

```
x0=[0   3   5   7   9];
y0=[1  1.2  1.7  2.0  2.1];
y10=[0  0.3  0.5  0.9  1.0];
hold on
plot(x0,y0,'o')
n=length(x0);
for i=1:n-1
    x1=[x0(i)  x0(i+1)];
    y1=[y0(i)  y0(i+1)];
    y11=[y10(i)  y10(i+1)];
    syms x;
    f=simplify(hermite(x1,y1,y11,x)); %%调用 hermite.m 程序
    sx=vpa(f,3)
    fprintf('sx=%s \t X∈(%d,%d)\n',char(sx),x0(i),x0(i+1));
    q=sym2poly(f);
    x2=x0(i):0.01:x0(i+1);
    y2=polyval(q,x2);
    hold on
    plot(x2,y2,'linewidth',2)
end
```

结果函数如下，图形如图5.14.

```
sx=0.0185*x^3-0.0333*x^2+1.0,          x∈(0,3)
sx=3.38*x-0.85*x^2+0.075*x^3-3.3,     x∈(3,5)
sx=28.4*x-4.85*x^2+0.275*x^3-53.3,   x∈(5,7)
sx=85.6*x-10.8*x^2+0.45*x^3-224.0,   x∈(7,9)
```

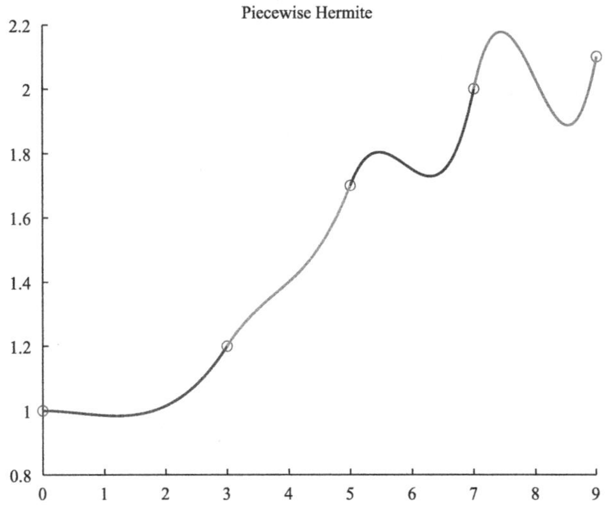

图 5.14　分段三次插值

5.6.9 三次样条插值 1 (边界条件：固支边界) (Cubic spline interpolation with clamped boundary)

```
function sx= Cubic_Sample_1(x0,y0,y_1,y_N,x)
syms x;
sx=0.0;
n=length(x0);
A=diag(2*ones(1,n));        %求解m的系数矩阵
A(1,2)=1;
A(n,n-1)=1;
d=zeros(n,1);
d(1)=6/(x0(2)-x0(1))*((y0(2)-y0(1))/(x0(2)-x0(1))-y_1);
d(n)=6/(x0(n)-x0(n-1))*(y_N-(y0(n)-y0(n-1))/(x0(n)-x0(n-1)));
for i=2:n-1
    u(i-1)=(x0(i)-x0(i-1))/(x0(i+1)-x0(i-1));
    lamda(i-1)=1-u(i-1);
    A(i,i-1)=u(i-1);
    A(i,i+1)=lamda(i-1);     %形成系数矩阵及向量c
    d(i)=6/(x0(i+1)-x0(i-1))*(y0(i+1)-y0(i))/(x0(i+1)-x0(i))-…
          6/(x0(i+1)-x0(i-1))*(y0(i)-y0(i-1))/(x0(i)-x0(i-1));
end
disp(A); %显示矩阵A
disp(d); %显示常数向量
M=LU_Algorithm(A,d);          %用LU法求解方程组
disp(M); %显示方程组的解
for i=1:n-1
    h=x0(i+1)-x0(i);    %第i个区间长度
    sx=1/(6*h)*((x-x0(i))^3*M(i+1)-(x-x0(i+1))^3*M(i)+…
       (6*y0(i+1)-h^2*M(i+1))*(x-x0(i))+…
       (h^2*M(i)-6*y0(i))*(x-x0(i+1)));%第i个区间的插值函数
    sx=simplify(sx);
    sx=vpa(sx,6);
    fprintf('sx=%s \t x∈(%4.1f,%4.1f) \n',char(sx),x0(i),x0(i+1));
    %分段输出函数
    %分段画图
    coeff=vpa(sym2poly(sx),6); %提取多项式的系数，并数值化
    q=double(coeff);  %将系数数值化
    x2=x0(i):0.01:x0(i+1);
```

```
    y2=polyval(q,x2);
    hold on
    plot(x0,y0,'o',x2,y2,'-'),title('Cubic Sample Interpolation_1')
end
```

附 LU 分解函数，参见 2.5.3.

例 5.25　已知五点，如下

x	0	3	5	7	9
y	1	1.2	1.7	2.0	2.1

用三次样条插值方法，构造分段函数插值通过以上五点，且 $y'(0)=0, y'(9)=1$.

```
%% 三次样条实验，第一类边界条件 (固支边界)
clc;clear all
x0=[0   3   5   7   9];
y0=[1  1.2  1.7  2.0  2.1];
y_1=0;
y_N=1;
x=sym('x');
sx=simplify(Cubic_Sample_1(x0,y0,y_1,y_N,x));
```

结果函数如下

```
sx=0.00833333*x^2+0.00462963*x^3+1.0      x∈(0.0,3.0)
sx=0.10625*x^2-0.29375*x-0.00625*x^3+1.29375     x∈(3.0,5.0)
sx=0.66875*x^2-3.10625*x-0.04375*x^3+5.98125     x∈(5.0,7.0)
sx=29.9688*x-4.05625*x^2+0.18125*x^3-71.1938     x∈(7.0,9.0)
```

插值图形如图 5.15.

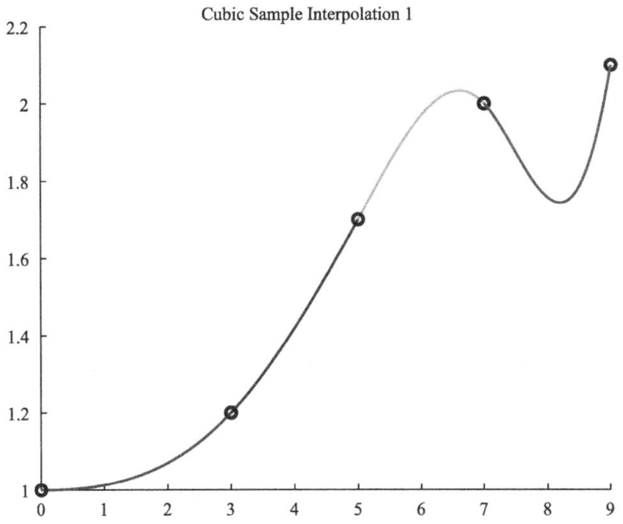

图 5.15　三次样条插值 (固支边界)

5.6.10 三次样条插值 2 (边界条件为自然边界) (Cubic spline interpolation with natural boundary)

```
function sx=Cubic_Sample_2(x0,y0,y2_1,y2_N,x)
syms x;
sx=0.0;
n=length(x0);
A=diag(2*ones(1,n));          %求解m的系数矩阵
d=zeros(n,1);
d(1)=2*y2_1;
d(n)=2*y2_N;
for i=2:n-1
    u(i-1)=(x0(i)-x0(i-1))/(x0(i+1)-x0(i-1));
    lamda(i-1)=1-u(i-1);
    A(i,i-1)=u(i-1);
    A(i,i+1)=lamda(i-1);      %形成系数矩阵及向量c
    d(i)=6/(x0(i+1)-x0(i-1))*(y0(i+1)-y0(i))/(x0(i+1)-x0(i))-…
        6/(x0(i+1)-x0(i-1))*(y0(i)-y0(i-1))/(x0(i)-x0(i-1));
end
disp(A);  %显示矩阵A
disp(d);  %显示常数向量
M=LU_Algorithm(A,d);              %用LU法求解方程组
disp(M);  %显示方程组的解
for i=1:n-1
    h=x0(i+1)-x0(i);    %第i个区间长度
    sx=1/(6*h)*((x-x0(i))^3*M(i+1)-(x-x0(i+1))^3*M(i)+…
       (6*y0(i+1)-h^2*M(i+1))*(x-x0(i))+…
       (h^2*M(i)-6*y0(i))*(x-x0(i+1)));%第i个区间的插值函数
    sx=collect(sx);
    sx=vpa(sx,6);
    fprintf('sx=%s \t x∈(%4.1f,%4.1f) \n',char(sx),x0(i), x0(i+1));
   %分段输出函数
    % 分段画图
    coeff=vpa(sym2poly(sx),6); %提取多项式的系数，并数值化
    q=double(coeff);  %将系数数值化
     x2=x0(i):0.01:x0(i+1);
    y2=polyval(q,x2);
    hold on
```

```
    plot(x0,y0,'o',x2,y2,'-'),title('Cubic Sample Interpolation 1')
end
```

例 5.26 已知五点，如下表，用三次样条插值方法，构造分段函数插值通过以上五点，且 $y'(0)=0, y'(9)=1$.

x	0	3	5	7	9
y	1	1.2	1.7	2.0	2.1

运行程序

```
clc;clear all
x0=[0   3   5   7   9];
y0=[1  1.2  1.7  2.0  2.1];
y2_1=0;
y2_N=0;
x=sym('x');
sx=simplify(Cubic_Sample_2(x0,y0,y2_1,y2_N,x));
```

结果函数为

```
sx=0.00223005*x+0.00715962*x^3+1.0   x∈(0.0,3.0)
sx=0.231866*x^2-0.693369*x-0.0186033*x^3+1.6956      x∈(3.0,5.0)
sx=0.970364*x-0.10088*x^2+0.00357981*x^3-1.07729     x∈(5.0,7.0)
sx=1.07388*x-0.115669*x^2+0.00428404*x^3-1.31884     x∈(7.0,9.0)
```

插值图形如图 5.16 所示.

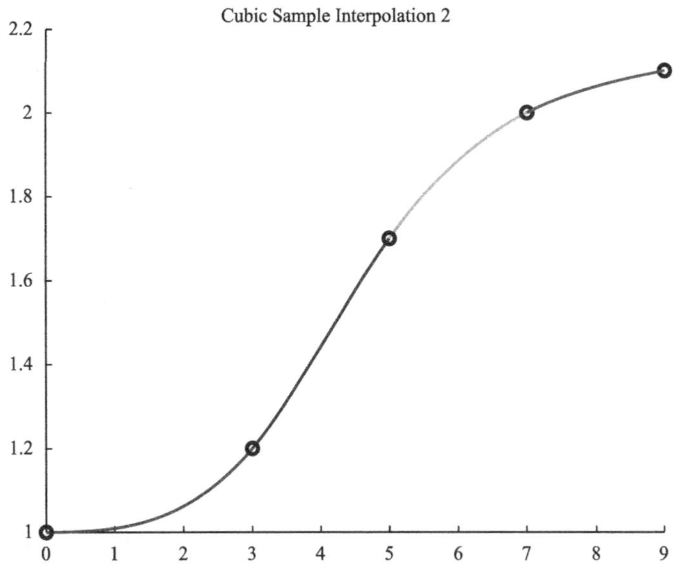

图 5.16 三次样条插值 (自然边界)

5.7 本章要点 (Hightlights)

1. 任意给定 $n+1$ 个支撑点 (support points) ，可以构造一个 n 次的插值多项式.
2. 带余项的 Lagrange 插值公式：

$$f(x)=\sum_{k=0}^{n}f(x_k)L_k(x)+f^{(n+1)}(\xi(x))\omega_n(x)/(n+1)!$$

其中，$L_k(x)=\omega_n(x)/[(x-x_k)\omega_n'(x_k)]$ ($\omega_n(x)=\prod_{i=0}^{n}(x-x_i)$) 是 Lagrange 基函数.

3. Lagrange 插值和 Neville 插值均要求插值结点 (支撑点) 互不相同.
4. 对于给定的插值结点 (支撑点) ，插值多项式是唯一的.
5. 如果 $f(x)$ 是不超过 n 次的多项式，则有 $f(x)=\sum_{k=0}^{n}f(x_k)L_k(x)$, 其中，$L_k(x)=\omega_n(x)/[(x-x_k)\omega_n'(x_k)]$ ($\omega_n(x)=\prod_{i=0}^{n}(x-x_i)$) 是 Lagrange 基函数.
6. 函数 $f(x)$ 的 n 阶差商 $f[x_0,x_1,\cdots,x_n]$ 与结点 $x_0,x_1,\cdots,x_n$ 的排序无关.
7. 函数 $f(x)$ 的 n 阶差商 $f[x_0,x_1,\cdots,x_n]$ 可以表示为 $f(x_0)$, $f(x_1),\cdots$, $f(x_n)$ 的线性组合，即：$f[x_0,x_1,\cdots,x_n]=\sum_{i=0}^{n}f(x_i)/\omega_n'(x_i)$ 这里 $\omega_n(x)=\prod_{i=0}^{n}(x-x_i)$.
8. 函数 $f(x)$ 的 n 阶差商 $f[x_0,x_1,\cdots,x_n]$ 与函数 $f(x)$ 的 n 阶导数之间有如下关系：$f[x_0,x_1,\cdots,x_n]=f^{(n)}(\xi)/n!$. 特别有 $f\left[\overbrace{x_0,x_0,\cdots,x_0}^{n+1}\right]=f^{(n)}(x_0)/n!$.
9. 带余项的 Newton 插值公式为

$$\begin{aligned}f(x)=f[x_0]+\sum_{k=1}^{n}f[x_0,x_1,\cdots,x_k]\cdot(x-x_0)(x-x_1)\cdots(x-x_{k-1})\\+f[x,x_0,x_1,\cdots,x_n](x-x_0)(x-x_1)\cdots(x-x_n).\end{aligned}$$

与 Lagrange 插值和 Neville 插值不同的是，Newton 插值中的插值结点 $x_0,x_1,\cdots,x_n$ 不必互异，当所有结点趋于同一点 x_0 时，带余项的 Newton 插值公式即为带余项的 Taylor 公式：

$$f(x)=f[x_0]+\sum_{k=1}^{n}\frac{f^{(k)}(x_0)}{k!}\cdot(x-x_0)^k+\frac{f^{(n+1)}(\xi(x))}{(n+1)!}(x-x_0)^{n+1}.$$

因此，Taylor 公式可视为 Newton 插值公式的极限形式.

10. Hermite 插值问题可以视为重结点的 Newton 插值问题，因此，Hermite 插值多项式的求解归结为重结点 Newton 插值多项式的构造，最终归结为重结点差商的计算.
11. 三次样条是一种 C^2 连续的分段三次多项式，其具体定义如下：给定区间 $[a,b]$ 的分割

$\Delta: a = x_0 < x_1 < \cdots < x_n = b$. 如果$S(x)$在每个子区间$[x_i, x_{i+1}]$上是一个三次多项式，且 $S(x) \in C^2[a,b]$，则称 $S(x)$ 为关于结点 $x_0, x_1, \cdots, x_n$ 的三次样条. 如果 $S(x_i) = f(x_i)$, $i = 0,1,\cdots,n$，则称$S(x)$为$f(x)$的三次插值样条. 给定固支边界条件：$S'(x_0) = f'(x_0), S'(x_n) = f'(x_n)$, 或自由边界(自然边界)条件：$S''(x_0) = 0, S''(x_n) = 0$，三次插值样条可通过求解对角占优的三对角方程组而得到. 对应于上述两种边界条件，有两种不同的方法建立方程组. 针对自由边界 (自然边界) 条件，通常采用 M 方法，即，将方程组的未知数假设为$M_i = S''(x_i), i = 0,1,\cdots,n$, 此时，由于$S''(x_0) = 0$, $S''(x_n) = 0$，自然得到：$M_0 = 0$, $M_n = 0$, 从而方程组由$n+1$阶降为$n-1$阶. 针对固支边界条件，通常采用 m 方法，即，将方程组的未知数假设为：$m_i = S'(x_i), i = 0,1,\cdots,n$, 此时，由于$S'(x_0) = f'(x_0), S'(x_n) = f'(x_n)$, 自然得到：$m_0 = f'(x_0), m_n = f'(x_n)$, 从而方程组由$n+1$阶降为$n-1$阶. 参见例 5.14.

5.8 问题讨论 (Questions for discussion)

1. Lagrange 插值多项式有什么不足之处？
2. 对任意给定的函数$f(x)$，设$P_n(x)$是$f(x)$的n次插值多项式，在什么条件下，n越大，$P_n(x)$与$f(x)$的误差会越小？
3. 设$f(x) = a_0 + a_1 x + a_2 x^2$ $(a_2 \neq 0)$，$f(x_i) = y_i, i = 1,2,\cdots,8$，且$x_i$，$i = 1,2,\cdots,8$，互异，张三和李四两位同学分别利用采样点$\{(x_i, y_i)\}_{i=1}^{4}$及$\{(x_i, y_i)\}_{i=5}^{8}$构造了插值多项式$P_3(x)$和$\overline{P}_3(x)$，问：$P_3(x), \overline{P}_3(x)$, $f(x)$三者之间有何关系？
4. 如何利用 Lagrange 插值 (Newton 插值) 和 Hermite 插值方法估计 ln2 的近似值？
5. 如何利用 Lagrange 插值 (Newton 插值) 和 Hermite 插值方法估计π的近似值？
6. 如何将 Lagrange 插值多项式推广到二元情形？
7. 如何定义二元差商？
8. 如何利用二元差商将 Newton 插值多项式推广到二元情形？
9. 如何构造双三次 Hermite 插值？
10. 如何构造双三次样条插值？
11. 与分段多项式逼近相比，样条函数逼近有哪些优势？
12. 试对分段三次 Hermite 插值与三次样条插值作比较分析.

5.9 关键术语 (Key terms)

linear (线性的)
quadratic (二次的)
polynomial (多项式)
interpolation (插值)

error (误差)
Lagrange basis functions (拉格朗日基函数)
Kronecker's symbol (克罗耐克记号)
Lagrange interpolating polynomial (拉格朗日插值多项式)
uniqueness (唯一性)
support points (支撑点)
system of linear equations (线性方程组)
Generalized Rolle's Theorem (广义罗尔定理)
Neville interpolation (Neville 插值)
algorithm (算法)
approximate (逼近)
divided differences (差商)
k th divided difference (k 阶差商)
Newton's interpolation (Newton 插值)
remainder term (余项)
error term (误差项)
Newton's interpolatory divided differenceformula (Newton 插值差商公式)
Newton's expansion of a function (函数的 Newton 展开)
leading coefficients (首项系数)
Newton's expansion with remainder term (带余项的 Newton 展开)
Taylor's expansion about $x = x_0$ with remainder term ($x = x_0$ 处带余项的 Taylor 展开)
Newton forward divided-difference formula (Newton 向前差商公式)
Newton forward difference formula (Newton 向前差分公式)
Newton backward divided-difference formula (Newton 向后差商公式)
Newton backward-difference formula (Newton 向后差分公式)
interpolating nodes (插值结点)
Hermite interpolation (Hermite 插值)
two-point Hermite interpolation (两点 Hermite 插值)
general Hermit interpolation (一般 Hermite 插值)
Runge phenomenon (龙格现象)
piecewise cubic interpolation (分段三次插值)
cubic splines (三次样条)
natural boundary (自然边界)
clamped boundary (固支边界)

5.10 延伸阅读 (Extending reading)

5.10.1 有理函数插值 (Interpolation by rational functions)

有理函数插值是指, 构造一个有理函数 $R_{m,n}(x)=\dfrac{p(x)}{q(x)}$, 其中 $p(x)$ 和 $q(x)$ 分别为 m 次和 n 次的多项式, 使得在给定的 $m+n+1$ 个点 $x_i, i=0,1,\cdots,m+n$, 有

$$R_{m,n}(x_i)=f(x_i), i=0,1,\cdots,m+n. \tag{5.10.1}$$

有理函数插值问题的求解显然要比多项式插值问题的求解复杂得多，通常将(5.10.1)转化为求解如下问题:

$$(fq-p)(x_i)=0, i=0,1,\cdots,m+n. \tag{5.10.2}$$

易知，当 $q(x_i)\neq 0, i=0,1,\cdots,m+n$ 时，(5.10.1) 与 (5.10.2) 等价. 设

$$p(x)=a_0+a_1x+\cdots+a_mx^m, q(x)=b_0+b_1x+\cdots+b_nx^n, \tag{5.10.3}$$

由 (5.10..2) 得到下述方程组

$$a_0+a_1x_i+\cdots+a_mx_i^m-b_0-b_1x_if_i-\cdots-b_nx_i^nf_i=0, i=0,1,\cdots,m+n, \tag{5.10.4}$$

其中 $f_i=f\left(x_i\right)$. (5.10.4) 是一个关于未知数 $a_0,a_1,\cdots,a_m,-b_0,-b_1,\cdots,-b_n$ 的 $(m+n+1)$ 个方程，故存在非平凡解.

令

$$M_{m,n}=\begin{pmatrix} 1 & x_0 & \cdots & x_0^m & f_0 & x_0f_0 & \cdots & x_0^nf \\ 1 & x_1 & \cdots & x_1^m & f_1 & x_1f_1 & \cdots & x_1^nf_1 \\ \vdots & \vdots & & \vdots & \vdots & \vdots & & \vdots \\ 1 & x_{m+n} & \cdots & x_{m+n}^m & f_{m+n} & x_{m+n}f_{m+n} & \cdots & x_{m+n}^nf_{m+n} \end{pmatrix}, \tag{5.10.5}$$

如果 $\operatorname{rank}M_{m,n}=m+n+1$，则 $M_{m,n}$ 中至少有一个 $m+n+1$ 阶子式不为零. 用 $M_{m,n}(j)$ 表示从 $M_{m,n}$ 中去掉第 $j+1$ 列后的矩阵，则至少存在一个 $j\in\{0,1,\cdots,m+n+1\}$ 使 $M_{m,n}(j)$ 非奇异，若 $0\leqslant j\leqslant m$，则

$$M_{m,n}(j)=\begin{pmatrix} 1 & x_0 & \cdots & x_0^{j-1} & x_0^{j+1} & \cdots & x_0^m & f_0 & x_0f_0 & \cdots & x_0^nf_0 \\ 1 & x_1 & \cdots & x_1^{j-1} & x_1^{j+1} & \cdots & x_1^m & f_1 & x_1f_1 & \cdots & x_1^nf_1 \\ \vdots & \vdots & & \vdots & \vdots & & \vdots & \vdots & \vdots & & \vdots \\ 1 & x_{m+n} & \cdots & x_{m+n}^{j-1} & x_{m+n}^{j+1} & \cdots & x_{m+n}^m & f_{m+n} & x_{m+n}f_{m+n} & \cdots & x_{m+n}^nf_{m+n} \end{pmatrix}, \tag{5.10.6}$$

此时，方程组(5.10.4) 可写成

$$M_{m,n}(j)\begin{pmatrix} a_0 \\ \vdots \\ a_{j-1} \\ a_{j+1} \\ \vdots \\ a_m \\ -b_0 \\ \vdots \\ -b_n \end{pmatrix} = -a_j \begin{pmatrix} x_0^j \\ x_1^j \\ \vdots \\ x_{m+n}^j \end{pmatrix}, \tag{5.10.7}$$

令 $a_j = 1$，有

$$(p - fq)(x) = a_0 + a_1 x + \cdots + a_{j-1}x^{j-1} + x^j + a_{j+1}x^{j+1} + \cdots + a_m x^m$$
$$-b_0 f(x) - b_1 x f(x) - \cdots - b_n x^n f(x),$$

或

$$a_0 + a_1 x + \cdots + a_{j-1}x^{j-1} + [(fq - p)(x) + x^j] + a_{j+1}x^{j+1} + \cdots + a_m x^m$$
$$-b_0 f(x) - b_1 x f(x) - \cdots - b_n x^n f(x) = 0, \tag{5.10.8}$$

联立求解(5.10.4) 和(5.10.7)，得

$$p(x) = -c_1 \begin{vmatrix} 1 & x_0 & \cdots & x_0^m & f_0 & x_0 f_0 & \cdots & x_0^n f_0 \\ 1 & x_1 & \cdots & x_1^m & f_1 & x_1 f_1 & \cdots & x_1^n f_1 \\ \vdots & \vdots & & \vdots & \vdots & \vdots & & \vdots \\ 1 & x_{m+n} & \cdots & x_{m+n}^m & f_{m+n} & x_{m+n} f_{m+n} & \cdots & x_{m+n}^n f_{m+n} \\ 1 & x & \cdots & x^m & 0 & 0 & \cdots & 0 \end{vmatrix}, \tag{5.10.9}$$

$$q(x) = c_1 \begin{vmatrix} 1 & x_0 & \cdots & x_0^m & f_0 & x_0 f_0 & \cdots & x_0^n f_0 \\ 1 & x_1 & \cdots & x_1^m & f_1 & x_1 f_1 & \cdots & x_1^n f_1 \\ \vdots & \vdots & & \vdots & \vdots & \vdots & & \vdots \\ 1 & x_{m+n} & \cdots & x_{m+n}^m & f_{m+n} & x_{m+n} f_{m+n} & \cdots & x_{m+n}^n f_{m+n} \\ 0 & 0 & \cdots & 0 & 1 & x & \cdots & x^n \end{vmatrix}, \tag{5.10.10}$$

其中 $c_1 = (-1)^{j+m+n} \big/ \det M_{m,n}(j)$. 若 $m+1 \leqslant j \leqslant m+n+1$ 使 $M_{m,n}(j)$ 非奇异，同样可以推出 (5.10.9) 和 (5.10.10).

通过行列式变换，可将 $p(x)$ 与 $q(x)$ 改写为

$$p(x) = c_2 \prod_{i=0}^{m+n}(x - x_i) \begin{vmatrix} f_0/(x-x_0) & f_0 & x_0 f_0 & \cdots & x_0^{n-1} f_0 & 1 & x_0 & \cdots & x_0^{m-1} \\ f_1/(x-x_1) & f_1 & x_1 f_1 & \cdots & x_1^{n-1} f_1 & 1 & x_1 & \cdots & x_1^{m-1} \\ \vdots & \vdots & \vdots & & \vdots & \vdots & \vdots & & \vdots \\ f_{m+n}/(x-x_{m+n}) & f_{m+n} & x_{m+n} f_{m+n} & \cdots & x_{m+n}^{n-1} f_{m+n} & 1 & x_{m+n} & \cdots & x_{m+n}^{m-1} \end{vmatrix}, \tag{5.10.11}$$

$$q(x)=c_2\prod_{i=0}^{m+n}(x-x_i)\begin{vmatrix}1/(x-x_0) & f_0 & x_0f_0 & \cdots & x_0^{n-1}f_0 & 1 & x_0 & \cdots & x_0^{m-1}\\ 1/(x-x_1) & f_1 & x_1f_1 & \cdots & x_1^{n-1}f_1 & 1 & x_1 & \cdots & x_1^{m-1}\\ \vdots & \vdots & \vdots & & \vdots & \vdots & \vdots & & \vdots\\ 1/(x-x_{m+n}) & f_{m+n} & x_{m+n}f_{m+n} & \cdots & x_{m+n}^{n-1}f_{m+n} & 1 & x_{m+n} & \cdots & x_{m+n}^{m-1}\end{vmatrix},$$

(5.10.12)

其中 $c_2=(-1)^{m(n+1)}c_1$.

$p(x)$ 与 $q(x)$ 还可以借助函数 $f(x)$ 的差商表示成如下更紧凑的行列式形式:

$$p(x)=\frac{1}{D}\begin{vmatrix}\sum_{i=0}^{m}c_{0i}B_i(x) & \sum_{i=1}^{m}c_{1i}B_i(x) & \cdots & \sum_{i=n}^{m}c_{ni}B_i(x)\\ c_{0,m+1} & c_{1,m+1} & \cdots & c_{n,m+1}\\ c_{0,m+2} & c_{1,m+2} & \cdots & c_{n,m+2}\\ \vdots & \vdots & & \vdots\\ c_{0,m+n} & c_{1,m+n} & \cdots & c_{n,m+n}\end{vmatrix}$$

$$q(x)=\frac{1}{D}\begin{vmatrix}1 & B_1(x) & \cdots & B_n(x)\\ c_{0,m+1} & c_{1,m+1} & \cdots & c_{n,m+1}\\ c_{0,m+2} & c_{1,m+2} & \cdots & c_{n,m+2}\\ \vdots & \vdots & & \vdots\\ c_{0,m+n} & c_{1,m+n} & \cdots & c_{n,m+n}\end{vmatrix},$$

其中,

$$B_0(x)=1,\quad B_j(x)=\prod_{l=1}^{j}(x-x_{l-1}),\ j\geqslant 1,\quad \begin{cases}c_{ij}=f[x_i,x_{i+1},\cdots,x_j],\ j\geqslant i;\\ c_{ij}=0,\ j<i.\end{cases}$$

5.10.2 Thiele 型连分式插值 (Interpolation by Thiele type continued fractions)

构造一般的有理插值函数需要求解线性方程组，这无疑给有理插值的表示和计算带来不便. 下面介绍一种利用连分式构造有理插值的方法.

定义 5.1 称下述形式的连分式

$$b_0+\frac{x-x_0}{b_1}\,\frac{}{+}\,\frac{x-x_1}{b_2}\,\frac{}{+\cdots+}\,\frac{x-x_{n-1}}{b_n}\,\frac{}{+\cdots} \tag{5.10.13}$$

为 Thiele 型连分式(Thiele type continued fractions).

定义 5.2 设 $X=\{x_i\,|\,i\in IN\}$ 是复平面上一点集，$f(x)$ 是定义在 $G(G\supset X)$ 上的函数. 令

$$\varphi[x_i]=f(x_i),\ i=0,1,2,\cdots, \tag{5.10.14}$$

$$\varphi[x_p,x_q]=\frac{x_q-x_p}{\varphi[x_q]-\varphi[x_p]}, \tag{5.10.15}$$

$$\varphi[x_i,\cdots,x_j,x_k,x_l]=\frac{x_l-x_k}{\varphi[x_i,\cdots,x_j,x_l]-\varphi[x_i,\cdots,x_j,x_k]},\tag{5.10.16}$$

称由上述公式确定的 $\varphi[x_0,x_1,\cdots,x_l]$ 为函数 $f(x)$ 在点 $x_0,x_1,\cdots,x_l$ 处的 l 阶逆差商或反差商(inverse differences; inverted differences).

定理 5.5 设

$$R_n(x)=b_0+\frac{x-x_0}{b_1}\,\frac{x-x_1}{+\ b_2}\,+\cdots+\frac{x-x_{n-1}}{b_n}=\frac{P_n(x)}{Q_n(x)},$$

则 $\deg P_n=\left[\frac{n+1}{2}\right],\deg Q_n=\left[\frac{n}{2}\right]$, 此处 $[\mu]$ 表示不超过 μ 的最大整数.

定理 5.6 设

$$R_n(x)=\varphi[x_0]+\frac{x-x_0}{\varphi[x_0,x_1]}\,\frac{x-x_1}{+\ \varphi[x_0,x_1,x_2]}\,+\cdots+\frac{x-x_{n-1}}{\varphi[x_0,x_1,\cdots,x_n]},\tag{5.10.17}$$

其中 $\varphi[x_0,x_1,\cdots,x_k]\neq 0,\infty,k=0,1,\cdots,n$, 为 $f(x)$ 在 $x_0,x_1,\cdots,x_k$ 处的 k 阶逆差商，则有

$$R_n(x_i)=f(x_i),\quad i=0,1,\cdots,n.$$

证明 反复利用逆差商的定义知:

$$\begin{aligned}
R_n(x_i)&=\varphi[x_0]+\frac{x_i-x_0}{\varphi[x_0,x_1]}\,\frac{x_i-x_1}{+\ \varphi[x_0,x_1,x_2]}\,+\cdots+\frac{x_i-x_{i-1}}{\varphi[x_0,x_1,\cdots,x_i]}\\
&=\varphi[x_0]+\frac{x_i-x_0}{\varphi[x_0,x_1]}+\cdots+\frac{x_i-x_{i-2}}{\varphi[x_0,\cdots,x_{i-1}]+\dfrac{x_i-x_{i-1}}{\dfrac{x_i-x_{i-1}}{\varphi[x_0,\cdots,x_{i-2},x_i]-\varphi[x_0,\cdots,x_{i-2},x_{i-1}]}}}\\
&=\varphi[x_0]+\frac{x_i-x_0}{\varphi[x_0,x_1]}+\cdots+\frac{x_i-x_{i-2}}{\varphi[x_0,\cdots,x_{i-2},x_i]}\\
&=\cdots\\
&=\varphi[x_0]+\frac{x_i-x_0}{\varphi[x_0,x_i]}\\
&=\varphi[x_i]=f(x_i),\quad i=0,1,\cdots,n.
\end{aligned}$$

定理 5.6 证毕.

定义 5.3 如果连分式

$$R_n(x)=b_0+\frac{x-x_0}{b_1}\,\frac{x-x_1}{+\ b_2}\,+\cdots+\frac{x-x_{n-1}}{b_n}$$

满足 $R_n(x_i)=f(x_i),i=0,1,\cdots,n$，则称该连分式为函数 $f(x)$ 的 $\left(\left[\frac{n+1}{2}\right]\Big/\left[\frac{n}{2}\right]\right)$ 阶 Thiele 型插值连分式.

定理 5.7 函数 $f(x)$ 的 $\left(\left[\frac{n+1}{2}\right]\Big/\left[\frac{n}{2}\right]\right)$ 阶 Thiele 型插值连分式是唯一的.

证明 设

$$R_n(x)=b_0+\frac{x-x_0}{b_1}\,{}_{+}\,\frac{x-x_1}{b_2}\,{}_{+\cdots+}\,\frac{x-x_{n-1}}{b_n}=\frac{P_n(x)}{Q_n(x)},$$

$$R_n^*(x)=b_0^*+\frac{x-x_0}{b_1^*}\,{}_{+}\,\frac{x-x_1}{b_2^*}\,{}_{+\cdots+}\,\frac{x-x_{n-1}}{b_n^*}=\frac{P_n^*(x)}{Q_n^*(x)}$$

是函数 $f(x)$ 的另一个 Thiele 型插值连分式, 则有

$$R_n^*(x_i)=R_n(x_i)=f(x_i),i=0,1,\cdots,n,$$

因此

$$P_n(x_i)Q_n^*(x_i)-Q_n(x_i)P_n^*(x_i)=0,\ i=0,1,\cdots,n,$$

注意到

$$\deg[P_n(x)Q_n^*(x)-Q_n(x)P_n^*(x)]\leqslant\left[\frac{n+1}{2}\right]+\left[\frac{n}{2}\right]=n,$$

有

$$P_n(x)Q_n^*(x)-Q_n(x)P_n^*(x)\equiv 0.$$

于是 $R_n(x)\equiv R_n^*(x)$.

唯一性即证.

关于 Thiele 型插值连分式，有如下误差估计.

定理 5.8 设 $\{x_0,x_1,\cdots,x_n\}\subset(a,b)$， $f(x)$ 在 (a,b) 上有直到 $n+1$ 阶的导数, 若

$$R_n(x)=b_0+\frac{x-x_0}{b_1}\,{}_{+}\,\frac{x-x_1}{b_2}\,{}_{+\cdots+}\,\frac{x-x_{n-1}}{b_n}=\frac{P_n(x)}{Q_n(x)}$$

满足 $R_n(x_i)=f(x_i),i=0,1,\cdots,n$, 则 $\forall x\in(a,b)$ 均 $\exists\xi\in I(x_0,x_1,\cdots,x_n,x)$,使得

$$f(x)-R_n(x)=\frac{\omega_{n+1}(x)}{Q_n(x)}\cdot\left.\frac{[f(x)Q_n(x)]^{(n+1)}}{(n+1)!}\right|_{x=\xi},$$

其中 $\omega_{n+1}(x)=(x-x_0)(x-x_1)\cdots(x-x_n)$, $I(x_0,x_1,\cdots,x_n,x)$ 表示包含 $x_0,x_1,\cdots,x_n$ 及 x 的最小开区间.

5.10.3 Padé 逼近 (Padé approximation)

定义 5.4 函数 $f(x)$ 的 (m,n) 阶 Padé 逼近问题，是指求解多项式

$$p(x)=\sum_{i=0}^{m}a_ix^i,\ q(x)=\sum_{i=0}^{n}b_ix^i, \tag{5.10.18}$$

使得

$$(fq-p)^{(k)}(0)=0,\ k=0,1,\cdots,m+n. \tag{5.10.19}$$

而称 p/q 的不可约形式 $r_{m,n}(x)=p_0(x)/q_0(x)=p(x)/q(x)$ 为 $f(x)$ 的 (m,n) 阶 Padé 逼近，记作 $[m/n]_{f(x)}$.

由定义不难看出，Padé 逼近实际上是一种切触有理插值. 设

$$f(x)=\sum_{i=0}^{\infty}c_i x^i ,$$

记

$$F_k(x)=\begin{cases}\sum_{i=0}^{k}c_i x^i, & k\geqslant 0,\\ 0, & k<0,\end{cases} \tag{5.10.20}$$

$$D_{m,n}=\begin{pmatrix} c_m & c_{m-1} & \cdots & c_{m+1-n}\\ c_{m+1} & c_m & \cdots & c_{m+2-n}\\ \vdots & \vdots & & \vdots\\ c_{m+n-1} & c_{m+n-2} & \cdots & c_m \end{pmatrix} \tag{5.10.21}$$

定理 5.9 设 f 的 (m,n) 阶 Padé 逼近为 $r_{m,n}(x)=p_0(x)/q_0(x)$，若 $D=\det D_{m,n}\neq 0$，则

$$p_0(x)=\frac{1}{D}\begin{vmatrix} F_m(x) & xF_{m-1}(x) & \cdots & x^n F_{m-n}(x)\\ c_{m+1} & & & \\ \vdots & & D_{m,n} & \\ c_{m+n} & & & \end{vmatrix},$$

$$q_0(x)=\frac{1}{D}\begin{vmatrix} 1 & x & \cdots & x^n\\ c_{m+1} & & & \\ \vdots & & D_{m,n} & \\ c_{m+n} & & & \end{vmatrix}.$$

5.10.4 数学家简介: 牛顿 (Newton)

艾萨克·牛顿 (Isaac Newton) 于 1642 年 12 月 25 日圣诞节出生 (根据当时在英国使用的 Julian calendar) 在英格兰林肯郡 (Lincolnshire) 的伍尔斯索普 (Woolsthorpe)，1727 年 3 月 31 日在英格兰米德尔塞克斯 (Middlesex) 的肯辛顿 (Kensington) 去世，终年 84 岁.

牛顿 12 岁时就读于格兰瑟姆国王学校 (King's School, Grantham)，在那里他学习古

典文学，17 岁时他以优等生的成绩毕业.

1661 年 6 月，18 岁的牛顿开始在剑桥大学三一学院 (Trinity College) 攻读法律学位. 1665 年，22 岁时，牛顿在数学领域做出了他人生中的第一个重大发现：广义二项式定理. 同年，他获得了学士学位. 此后，他开始发展一套新的数学理论，也就是后来广为人知的微积分学. 1669 年，26 岁的牛顿被授予剑桥大学三一学院卢卡斯数学教授席位 (the Lucasian Professor of Mathematics) .

牛顿的主要科学成就：

- 推广了二项式定理.
- 证明了阳光是由彩虹的所有颜色组成的. 他用一个玻璃棱镜将一束阳光分成不同的颜色，然后用另一个棱镜将彩虹的颜色重新组合成一束白光.
- 建造了世界上第一台反射望远镜.
- 创立了微积分.
- 出版了《自然哲学的数学原理》，这是有史以来最重要的科学著作之一，在该书中，牛顿使用数学来解释重力和运动.
- 发现了万有引力定律.
- 创立了三大运动定律.
- 指出开普勒 (Kepler) 行星运动定律是牛顿万有引力的特例.
- 证明了在空间运动的所有物体在重力作用下，其运动轨迹的形状必定是一条二次曲线 (圆、椭圆或抛物线)，从而解释了所有行星和彗星的运行轨迹.
- 指出潮汐是由地球、月球和太阳之间的引力相互作用引起的.
- 正确地预测地球不是完美的球形，而是扁球体, 赤道周围比两极周围大.
- 给出了方程求根的方法.

5.10.5 数学家简介：拉格朗日 (Lagrange)

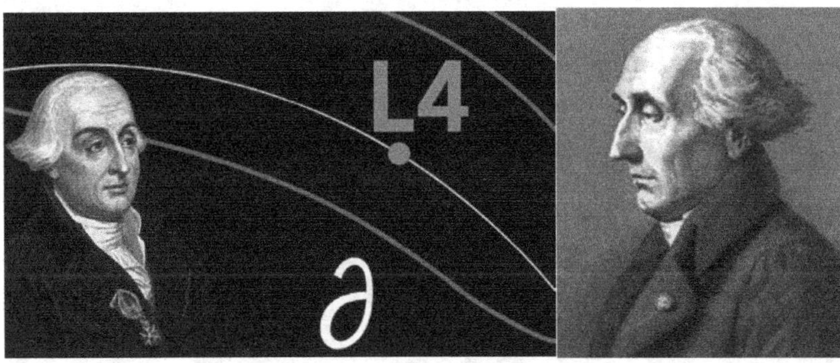

约瑟夫・路易斯・拉格朗日 (Joseph-Louis Lagrange) ，1736 年 1 月 25 日出生于意大利都灵(Turin) 的朱塞佩・洛多维科・拉格拉贾(Giuseppe Lodovico Lagrangia) ，1813 年 4 月 10 日在巴黎去世，终年 77 岁. 拉格朗日在数学、力学和天文学三个学科领域中都有历史性的贡献，其中尤以数学方面的成就最为突出. 其主要贡献涉及物理学、天体力学、微积分、代数、数论、群论等. 拉格朗日没有获得任何学位，基本上是靠自学成

才. 拉格朗日是变分法的创始人. 我们熟悉的函数一阶导数记号 $f'(x)$ 和二阶导数记号 $f''(x)$ 就是拉格朗日引进的.

拉格朗日的主要贡献：

- 在莱昂哈德·欧拉 (Leonhard Euler) 早期工作基础上，创立了变分法.
- 引进了偏导数记号"∂"，并创建了第一个偏微分方程.
- 在虚功概念基础上，利用拉格朗日函数，为固体和流体创建了一个全新的力学领域：拉格朗朗日力学.
- 引进了广义坐标的概念. 拉格朗日力学可以在任何坐标系中使用——通过选择合适的坐标系可以使问题得到简化.
- 创建了势的概念. 例如，引力场就是势场.
- 发现了拉格朗日轨道. 建立了各类天体的运动方程，特别根据他在微分方程解法中使用的常数变异法，建立了以天体椭圆轨道根数为基本变量的运动方程，称作拉格朗日行星运动方程.
- 提出了拉格朗日插值法和求解多元函数极值的拉格朗日乘子法.
- 解决了费马提出的困扰数学家们百年之久的数论难题.
- 群论的奠基人.

5.10.6　数学家简介：埃尔米特 (Hermite)

查尔斯·埃尔米特 (Charles Hermite) 1822年12月24日出生于法国摩泽尔 (Moselle) 的迪尤兹 (Dieuze) ，1901 年 1 月 14 日在巴黎去世，终年 79 岁. 曾任法兰西学院、巴黎高等师范学校、巴黎大学教授，法兰西科学院院士. 在数论、椭圆函数、二次型、不变理论、正交多项式、代数等方面都有重要发现. 以他的名字命名的数学概念有：埃尔米特多项式 (Hermite polynomials) 、埃尔米特插值 (Hermite interpolation) 、埃尔米特标准

型 (Hermite normal form) 、埃尔米特算子 (Hermitian operators) 、三次埃尔米特样条 (cubic Hermite splines) 等. 埃尔米特首次证明了自然对数的底 e 是一个超越数 (transcendental number) . 他的方法后来被费迪南德・冯・林德曼(Ferdinand von Lindemann) 用来证明 π 是超越数. 埃尔米特培养了一个杰出学生，他的名字叫做：儒勒・亨利・庞加莱 (Jules Henri Poincaré，1854 年 4 月 29 日～1912 年 7 月 17 日，法国数学家、天体力学家、数学物理学家、科学哲学家) .

5.11 习题 (Exercises)

1. Suppose $x_0, x_1, \cdots, x_n$ are distinct numbers. Let $\omega(x) = \prod_{k=0}^{n}(x - x_k)$, and $L_i(x) = \dfrac{\omega(x)}{(x - x_i)\omega'(x_i)}$, $i = 0,1,\cdots,n$. Show that the following equality holds true for any polynomial $p(x)$ of degree not exceeding n

$$\sum_{i=0}^{n} p(x_i)L_i(x) = p(x).$$

2. Suppose $\omega_n(x) = \prod_{i=0}^{n}(x - x_i)$, show $\sum_{i=0}^{n} 1/\omega_n'(x_i) = 0$.

3. Suppose $p_3(x)$ is the monic cubic interpolating polynomial for the data $(0,0)$, $(0.5, y)$, $(1,3)$ and $(2,2)$. Find y.

4. Construct a cubic polynomial $p_3(x)$ interpolating $f(x) = \sqrt{x}$ at $x_0 = 1.69$, $x_1 = 1.96$, $x_2 = 2.25$, $x_3 = 2.56$. Compute $p_3(2)$ and estimate the error $|f(2) - p_3(2)|$.

5. Show that the Lagrange basis function $L_i(x)$ can be expressed as follows

$$L_i(x) = (-1)^i \begin{vmatrix} 1 & x & \cdots & x^{n-1} & x^n \\ 1 & x_0 & \cdots & x_0^{n-1} & x_0^n \\ \vdots & \vdots & & \vdots & \vdots \\ 1 & x_{i-1} & \cdots & x_{i-1}^{n-1} & x_{i-1}^n \\ 1 & x_{i+1} & \cdots & x_{i+1}^{n-1} & x_{i+1}^n \\ \vdots & \vdots & & \vdots & \vdots \\ 1 & x_n & \cdots & x_n^{n-1} & x_n^n \end{vmatrix} \cdot \frac{1}{\begin{vmatrix} 1 & x_0 & \cdots & x_0^{n-1} & x_0^n \\ 1 & x_1 & \cdots & x_1^{n-1} & x_1^n \\ \vdots & \vdots & & \vdots & \vdots \\ 1 & x_n & \cdots & x_n^{n-1} & x_n^n \end{vmatrix}} \quad \text{for } i = 0,1,\cdots,n.$$

6. Neville's method is used to approximate $f(0.5)$ through the following table.

$x_0 = 0$	$P_0 = 0$		
$x_1 = 0.4$	$P_1 = 2.8$	$P_{0,1} = 3.5$	
$x_0 = 0.7$	$P_2 = ?$	$P_{1,2} = ?$	$P_{0,1,2} = 27/7$

Dertermine P_2 and $P_{1,2}$.

7. Suppose $x_j = j, j = 0,1,2,3$ and it is known that

$$P_{0,1}(x) = 2x+1,\ P_{0,2}(x) = x+1, \text{and}\ P_{1,2,3}(2.5) = 3.$$

Find $P_{0,1,2,3}(2.5)$.

8. Suppose $x_j = j$, for $j = 0,1,2,3$, and it is known that $P_{0,1}(x) = x+1$, $P_{1,2}(x) = 3x-1$, and $P_{1,2,3}(1.5) = 4$. Find $P_{0123}(1.5)$, $P_{0123}(0)$, $P_{0123}(1)$, $P_{0123}(2)$ and $P_{0123}(3)$.

9. Suppose $f(x) \in C^1[a,b]$. Show that $f'(x) \geqslant (\leqslant) 0$ if and only if

$$f[x_1, x_2] \geqslant (\leqslant) 0,\ \forall x_1, x_2 \in [a,b].$$

10. Suppose $f(x) \in C^2[a,b]$. Show that $f''(x) \geqslant (\leqslant) 0$ if and only if

$$f[x_1, x_2, x_3] \geqslant (\leqslant) 0,\ \forall x_1, x_2, x_3 \in [a,b].$$

11. Suppose three distinct points $A(x_0, f(x_0))$, $B(x_1, f(x_1))$ and $C(x_2, f(x_2))$ lie on the same straight line. Show that the second divided difference $f[x_0, x_1, x_2]$ of $f(x)$ with respect to x_0, x_1, x_2 must be zero.

12. Suppose $x_0 < x_1 < x_2$, $f(x_1) < \min\{f(x_0), f(x_2)\}$. Show that the conic curve $y = p(x)$ passing through the three points $A(x_0, f(x_0))$, $B(x_1, f(x_1))$ and $C(x_2, f(x_2))$ satisfies $p''(x) > 0$.

13. Suppose

$$L(x_0, x_1, \cdots, x_n; f) = \begin{vmatrix} 1 & x_0 & \cdots & x_0^{n-1} & f(x_0) \\ 1 & x_1 & \cdots & x_1^{n-1} & f(x_1) \\ \vdots & \vdots & & \vdots & \vdots \\ 1 & x_n & \cdots & x_n^{n-1} & f(x_n) \end{vmatrix}.$$

Show that the nth divided difference $f[x_0, x_1, \cdots, x_n]$ of $f(x)$ with respect to $x_0, x_1, \cdots, x_n$ can be expressed as

$$f[x_0, x_1, \cdots, x_n] = \frac{L(x_0, x_1, \cdots, x_n; f)}{L(x_0, x_1, \cdots, x_n; x^n)}.$$

14. Show $f[x, x+h] = f[x,x] + hf[x,x,x+h]$.

15. Show $f[x, x+h] = f[x,x] + hf[x,x,x] + h^2 f[x,x,x,x+h]$.

16. Show $f[x, x+h] = \sum_{k=1}^{n} h^{k-1} f[\overbrace{x, \cdots, x}^{k+1}] + h^n f[\overbrace{x, \cdots, x}^{n+1}, x+h]$.

17. Show $f(x+h) = \sum_{k=0}^{n} \frac{f^{(k)}(x)}{k!} h^k + h^{n+1} f[\overbrace{x, \cdots, x}^{n+1}, x+h]$.

18. Show $f[x-h, x] = f[x,x] - hf[x,x,x] + h^2 f[x-h, x, x, x]$.

19. Show $f[x-h, x, x+h] = f[x,x,x] + h^2 f[x-h, x, x, x, x+h]$.

20. Show $f[x-h, x, x, x, x+h] = f[x,x,x,x,x] + h^2 f[x-h, x, x, x, x, x, x+h]$.

21. Show $f[x-h, \overbrace{x, \cdots, x}^{n}, x+h] = f[\overbrace{x, \cdots, x}^{n+2}] + h^2 f[x-h, \overbrace{x, \cdots, x}^{n+2}, x+h]$.

22. Suppose $f(x) \in C^4[x-h, x+h]$. Show that there exists a number $\xi \in (x-h, x+h)$ such

that

$$2f[x-h,x,x+h]=f''(x)+\frac{h^2}{12}f^{(4)}(\xi).$$

23. Suppose $f(x)\in C^6[x-h,x+h]$. Show that there exists a number $\eta\in(x-h,x+h)$ such that

$$2f[x-h,x,x+h]=f''(x)+\frac{h^2}{12}f^{(4)}(x)+\frac{h^4}{360}f^{(6)}(\eta).$$

24. Suppose $f(x)\in C^{2n+2}[x-h,x+h]$. Show that there exists a number $\xi\in(x-h,x+h)$ such that

$$f[x-h,x,x+h]=\sum_{k=1}^{n}\frac{h^{2k-2}}{(2k)!}f^{(2k)}(x)+\frac{h^{2n}}{(2n+2)!}f^{(2n+2)}(\xi).$$

25. Let $f(x)=x^7$. Compute the fifth divided difference $f[0,1,1,1,2,2]$ of $f(x)$. It is known that this divided difference is expressible in terms of the fifth derivative of $f(x)$ evaluated at some ξ, $0<\xi<2$. Determine ξ.

26. Let $f(x)=\mathrm{e}^x$. Prove that for any real number t, one has

$$f[t,t+1,\cdots,t+n]=\frac{(\mathrm{e}-1)^n}{n!}\mathrm{e}^t.$$

{*Hint*: use induction on n. }

27. Let $f(x)=\mathrm{e}^x$. We know that

$$f[0,1,\cdots,n]=\frac{f^{(n)}(\xi)}{n!},\quad 0<\xi<n.$$

Use the result in the above problem to determine ξ. Is ξ located to the left or to the right of the midpoint $n/2$?

28. (Euler, 1734) Let $x_k=10^k, k=0,1,2,\cdots$, and $f(x)=\log_{10}x$.

a. Show that

$$f[x_0,x_1,\cdots,x_n]=\frac{(-1)^{n-1}}{10^{n(n-1)/2}(10^n-1)},\quad n=1,2,3,\cdots.$$

(*Hint*: prove more generally

$$f[x_r,x_{r+1},\cdots,x_{r+n}]=\frac{(-1)^{n-1}}{10^{rn+n(n-1)/2}(10^n-1)},\quad r\geqslant 0, n=1,2,3,\cdots$$

by induction on n.)

b. Use Newton's interpolatory divided difference formula to determine the polynomial $p_n(x)$ of degree n, which interpolates $f(x)$ at $x_0,x_1,\cdots,x_n$. Show that $\lim\limits_{n\to\infty}p_n(x)$ exists for $1\leqslant x<10$. Is the limit equal to $\log_{10}x$?

29. Show that

$$\frac{\partial}{\partial x_0} f[x_0, x_1, \cdots, x_n] = f[x_0, x_0, x_1, \cdots, x_n],$$

assuming $f(x)$ is differentiable at x_0. What about the partial derivative with respect to one of the other variables?

30. Suppose $g_j(x) = x - x_j$. Show that

$$(fg_j)[x_0, x_1, \cdots, x_n] = f[x_0, x_1, \cdots, x_{j-1}, x_{j+1}, \cdots, x_n].$$

31. (Mikeladze, 1941) Assuming that $x_0, x_1, \cdots, x_n$ are mutually distinct, show that

$$f[\overbrace{x_0, x_0, \cdots, x_0}^{m \text{ times}}, x_1, \cdots, x_n]$$

$$= \frac{f[\overbrace{x_0, x_0, \cdots, x_0}^{m \text{ times}}]}{\prod_{i=1}^{n}(x_0 - x_i)} + \sum_{j=1}^{n} \frac{f[\overbrace{x_0, x_0, \cdots, x_0}^{(m-1) \text{ times}}, x_j]}{\prod_{\substack{i=1 \\ i \neq j}}^{n}(x_j - x_i)}.$$

{*Hint:* use induction on m.}

32. Let $f(x) = \dfrac{1}{a-x}$, and $a, x_0, x_1, \cdots, x_n$ are mutually distinct numbers. Show that

$$f[x_0, x_1, \cdots, x_k] = \frac{1}{(a-x_0)(a-x_1)\cdots(a-x_k)}, \quad k = 1, 2, \cdots, n.$$

33. Let $f(x) = a_0 + a_1 x + \cdots + a_n x^n$ and $f(x_i) = 0, i = 1, 2, \cdots, n$, where $a_n \neq 0$, and $x_1, x_2, \cdots, x_n$ are mutually distinct numbers. Show that

$$\sum_{i=1}^{n} \frac{x_i^k}{f'(x_i)} = \begin{cases} 0, & 0 \leqslant k \leqslant n-2; \\ \dfrac{1}{a_n}, & k = n-1. \end{cases}$$

34. Suppose $x_1, x_2, \cdots, x_n$ are mutually distinct numbers, and $F(x) = \prod_{i=1}^{n}(x - x_i)$. Show that

$$\frac{x^k}{F(x)} = \begin{cases} 1 + \displaystyle\sum_{i=1}^{n} \frac{x_i^n}{(x-x_i)F'(x_i)}, & k = n; \\ \displaystyle\sum_{i=1}^{n} \frac{x_i^k}{(x-x_i)F'(x_i)}, & 0 \leqslant k \leqslant n-1. \end{cases}$$

35. Use the Newton's interpolatory divided-difference formula based on the following data to construct a cubic polynomial $P_3(x)$ such that $P_3(x_i) = f(x_i)$ for $i = 0, 1, 2, 3$, and give the remainder term.

i	0	1	2	3
x_i	1	2	3	4
$f(x_i)$	3	10	29	66

36. For a function $f(x)$, the forward divided differences are given in the following table

$x_0=0$	$f(x_0)=1$			
$x_1=1$	$f(x_1)=A$	$f[x_0,x_1]=1$		
$x_2=2$	$f(x_2)=B$	$f[x_1,x_2]=D$	$f[x_0,x_1,x_2]=3$	
$x_3=3$	$f(x_3)=C$	$f[x_2,x_3]=E$	$f[x_1,x_2,x_3]=F$	$f[x_0,x_1,x_2,x_3]=1$

Determine A,B,C,D,E, and F in the table.

37. For a function $f(x)$, the forward divided differences are given in the following table

$x_0=0$	$f(x_0)=1$			
$x_0=0$	$f(x_0)=1$	$f[x_0,x_0]=3$		
$x_1=1$	$f(x_1)=A$	$f[x_0,x_1]=B$	$f[x_0,x_0,x_1]=C$	
$x_1=1$	$f(x_1)=A$	$f[x_1,x_1]=12$	$f[x_0,x_1,x_1]=D$	$f[x_0,x_0,x_1,x_1]=1$

a. Determine A,B,C and D in the table.

b. Use the data in the above table to construct a cubic polynomial $P_3(x)$ such that
$P_3(x_0)=f(x_0),\ P_3(x_1)=f(x_1),\ P_3'(x_0)=f'(x_0),\ P_3'(x_1)=f'(x_1).$

c. Estimate the error $R(x)=f(x)-P_3(x)$.

38. Suppose $A(x_0,f(x_0))$, $B(x_1,f(x_1))$, $C(x_0-h,f(x_0-h))$, $D(x_1+h,f(x_1+h))$ are four vertices of an isosceles trapezoid. Show that the four vertices lie on a parabola.

39. If $f(x)=\mathrm{e}^x$, find out $\lim\limits_{x_0,x_1,\cdots,x_n\to 0} f[x_0,x_1,\cdots,x_n]$.

40. Use the formula

$$(fg)[x_0,x_1,\cdots,x_k]=\sum_{i=0}^{k} f[x_0,x_1,\cdots,x_i]g[x_i,x_{i+1},\cdots,x_k]$$

to derive the Leibniz rule of derivatives concerning the nth derivative of the product of two functions

$$[f(x)g(x)]^{(k)}=\sum_{i=0}^{k}\binom{k}{i}f^{(i)}(x)g^{(k-i)}(x).$$

41. Suppose $f(x)=x^m$. Find the mth divided difference $f[0,1,\cdots,m]$.

42. Suppose $f(x)=ax^3+bx^2+cx+d$. Then the third divided difference $f[0,1,2,3]=$______, the fourth divided difference $f[0,1,2,3,4]=$______.

43. Define the first-order forward difference of $P(x)$ at x as: $\Delta P(x)=P(x+1)-P(x)$ and denote by $P[x_0,x_1,\cdots,x_n]$ the nth divided difference of $P(x)$ with respect to $x_0,x_1,\cdots,x_n$. Suppose all the third-order forward differences of a polynomial $P(x)$ are 1, and $P(0)=2$, $P(1)=-1$, $P(2)=4$. Show $P[x,x+1,x+2,x+3]=\frac{1}{3!}\Delta^3P(x)$.

44. Construct a two-point cubic Hermite polynomial $H_3(x)$ interpolating $f(x)=\sqrt{x}$ at $x_0 = 1.96$, and $x_1 = 2.25$. Compute $H_3(2)$ and estimate the error $|f(2)-H_3(2)|$.

45. Use the following data to construct an interpolating polynomial $P_3(x)$ of degree three so that $P_3(x_i) = f(x_i)$ for $i = 0,1,2$ and $P_3'(x_0) = f'(x_0)$.

x_i	0	1	2
$f(x_i)$	1	2	9
$f'(x_i)$	0		

46. Suppose $n \geqslant 1$. The function $f(x)$ and its derivatives of order up to and including $2n+1$ are continuous on $[a,b]$. The points $x_i, i = 0,1,\cdots,n,$ are distinct and lie in $[a,b]$. Construct polynomials $l_0(x), h_i(x), k_i(x), i = 1,2,\cdots,n,$ of degree $2n$ such that the polynomial

$$p_{2n}(x) = l_0(x)f(x_0) + \sum_{i=1}^{n}[h_i(x)f(x_i) + k_i(x)f'(x_i)]$$

satisfies the conditions

$$p_{2n}(x_i) = f(x_i), \quad i = 0,1,\cdots,n,$$

and

$$p_{2n}'(x_i) = f'(x_i), \quad i = 1,2,\cdots,n.$$

Show also that for each value of x in $[a,b]$ there is a number η, depending on x, such that

$$f(x) - p_{2n}(x) = \frac{(x-x_0)\prod_{i=1}^{n}(x-x_i)^2}{(2n+1)!} f^{(2n+1)}(\eta).$$

47. Suppose that $n \geqslant 2$. The function $f(x)$ and its derivatives of order up to and including $2n$ are continuous on $[a,b]$. The points $x_i, i = 0,1,\cdots,n,$ are distinct and lie in $[a,b]$. Explain how to construct polynomials $l_0(x),\ l_n(x),\ h_i(x),\ k_i(x),\ i = 1,2,\cdots,n-1,$ of degree $2n-1$ such that the polynomial

$$p_{2n-1}(x) = l_0(x)f(x_0) + l_n(x)f(x_n) + \sum_{i=1}^{n-1}[h_i(x)f(x_i) + k_i(x)f'(x_i)]$$

satisfies the conditions

$$p_{2n-1}(x_i) = f(x_i), \quad i = 0,1,\cdots,n,$$

and

$$p_{2n-1}'(x_i) = f'(x_i), \quad i = 1,2,\cdots,n-1.$$

Show also that for each value of x in $[a,b]$, there is a number η, depending on x, such that

$$f(x)-p_{2n-1}(x)=\frac{(x-x_0)(x-x_n)\prod_{i=1}^{n-1}(x-x_i)^2}{(2n)!}f^{(2n)}(\eta).$$

48. Let $n\geqslant 1$. The interpolation points x_j, $j=0,1,\cdots,2n-1$, are distinct, and $x_{n+j}=x_j+\varepsilon$ for each $j=0,1,\cdots,n-1$. The Lagrange polynomial of degree $2n-1$ for the function $f(x)$ using these points is denoted by $p_{2n-1}(x)$. Show that the terms involving $f(x_j)$ and $f(x_{n+j})$ in $p_{2n-1}(x)$ may be written

$$\frac{\varphi_j(x)\varphi_j(x-\varepsilon)}{\varepsilon\varphi_j(x_j)}\left\{\frac{x-x_j}{\varphi_j(x_j+\varepsilon)}f(x_j+\varepsilon)-\frac{x-x_j-\varepsilon}{\varphi_j(x_j-\varepsilon)}f(x_j)\right\},$$

where

$$\varphi_j(x)=\prod_{\substack{i=0\\ i\neq j}}^{n-1}(x-x_i).$$

Find the limit of this expression as $\varepsilon\to 0$, and deduce that $p_{2n-1}(x)-q_{2n-1}(x)\to 0$ as $\varepsilon\to 0$, where $q_{2n-1}(x)$ is the Hermite interpolation polynomial for $f(x)$, using the points x_j, $j=0,1,\cdots,n-1$.

49. A natural cubic spline $S(x)$ on $[0,2]$ is defined by

$$S(x)=\begin{cases} S_0(x)=1+2x-x^3, & 0\leqslant x<1;\\ S_1(x)=2+b(x-1)+c(x-1)^2+d(x-1)^3, & 1\leqslant x\leqslant 2.\end{cases}$$

Find b, c and d.

50. A clamped cubic spline $S(x)$ for a function $f(x)$ is defined by

$$S(x)=\begin{cases} S_0(x)=1+Bx+2x^2-2x^3, & 0\leqslant x<1,\\ S_1(x)=1+b(x-1)-4(x-1)^2+7(x-1)^3, & 1\leqslant x\leqslant 2.\end{cases}$$

Find $f'(0)$ and $f'(2)$.

51. For what value of k is the following a spline function?

$$q(x)=\begin{cases} kx^2+\dfrac{3}{2}, & 0\leqslant x\leqslant 1;\\ x^2+x+\dfrac{1}{2}, & 1\leqslant x\leqslant 2.\end{cases}$$

52. For what value of k is the following a spline function?

$$q(x)=\begin{cases} x^3-x^2+kx+1, & 0\leqslant x\leqslant 1;\\ -x^3+(k+2)x^2-kx+3, & 1\leqslant x\leqslant 2.\end{cases}$$

53. For what value of k is the following a spline function?

$$q(x)=\begin{cases} x^3+3x^2+1, & -1\leqslant x\leqslant 0;\\ -x^3+kx^2+1, & 0\leqslant x\leqslant 1.\end{cases}$$

54. Suppose $f(x) \in C^2[a,b]$, $S(x)$ is a clamped cubic interpolation spline defined on the partition $\Delta: a = x_0 < x_1 < \cdots < x_n = b$. Show that

$$\int_a^b S''(x)[f''(x) - S''(x)]\mathrm{d}x = 0.$$

55. Suppose $f(x) \in C^2[a,b]$, $S(x)$ is a clamped cubic interpolation spline defined on the partition $\Delta: a = x_0 < x_1 < \cdots < x_n = b$. Show that

$$\int_a^b [f''(x) - S''(x)]^2\mathrm{d}x = \int_a^b [f''(x)]^2\mathrm{d}x - \int_a^b [S''(x)]^2\mathrm{d}x.$$

56. Suppose $f(x) \in C^4[a,b]$, $S(x)$ is a clamped cubic interpolation spline defined on the partition $\Delta: a = x_0 < x_1 < \cdots < x_n = b$. Show that

$$\int_a^b [f''(x) - S''(x)]^2\,\mathrm{d}x = \int_a^b [f(x) - S(x)]f^{(4)}(x)\,\mathrm{d}x.$$

Chapter 6 Approximation Theory

(逼近论)

本章内容 (Contents)

学习目标 (Learning Objectives)

- 掌握最佳拟合准则
- 熟练应用最小二乘法进行直线拟合 (线性回归)
- 掌握多项式拟合 (多项式回归)
- 了解指数拟合
- 了解函数逼近的基本思想
- 知悉最小平方逼近多项式的存在性
- 掌握最小平方逼近多项式的求解方法
- 熟悉函数系的线性无关性和正交性概念
- 能熟练应用 Gram-Schmidt 正交化过程构造正交多项式
- 掌握 Chebyshev 多项式的递推公式和正交性
- 掌握 Chebyshev 多项式的零点和极值点
- 掌握 Chebyshev 多项式的极小性质
- 了解 Chebyshev 多项式的零点在多项式插值误差分析中的特殊作用
- 会用 Chebyshev 极小性质构造容许误差下的降阶逼近多项式

6.1 Discrete least squares approximation (离散最小二乘逼近)

- Linear regression (线性回归)

- Criteria for the "best" fit (最佳拟合准则)
- Least squares fit of a straight line (最小二乘直线拟合)
- Polynomial fitting (polynomial regression) (多项式拟合(多项式回归))
- Exponential fitting (指数拟合)

6.1.1 Linear regression (线性回归)

The simplest example of a least-squares approximation (最小平方逼近) is fitting (拟合) a straight line to a set of paired observations:

$$(x_1, y_1), (x_2, y_2), \cdots, (x_m, y_m).$$

The mathematical expression for the straight line is

$$y = a_0 + a_1 x + e, \tag{6.1.1}$$

where a_0, a_1 are coefficients representing the intercept (截距) and the slope (斜率), respectively, and e is the error, or residual (余数), between the model and the observations, which can be represented by rearranging (6.1.1) as

$$e = y - a_0 - a_1 x.$$

Thus, the error, or residual, is the discrepancy (差异) between the true value of y and approximate value, $a_0 + a_1 x$, predicted by the linear equation.

6. 1.2 Criteria for the "best" fit (最佳拟合准则)

One strategy for fitting the "best" line through the data would be to minimize the sum of the residual errors for all the available data, as in

$$\sum_{i=1}^{m} e_i = \sum_{i=1}^{m} (y_i - a_0 - a_1 x_i). \tag{6.1.2}$$

This is an inadequate (不充分的) criterion. Obviously, the best fit is the line connecting the points. However, any straight line passing through the midpoint of the connecting line (except a perfectly vertical line) results in a minimum value of (6.1.2) equal to zero because the errors cancel.

Therefore, another logical criterion might be to minimize the sum of the absolute values of the discrepancies, as in

$$\sum_{i=1}^{m} |e_i| = \sum_{i=1}^{m} |y_i - a_0 - a_1 x_i|.$$

This quantity is called the absolute deviation (绝对偏差).

To minimize a function of two variables, we need to set its partial derivatives (偏导数) to zero and simultaneously solve the resulting equation.

In the case of the absolute deviation, we need to find a_0 and a_1 with

$$0=\frac{\partial}{\partial a_0}\sum_{i=1}^{m}|y_i - a_0 - a_1x_i|$$

and

$$0=\frac{\partial}{\partial a_1}\sum_{i=1}^{m}|y_i - a_0 - a_1x_i|.$$

The difficulty is that the absolute value function is not differentiable (可微的) at zero, and we may not be able to find solutions to this pair of equations.

6.1.3 Least squares fit of a straight line (最小二乘直线拟合)

The least squares approach to this problem involves determining the best approximating line when the error involved is the sum of the squares of the differences between the y-values on the approximating line and the given y-values.

The general problem of fitting the best least squares line to a collection of data $\{(x_i, y_i)\}_{i=1}^{m}$ involves minimizing the total error,

$$E = E(a_0, a_1) = \sum_{i=1}^{m}(y_i - a_0 - a_1x_i)^2,$$

with respect to the parameters a_0 and a_1.

For a minimum to occur, we need

$$0=\frac{\partial}{\partial a_0}\sum_{i=1}^{m}(y_i - a_0 - a_1x_i)^2 = -2\sum_{i=1}^{m}(y_i - a_0 - a_1x_i)$$

and

$$0=\frac{\partial}{\partial a_1}\sum_{i=1}^{m}(y_i - a_0 - a_1x_i)^2 = -2\sum_{i=1}^{m}(y_i - a_0 - a_1x_i)x_i.$$

Simplifying these equations produces the normal equations:

$$ma_0 + \left(\sum_{i=1}^{m}x_i\right)a_1 = \sum_{i=1}^{m}y_i,$$

$$\left(\sum_{i=1}^{m}x_i\right)a_0 + \left(\sum_{i=1}^{m}x_i^2\right)a_1 = \sum_{i=1}^{m}x_iy_i.$$

Solving this system of equations for a_0 and a_1 gives

$$a_0 = \frac{\sum_{i=1}^{m}x_i^2\sum_{i=1}^{m}y_i - \sum_{i=1}^{m}x_iy_i\sum_{i=1}^{m}x_i}{m\left(\sum_{i=1}^{m}x_i^2\right) - \left(\sum_{i=1}^{m}x_i\right)^2}$$

and

$$a_1 = \frac{m\sum_{i=1}^{m} x_i y_i - \sum_{i=1}^{m} x_i \sum_{i=1}^{m} y_i}{m\left(\sum_{i=1}^{m} x_i^2\right) - \left(\sum_{i=1}^{m} x_i\right)^2}.$$

6.1.4 Polynomial fitting (polynomial regression) (多项式拟合(多项式回归))

Similarly a polynomial

$$P_n(x) = a_0 + a_1 x + \cdots + a_n x^n$$

of degree $n(n < m-1)$ can be used to approximate a set of data $\{(x_i, y_i)\}_{i=1}^{m}$ in the sense of the least squares.

This problem is to determine the constants $a_0, a_1, \cdots, a_n$ to minimize the least squares error

$$E = \sum_{i=1}^{m} (y_i - P_n(x_i))^2 = \sum_{i=1}^{m} \left(y_i - \sum_{j=0}^{n} a_j x_i^j \right)^2.$$

It follows

$$\frac{\partial E}{\partial a_k} = -2\sum_{i=1}^{m} \left(y_i - \sum_{j=0}^{n} a_j x_i^j \right) x_i^k, \ k = 0, 1, \cdots, n.$$

As in the linear case, for E to be minimized, it is necessary that

$$\frac{\partial E}{\partial a_k} = 0, \ k = 0, 1, \cdots, n.$$

Thus, for each k,

$$0 = \frac{\partial E}{\partial a_k} = -2\sum_{i=1}^{m} y_i x_i^k + 2\sum_{j=0}^{n} a_j \sum_{i=1}^{m} x_i^{j+k}.$$

This gives $n+1$ normal equations in the $n+1$ unknowns a_k

$$\sum_{j=0}^{n} \left(\sum_{i=1}^{m} x_i^{j+k} \right) a_j = \sum_{i=1}^{m} y_i x_i^k, \ k = 0, 1, \cdots, n.$$

Let us now prove that the above equations have a unique solution provided that the x_i are distinct.

To prove

$$\sum_{j=0}^{n} \left(\sum_{i=1}^{m} x_i^{j+k} \right) a_j = \sum_{i=1}^{m} y_i x_i^k, \ k = 0, 1, \cdots, n \tag{6.1.3}$$

have a unique solution is to prove

$$\sum_{j=0}^{n} \left(\sum_{i=1}^{m} x_i^{j+k} \right) a_j = 0, \ k = 0, 1, \cdots, n \tag{6.1.4}$$

have only zero solution.

Clearly, (6.1.4) can be written as

$$\sum_{i=1}^{m}\left(\sum_{j=0}^{n}a_j x_i^j\right)x_i^k = 0, \quad k = 0,1,\cdots,n.$$

Multiplying the both sides of the above equation by a_k and summing over all k for $k = 0,1,\cdots, n,$ gives

$$\sum_{i=1}^{m}\left(\sum_{j=0}^{n}a_j x_i^j\right)\sum_{k=0}^{n}a_k x_i^k = 0,$$

namely,

$$\sum_{i=1}^{m}\left(\sum_{j=0}^{n}a_j x_i^j\right)^2 = 0.$$

Therefore

$$\sum_{j=0}^{n}a_j x_i^j = 0, \quad i = 1,2,\cdots,m.$$

This means that the polynomial of degree n

$$P_n(x) = \sum_{j=0}^{n}a_j x^j$$

has m distinct zeros $x_i, i = 1,2,\cdots,m.$ $m > n+1$ implies $P_n(x) \equiv 0,$ which leads to

$$a_j = 0, \quad j = 0,1,\cdots,n,$$

as asserted.

Remark One can use Theorem 2.12 and Theorem 2.13 in the Extending Reading part of Chapter 2 to find out the solution to the polynomial fitting problem.

6.1.5 Exponential fitting (指数拟合)

There occasionally occurs the situation when the data are possibly exponentially distributed. In this case, the approximating function is required to be assumed as the form

$$y = b\mathrm{e}^{ax}$$

for some constants a and b.

The difficulty with applying the least squares procedure in a situation of this type comes from attempting to minimize the quantity

$$E = \sum_{i=1}^{m}\left(y_i - b\mathrm{e}^{ax_i}\right)^2.$$

Differentiating E with respect to a and b, respectively, gives

$$0 = \frac{\partial E}{\partial b} = 2\sum_{i=1}^{m}\left(y_i - b\mathrm{e}^{ax_i}\right)\left(-\mathrm{e}^{ax_i}\right)$$

and

$$0=\frac{\partial E}{\partial a}=2\sum_{i=1}^{m}\left(y_i-be^{ax_i}\right)\left(-bx_ie^{ax_i}\right),$$

which results in solving a system of nonlinear equations. The common practice to get out the tight corner is to consider the logarithm (对数) of the approximating equation: $\ln y=\ln b+ax$. Let $Y=\ln y$, $B=\ln b$. Then one can use the method for linear regression to get the solution a and B, and hence $b=e^{B}$.

6.2 Orthogonal polynomials and least squares approximation (正交多项式和最小平方逼近)

- Basic ideas (基本思想)
- Linearly independent functions (线性无关函数)
- Orthogonal functions (正交函数)
- Gram-Schmidt process (Gram-Schmidt 正交化)

6.2.1 Basic ideas (基本思想)

Suppose $f\in C[a,b]$. Find a polynomial $P_n(x)$ of degree at most n so that the error

$$\int_a^b\left[f(x)-P_n(x)\right]^2\mathrm{d}x$$

is minimized.

The polynomial $P_n(x)$ which minimizes the above error is called least squares approximating polynomial (最小平方逼近多项式).

To determine the least squares approximating polynomial $P_n(x)$, let

$$P_n(x)=a_0+a_1x+\cdots+a_nx^n=\sum_{k=0}^{n}a_kx^k$$

and define

$$E=E(a_0,a_1,\cdots,a_n)=\int_a^b\left[f(x)-\sum_{k=0}^{n}a_kx^k\right]^2\mathrm{d}x.$$

Then

$$\frac{\partial E}{\partial a_j}=\int_a^b2\left[\sum_{k=0}^{n}a_kx^k-f(x)\right]x^j\mathrm{d}x.$$

A necessary condition for the numbers $a_0,a_1,\cdots,a_n$ to minimize E is that

$$\frac{\partial E}{\partial a_j}=0,\quad j=0,1,\cdots,n$$

which results in

$$\int_a^b 2\left[\sum_{k=0}^n a_k x^k - f(x)\right]x^j \mathrm{d}x = 0,\ j = 0,1,\cdots,n,$$

namely,

$$\sum_{k=0}^n a_k \int_a^b x^{j+k}\mathrm{d}x = \int_a^b x^j f(x)\mathrm{d}x,\ j = 0,1,\cdots,n. \tag{6.2.1}$$

We now prove the linear system (6.2.1) has a unique solution provided $f(x) \in C[a,b]$. To prove (6.2.1) has a unique solution is to prove that the coefficients matrix with respect to $a_0, a_1, \cdots, a_n$ is nonsingular, which is equivalent to showing that the corresponding homogeneous equations (齐次方程)

$$\sum_{k=0}^n a_k \int_a^b x^{j+k}\mathrm{d}x = 0,\ j = 0,1,\cdots,n \tag{6.2.2}$$

have only zero solution.

Multiplying both sides of (6.2.2) by a_j and summing over all j for $j = 0,1,\cdots,n$ produces

$$\sum_{j=0}^n \sum_{k=0}^n \int_a^b a_k x^k \cdot a_j x^j \mathrm{d}x = 0.$$

Interchanging the integral sign and the summation signs gives

$$\int_a^b \left[\sum_{j=0}^n a_j x^j\right] \cdot \left[\sum_{k=0}^n a_k x^k\right] \mathrm{d}x = 0$$

or

$$\int_a^b \left[\sum_{k=0}^n a_k x^k\right]^2 \mathrm{d}x = 0,$$

which implies

$$\sum_{k=0}^n a_k x^k \equiv 0,\ x \in [a,b].$$

Therefore $a_k = 0$ for each $k = 0,1,\cdots,n$.

Example 6.1 Find the least squares approximating polynomial of degree two for the function $f(x) = \sin \pi x$ on the interval $[0,1]$.

Solution. To determine the least squares approximating polynomial $P_2(x)$, let

$$P_2(x) = a_0 + a_1 x + a_2 x^2$$

and define

$$E = E(a_0, a_1, a_2) = \int_0^1 \left[a_0 + a_1 x + a_2 x^2 - \sin \pi x\right]^2 \mathrm{d}x.$$

Then

$$\frac{\partial E}{\partial a_0} = \int_0^1 2\left(a_0 + a_1 x + a_2 x^2 - \sin \pi x\right)\mathrm{d}x,$$

$$\frac{\partial E}{\partial a_1}=\int_0^1 2\left(a_0+a_1x+a_2x^2-\sin\pi x\right)x\,\mathrm{d}x,$$

$$\frac{\partial E}{\partial a_2}=\int_0^1 2\left(a_0+a_1x+a_2x^2-\sin\pi x\right)x^2\,\mathrm{d}x.$$

Setting

$$\frac{\partial E}{\partial a_j}=0,\ j=0,1,2,$$

we get the normal equations for $P_2(x)=a_0+a_1x+a_2x^2$ as follows

$$\begin{cases} a_0\int_0^1 \mathrm{d}x+a_1\int_0^1 x\mathrm{d}x+a_2\int_0^1 x^2\mathrm{d}x=\int_0^1 \sin\pi x\,\mathrm{d}x, \\ a_0\int_0^1 x\mathrm{d}x+a_1\int_0^1 x^2\mathrm{d}x+a_2\int_0^1 x^3\mathrm{d}x=\int_0^1 x\sin\pi x\,\mathrm{d}x, \\ a_0\int_0^1 x^2\mathrm{d}x+a_1\int_0^1 x^3\mathrm{d}x+a_2\int_0^1 x^4\mathrm{d}x=\int_0^1 x^2\sin\pi x\,\mathrm{d}x. \end{cases}$$

Performing the integration yields

$$\begin{cases} a_0+\frac{1}{2}a_1+\frac{1}{3}a_2=\frac{2}{\pi}, \\ \frac{1}{2}a_0+\frac{1}{3}a_1+\frac{1}{4}a_2=\frac{1}{\pi}, \\ \frac{1}{3}a_0+\frac{1}{4}a_1+\frac{1}{5}a_2=\frac{\pi^2-4}{\pi^3}. \end{cases}$$

Solving these equations for a_0, a_1 and a_2 gives

$$a_0=\frac{12\pi^2-120}{\pi^3}\approx-0.050465,\quad a_1=\frac{720-60\pi^2}{\pi^3}\approx 4.12251,\quad a_2=-a_1\approx-4.12251.$$

Therefore the least squares approximating polynomial of degree two for the function $f(x)=\sin\pi x$ on the interval $[0,1]$ is

$$P_2(x)=\frac{12\pi^2-120}{\pi^3}+\frac{720-60\pi^2}{\pi^3}x-\frac{720-60\pi^2}{\pi^3}x^2.$$

Example 6.1 illustrates how difficult it is to obtain a least squares polynomial approximation, even if only three parameters are involved.

6.2.2 Linearly independent functions (线性无关函数)

Definition 6.1 The set of functions $\phi_0,\cdots,\phi_n$ is said to be linearly independent on $[a,b]$, if whenever

$$c_0\phi_0(x)+c_1\phi_1(x)+\cdots+c_n\phi_n(x)=0$$

for all $x\in[a,b]$, we have

$$c_0=c_1=\cdots=c_n=0.$$

Otherwise the set of functions is said to be linearly dependent (线性相关).

Theorem 6.1 If $\phi_j(x)$ is a polynomial of exactly degree j for $j=0,1,\cdots,n$, then $\{\phi_0(x),\phi_1(x),\cdots,\phi_n(x)\}$ is linearly independent on any interval $[a,b]$.

Proof. Suppose

$$\phi_i(x)=a_{0,i}+a_{1,i}x+\cdots+a_{i,i}x^i, a_{ii}\neq 0,\ i=0,1,\cdots,n,$$

and $c_0,c_1,\cdots,c_n$ are real numbers for which

$$p(x)=c_0\phi_0(x)+c_1\phi_1(x)+\cdots+c_n\phi_n(x)=0,\ x\in[a,b].$$

Then $p(x)$ can be written as

$$p(x)=\begin{pmatrix}\phi_0(x) & \phi_1(x) & \cdots & \phi_n(x)\end{pmatrix}\begin{pmatrix}c_0\\ c_1\\ \vdots\\ c_n\end{pmatrix}$$

$$=(1\ x\ \cdots\ x^n)\begin{pmatrix}a_{00} & a_{01} & a_{02} & \cdots & a_{0n}\\ & a_{11} & a_{12} & \cdots & a_{1n}\\ & & a_{22} & \cdots & a_{2n}\\ & & & \ddots & \vdots\\ & & & & a_{nn}\end{pmatrix}\cdot\begin{pmatrix}c_0\\ c_1\\ \vdots\\ c_n\end{pmatrix}.$$

$p(x)=0$ for all $x\in[a,b]$ implies that for any $n+1$ distinct points $x_i, i=0,1,\cdots,n$ in $[a,b]$, $p(x_i)=0$, namely,

$$\begin{pmatrix}1 & x_0 & \cdots & x_0^n\\ 1 & x_1 & \cdots & x_1^n\\ \vdots & \vdots & & \vdots\\ 1 & x_n & \cdots & x_n^n\end{pmatrix}\cdot\begin{pmatrix}a_{00} & a_{01} & \cdots & a_{0n}\\ & a_{11} & \cdots & a_{1n}\\ & & \ddots & \vdots\\ & & & a_{nn}\end{pmatrix}\cdot\begin{pmatrix}c_0\\ c_1\\ \vdots\\ c_n\end{pmatrix}=0$$

from which follows

$$c_0=c_1=\cdots=c_n=0$$

since both the matrices are nonsingular.

Hence $\{\phi_0,\cdots,\phi_n\}$ is linearly independent on any interval $[a,b]$.

The proof of Theorem 6.1 is completed.

Theorem 6.2 Denote by Π_n the set of all polynomials of degree at most n. If $\{\phi_0(x),\phi_1(x),\cdots,\phi_n(x)\}$ is a collection of linearly independent polynomials in Π_n, then any polynomial in Π_n can be written uniquely as a linear combination of $\phi_0(x),\phi_1(x),\cdots,\phi_n(x)$.

Proof. Suppose for $i=0,1,\cdots,n$,

$$\phi_i(x)=a_{0,i}+a_{1,i}x+\cdots+a_{n,i}x^n.$$

Denote by A the coefficients matrix of $\phi_0(x),\phi_1(x),\cdots,\phi_n(x)$, that is,

$$A=\begin{pmatrix} a_{00} & a_{10} & \cdots & a_{n0} \\ a_{01} & a_{11} & \cdots & a_{n1} \\ \vdots & \vdots & & \vdots \\ a_{0n} & a_{1n} & \cdots & a_{nn} \end{pmatrix}.$$

Then we have

$$\begin{pmatrix} \phi_0(x) \\ \phi_1(x) \\ \vdots \\ \phi_n(x) \end{pmatrix}=\begin{pmatrix} a_{00} & a_{10} & \cdots & a_{n0} \\ a_{01} & a_{11} & \cdots & a_{n1} \\ \vdots & \vdots & & \vdots \\ a_{0n} & a_{1n} & \cdots & a_{nn} \end{pmatrix}\begin{pmatrix} 1 \\ x \\ \vdots \\ x^n \end{pmatrix}.$$

Now we prove that A is nonsingular. Assume that A is singular, then the linear system of equations

$$(c_0,c_1,\cdots,c_n)\begin{pmatrix} a_{00} & a_{10} & \cdots & a_{n0} \\ a_{01} & a_{11} & \cdots & a_{n1} \\ \vdots & \vdots & & \vdots \\ a_{0n} & a_{1n} & \cdots & a_{nn} \end{pmatrix}=(0,0,\cdots,0)$$

has nonzero solution $(c_0,c_1,\cdots,c_n)$, resulting in that the numbers $c_0,c_1,\cdots,c_n$, which are not all zero, satisfy

$$(c_0,c_1,\cdots,c_n)\begin{pmatrix} \phi_0(x) \\ \phi_1(x) \\ \vdots \\ \phi_n(x) \end{pmatrix}=(c_0,c_1,\cdots,c_n)\begin{pmatrix} a_{00} & a_{10} & \cdots & a_{n0} \\ a_{01} & a_{11} & \cdots & a_{n1} \\ \vdots & \vdots & & \vdots \\ a_{0n} & a_{1n} & \cdots & a_{nn} \end{pmatrix}\begin{pmatrix} 1 \\ x \\ \vdots \\ x^n \end{pmatrix}=0$$

contradicting the fact that $\{\phi_0(x),\phi_1(x),\cdots,\phi_n(x)\}$ is a collection of linearly independent polynomials in Π_n. Therefore the matrix A must be nonsingular. And we can write

$$\begin{pmatrix} 1 \\ x \\ \vdots \\ x^n \end{pmatrix}=\begin{pmatrix} a_{00} & a_{10} & \cdots & a_{n0} \\ a_{01} & a_{11} & \cdots & a_{n1} \\ \vdots & \vdots & & \vdots \\ a_{0n} & a_{1n} & \cdots & a_{nn} \end{pmatrix}^{-1}\begin{pmatrix} \phi_0(x) \\ \phi_1(x) \\ \vdots \\ \phi_n(x) \end{pmatrix}=A^{-1}\begin{pmatrix} \phi_0(x) \\ \phi_1(x) \\ \vdots \\ \phi_n(x) \end{pmatrix}$$

If $p(x)$ is a polynomial in Π_n, then $p(x)$ can be written uniquely as

$$p(x)=(a_0,a_1,\cdots,a_n)\begin{pmatrix} 1 \\ x \\ \vdots \\ x^n \end{pmatrix}=(a_0,a_1,\cdots,a_n)A^{-1}\begin{pmatrix} \phi_0(x) \\ \phi_1(x) \\ \vdots \\ \phi_n(x) \end{pmatrix},$$

which shows that $p(x)$ can be written uniquely as a linear combination of $\phi_0(x),\phi_1(x),\cdots,\phi_n(x)$.

The proof of Theorem 6.2 is completed.

6.2.3 Orthogonal functions (正交函数)

Definition 6.2 (weight function (权函数)) An integrable function $w(x)$ is called a weight function on the interval I if $w(x) \geqslant 0$ for all x in I, and $w(x) \neq 0$ on any subinterval of I.

Definition 6.3 (orthogonality (正交性)) $\{\phi_0(x), \phi_1(x), \cdots, \phi_n(x)\}$ is said to be an orthogonal set of functions (正交函数系) for the interval $[a,b]$ with respect to the weight function $w(x)$ if

$$\int_a^b w(x)\phi_j(x)\phi_k(x)\mathrm{d}x = \begin{cases} 0, & \text{when } j \neq k; \\ \alpha_k > 0, & \text{when } j = k. \end{cases}$$

If, in addition, $\alpha_k = 1$ for each $k = 0,1,\cdots,n$, the set is said to be orthonormal (规范正交的).

Theorem 6.3 If $\{\phi_0(x), \phi_1(x), \cdots, \phi_n(x)\}$ is an orthogonal set of functions on an interval $[a,b]$ with respect to the weight function $w(x)$, then the least squares approximation to $f(x)$ on $[a,b]$ with respect to $w(x)$ is

$$p(x) = \sum_{k=0}^{n} a_k \phi_k(x)$$

which minimizes the error

$$E = E(a_0, a_1, \cdots, a_n) = \int_a^b w(x)[f(x) - \sum_{k=0}^{n} a_k \phi_k(x)]^2 \mathrm{d}x,$$

where, for each $k = 0,1,\cdots,n$,

$$a_k = \frac{\int_a^b w(x)\phi_k(x)f(x)\mathrm{d}x}{\int_a^b w(x)[\phi_k(x)]^2 \mathrm{d}x}.$$

Proof. From

$$0 = \frac{\partial E}{\partial a_j} = 2\int_a^b w(x)[f(x) - \sum_{k=0}^{n} a_k \phi_k(x)]\phi_j(x)\mathrm{d}x, \quad j = 0,1,\cdots,n,$$

it follows

$$\int_a^b w(x)\phi_j(x)f(x)\mathrm{d}x = \sum_{k=0}^{n} a_k \int_a^b w(x)\phi_k(x)\phi_j(x)\mathrm{d}x = a_j \int_a^b w(x)[\phi_j(x)]^2 \mathrm{d}x, \; j = 0,1,\cdots,n.$$

Therefore

$$a_j = \frac{\int_a^b w(x)\phi_j(x)f(x)\mathrm{d}x}{\int_a^b w(x)[\phi_j(x)]^2 \mathrm{d}x}, \quad j = 0,1,\cdots,n.$$

6.2.4 Gram-Schmidt process (Gram-Schmidt 正交化)

Theorem 6.4 The set of polynomial functions $\{\phi_0(x), \phi_1(x), \cdots, \phi_n(x)\}$ defined in the following way is orthogonal on $[a,b]$ with respect to the weight function $w(x)$,

$$\phi_0(x) \equiv 1, \quad \phi_1(x) = x - B_1, \quad x \in [a,b],$$

where

$$B_1 = \frac{\int_a^b x w(x)[\phi_0(x)]^2 \mathrm{d}x}{\int_a^b w(x)[\phi_0(x)]^2 \mathrm{d}x}$$

and when $k \geqslant 2$,

$$\phi_k(x) = (x - B_k)\phi_{k-1}(x) - C_k \phi_{k-2}(x), \quad x \in [a,b]$$

where

$$B_k = \frac{\int_a^b x w(x)[\phi_{k-1}(x)]^2 \mathrm{d}x}{\int_a^b w(x)[\phi_{k-1}(x)]^2 \mathrm{d}x}$$

and

$$C_k = \frac{\int_a^b x w(x)\phi_{k-1}(x)\phi_{k-2}(x) \mathrm{d}x}{\int_a^b w(x)[\phi_{k-2}(x)]^2 \mathrm{d}x}.$$

Corollary 6.5 For any $n > 0$, the set of polynomial functions $\{\phi_0(x), \phi_1(x), \cdots, \phi_n(x)\}$ given in Theorem 6.4 is linearly independent on $[a,b]$ and

$$\int_a^b w(x)\phi_n(x)Q_k(x) \mathrm{d}x = 0$$

for any polynomial $Q_k(x)$ of degree $k < n$.

Proof. Since $\phi_n(x)$ is a polynomial of degree n, Theorem 6.1 implies that $\{\phi_0(x), \phi_1(x), \cdots, \phi_n(x)\}$ is a linearly independent set.

Let $Q_k(x)$ be a polynomial of degree k. By Theorem 6.2 there exist numbers $c_0, c_1, \cdots, c_k$ such that

$$Q_k(x) = \sum_{j=0}^{k} c_j \phi_j(x).$$

Thus,

$$\int_a^b w(x)\phi_n(x)Q_k(x) \mathrm{d}x = \sum_{j=0}^{k} c_j \int_a^b w(x)\phi_n(x)\phi_j(x) \mathrm{d}x = 0,$$

since $\phi_n(x)$ is orthogonal to $\phi_j(x)$ for each $j = 0, 1, \cdots, k$.

Example 6.2 Using the recursive procedure of Theorem 6.4 with

$$P_0(x) \equiv 1,\ w(x) \equiv 1$$

on $[-1,1]$ gives

$$B_1 = \frac{\int_{-1}^{1} x\mathrm{d}x}{\int_{-1}^{1} \mathrm{d}x} = 0.$$

And

$$P_1(x) = (x - B_1)P_0(x) = x,$$

$$B_2 = \frac{\int_{-1}^{1} x[P_1(x)]^2 \mathrm{d}x}{\int_{-1}^{1} [P_1(x)]^2 \mathrm{d}x} = 0, \quad C_2 = \frac{\int_{-1}^{1} x^2 \mathrm{d}x}{\int_{-1}^{1} \mathrm{d}x} = \frac{1}{3}.$$

So

$$P_2(x) = (x - B_2)P_1(x) - C_2 P_0(x) = x^2 - \frac{1}{3}$$

and

$$B_3 = \frac{\int_{-1}^{1} x[P_2(x)]^2 \mathrm{d}x}{\int_{-1}^{1} [P_2(x)]^2 \mathrm{d}x} = 0,$$

$$C_3 = \frac{\int_{-1}^{1} xP_2(x)P_1(x)\mathrm{d}x}{\int_{-1}^{1} [P_1(x)]^2 \mathrm{d}x} = \frac{\int_{-1}^{1} x^2\left(x^2 - \frac{1}{3}\right)\mathrm{d}x}{\int_{-1}^{1} x^2 \mathrm{d}x} = \frac{4}{15}.$$

So

$$\begin{aligned} P_3(x) &= (x - B_3)P_2(x) - C_3 P_1(x) \\ &= x\left(x^2 - \frac{1}{3}\right) - \frac{4}{15}x = x^3 - \frac{3}{5}x. \end{aligned}$$

6.3 Chebyshev polynomials and economization of power series (Chebyshev 多项式与幂级数约化)

- Definition of Chebyshev polynomial $T_n(x)$ (Chebyshev 多项式 $T_n(x)$ 的定义)
- Orthogonality of the Chebyshev polynomials (Chebyshev 多项式的正交性)
- The zeros and extreme points of $T_n(x)$ ($T_n(x)$ 的零点与极值点)
- Minimization property (极小性质)
- Application of minimization property in polynomial interpolation (极小性质在多项式插值中的应用)

- Economization of power series (幂级数的约化)

6.3.1 Definition of Chebyshev polynomial $T_n(x)$ (Chebyshev 多项式 $T_n(x)$ 的定义)

For $x \in [-1,1]$, define

$$T_n(x) = \cos(n \arccos x),\ n \geqslant 0. \tag{6.3.1}$$

We call $\{T_n(x)\}$ the Chebyshev polynomials, which are orthogonal on $[-1,1]$ with respect to the weight function

$$w(x) = \frac{1}{\sqrt{1-x^2}},\quad x \in (-1,1).$$

Although they can be derived by the Gram-Schmidt process in the previous section, it is easier to give their definition and then show that they satisfy the required orthogonality properties.

Let us now show that $T_n(x)$ is a polynomial in x. First note that

$$T_0(x) = 1,\ T_1(x) = \cos(\arccos x) = x.$$

For $n \geqslant 1$, let $\theta = \arccos x$. Then

$$T_n(x) = \cos n\theta,$$

where $\theta \in [0,\pi]$. Using the trigonometric identity (三角恒等式)

$$\cos(n+1)\theta + \cos(n-1)\theta = 2\cos n\theta \cos\theta$$

gives

$$T_{n+1}(x) = 2xT_n(x) - T_{n-1}(x) \tag{6.3.2}$$

Since $T_0(x) = 1$ and $T_1(x) = x$, the above recurrence relation (递推关系) implies that $T_n(x)$ is a polynomial of degree n with leading coefficient 2^{n-1}, when $n \geqslant 1$.

The next three Chebyshev polynomials are

$$\begin{aligned} T_2(x) &= 2xT_1(x) - T_0(x) = 2x^2 - 1, \\ T_3(x) &= 2xT_2(x) - T_1(x) = 4x^3 - 3x, \\ T_4(x) &= 2xT_3(x) - T_2(x) = 8x^4 - 8x^2 + 1. \end{aligned}$$

6.3.2 Orthogonality of the Chebyshev polynomials (Chebyshev 多项式的正交性)

By the definition of orthogonality, the definition of Chebyshev polynomials and the trigometric identity we have

$$\int_{-1}^{1} \frac{1}{\sqrt{1-x^2}} T_n(x) T_m(x) \mathrm{d}x \quad (x = \cos\theta)$$

$$= \int_{\pi}^{0} \frac{1}{\sin\theta} \cos n\theta \cos m\theta (-\sin\theta) \mathrm{d}\theta = \int_0^{\pi} \cos n\theta \cos m\theta \mathrm{d}\theta$$

$$= \frac{1}{2}\int_0^{\pi} [\cos(n+m)\theta + \cos(n-m)\theta] \mathrm{d}\theta$$

$$
=\begin{cases}\dfrac{1}{2}\left[\dfrac{1}{n+m}\sin(n+m)\theta\Big|_0^{\pi}+\dfrac{1}{n-m}\sin(n-m)\theta\Big|_0^{\pi}\right], n\neq m;\\[2ex] \dfrac{1}{2}\left[\dfrac{1}{2n}\sin 2n\theta\Big|_0^{\pi}+\pi\right], \quad n=m\neq 0;\\[2ex] \dfrac{1}{2}(\pi+\pi), \quad n=m=0.\end{cases}
$$

$$
=\begin{cases}0, \quad n\neq m;\\ \dfrac{\pi}{2}, \quad n=m\neq 0;\\ \pi, \quad n=m=0.\end{cases}
$$

which means that the Chebyshev polynomials $\{T_n(x)\}$ are orthogonal on $(-1,1)$ with respect to the weight function $w(x)=\dfrac{1}{\sqrt{1-x^2}}$.

6.3.3 The zeros and extreme points of $T_n(x)$ ($T_n(x)$ 的零点与极值点)

Theorem 6.6 The Chebyshev polynomial $T_n(x)$ of degree $n\geqslant 1$ has n simple zeros in $[-1,1]$ at

$$
x_k=\cos\frac{2k-1}{2n}\pi,
$$

for each $k=1,2,\cdots,n$. And $T_n(x)$ assumes its absolute extreme at

$$
x_k' = \cos\frac{k\pi}{n},
$$

with

$$
T_n(x_k') = (-1)^k, k=0,1,\cdots,n. \tag{6.3.3}
$$

Proof. Since

$$
\begin{aligned}
T_n(x_k)&=\cos(n\arccos x_k)=\cos\left(n\arccos\left(\cos\frac{2k-1}{2n}\pi\right)\right)\\
&=\cos\left(k\pi-\frac{\pi}{2}\right)=0, \quad k=1,2,\cdots,n,
\end{aligned}
$$

this shows that $T_n(x)$ has n distinct zeros $x_k=\cos\dfrac{2k-1}{2n}\pi,\ k=1,2,\cdots,n$, in $[-1,1]$. Note that

$$
\begin{aligned}
T_n'(x)&=\frac{\mathrm{d}}{\mathrm{d}x}\cos(n\arccos x)=-\sin(n\arccos x)\cdot n\cdot\left(-\frac{1}{\sqrt{1-x^2}}\right)\\
&=\frac{n\sin(n\arccos x)}{\sqrt{1-x^2}}
\end{aligned}
$$

and for each $k=1,2,\cdots,n-1$ ($k=0,n$, $\sin\frac{k\pi}{n}=0$, which results in that $T_n'\left(x_k'\right)$ is undefined)

$$T_n'\left(x_k'\right)=\frac{n\sin\left(n\arccos\left(\cos\frac{k\pi}{n}\right)\right)}{\sqrt{1-\left(\cos\frac{k\pi}{n}\right)^2}}=0.$$

Since $T_n(x)$ is a polynomial of degree n, its derivative $T_n'(x)$ is a polynomial of degree $n-1$, and all the zeros of $T_n'(x)$ occur at these $n-1$ points. The only other possibilities for extreme of $T_n(x)$ occur at the endpoints of the interval $[-1,1]$, that is, at $x_0'=1$ and $x_n'=-1$. Since for any $k=0,1,\cdots,n$ we have

$$T_n\left(x_k'\right)=\cos\left(n\arccos\left(\cos\frac{k\pi}{n}\right)\right)=\cos k\pi=(-1)^k.$$

A maximum occurs at each even value of k and a minimum at each odd value of k.

The proof of Theorem 6.6 is completed.

Denote by $\tilde{T}_n(x)$ the monic (polynomial with leading coefficient 1) Chebyshev polynomials. Then for each $n\geqslant 1$

$$\tilde{T}_0(x)=1,\quad \tilde{T}_n(x)=\frac{1}{2^{n-1}}T_n(x).$$

Because $\tilde{T}_n(x)$ is just a multiple of $T_n(x)$, $\tilde{T}_n(x)$ has n simple zeros at $x_k=\cos\frac{2k-1}{2n}\pi$, for each $k=1,2,\cdots,n$ and the extreme values of $\tilde{T}_n(x)$ for $n\geqslant 1$, occur at

$$x_k'=\cos\frac{k\pi}{n},$$

with

$$\tilde{T}_n\left(x_k'\right)=\frac{(-1)^k}{2^{n-1}},\ k=0,1,\cdots,n. \tag{6.3.4}$$

6.3.4 Minimization property (极小性质)

Let $\tilde{\Pi}_n$ denote the set of all monic polynomials of degree n. The relation expressed in Eq. (6.3.4) leads to an important minimization property that distinguishes $\tilde{T}_n(x)$ from the other polynomials in $\tilde{\Pi}_n$.

Theorem 6.7. When $n\geqslant 1$, the polynomials of the form $\tilde{T}_n(x)$ have the property that

$$\frac{1}{2^{n-1}}=\max_{x\in[-1,1]}\left|\tilde{T}_n(x)\right|\leqslant\max_{x\in[-1,1]}\left|P_n(x)\right|,$$
$$P_n(x)\in\tilde{\Pi}_n.$$

Moreover, the equality occurs only if $P_n\equiv\tilde{T}_n$.

Proof. Suppose $P_n(x)\in\tilde{\Pi}_n$ and

$$\max_{x\in[-1,1]}\left|P_n(x)\right| < \frac{1}{2^{n-1}} = \max_{x\in[-1,1]}\left|\tilde{T}_n(x)\right|.$$

Let $Q(x)=\tilde{T}_n(x)-P_n(x)$. Then $Q(x)\neq 0$. Since $\tilde{T}_n(x)$ and $P_n(x)$ are both monic (首项系数为 1) polynomials of degree n, $Q(x)$ is a polynomial of degree at most $n-1$. Moreover, at the extreme points of $\tilde{T}_n(x)$ since $\left|P_n(x)\right| < \frac{1}{2^{n-1}}$, for each $k=0,1,\cdots,n$

$$Q(x_k')=\tilde{T}_n(x_k')-P_n(x_k')=\frac{(-1)^k}{2^{n-1}}-P_n(x_k')=\begin{cases}<0, & k \text{ is odd;}\\ >0, & k \text{ is even.}\end{cases}$$

The Intermediate Value Theorem (介值定理) implies that the polynomial $Q(x)$ has at least one zero between x_j' and x_{j+1}' for each $j=0,1,\cdots,n-1$, thus, $Q(x)$ has at least n zeros in the interval $[-1,1]$. But the degree of $Q(x)$ is less than n, so $Q(x)\equiv 0$. This implies that $P_n(x)\equiv\tilde{T}_n(x)$.

The proof of Theorem 6.7 is completed.

6.3.5 Application of minimization property in polynomial interpolation (极小性质在多项式插值中的应用)

Theorem 6.8 If $P(x)$ is the interpolating polynomial of degree at most n with nodes at the roots of $T_{n+1}(x)$, then for each $f\in C^{n+1}[-1,1]$

$$\max_{x\in[-1,1]}|f(x)-P(x)| \leqslant \frac{1}{2^n(n+1)!}\max_{x\in[-1,1]}\left|f^{(n+1)}(x)\right|.$$

Proof. Suppose $x_0,x_1,\cdots,x_n$ are the roots of $T_{n+1}(x)$. Then

$$\tilde{T}_{n+1}(x)=(x-x_0)(x-x_1)\cdots(x-x_n)=\frac{1}{2^n}T_{n+1}(x).$$

By the error formula of interpolating polynomial,

$$\begin{aligned}f(x)-P(x)&=\frac{f^{(n+1)}(\xi)}{(n+1)!}(x-x_0)(x-x_1)\cdots(x-x_n)\\&=\frac{f^{(n+1)}(\xi)}{(n+1)!}\tilde{T}_{n+1}(x),\xi\in(-1,1).\end{aligned}$$

From Theorem 6.7 immediately follows

$$\max_{x\in[-1,1]}|f(x)-P(x)| \leqslant \frac{1}{2^n(n+1)!}\max_{x\in[-1,1]}\left|f^{(n+1)}(x)\right|.$$

The proof of Theorem 6.8 is completed.

This technique for choosing points to minimize the interpolating error can be extended to a general closed interval $[a,b]$ by using the change of variables

$$t=\frac{1}{2}[(b-a)x+(a+b)]$$

to transform the numbers x_k in the interval $[-1,1]$ into the corresponding number t_k in the interval $[a,b]$.

6.3.6 Economization of power series (幂级数的约化)

Chebyshev polynomials can also be used to reduce the degree of an approximating polynomial with a minimal loss of accuracy. Because the Chebyshev polynomials have a minimum maximum absolute value that is spread uniformly on an interval, they can be used to reduce the degree of an approximation polynomial without exceeding the error tolerance (误差容许).

Let us consider approximating an arbitrary nth-degree polynomial

$$P_n(x) = a_n x^n + a_{n-1}x^{n-1} + \cdots + a_1 x + a_0, \quad x \in [-1,1]$$

with a polynomial of degree at most $n-1$. We want to choose a polynomial $P_{n-1}(x)$ in Π_{n-1} so that

$$\max_{x\in[-1,1]} |P_n(x) - P_{n-1}(x)|$$

is as small as possible.

Since $(P_n(x) - P_{n-1}(x))/a_n$ is a monic polynomial of degree n, applying Theorem 6.7 gives

$$\max_{x\in[-1,1]} \left| \frac{1}{a_n}(P_n(x) - P_{n-1}(x)) \right| \geqslant \frac{1}{2^{n-1}},$$

where equality occurs precisely when

$$\frac{1}{a_n}(P_n(x) - P_{n-1}(x)) = \tilde{T}_n(x).$$

This means that we should choose

$$P_{n-1}(x) = P_n(x) - a_n \tilde{T}_n(x),$$

and with this choice we have the minimum value of

$$\max_{x\in[-1,1]} |P_n(x) - P_{n-1}(x)| = |a_n| \max_{x\in[-1,1]} |\tilde{T}_n(x)| = \frac{|a_n|}{2^{n-1}}.$$

Example 6.3 The function $f(x) = \mathrm{e}^x$ is approximated on the interval $[-1,1]$ by the fourth Maclaurin polynomial

$$P_4(x) = 1 + x + \frac{1}{2}x^2 + \frac{1}{6}x^3 + \frac{1}{24}x^4,$$

which has truncation error (截断误差) for $x \in [-1,1]$

$$|R_4(x)| = \frac{1}{120}|f^{(5)}(\xi)| \cdot |x^5| \leqslant \frac{\mathrm{e}}{120} \approx 0.0226523.$$

Suppose that an error of 0.05 is tolerable and that we would like to reduce the degree of the approximating polynomial while staying within this bound.

The polynomial of degree three or less that best uniformly approximates $P_4(x)$ on $[-1,1]$ is

$$P_3(x)=P_4(x)-a_4\tilde{T}_4(x)$$
$$=1+x+\frac{1}{2}x^2+\frac{1}{6}x^3+\frac{1}{24}x^4-\frac{1}{24}\left(x^4-x^2+\frac{1}{8}\right)=\frac{191}{192}+x+\frac{13}{24}x^2+\frac{1}{6}x^3.$$

With this choice, we have

$$\left|P_4(x)-P_3(x)\right|=\left|a_4\tilde{T}_4(x)\right|\leqslant\frac{1}{24}\cdot\frac{1}{2^3}=\frac{1}{192}\approx 0.00520833.$$

Adding this error bound to the bound for the Maclaurin truncation error gives

$$\left|f(x)-P_3(x)\right|\leqslant\left|f(x)-P_4(x)\right|+\left|P_4(x)-P_3(x)\right|$$
$$=\left|R_4(x)\right|+\left|a_4\tilde{T}_4(x)\right|=\frac{\mathrm{e}}{120}+\frac{1}{192}\approx 0.02786063$$

which is still within the permissible error of 0.05.

The polynomial of degree two or less that best uniformly approximates $P_3(x)$ on $[-1,1]$ is

$$P_2(x)=P_3(x)-a_3\tilde{T}_3(x)$$
$$=\frac{191}{192}+x+\frac{13}{24}x^2+\frac{1}{6}x^3-\frac{1}{6}\left(x^3-\frac{3}{4}x\right)=\frac{191}{192}+\frac{9}{8}x+\frac{13}{24}x^2.$$

However

$$\left|P_3(x)-P_2(x)\right|=\left|\frac{1}{6}\tilde{T}_3(x)\right|=\frac{1}{6}\cdot\frac{1}{2^2}=\frac{1}{24}\approx 0.042.$$

which exceeds the tolerance of 0.05 when added to the already accumulated error bound of 0.0283.

Consequently, the polynomial of least degree that best approximates e^x on $[-1,1]$ with an error bound of less than 0.05 is

$$P_3(x)=\frac{191}{192}+x+\frac{13}{24}x^2+\frac{1}{6}x^3.$$

6.4 本章算法程序及实例 (Algorithms and examples)

6.4.1 最小二乘法 (Discrete least squares approximation)

```
%最小二乘法的函数逼近数据拟合，分别用一、二、三次多项式拟合
function [f,q,e]=LSM(x0,y0,n)
m=length(x0);
A=zeros(n+1,n+1);
for i=1:n+1
    for j=1:n+1
        sx=x0.^(i+j-2);
        A(i,j)=sum(sx);%最小二乘法的系数矩阵
```

```
        end
    end
    B=zeros(n+1,1);
    for i=1:n+1
        sy=x0.^(i-1);
        B(i)=sy*y0';     %常数列向量
    end
    a=inv(A)*B;    %计算多项式的系数
    a=a';
    syms x;
    f=0;
    for i=1:n+1
        f=f+a(i)*x^(i-1);
    end
    f=vpa(f,6);            %输出拟合多项式
    q=fliplr(a);
    y1=zeros(1,m);
    for i=1:m
        y1(i)=polyval(q,x0(i));      %计算多项式的函数值
    end
    e0=(y1-y0).^2;    %计算每个结点的因变量的误差平方
    e=sum(e0);    %输出总误差
```

例 6.4 已知数据点

x	0	1	2	3	4
y	1	1.2	1.7	2.0	2.1

分别用最小二乘法构成二、三次多项式对数据进行拟合.

运行程序:

```
clc,clear
x0=[0 1 2 3 4];y0=[1 1.2 1.7 2.0 2.1];
[f2,q2,e2]=LSM(x0,y0,2);  %输出2次拟合多项式
disp(f2);  %输出多项式函数
disp(e2);  %输出误差
plot(x0,y0,'*')
m=length(x0);
x1=x0(1):0.1:x0(m);
y1=polyval(q2,x1);
hold on
```

```
plot(x1,y1,'-.','linewidth',2)
%%再进行三次多项式拟合
[f3,q3,e3]=LSM(x0,y0,3);  %输出3次拟合多项式
disp(f3);  %输出多项式函数
disp(e3);  %输出误差
x2=x0(1):0.1:x0(m);
y2=polyval(q3,x1);
hold on
plot(x2,y2,'--','linewidth',2), title('最小二乘法拟合曲线')
```

结果：多项式函数如下，图形如图6.1所示.

s2=- 0.0285714*x^2 + 0.414286*x + 0.942857，误差为Erorrs2= 0.0286

s3=- 0.0416667*x^3 + 0.221429*x^2 + 0.0559524*x + 0.992857，误差为Erorrs3= 0.0036

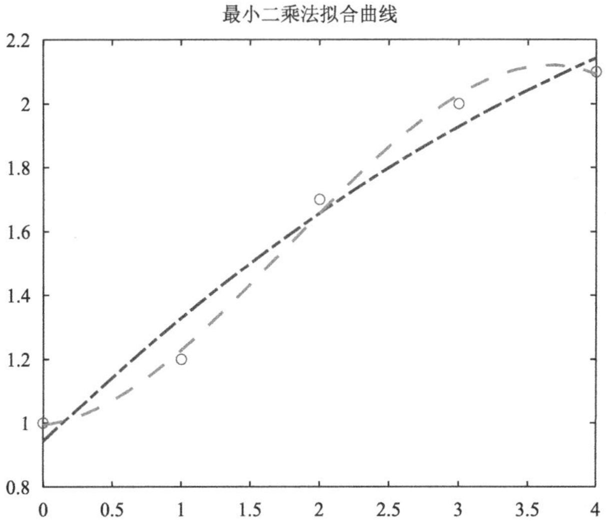

图 6.1 多项式拟合

6.4.2 指数拟合 (Exponential fitting)

例 6.5 已知数据点

x	0	1	2	3	4
y	1	1.2	1.7	2.0	2.1

用指数拟合方法 ($y = ae^{bx}$) 对数据进行拟合.

直接运行程序:

```
clc; clear all
x0=[0  1  2  3  4];
```

```
y0=[1  1.2  1.7  2.0  2.1];
plot(x0,y0,'*');
hold on;
z0=log(y0);   %y=a*exp(b*x)转化为z=b*x+ln(a);
[f,q,e]=LSM(x0,z0,1);  %对z=bx+ln(a)进行线性拟合，通过LSM
%% 画图
m=length(x0);
x1=x0(1):0.1:x0(m);
z1=polyval(q,x1);
y1=exp(z1);  %转化为y值
plot(x1,y1,'r','Linewidth',2);title('Exponential fitting')
%% 输出拟合函数
syms x
a=exp(q(2));
b=q(1);
g=vpa(a*exp(b*x),4);
disp(g);
%% 统计误差平方误差
z2=polyval(q,x0);
y2=exp(z2);
Errors=sum((y0-y2).^2)
```

结果拟合函数为

f=1.031*exp(0.1995*x)，误差为Errors=0.0826.

拟合图形见图 6.2.

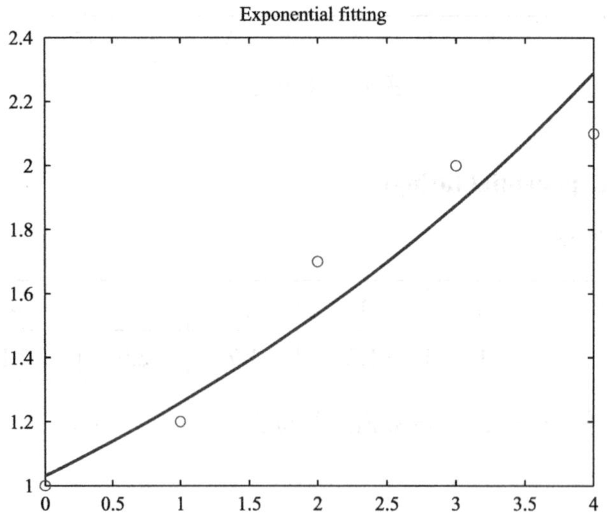

图 6.2　指数拟合

6.4.3 Gram-Schmidt 正交化 (Gram-Schmidt process)

```
function q=Gram_Schmidt(w_x,a,b,k)
syms x;
%% 初始多项式
p=sym('p',[1,k]); %%定义一个初始多项式函数向量
for i=1:k  %%计算前k个正交多项式函数
   p(i)= x^(i-1);
end
%% 正交化
q=sym('q',[1,k]); %%定义一个正交后函数向量
q(1)=1;
for i=2:k
   for j=2:i
        p(i)=p(i)-(int(w_x*q(j-1)*p(i),x,a,b))/(int(w_x*q(j-1)*q(j-1),…
        x,a,b))*q(j-1);
   end
   q(i)=p(i);
end
disp(q)
```

例 6.6 已知初始多项式为$1,x,x^2,x^3$，权函数为$w(x)=x^2$，区间$[-1,1]$，求区间上的带权$w(x)$的正交多项式.

运行程序：

```
syms x;
q=Gram_Schmidt(x^2,-1,1,4);
```

结果：正交多项式系

```
q=[1,x,x^2-3/5,x^3-(5*x)/7].
```

6.4.4 勒让德正交多项式 (Legendre orthogonal polynomials)

例 6.7 求 5 个勒让德正交多项式函数，其权函数为$w(x)=x^2$，区间$[-1,1]$.

```
%%方法一可以直接由Gram-Schmidt计算，权函数w_x=1; 区间[a,b]=[-1,1], k=5
symsx;
q=Gram_Schmidt(1,-1,1,5);
```

结果

```
     q=[ 1, x, x^2 - 1/3, x^3 - (3*x)/5, x^4 - (6*x^2)/7 + 3/35]
%%方法二直接运行以下程序 (递推公式计算)
```

```
clear all;clc;
syms x;
P_0=1;
P_1=x;
P=sym(zeros(1,3));      %除初始两个函数外，再加上新生产的3个多项式
for n=1:3
  Pn=collect(x*(2*n+1)/(n+1)*P_1-n/(n+1)*P_0)
  P_0=P_1;
  P_1=Pn;
  P(n)=Pn;
end
```

运行结果

```
p1=1,
p2=x,
p3=(3*x^2)/2-1/2,
p4=(5*x^3)/2-(3*x)/2,
p5=(35*x^4)/8-(15*x^2)/4+3/8.
```

注：两种方法求出的正交多项式是成比例的.

6.4.5 切比雪夫正交多项式 (Chebyshev orthogonal polynomials)

例 6.8 求 5 个切比雪夫正交多项式函数，其权函数为 $w(x)=\dfrac{1}{\sqrt{1-x^2}}$，区间 $[-1,1]$.

```
%%方法一由Gram-Schmidt计算，权函数w_x=1/sqrt(1-x^2)，区间[a,b]=[-1,1]，k=5
syms x;
q=Gram_Schmidt(1/(1-x^2)^(1/2),-1,1,5);
```

结果

```
        q=[ 1, x, x^2 - 1/2, x^3 - (3*x)/4, x^4 - x^2 + 1/8]
```

```
%%方法二运行以下程序 (递推公式计算)
clear all; clc;
syms x;
T_0=1
T_1=x
T=sym(zeros(1,3));      %除初始两个函数外，再加上新生产的3个多项式
%切比雪夫正交多项式序列前5项求取
for n=1:3
Tn=collect(2*x*T_1-T_0)
T_0=T_1;
T_1=Tn;
```

```
T(n)=Tn;
end
```

结果

```
T1=1,
T2=x,
T3=2*x^2-1,
T4=x*(4*x^2-3),
T5=8*x^4-8*x^2+1.
```

6.4.6 最佳平方逼近 (Least squares approximation)

```
function x = Optimal_Square_Approximation(f,n,I_l,I_r)
%% f为待逼近函数，n为最佳逼近多项式次数，[I_l,I_r]为积分区间
format long;
rightfun=@(i) (@(x) x.^i .* f(x)); %定义x^i*f(x)的匿名函数用于后续求积分
leftfun=@(i,j) (@(x) x.^i .* x.^j); %定义左端的系数矩阵基本形式
%% 定义线性方程的左右两端矩阵
A=zeros(n+1,n+1);
b=zeros(n+1,1);
for p=0:n
   for q=0:n
      A(p+1,q+1)=integral(leftfun(p,q),I_l,I_r); %计算左端系数矩阵的每项积分
   end
   % 计算右端矩阵的每项
   b(p+1,1)=integral(rightfun(p),I_l,I_r);
end
x=inv(A)*b;
```

例 6.9 求函数 $y=\ln x$ 在区间 $[5,10]$ 上的最佳一次逼近和最佳二次逼近多项式.

运行以下程序:

```
clc; clear all
format long;
%% 最佳逼近
f=@(x) log(x);
f1=Optimal_Square_Approximation(f,1,5,10);    %获取最佳一次逼近多项式系数
f2=Optimal_Square_Approximation(f,2,5,10);    %获取最佳二次逼近多项式系数
%% 对多项式系数按从高次到低次排列
f1=flip(f1);
f2=flip(f2);
x0=5:0.01:10;
```

```
y0=f(x0);
%% 画图
figure;
plot(x0,polyval(f1,x0),'b--',x0,polyval(f2,x0),'r--',x0,y0,'k-','Linew
idth',2),title('Square Approximation')
xlabel('x轴');ylabel('y轴');
legend('最佳一次逼近多项式','最佳二次逼近多项式','原曲线');
format short;
```

结果如图 6.3 所示.

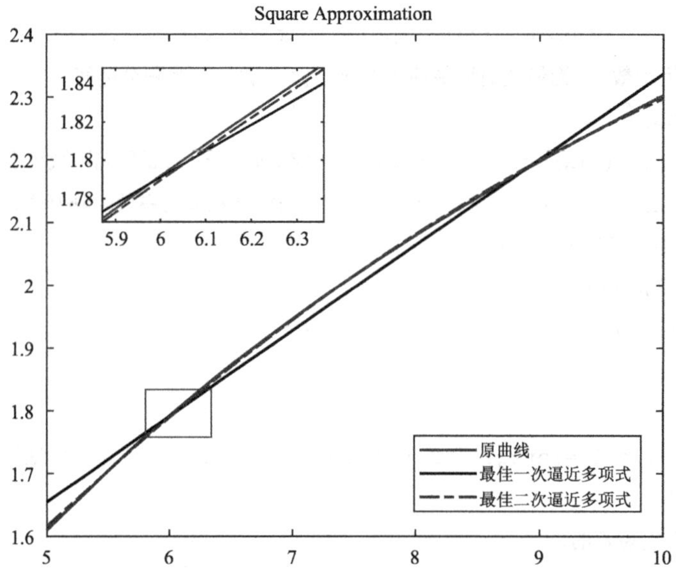

图 6.3 最佳平方逼近

6.5 本章要点 (Hightlights)

1. 最小二乘法，也称最小二乘拟合，或离散最小二乘逼近，是基于误差平方 $\sum_{i=1}^{m}[y_i - P_n(x_i)]^2$ 最小化而提出的一种拟合方法. 该方法的本质是利用多项式逼近离散数据 $\{(x_i, y_i)\}\big|_{i=1}^{m}$，该方法的应用场景是离散数据量较大，而用于拟合的多项式次数 n 远小于离散数据的个数 m.
2. 最小二乘法中多项式系数的求解实际上归结为一个线性方程组的求解问题.
3. 函数的最小平方逼近，是基于加权误差平方 $\int_a^b w(x)[f(x) - p_n(x)]^2 \mathrm{d}x$ 最小化而提出的一种逼近方法，该方法的本质是利用多项式逼近连续函数.

4. 如果$\phi_j(x)$表示 j 次多项式，则$\left\{\phi_0(x),\cdots,\phi_n(x)\right\}$在区间$[a,b]$内线性无关.
5. 一组线性无关的多项式可通过 Gram-Schmidt 正交化方法转换为一组正交多项式.
6. 利用一组正交多项式作为基底，可将多项式表示为其线性组合，从而不用求解线性方程组就可直接求出最小平方逼近多项式.
7. Chebyshev 多项式$T_n(x)$的最高次项的系数为2^{n-1}，并有递推公式

$$T_{n+1}(x)=2xT_n(x)-T_{n-1}(x).$$

8. Chebyshev 多项式$T_n(x)$在$[-1,1]$上有 n 个零点和 $n+1$ 个极值点，并且极大值、极小值点交替出现.
9. Chebyshev 多项式$T_n(x)$的极小值性质表明：在所有最高次系数为 1 的 n 次多项式中，$T_n(x)$与 0 的偏差最小，并且最小值为$\dfrac{1}{2^{n-1}}$.
10. 利用 Chebyshev 多项式可以降低用于函数逼近的多项式的次数，并使得增加的偏差尽可能小.

6.6 问题讨论 (Questions for discussion)

1. 多项式拟合相较于多项式插值有哪些优势？采用 $m-1$ 次多项式对离散数据$\{(x_i,y_i)\}\big|_{i=1}^{m}$进行拟合，则求解出的多项式具有什么特性？
2. 将多项式拟合写成矩阵形式，并讨论多项式拟合中系数的存在性.
3. 设 $f(x)\in C[a,b]$, $p(x)=c_0\varphi_0(x)+c_1\varphi_1(x)+\cdots+c_n\varphi_n(x)$, $w(x)$ 为权函数，则 $p(x)$ 是 $f(x)$ 的最小平方逼近多项式的充要条件为

$$\int_a^b w(x)[f(x)-p(x)]\varphi_j(x)\,\mathrm{d}x=0,\quad j=0,1,\cdots,n.$$

4. 证明 Chebyshev 多项式$T_n(x)$满足微分方程

$$(1-x^2)T_n''(x)-xT_n'(x)+n^2T_n(x)=0.$$

5. 设 $\varphi_j(x)$, $j=0,1,\cdots$, 是区间 (a,b) 上关于权函数 $w(x)$ 正交的多项式系，且 $\varphi_j(x)$ 是 j 次多项式, $j=0,1,\cdots$, 证明：存在常数 C_j, 使得

$$\varphi_{j+1}(x)-C_jx\varphi_j(x)=\sum_{k=0}^{j}\alpha_{jk}\varphi_k(x),\quad \alpha_{jk}\in\mathbf{R}.$$

利用正交性质证明当 $k<j-1$ 时，$\alpha_{jk}=0$, 从而推出下述递推关系式：

$$\varphi_{j+1}(x)-(C_jx+D_j)\varphi_j(x)+E_j\varphi_{j-1}(x)=0,\quad j\geqslant 1.$$

6. 如果对上题中的 $\varphi_j(x)$, $j=0,1,\cdots$, 进行规范化，使得 $\varphi_j(x)$ 的首项系数大于零，则有 ① $C_j>0$, $j\geqslant 1$; ② $\int_a^b w(x)x\varphi_{j-1}(x)\varphi_j(x)\mathrm{d}x>0$; ③ $E_j>0$, $j\geqslant 1$; ④ $\varphi_j(x)$ 的零点与 $\varphi_{j-1}(x)$ 的零点相互交错.

7. 设权函数 $w(x)$ 在区间 $(-a,a)$ 上是偶函数，$\varphi_j(x), j=0,1,\cdots,n$, 是 $(-a,a)$ 上关于权函数 $w(x)$ 的一组正交多项式，则 $\varphi_j(x)$ 的奇偶性与 j 的奇偶性有何关系？若 $f(x)$ 在 $(-a,a)$ 上的最小平方逼近多项式表示为

$$p_n(x)=\alpha_0\varphi_0(x)+\alpha_1\varphi_1(x)+\cdots+\alpha_n\varphi_n(x),$$

则当 $f(x)$ 为偶函数时，所有的奇系数 $\alpha_1,\alpha_3,\alpha_5,\cdots$ 为零；当 $f(x)$ 为奇函数时，所有的偶系数 $\alpha_2,\alpha_4,\alpha_6,\cdots$ 为零.

8. 在区间 $[-1,1]$ 上，用 $f(x)=\sin x$ 的 7 次 Maclaurin 多项式求 $f(x)=\sin x$ 的 5 次约化幂级数，并估计误差.

6.7 关键术语 (Key terms)

discrete least squares approximation (离散最小二乘逼近)
regression (回归)
fit (拟合)
approximation (逼近)
least-squares (最小二乘)
intercept (截距)
slope (斜率)
residual (余数)
discrepancy (差异)
interpolation (插值)
error (误差)
minimize (最小化)
partial derivatives (偏导数)
absolute deviation (绝对偏差)
least squares fit of a straight line (最小二乘直线拟合)
polynomial fitting (多项式拟合)
exponential fitting (指数拟合)
least squares approximating polynomial (最小平方逼近多项式)
linearly independent (线性无关)
linear combination (线性组合)
singular (奇异的)
nonsingular (非奇异的)
orthogonal functions (正交函数系)
weight function (权函数)
orthogonal (正交的)
orthogonality (正交性)

orthonormal (规范正交的)
Chebyshev polynomial (Chebyshev 多项式)
recurrence relation (递推关系)
extreme point (极值点)
minimization property (极小性质)
polynomial interpolation (多项式插值)
power series (幂级数)
economization (约化)
truncation error (截断误差)

6.8 延伸阅读 (Extending reading)

最佳(一致)逼近 (Best uniform approximation)

在 6.2 节中，介绍了最小平方逼近 (least squares approximation) 的概念，还有一种逼近叫做最佳(一致)逼近 (best uniform approximation)，这两种逼近的区别在于它们使用不同的度量，亦即不同的范数. 设 p 是正整数，如果积分

$$\left(\int_a^b |f(x)|^p \mathrm{d}x\right)$$

存在，则称 $f(x)$ 在 $[a,b]$ 上 L^p 可积，记作 $f(x)\in L^p[a,b]$，并称

$$\|f\|_p=\left(\int_a^b |f(x)|^p \mathrm{d}x\right)^{1/p}$$

为 $f(x)$ 的 p–范数. 显然，若 $f(x)$ 在 $[a,b]$ 上连续，即，$f(x)\in C[a,b]$，则 $f(x)\in L^p[a,b]$. 不难证明

$$\lim_{p\to\infty}\|f\|_p=\max_{x\in[a,b]}|f(x)|=\|f\|_\infty .$$

所谓的**最小平方逼近**问题实际上是求解 n 次多项式 $p^*(x)$，实现下述 2-范数意义下的极小化问题：

$$\min_{p(x)\in\Pi_n}\int_a^b\left[f(x)-p(x)\right]^2\mathrm{d}x=\min_{p(x)\in\Pi_n}\|f-p\|_2^2=\left\|f-p^*\right\|_2^2 .$$

上述问题显然等价于

$$\left\|f-p^*\right\|_2=\min_{p(x)\in\Pi_n}\|f-p\|_2 .$$

因此，最小平方逼近也称作**最佳 2-范数逼近** (best approximation in the 2-norm).

最佳 (一致) 逼近 问题则是求解 n 次多项式 $p^*(x)$，实现下述无穷范数意义下的极小化问题：

$$\left\|f-p^*\right\|_\infty=\min_{p(x)\in\Pi_n}\|f-p\|_\infty .$$

因此，最佳 (一致) 逼近也称作是**最佳∞-范数逼近** (best approximation in the ∞-norm)

或极小极大逼近 (minimax approximation).

首先介绍逼近论中一个著名的定理。

定理 6.9 (Weierstrass 逼近定理) 设 $f(x)$ 在 $[a,b]$ 上连续，$\forall \varepsilon > 0$, 则存在多项式 $p(x)$，使得

$$\|f-p\|_\infty < \varepsilon,$$

其中，$\|f\|_\infty = \max_{x\in[a,b]} |f(x)|$.

证明 考虑 Bernstein 多项式

$$b_{nk}(x) = \binom{n}{k} x^k (1-x)^{n-k}, \quad x\in[0,1]$$

令

$$u(t) = (1-x+tx)^n,$$

则由

$$u(t) = (1-x+tx)^n = \sum_{k=0}^{n} b_{nk}(x) t^k$$

知

$$u'(t) = nx(1-x+tx)^{n-1} = \sum_{k=0}^{n} k b_{nk}(x) t^{k-1},$$

$$u''(t) = n(n-1)x^2(1-x+tx)^{n-2} = \sum_{k=0}^{n} k(k-1) b_{nk}(x) t^{k-2},$$

从而，由 $u(1)=1, u'(1)=nx, u''(1)=n(n-1)x^2$ 得到如下 3 个恒等式：

$$\sum_{k=0}^{n} b_{nk}(x) = 1,$$

$$\sum_{k=0}^{n} k b_{nk}(x) = nx,$$

$$\sum_{k=0}^{n} k(k-1) b_{nk}(x) = n(n-1)x^2.$$

于是，

$$\begin{aligned}\sum_{k=0}^{n}\left(x-\frac{k}{n}\right)^2 b_{nk}(x) &= \sum_{k=0}^{n}\left(x^2 - 2\frac{k}{n}x + \frac{k^2}{n^2}\right) b_{nk}(x) \\ &= x^2\sum_{k=0}^{n} b_{nk}(x) - \frac{2x}{n}\sum_{k=0}^{n} k b_{nk}(x) + \frac{1}{n^2}\sum_{k=0}^{n} k^2 b_{nk}(x) \\ &= x^2 - \frac{2x}{n} nx + \frac{1}{n^2}\sum_{k=0}^{n}[k(k-1)b_{nk}(x) + k b_{nk}(x)]\end{aligned}$$

$$=x^2-\frac{2x}{n}nx+\frac{1}{n^2}[n(n-1)x^2+nx]$$

$$=x^2-2x^2+\left(1-\frac{1}{n}\right)x^2+\frac{x}{n}$$

$$=\frac{x(1-x)}{n}.$$

令

$$g(x)=f[a+(b-a)x],$$

则由 $f(x)\in C[a,b]$ 知 $g(x)\in C[0,1]$. 令

$$p(x)=\sum_{k=0}^{n}g\left(\frac{k}{n}\right)b_{nk}(x),$$

则 $p(x)$ 是一个 n 次多项式，且

$$g(x)-p(x)=\sum_{k=0}^{n}\left[g(x)-g\left(\frac{k}{n}\right)\right]b_{nk}(x).$$

因 $g(x)\in C[0,1]$, 故 $g(x)$ 在 $[0,1]$ 上一致连续，于是，$\forall\varepsilon>0,\exists\delta>0$, $\forall x,y\in[0,1]$, 且 $|x-y|<\delta$, 恒有 $|g(x)-g(y)|<\varepsilon/2$. 令 $M=\max\limits_{x\in[0,1]}|g(x)|$，则有

$$\begin{aligned}|g(x)-p(x)|&\leqslant\sum_{k=0}^{n}\left|g(x)-g\left(\frac{k}{n}\right)\right|b_{nk}(x)\\&=\sum_{\left|x-\frac{k}{n}\right|<\delta}\left|g(x)-g\left(\frac{k}{n}\right)\right|b_{nk}(x)+\sum_{\left|x-\frac{k}{n}\right|\geqslant\delta}\left|g(x)-g\left(\frac{k}{n}\right)\right|b_{nk}(x)\\&<\sum_{\left|x-\frac{k}{n}\right|<\delta}\frac{\varepsilon}{2}b_{nk}(x)+\sum_{\left|x-\frac{k}{n}\right|\geqslant\delta}2M\frac{\left(x-\frac{k}{n}\right)^2}{\delta^2}b_{nk}(x)\\&<\sum_{k=0}^{n}\frac{\varepsilon}{2}b_{nk}(x)+\sum_{k=0}^{n}\frac{2M}{\delta^2}\left(x-\frac{k}{n}\right)^2b_{nk}(x)\\&=\frac{\varepsilon}{2}+\frac{2M}{\delta^2}\frac{x(1-x)}{n}.\end{aligned}$$

取 $N=\left[\frac{M}{\delta^2\varepsilon}\right]$，这里[]表示取整，则当 $n>N$ 时，有 $n>\frac{M}{\delta^2\varepsilon}$，由 $x\in[0,1]$, 知 $x(1-x)\leqslant\frac{1}{4}$, 从而

$$\frac{2M}{\delta^2}\frac{x(1-x)}{n}\leqslant\frac{M}{2n\delta^2}<\frac{\varepsilon}{2}.$$

因此，当 $n>N$ 时，$\forall x\in[0,1]$, 有

$$|g(x)-p(x)|<\varepsilon.$$

于是，当 $n > N$ 时，$\forall x \in [a,b]$, 有

$$|f(x) - p(x)| < \varepsilon.$$

即

$$\|f - p\|_\infty < \varepsilon.$$

定理 6.9 证毕.

定义 6.4 令

$$\Delta(p) = \max_{a \leqslant x \leqslant b} |f(x) - p(x)| = \|f - p\|_\infty,$$

则称 $\Delta(p)$ 为 $p(x)$ 与 $f(x)$ 的偏差.

定义 6.5 令

$$E_n = E_n(f) = \inf_{p \in \Pi_n} \Delta(p) = \inf_{p \in \Pi_n} \|f - p\|_\infty,$$

其中，Π_n 表示次数不超过 n 的实系数多项式集合，则称 $E_n(f)$ 为 Π_n 对给定函数 $f(x)$ 的最小偏差或最佳(一致)逼近或最佳 ∞-范数逼近或极小极大逼近.

由 $E_n(f)$ 的定义易知：

$$E_n(f) \geqslant E_{n+1}(f) \geqslant E_{n+2}(f) \geqslant \cdots$$

再由定理 6.9 (Weierstrass 逼近定理) 推得

$$\lim_{n \to \infty} E_n(f) = 0.$$

定理 6.10 (Borel 存在定理) $\forall f(x) \in C[a,b]$, $\exists p(x) \in \Pi_n$, 使得

$$\Delta(p) = E_n(f).$$

证明 由 $E_n(f)$ 是 $\Delta(p)$ 下确界知，$\forall \varepsilon > 0$, $\exists p_\varepsilon(x) \in \Pi_n$, 使得

$$E_n(f) \leqslant \Delta(p_\varepsilon) < E_n(f) + \varepsilon.$$

任取正整数 m，取 $\varepsilon = \dfrac{1}{m}$, 则 $\exists p_m(x) \in \Pi_n$, 使得

$$E_n(f) \leqslant \Delta(p_m) < E_n(f) + \frac{1}{m}.$$

因

$$\begin{aligned} |p_m(x)| &\leqslant |f(x) - p_m(x)| + |f(x)| \\ &\leqslant \Delta(p_m) + \max_{a \leqslant x \leqslant b} |f(x)| \\ &< E_n(f) + \frac{1}{m} + \max_{a \leqslant x \leqslant b} |f(x)| \\ &\leqslant E_n(f) + 1 + \max_{a \leqslant x \leqslant b} |f(x)|, \end{aligned}$$

因此 $\{p_m(x)\}$ 有界. 设 $x_0, x_1, \cdots, x_n$ 是 $[a,b]$ 中 $n+1$ 个互不相同的点，利用 Newton 插值多项式，可将每一个 $p_m(x) \in \Pi_n$ 写成如下形式：

$$p_m(x) = b_0^{(m)} + b_1^{(m)}(x - x_0) + b_2^{(m)}(x - x_0)(x - x_1) + \cdots + b_n^{(m)}(x - x_0)\cdots(x - x_{n-1}),$$

其中 $b_i^{(m)} = p_m[x_0, x_1, \cdots, x_i], i = 0,1,\cdots,n$, 表示 $p_m(x)$ 在结点 $x_0, x_1, \cdots, x_i$ 处的 i 阶差商. 由差商的性质知

$$b_i^{(m)} = p_m[x_0, x_1, \cdots, x_i] = \sum_{j=0}^{i} \frac{p_m(x_j)}{\prod\limits_{k=0, k\neq j}^{i} (x_j - x_k)}, \quad i = 0, 1, \cdots, n,$$

因此 $\{p_m(x)\}$ 有界蕴含了 $\{b_i^{(m)}\}$, $i = 0, 1, \cdots, n$, 有界.由致密性定理 (Bolzano-Weierstrass 定理)，可选出 $n+1$ 个同时收敛的子列 $\{b_i^{(m_j)}\}$, $i = 0, 1, \cdots, n$, 使得

$$\lim_{j\to\infty} b_i^{(m_j)} = b_i, \quad i = 0, 1, \cdots, n.$$

令

$$p(x) = b_0 + b_1(x - x_0) + b_2(x - x_0)(x - x_1) + \cdots + b_n(x - x_0)\cdots(x - x_{n-1}),$$

则有

$$\lim_{j\to\infty} p_{m_j}(x) = p(x), \quad \forall x \in [a,b].$$

因此，

$$\lim_{j\to\infty} \Delta(p_{m_j}) = \lim_{j\to\infty} \max_{a\leqslant x\leqslant b} \left|f(x) - p_{m_j}(x)\right| = \max_{a\leqslant x\leqslant b} \left|f(x) - p(x)\right| = \Delta(p).$$

从而，利用不等式

$$E_n(f) \leqslant \Delta(p_{m_j}) < E_n(f) + \frac{1}{m_j}$$

推得

$$\lim_{j\to\infty} \Delta(p_{m_j}) = \Delta(p) = E_n(f).$$

定理 6.10 证毕.

定理 6.10 中的多项式 $p(x) \in \Pi_n$ 称作 $f(x)$ 在 Π_n 中的**最佳逼近多项式**, 有时也称为**极小极大多项式**.

定理 6.11　(De la Vall′ee Poussin 定理) 设 $f(x) \in C[a,b]$, $p(x) \in \Pi_n$. 如果区间 $[a,b]$ 内存在 $n+2$ 个不同的点 $x_0 < x_1 < \cdots < x_{n+1}$, 使得

$$\mathrm{sign}[f(x_i) - p(x_i)] \cdot \mathrm{sign}[f(x_{i+1}) - p(x_{i+1})] = -1, \quad i = 0, 1, \cdots, n,$$

其中，sign 表示符号函数，即

$$\mathrm{sign}\,(x) = \begin{cases} 1, & x > 0; \\ 0, & x = 0; \\ -1, & x < 0, \end{cases}$$

则有

$$\min_{q\in\Pi_n} \|f - q\|_\infty \geqslant \min_{i=0,1,\cdots,n+1} |f(x_i) - p(x_i)|.$$

证明　令

$$\lambda = \min_{i=0,1,\cdots,n+1} |f(x_i) - p(x_i)|,$$

则 $\lambda \geqslant 0$. 当 $\lambda = 0$ 时，定理的结论显然成立. 下设 $\lambda > 0$. 采用反证法. 假设有一个最佳逼

近多项式 $p_n(x)\in\Pi_n$，满足

$$\|f-p_n\|_\infty=\min_{q\in\Pi_n}\|f-q\|_\infty<\lambda,$$

则由

$$p_n(x_i)-p(x_i)=f(x_i)-p(x_i)-[f(x_i)-p_n(x_i)]>f(x_i)-p(x_i)-\lambda$$

知

$$\operatorname{sign}[p_n(x_i)-p(x_i)]=\operatorname{sign}[f(x_i)-p(x_i)].$$

从而，由定理 6.11 的条件推得

$$\operatorname{sign}[p_n(x_i)-p(x_i)]\cdot\operatorname{sign}[p_n(x_{i+1})-p(x_{i+1})]=-1,\quad i=0,1,\cdots,n.$$

上式表明，连续可微函数 $p_n(x)-p(x)$ 在 $n+2$ 个不同的点 $x_0<x_1<\cdots<x_{n+1}$ 处的函数值异号，根据零点定理，$p_n(x)-p(x)$ 在 (a,b) 内有 $n+1$ 个零点，然而 $p_n(x)\in\Pi_n$，$p_n(x)\in\Pi_n$，蕴含 $p_n(x)-p(x)\in\Pi_n$ 矛盾. 因此，必有

$$\min_{q\in\Pi_n}\|f-q\|_\infty\geqslant\min_{i=0,1,\cdots,n+1}|f(x_i)-p(x_i)|.$$

定理 6.11 证毕.

De la Vall′ee Poussin 定理给出了最佳逼近误差的一个下界估计.

定理 6.12 (Chebyshev 等幅振荡定理 (Chebyshev Equioscillation Theorem)) 设 $f(x)\in C[a,b]$，n 次多项式 $p(x)$ 是 $f(x)$ 在 $[a,b]$ 上的最佳逼近多项式，当且仅当在 $[a,b]$ 内存在至少 $n+2$ 个点 x_i，$i=0,1,\cdots,n+1$，满足 $a\leqslant x_0<x_1<\cdots<x_{n+1}\leqslant b$，使得

$$|f(x_i)-p(x_i)|=\|f-p\|_\infty,\quad i=0,1,\cdots,n+1,$$

且

$$\operatorname{sign}[f(x_i)-p(x_i)]\cdot\operatorname{sign}[f(x_{i+1})-p(x_{i+1})]=-1,\quad i=0,1,\cdots,n.$$

等幅振荡定理可表述为：n 次多项式 $p(x)$ 是 $f(x)$ 在 $[a,b]$ 上的最佳逼近多项式，当且仅当 $f(x)-p(x)$ 在 $[a,b]$ 上的 $n+2$ 个点 x_i，$i=0,1,\cdots,n+1$，交替取得具有相同绝对值的极大值和极小值. 这 $n+2$ 个点 x_i，$i=0,1,\cdots,n+1$，称作临界点 (critical points).

证明 如果 $f(x)\in\Pi_n$，定理的结论显然成立. 以下假设 $f(x)\notin\Pi_n$.

充分性. 设 $f(x)-p(x)$ 在 $[a,b]$ 上的 $n+2$ 个点 x_i，$i=0,1,\cdots,n+1$，交替取得具有相同绝对值 L 的极大值和极小值，则 $L=\|f-p\|_\infty$. 一方面，根据 De la Vall′ee Poussin 定理知，$E_n(f)=\min\limits_{q\in\Pi_n}\|f-q\|_\infty\geqslant L$. 另一方面，$E_n(f)$ 的定义蕴含了，$E_n(f)\leqslant\|f-p\|_\infty=L$. 因此，$E_n(f)=L=\|f-p\|_\infty$，即，$n$ 次多项式 $p(x)$ 是 $f(x)$ 在 $[a,b]$ 上的最佳逼近多项式.

必要性. 设 n 次多项式 $p(x)$ 是 $f(x)$ 在 $[a,b]$ 上的最佳逼近多项式. 令

$$e_n(x)=f(x)-p(x).$$

采用反证法. 假设 $e_n(x)=f(x)-p(x)$ 在 $[a,b]$ 上至多有 $n+1$ 个点 x_i，$i=0,1,\cdots,n$，交替取得具有相同绝对值的极大值和极小值，即 $|e_n(x_i)|=E_n(f), i=0,1,\cdots,n$，且 $e_n(x_i)=-e_n(x_{i+1})$，$i=0,1,\cdots,n-1$. 从而存在正整数 m，使得 $e_n(x_i)=(-1)^{m+i}E_n(f)$, $i=0,1,\cdots,n$, 不妨设 $a\leqslant x_0$

$< x_1 < \cdots < x_n \leqslant b$, 由连续函数的介值定理知，$\exists u_i \in (x_i, x_{i+1}), i = 0,1,\cdots,n-1$, 使得 $e_n(u_i) = 0, i = 0,1,\cdots,n-1$. 令

$$q(x) = (-1)^{m-n}(x-u_0)(x-u_1)\cdots(x-u_{n-1}),$$

则 $q(x) \in \Pi_n$, 且

$$\begin{aligned}\operatorname{sign}[q(x_i)] &= \operatorname{sign}[(-1)^{m-n}(x_i-u_0)(x_i-u_1)\cdots(x_i-u_i)\cdots(x_i-u_{n-1})] \\ &= (-1)^{m-n}(-1)^{n-i} = (-1)^{m-i} = (-1)^{m+i},\ i = 0,1,\cdots,n-1.\end{aligned}$$

对任意的 $\alpha > 0$, 显然有 $\operatorname{sign}[\alpha q(x_i)] = (-1)^{m+i}$. 从而，

$$\operatorname{sign}[\alpha q(x_i)] = (-1)^{m+i} = \operatorname{sign}[e_n(x_i)],\ i = 0,1,\cdots,n-1.$$

由 $e_n(x)$ 和 $\alpha q(x)$ 的连续性知，存在 x_i 的邻域 I_i，使得

$$\operatorname{sign}[\alpha q(x)] = (-1)^{m+i} = \operatorname{sign}[e_n(x)],\ \forall x \in I_i, i = 0,1,\cdots,n-1.$$

显然这些 I_i，$i = 0,1,\cdots,n-1$，互不相交. 令

$$I = I_0 \cup I_1 \cup \cdots \cup I_{n-1},$$

则有

$$\operatorname{sign}[\alpha q(x)] = \operatorname{sign}[e_n(x)], \quad \forall x \in I.$$

令 $\mu = \min\limits_{x\in I}|e_n(x)|$, 由 $\operatorname{sign}[e_n(x)] = (-1)^{m+i}, \forall x \in I_i$，$i = 0,1,\cdots,n-1$, 知 $\mu > 0$.

选取 α, 使

$$0 < \alpha < \frac{\mu}{\|q\|_\infty},$$

则 $\forall x \in I$,

$$|\alpha q(x)| < \frac{\mu}{\|q\|_\infty}|q(x)| \leqslant \mu \leqslant |e_n(x)|.$$

于是，$\forall x \in I$，有

$$|f(x) - [p(x) + \alpha q(x)]| = |e_n(x) - \alpha q(x)| = |e_n(x)| - |\alpha q(x)| < |e_n(x)| \leqslant E_n(f).$$

记 $J = [a,b] \setminus I$, 令 $d = \max\limits_{x\in J}|e_n(x)|$, 则

$$d = \max_{x\in J}|e_n(x)| < \max_{x\in[a,b]}|e_n(x)| = \max_{x\in[a,b]}|f(x) - p(x)| = \|f - p\|_\infty = E_n(f).$$

选取 α, 使

$$0 < \alpha < \frac{1}{2\|q\|_\infty}(E_n(f) - d),$$

则 $\forall x \in J$,

$$\begin{aligned}|f(x) - [p(x) + \alpha q(x)]| &= |e_n(x) - \alpha q(x)| \\ &\leqslant |e_n(x)| + \alpha|q(x)| \\ &< d + \frac{|q(x)|}{2\|q\|_\infty}(E_n(f) - d)\end{aligned}$$

$$\begin{aligned}&\leqslant d+\frac{1}{2}(E_n(f)-d)\\&=\frac{1}{2}d+\frac{1}{2}E_n(f)\\&\leqslant\frac{1}{2}E_n(f)+\frac{1}{2}E_n(f)\\&=E_n(f).\end{aligned}$$

综合以上分析，现取

$$0<\alpha<\min\left\{\frac{1}{2\|q\|_\infty}(E_n(f)-d),\ \frac{\mu}{\|q\|_\infty}\right\},$$

则 $\forall x\in[a,b]=I\cup J$, 均有

$$|f(x)-[p(x)+\alpha q(x)]|<E_n(f)=\|f-p\|_\infty.$$

由于 $p(x)+\alpha q(x)$ 是 n 次多项式，上述不等式表明，$p(x)+\alpha q(x)$ 是比 $p(x)$ 更优的对 $f(x)$ 的逼近，这与 n 次多项式 $p(x)$ 是 $f(x)$ 在 $[a,b]$ 上的最佳逼近多项式矛盾. 必要性得证. 定理 6.12 证毕.

定理 6.13 (最佳逼近唯一性定理) 设 $f(x)\in C[a,b]$, 则 Π_n 中对 $f(x)$ 的最佳一致逼近 $p_n^*(x)$ 是唯一的.

证明 设 $p(x)\in\Pi_n$ 和 $q(x)\in\Pi_n$ 都是 $f(x)$ 的最佳一致逼近，即

$$\|f-p\|_\infty=E_n(f)=\|f-q\|_\infty.$$

令

$$r(x)=\frac{p(x)+q(x)}{2},$$

则 $r(x)\in\Pi_n$，且

$$\|f-r\|_\infty=\left\|\frac{f-p}{2}+\frac{f-q}{2}\right\|_\infty\leqslant\frac{1}{2}\|f-p\|_\infty+\frac{1}{2}\|f-q\|_\infty=E_n(f),$$

又显然，$\|f-r\|_\infty\geqslant E_n(f)$. 因此，$\|f-r\|_\infty=E_n(f)$. 根据 Chebyshev 等幅振荡定理，在 $[a,b]$ 中至少存在 $n+2$ 个点 $x_0<x_1<\cdots<x_{n+1}$, 使得

$$f(x_i)-r(x_i)=(-1)^{m+i}E_n(f),\quad i=0,1,\cdots,n+1,$$

其中，m 是某个正整数. 因此，有

$$\frac{f(x_i)-p(x_i)}{2}+\frac{f(x_i)-q(x_i)}{2}=(-1)^{m+i}E_n(f),\quad i=0,1,\cdots,n+1.$$

从而

$$\frac{1}{2}|f(x_i)-p(x_i)|+\frac{1}{2}|f(x_i)-q(x_i)|\leqslant E_n(f),\quad i=0,1,\cdots,n+1.$$

由 $\|f-p\|_\infty=E_n(f)=\|f-q\|_\infty$，推得

$$\frac{1}{2}|f(x_i)-p(x_i)|=\frac{1}{2}|f(x_i)-q(x_i)|=E_n(f),\quad i=0,1,\cdots,n+1,$$

于是有

$$f(x_i)-p(x_i)=f(x_i)-q(x_i)=(-1)^{m+i}E_n(f),\quad i=0,1,\cdots,n+1.$$

上式蕴含

$$p(x_i)-q(x_i)=0,\quad i=0,1,\cdots,n+1,$$

从而，由 $p(x)\in\Pi_n$，$q(x)\in\Pi_n$ 推得：$p(x)\equiv q(x)$.

定理 6.13 证毕.

6.9 习题 (Exercises)

1. Suppose $f(0)=1$, $f(1)=2$, $f(3)=4$, $f(5)=8$, find the least squares fit of a straight line to the given data.
2. Find the least squares polynomial of degrees 1, 2, and 3 for the data in the following table. Compute the error E in each case. Graph the data and the polynomials.

x_i	1.0	1.1	1.3	1.5	1.9	2.1
y_i	1.84	1.96	2.21	2.45	2.94	3.18

3. Find the linear least squares polynomial approximation to $f(x)=x^2+3x+2$ on the interval $[0,1]$.
4. Find the linear least squares polynomial approximation to $f(x)=x^3$ on the interval $[0,2]$.
5. Find the linear least squares polynomial approximation to $f(x)=\mathrm{e}^x$ on the interval $[0,2]$.
6. Find the linear least squares polynomial approximation to $f(x)=1/x$ on the interval $[1,3]$.
7. Find the linear least squares polynomial approximation to $f(x)=\frac{1}{2}\cos x+\frac{1}{3}\sin 2x$ on the interval$[0,1]$.
8. Find the linear least squares polynomial approximation to $f(x)=x\ln x$ on the interval $[1,3]$.
9. Let

$$f(x)=\begin{cases}0, & 0\leqslant x\leqslant\dfrac{1}{2};\\[2mm] 1, & \dfrac{1}{2}\leqslant x\leqslant 1.\end{cases}$$

a. Find the linear least squares approximant to $f(x)$ on $[0,1]$.

b. Can you do better with continuous piecewise linear functions relative to the partition $[0,1]=\left[0,\frac{1}{2}\right]\cup\left[\frac{1}{2},1\right]$?

10. Construct the orthogonal polynomials $\varphi_k(x)$, $k=0,1,2,3$, with respect to the weight

function $w(x) = x^2$ on $[-1,1]$.

11. Construct orthogonal polynomials of degrees 0, 1 and 2 on the interval $(0,1)$ with the weight function $w(x) = -\ln x$.
12. Let the polynomials $\varphi_j(x)$, $j = 0,1,\cdots$, form an orthogonal system on the interval $(-1,1)$ with respect to the weight function $w(x) \equiv 1$. Show that the polynomials $\varphi_j((2x-a-b)/(b-a))$, $j = 0,1,\cdots$, represent an orthogonal system for the interval (a,b) and the same weight function.
13. Use the Gram-Schmidt procedure to calculate L_1, L_2 and L_3, where $\{L_0(x), L_1(x), L_2(x), L_3(x)\}$ is an orthogonal set of polynomials on $(0,\infty)$ with respect to the function $w(x) = e^{-x}$ and $L_0(x) \equiv 1$. The polynomials obtained from this procedure are called the Laguerre polynomials.
14. Suppose that the polynomials $\varphi_j(x)$, $j = 0,1,\cdots$, form an orthogonal system on the interval $(0,1)$ with respect to the weight function $w(x) = x^\alpha, \alpha > 0$. Find, in terms of $\varphi_j(x)$, $j = 0,1,\cdots$, a system of orthogonal polynomials for the interval $(0,b)$ and the same weight function.
15. Show that for any positive integers i and j with $i > j$, we have

$$T_i(x)T_j(x) = \frac{1}{2}[T_{i+j}(x) + T_{i-j}(x)],$$

where $\{T_n(x)\}$ are the Chebyshev polynomials.

16. Show, by induction or otherwise, that, for $0 \leqslant k \leqslant n$,

$$\frac{\mathrm{d}^k}{\mathrm{d}x^k}(1-x^2)^n = (1-x^2)^{n-k} q_k(x),$$

where $q_k(x)$ is a polynomial of degree k. Deduce that all the derivatives of the function $(1-x^2)^n$ of order less than n vanish at $x = \pm 1$. Define

$$\varphi_j(x) = \frac{\mathrm{d}^j}{\mathrm{d}x^j}(1-x^2)^j.$$

Show by repeated integration by parts that

$$\int_{-1}^{1} \varphi_k(x)\varphi_j(x)\mathrm{d}x = 0, \quad 0 \leqslant k < j.$$

17. Define the Legendre polynomials: $p_0(x) = 1$, $p_n(x) = \frac{1}{2^n n!}\frac{\mathrm{d}^n}{\mathrm{d}x^n}(x^2-1)^n$, $n = 1,2,\cdots$. Show that $p_n(x)$ is a polynomial of degree n with leading coefficient $\frac{(2n)!}{2^n (n!)^2}$, and

$$\int_{-1}^{1} p_n(x)p_m(x)\mathrm{d}x = \begin{cases} 0, & m \neq n; \\ \dfrac{2}{2n+1}, & m = n. \end{cases}$$

18. Prove that the Legendre polynomials: $p_0(x)=1,\ p_n(x)=\dfrac{1}{2^n n!}\dfrac{\mathrm{d}^n}{\mathrm{d}x^n}(x^2-1)^n,\ n=1,2,\cdots,$ satisfy the following recursion relation:

$$(n+1)p_{n+1}(x)=(2n+1)xp_n(x)-np_{n-1}(x),\quad n=1,2,\cdots.$$

19. Prove that the Legendre polynomials: $p_0(x)=1,\ p_n(x)=\dfrac{1}{2^n n!}\dfrac{\mathrm{d}^n}{\mathrm{d}x^n}(x^2-1)^n,\ n=1,2,\cdots,$ satisfy the following differential equation:

$$(1-x^2)p_n''(x)-2xp_n'(x)+n(n+1)p_n(x)=0.$$

20. Prove that the Legendre polynomials: $p_0(x)=1,\ p_n(x)=\dfrac{1}{2^n n!}\dfrac{\mathrm{d}^n}{\mathrm{d}x^n}(x^2-1)^n,\ n=1,2,\cdots,$ satisfy the following differential equation:

$$p_n'(x)=xp_{n-1}'(x)+np_{n-1}(x).$$

21. Prove that the Legendre polynomials: $p_0(x)=1,\ p_n(x)=\dfrac{1}{2^n n!}\dfrac{\mathrm{d}^n}{\mathrm{d}x^n}(x^2-1)^n,\ n=1,2,\cdots,$ satisfy the following differential equation:

$$xp_n'(x)-p_{n-1}'(x)=np_n(x).$$

22. Let $p_0(x)=1,\ \tilde{p}_n(x)=\dfrac{n!}{(2n)!}\dfrac{\mathrm{d}^n}{\mathrm{d}x^n}(x^2-1)^n,\ n=1,2,\cdots$. Then $\tilde{p}_n(x)\in\tilde{\Pi}_n$, and for any polynomial $Q_n(x)\in\tilde{\Pi}_n$, there holds

$$\|Q_n\|_2^2=\int_{-1}^{1}Q_n^2(x)\mathrm{d}x\geqslant\int_{-1}^{1}\tilde{p}_n^2(x)\mathrm{d}x=\|\tilde{p}_n\|_2^2.$$

23. Show, by induction or otherwise, that, for $0\leqslant k\leqslant j$,

$$\frac{\mathrm{d}^k}{\mathrm{d}x^k}(x^j\mathrm{e}^{-x})=x^{j-k}q_k(x)\mathrm{e}^{-x},$$

where $q_k(x)$ is a polynomial of degree k. Define

$$L_j(x)=\mathrm{e}^x\frac{\mathrm{d}^j}{\mathrm{d}x^j}(x^j\mathrm{e}^{-x}).$$

Show that, for each $j\geqslant 0$, $L_j(x)$ is a polynomial of degree j, known as the Laguerre polynomials, and that these polynomials form an orthogonal system on the interval $(0,\infty)$ with respect to the weight function $w(x)=\mathrm{e}^{-x}$. Write down the polynomials with $j=0,1,2$ and 3.

24. Show that the Laguerre polynomials $L_j(x)=\mathrm{e}^x\dfrac{\mathrm{d}^j}{\mathrm{d}x^j}(x^j\mathrm{e}^{-x})$ have the following property:

$$\int_0^{+\infty}\mathrm{e}^{-x}L_m(x)L_n(x)\,\mathrm{d}x=\begin{cases}0, & m\neq n;\\ (n!)^2, & m=n.\end{cases}$$

25. Show that the Laguerre polynomials $L_j(x)=\mathrm{e}^x\dfrac{\mathrm{d}^j}{\mathrm{d}x^j}(x^j\mathrm{e}^{-x})$ satisfy the recursion relation:

$$L_{n+1}(x)=(1+2n-x)L_n(x)-n^2L_{n-1}(x),\quad n=1,2,\cdots.$$

26. Find the sixth Maclaurin polynomial for $x\mathrm{e}^x$, and use Chebyshev economization to obtain a lesser-degree polynomial approximation while keeping the error less than 0.01 on $[-1, 1]$.

27. Find the sixth Maclaurin polynomial for $\sin x$, and use Chebyshev economization to obtain a lesser-degree polynomial approximation while keeping the error less than 0.01 on $[-1, 1]$.

28. Define the Hermite polynomials: $H_0(x)=1,\ \mathrm{H}_n(x)=(-1)^n\mathrm{e}^{x^2}\dfrac{\mathrm{d}^n}{\mathrm{d}x^n}\mathrm{e}^{-x^2}, n=1,2,\cdots$. Show that $H_n(x)$ is a polynomial of degree n and

$$\int_0^{+\infty}H_m(x)H_n(x)\,\mathrm{d}x=\begin{cases}0, & m\neq n;\\ 2^n n!\sqrt{\pi}, & m=n.\end{cases}$$

29. Show that the Hermite polynomials: $H_0(x)=1,\ \mathrm{H}_n(x)=(-1)^n\mathrm{e}^{x^2}\dfrac{\mathrm{d}^n}{\mathrm{d}x^n}\mathrm{e}^{-x^2}, n=1,2,\cdots$, satisfy the following recursion relation:

$$H_{n+1}(x)=2xH_n(x)-2nH_{n-1}(x),\quad n=1,2,\cdots.$$

Chapter 7 Numerical Differentiation and Integration

(数值微分与数值积分)

本章内容 (Contents)

学习目标 (Learning Objectives)

- 了解数值微分的概念
- 掌握向前、向后差分公式
- 掌握三点公式、五点公式，并会用其估算导数
- 了解高阶导数逼近公式，并会用公式计算高阶导数
- 理解舍入误差与截断误差对整体误差的影响
- 掌握 Richardson 外推法的基本思想
- 会用外推法构造指定精度的数值微分公式
- 掌握中点公式、梯形公式、Simpson 公式以及 Newton-Cotes 公式
- 掌握以上中点公式、梯形公式和 Simpson 公式的代数精度
- 掌握数值积分代数精度的定义，并会利用代数精度推导积分公式
- 掌握复化中点公式、复化梯形公式、复化 Simpson 公式
- 理解复化公式的稳定性，会用累积截断误差来判断稳定性
- 知悉用外推法推导 Romberg 积分的思想
- 理解高斯求积、高斯点的基本思想和概念
- 掌握高斯点的性质
- 掌握高斯求积公式中高斯点和高斯系数的求法

7.1 Numerical differentiation (数值微分)

- Forward-difference formula (向前差分公式)
- Backward-difference formula (向后差分公式)
- Three-point formula (三点公式)
- Five-point formula (五点公式)
- Approximation to higher derivatives (高阶导数逼近)
- Effect of round-off error (舍入误差影响)
- Examples (例题)

As we know, polynomials belong to the class of the simplest functions, which can be used to approximate any continuous function defined on a closed interval. And the derivatives and integrals of polynomials are easily obtained and evaluated. It should not be surprising, then, that many procedures for approximating derivatives and integrals use the polynomials that approximate the function.

7.1.1 Forward-difference formula (向前差分公式)

As everyone knows, the derivative of the function $f(x)$ at x_0 is defined by

$$f'(x_0)=\lim_{h\to 0}\frac{f(x_0+h)-f(x_0)}{h}.$$

This definition tells us that the quotient

$$\frac{f(x_0+h)-f(x_0)}{h}$$

which is just the first divided difference of $f(x)$ with respect to x_0 and x_0+h, provides an approximation to $f'(x_0)$ when h is small enough.

Suppose that $f(x)\in C^2[a,b]$, $x_0\in(a,b)$ and $h>0$ is sufficiently small to ensure that $x_1=x_0+h\in[a,b]$. To approximate $f'(x_0)$, we construct a linear interpolating polynomial $P_1(x)$ for f determined by x_0 and x_1, with remainder term:

$$f(x)=P_1(x)+\frac{1}{2!}(x-x_0)(x-x_1)f''(\xi(x)), x\in[a,b], \tag{7.1.1}$$

where $\xi(x)$ lies among x_0, x_1 and x, $P_1(x)$ is given by Newton's interpolatory (forward) divided-difference formula as

$$P_1(x)=f(x_0)+(x-x_0)f[x_0,x_1]=f(x_0)+(x-x_0)f[x_0,x_0+h].$$

Differentiating (7.1.1) gives

$$\begin{aligned}f'(x)&=P_1'(x)+\frac{\mathrm{d}}{\mathrm{d}x}\left[\frac{1}{2}(x-x_0)(x-x_1)f''(\xi(x))\right]\\&=f[x_0,x_1]+\frac{2x-x_0-x_1}{2}f''(\xi(x))+\frac{1}{2}(x-x_0)(x-x_1)\frac{\mathrm{d}}{\mathrm{d}x}f''(\xi(x)).\end{aligned}$$

So

$$f'(x_0)=f[x_0,x_1]-\frac{h}{2}f''(\xi)=\frac{f(x_0+h)-f(x_0)}{h}-\frac{h}{2}f''(\xi),$$

or

$$\left|f'(x_0)-\frac{f(x_0+h)-f(x_0)}{h}\right|\leqslant\frac{h}{2}M.$$

Where M is a bound on $|f''(x)|$ for $x\in[a,b]$,which means that for small values of h, the difference quotient $\frac{f(x_0+h)-f(x_0)}{h}$ can be used to approximate $f'(x_0)$ with an error bounded by $Mh/2$.

The formula

$$f'(x_0)=\frac{f(x_0+h)-f(x_0)}{h}-\frac{h}{2}f''(\xi) \tag{7.1.2}$$

is known as **the forward-difference formula**.

7.1.2 Backward-difference formula (向后差分公式)

Suppose $h>0$ is sufficiently small to ensure that $x_{-1}=x_0-h\in[a,b]$. Then by Newton backward divided-difference formula

$$f(x)=P_1(x)+\frac{1}{2!}(x-x_{-1})(x-x_0)f''(\xi(x))$$
$$=f(x_0)+(x-x_0)f[x_0,x_{-1}]+\frac{1}{2!}(x-x_{-1})(x-x_0)f''(\xi(x)).$$

Differentiating the above expression gives

$$f'(x_0)=f[x_0,x_{-1}]+\frac{x_0-x_{-1}}{2}f''(\xi)=\frac{f(x_0)-f(x_0-h)}{h}+\frac{h}{2}f''(\xi).$$

The approximation

$$f'(x_0)=\frac{f(x_0)-f(x_0-h)}{h}+\frac{h}{2}f''(\xi) \tag{7.1.3}$$

is known as **the backward-difference formula** (向后差分公式), which generates an error also bounded by $Mh/2$.

7.1.3 Three-point formula (三点公式)

Suppose t_0, t_1, t_2 are equally spaced and

$$t_1=t_0+h,\ t_2=t_0+2h.$$

Then using the quadratic Newton's interpolatory divided difference formula with remainder term gives

$$f(x)=f(t_0)+(x-t\ )f[t_0,t_1]+(x-t_0)(x-t_1)f[t_0,t_1,t_2]+(x-t_0)(x-t_1)(x-t_2)f[t_0,t_1,t_2,x],$$

which leads to

$$f'(t_j) = f[t_0,t_1] + (2t_j - t_0 - t_1)f[t_0,t_1,t_2] + f[t_0,t_1,t_2,t_j] \prod_{k=0,k\neq j}^{2} (t_j - t_k)$$
$$= f[t_0,t_1] + (2t_j - t_0 - t_1)f[t_0,t_1,t_2] + \frac{f^{(3)}(\xi_j)}{3!} \prod_{k=0,k\neq j}^{2} (t_j - t_k).$$

Namely,

$$\begin{aligned} f'(t_0) &= f[t_0,t_1] - hf[t_0,t_1,t_2] + \frac{h^2}{3} f^{(3)}(\xi_0) \\ &= \frac{1}{h}\left[-\frac{3}{2}f(t_0) + 2f(t_1) - \frac{1}{2}f(t_2)\right] + \frac{h^2}{3} f^{(3)}(\xi_0). \end{aligned} \tag{7.1.4}$$

$$\begin{aligned} f'(t_1) &= f[t_0,t_1] + hf[t_0,t_1,t_2] - \frac{h^2}{6} f^{(3)}(\xi_1) \\ &= \frac{1}{h}[-\frac{1}{2}f(t_0) + \frac{1}{2}f(t_2)] - \frac{h^2}{6} f^{(3)}(\xi_1). \end{aligned} \tag{7.1.5}$$

$$\begin{aligned} f'(t_2) &= f[t_0,t_1] + 3hf[t_0,t_1,t_2] + \frac{h^2}{3} f^{(3)}(\xi_2) \\ &= \frac{1}{h}[\frac{1}{2}f(t_0) - 2f(t_1) + \frac{3}{2}f(t_2)] + \frac{h^2}{3} f^{(3)}(\xi_2). \end{aligned} \tag{7.1.6}$$

Setting $t_0 = x_0,\ t_1 = x_0 + h,\ t_2 = x_0 + 2h$ in (7.1.4) produces

$$f'(x_0) = \frac{1}{2h}[-3f(x_0) + 4f(x_0+h) - f(x_0+2h)] + \frac{h^2}{3} f^{(3)}(\xi_0). \tag{7.1.4*}$$

Setting $t_1 = x_0,\ t_0 = x_0 - h,\ t_2 = x_0 + h$ in (7.1.5) produces

$$f'(x_0) = \frac{1}{2h}[-f(x_0-h) + f(x_0+h)] - \frac{h^2}{6} f^{(3)}(\xi_1). \tag{7.1.5*}$$

Setting $t_2 = x_0,\ t_1 = x_0 - h,\ t_0 = x_0 - 2h$ in (7.1.6) produces

$$f'(x_0) = \frac{1}{2h}[f(x_0-2h) - 4f(x_0-h) + 3f(x_0)] + \frac{h^2}{3} f^{(3)}(\xi_2). \tag{7.1.6*}$$

Note that since the equation (7.1.6)* can be obtained from (7.1.4)* by simply replacing h with $-h$, there are actually only two formulas

$$f'(x_0) = \frac{1}{2h}[-3f(x_0) + 4f(x_0+h) - f(x_0+2h)] + \frac{h^2}{3} f^{(3)}(\xi_0). \tag{7.1.7}$$

$$f'(x_0) = \frac{1}{2h}[f(x_0+h) - f(x_0-h)] - \frac{h^2}{6} f^{(3)}(\xi_1). \tag{7.1.8}$$

where ξ_0 lies between x_0 and x_0+2h, and ξ_1 lies between x_0-h and x_0+h.

(7.1.7) and (7.1.8) are called **three-point formulas**.

7.1.4 Five-point formula (五点公式)

Suppose $t_i, i = 0,1,2,3,4,$ are equally spaced and

$$t_i = t_0 + ih,\ i = 1,2,3,4.$$

Applying the quartic Newton's interpolatory divided difference formula gives

$$f(x) = f(t_0) + (x-t_0)f[t_0,t_1] + (x-t_0)(x-t_1)f[t_0,t_1,t_2] + (x-t_0)(x-t_1)(x-t_2)f[t_0,t_1,t_2,t_3]$$
$$+ (x-t_0)\cdots(x-t_3)f[t_0,t_1,t_2,t_3,t_4] + (x-t_0)\cdots(x-t_4)f[t_0,t_1,t_2,t_3,t_4,x].$$

Differentiating the both sides of the above expression yields

$$f'(t_2) = f[t_0,t_1] + (2t_2 - t_0 - t_1)f[t_0,t_1,t_2] + (t_2-t_0)(t_2-t_1)f[t_0,t_1,t_2,t_3]$$
$$+ (t_2-t_0)(t_2-t_1)(t_2-t_3)f[t_0,t_1,t_2,t_3,t_4] + (t_2-t_0)\cdots(t_2-t_4)f[t_0,t_1,t_2,t_3,t_4,t_2]$$
$$= f[t_0,t_1] + 3hf[t_0,t_1,t_2] + 2h^2 f[t_0,t_1,t_2,t_3] - 2h^3 f[t_0,t_1,t_2,t_3,t_4] + 4h^4 \frac{f^{(5)}(\xi)}{5!},$$

where ξ is located between t_0 and $t_0 + 4h$. Since

$$f[t_0,t_1] = \frac{1}{h}[f(t_1) - f(t_0)],$$

$$f[t_0,t_1,t_2] = \frac{f(t_0)}{(t_0-t_1)(t_0-t_2)} + \frac{f(t_1)}{(t_1-t_0)(t_1-t_2)} + \frac{f(t_2)}{(t_2-t_0)(t_2-t_1)}$$
$$= \frac{1}{2h^2}f(t_0) - \frac{1}{h^2}f(t_1) + \frac{1}{2h^2}f(t_2),$$

$$f[t_0,t_1,t_2,t_3] = \frac{f(t_0)}{(t_0-t_1)(t_0-t_2)(t_0-t_3)} + \frac{f(t_1)}{(t_1-t_0)(t_1-t_2)(t_1-t_3)}$$
$$+ \frac{f(t_2)}{(t_2-t_0)(t_2-t_1)(t_2-t_3)} + \frac{f(t_3)}{(t_3-t_0)(t_3-t_1)(t_3-t_2)}$$
$$= -\frac{1}{6h^3}f(t_0) + \frac{1}{2h^3}f(t_1) - \frac{1}{2h^3}f(t_2) + \frac{1}{6h^3}f(t_3),$$

$$f[t_0,t_1,t_2,t_3,t_4] = \frac{f(t_0)}{(t_0-t_1)(t_0-t_2)(t_0-t_3)(t_0-t_4)} + \frac{f(t_1)}{(t_1-t_0)(t_1-t_2)(t_1-t_3)(t_1-t_4)}$$
$$+ \frac{f(t_2)}{(t_2-t_0)(t_2-t_1)(t_2-t_3)(t_2-t_4)} + \frac{f(t_3)}{(t_3-t_0)(t_3-t_1)(t_3-t_2)(t_3-t_4)}$$
$$+ \frac{f(t_4)}{(t_4-t_0)(t_4-t_1)(t_4-t_2)(t_4-t_3)}$$
$$= \frac{1}{24h^4}f(t_0) - \frac{1}{6h^4}f(t_1) + \frac{1}{4h^4}f(t_2) - \frac{1}{6h^4}f(t_3) + \frac{1}{24h^4}f(t_4),$$

where the property of divided difference

$$f[t_0,t_1,\cdots,t_n] = \sum_{i=0}^{n} \frac{f(t_i)}{(t_i-t_0)\cdots(t_i-t_{i-1})(t_i-t_{i+1})\cdots(t_i-t_n)}$$

is used. We have

$$f'(t_2) = \frac{1}{h}[f(t_1) - f(t_0)] + \frac{1}{h}\left[\frac{3}{2}f(t_0) - 3f(t_1) + \frac{3}{2}f(t_2)\right] + \frac{1}{h}\left[-\frac{1}{3}f(t_0) + f(t_1) - f(t_2) + \frac{1}{3}f(t_3)\right]$$
$$-\frac{1}{h}\left[\frac{1}{12}f(t_0) - \frac{1}{3}f(t_1) + \frac{1}{2}f(t_2) - \frac{1}{3}f(t_3) + \frac{1}{12}f(t_4)\right] + \frac{h^4}{30}f^{(5)}(\xi)$$

$$=\frac{1}{12h}[f(t_0)-8f(t_1)+8f(t_3)-f(t_4)]+\frac{h^4}{30}f^{(5)}(\xi).$$

Setting

$$t_2=x_0, t_1=x_0-h, t_0=x_0-2h, t_3=x_0+h, t_4=x_0+2h$$

produces

$$f'(x_0)=\frac{1}{12h}[f(x_0-2h)-8f(x_0-h)+8f(x_0+h)-f(x_0+2h)]+\frac{h^4}{30}f^{(5)}(\xi) \tag{7.1.9}$$

where ξ is located between x_0-2h and x_0+2h. (7.1.9) is called **five-point formula**.

Similarly one can obtain four other five-point formulas as follows

$$f'(x_0)=\frac{1}{12h}[-25f(x_0)+48f(x_0+h)-36f(x_0+2h)+16f(x_0+3h)-3f(x_0+4h)]+\frac{h^4}{5}f^{(5)}(\xi_1), \tag{7.1.10}$$

where ξ_1 is located between x_0 and x_0+4h.

$$f'(x_0)=\frac{1}{12h}[-3f(x_0-h)-10f(x_0)+18f(x_0+h)-6f(x_0+2h)+f(x_0+3h)]-\frac{h^4}{20}f^{(5)}(\xi_2), \tag{7.1.11}$$

where ξ_2 is located between x_0-h and x_0+3h.

$$f'(x_0)=\frac{1}{12h}[-f(x_0-3h)+6f(x_0-2h)-18f(x_0-h)+10f(x_0)+3f(x_0+h)]-\frac{h^4}{20}f^{(5)}(\xi_3), \tag{7.1.12}$$

where ξ_3 is located between x_0-3h and x_0+h.

$$f'(x_0)=\frac{1}{12h}[3f(x_0-4h)-16f(x_0-3h)+36f(x_0-2h)-48f(x_0-h)+25f(x_0)]+\frac{h^4}{5}f^{(5)}(\xi_4), \tag{7.1.13}$$

where ξ_4 is located between x_0-4h and x_0.

It is not difficult to see that the five-point formulas (7.1.10) and (7.1.11) can be obtained from (7.1.13) and (7.1.12) by replacing h with $-h$ respectively.

To obtain general derivative approximation formulas, suppose that $\{x_0, x_1, \cdots, x_n\}$ are $(n+1)$ distinct numbers in some interval I and that $f(x)\in C^{n+1}(I)$. By the Lagrange interpolation formula with remainder term,

$$f(x)=\sum_{i=0}^{n}L_i(x)f(x_i)+\frac{(x-x_0)(x-x_1)\cdots(x-x_n)}{(n+1)!}f^{(n+1)}(\xi),$$

for some ξ in I, where $L_i(x)$ denotes the i th Lagrange basis functions for $f(x)$ at $x_0, x_1, \cdots, x_n$.

Differentiating this expression gives

$$f'(x)=\sum_{i=0}^{n}L_i'(x)f(x_i)+\frac{f^{(n+1)}(\xi)}{(n+1)!}\cdot\frac{\mathrm{d}}{\mathrm{d}x}[(x-x_0)(x-x_1)\cdots(x-x_n)]$$
$$+\frac{(x-x_0)(x-x_1)\cdots(x-x_n)}{(n+1)!}\cdot\frac{\mathrm{d}}{\mathrm{d}x}f^{(n+1)}(\xi).$$

When x is one of the numbers x_j, the coefficient of $\frac{\mathrm{d}}{\mathrm{d}x}f^{(n+1)}(\xi)$ is 0, and the formula becomes

$$f'(x_j)=\sum_{i=0}^{n}L_i'(x_j)f(x_i)+\frac{f^{(n+1)}(\xi)}{(n+1)!}(x_j-x_0)\cdots(x_j-x_{j-1})(x_j-x_{j+1})\cdots(x_j-x_n),\tag{7.1.14}$$

which is called an $(n+1)$ **-point formula** to approximate $f'(x_j)$.

If the nodes are equally spaced, that is, when $x_j=x_0+jh,\ j=1,2,\cdots,n,$ for some $h\neq 0$ (7.1.14) can be rewritten as

$$f'(x_j)=\sum_{i=0}^{n}L_i'(x_0+jh)f(x_0+ih)+(-1)^{n-j}\frac{f^{(n+1)}(\xi)}{(n+1)\binom{n}{j}}h^n.\tag{7.1.15}$$

7.1.5 Approximation to higher derivatives (高阶导数逼近)

Suppose $t_i, i=0,1,2,$ are equally spaced and

$$t_1=t_0+h, t_2=t_0+2h.$$

Then

$$f(x)=f(t_0)+(x-t_0)f[t_0,t_1]+(x-t_0)(x-t_1)f[t_0,t_1,t_2]+(x-t_0)(x-t_1)(x-t_2)f[t_0,t_1,t_2,x].$$

Differentiating twice gives

$$f''(x)=2f[t_0,t_1,t_2]+\frac{\mathrm{d}^2}{\mathrm{d}x^2}\{(x-t_0)(x-t_1)(x-t_2)\cdot f[t_0,t_1,t_2,x]\}$$
$$=2f[t_0,t_1,t_2]+\sum_{i=0}^{2}\binom{2}{i}\cdot[(x-t_0)(x-t_1)(x-t_2)]^{(i)}(f[t_0,t_1,t_2,x])^{(2-i)},$$

which implies

$$f''(t_1)=2f[t_0,t_1,t_2]+\binom{2}{1}[(x-t_0)(x-t_1)(x-t_2)]'\big|_{x=t_1}\cdot f[t_0,t_1,t_2,x,x]\big|_{x=t_1}$$
$$+\binom{2}{2}[(x-t_0)(x-t_1)(x-t_2)]''\big|_{x=t_1}\cdot f[t_0,t_1,t_2,t_1]$$
$$=2f[t_0,t_1,t_2]+2(t_1-t_0)(t_1-t_2)\cdot f[t_0,t_1,t_2,t_1,t_1]+[(x-t_0)(x-t_1)(x-t_2)]''\big|_{x=t_1}\cdot f[t_0,t_1,t_2,t_1]$$
$$=2f[t_0,t_1,t_2]+2(t_1-t_0)(t_1-t_2)\cdot f[t_0,t_1,t_2,t_1,t_1]+\{(x-t_1)[(x-t_1)^2-h^2]\}''\big|_{x=t_1}\cdot f[t_0,t_1,t_2,t_1]$$
$$=2f[t_0,t_1,t_2]-2h^2\frac{f^{(4)}(\xi)}{4!}$$

$$=2\cdot\frac{1}{2h^2}[f(t_0)-2f(t_1)+f(t_2)]-\frac{h^2}{12}f^{(4)}(\xi)$$

$$=\frac{1}{h^2}[f(t_0)-2f(t_1)+f(t_2)]-\frac{h^2}{12}f^{(4)}(\xi).$$

Now replacing t_0, t_1 and t_2 by x_0-h, x_0 and x_0+h respectively yields:

$$f''(x_0)=\frac{1}{h^2}[f(x_0-h)-2f(x_0)+f(x_0+h)]-\frac{h^2}{12}f^{(4)}(\xi). \tag{7.1.16}$$

Eq. (7.1.16) can also be derived in a way as below.

Expand $f(x)$ in a third Taylor polynomial about the point x_0 and evaluate at x_0-h and x_0+h, respectively. Then

$$f(x_0+h)=f(x_0)+hf'(x_0)+\frac{h^2}{2}f''(x_0)+\frac{h^3}{6}f'''(x_0)+\frac{h^4}{24}f^{(4)}(\xi_1),$$

$$f(x_0-h)=f(x_0)-hf'(x_0)+\frac{h^2}{2}f''(x_0)-\frac{h^3}{6}f'''(x_0)+\frac{h^4}{24}f^{(4)}(\xi_2).$$

If we add these equations, the terms involving $f'(x_0)$ and $f'''(x_0)$ cancel and we obtain

$$f(x_0+h)+f(x_0-h)=2f(x_0)+f''(x_0)h^2+\frac{1}{24}[f^{(4)}(\xi_1)+f^{(4)}(\xi_2)]h^4.$$

Solving this equation for $f''(x_0)$ gives

$$f''(x_0)=\frac{1}{h^2}[f(x_0+h)+f(x_0-h)-2f(x_0)]-\frac{h^2}{24}[f^{(4)}(\xi_1)+f^{(4)}(\xi_2)]. \tag{7.1.17}$$

Suppose $f^{(4)}(x)$ is continuous on (x_0-h, x_0+h), then the Intermediate Value Theorem implies that a number ξ exists between ξ_1 and ξ_2, and hence in (x_0-h, x_0+h) with

$$f^{(4)}(\xi)=\frac{1}{2}[f^{(4)}(\xi_1)+f^{(4)}(\xi_2)].$$

This permits us to rewrite Eq. (7.1.17) as

$$f''(x_0)=\frac{1}{h^2}[f(x_0+h)+f(x_0-h)-2f(x_0)]-\frac{h^2}{12}f^{(4)}(\xi). \tag{7.1.18}$$

Similarly one can obtain other two three-point formulas for second derivative $f''(x_0)$

$$f''(x_0)=\frac{1}{h^2}[f(x_0)-2f(x_0+h)+f(x_0+2h)]+hf'''(\xi), \tag{7.1.19}$$

$$f''(x_0)=\frac{1}{h^2}[f(x_0-2h)-2f(x_0-h)+f(x_0)]-hf'''(\xi), \tag{7.1.20}$$

and five five-point formulas used to estimate the second derivative $f''(x_0)$ as follows

$$f''(x_0)=\frac{1}{12h^2}[35f(x_0)-104f(x_0+h)+114f(x_0+2h)-56f(x_0+3h)+11f(x_0+4h)]$$
$$-\frac{5h^3}{6}f^{(5)}(\xi).$$

$$f''(x_0)=\frac{1}{12h^2}[11f(x_0-h)-20f(x_0)+6f(x_0+h)+4f(x_0+2h)-f(x_0+3h)]$$
$$+\frac{h^3}{12}f^{(5)}(\xi).$$

$$f''(x_0)=-\frac{1}{12h^2}[f(x_0-2h)-16f(x_0-h)+30f(x_0)-16f(x_0+h)+f(x_0+2h)]$$
$$+\frac{h^4}{90}f^{(6)}(\xi).$$

$$f''(x_0)=\frac{1}{12h^2}[-f(x_0-3h)+4f(x_0-2h)+6f(x_0-h)-20f(x_0)+11f(x_0+h)]$$
$$-\frac{h^3}{12}f^{(5)}(\xi).$$

$$f''(x_0)=\frac{1}{12h^2}[11f(x_0-4h)-56f(x_0-3h)+114f(x_0-2h)-104f(x_0-h)+35f(x_0)]$$
$$+\frac{5h^3}{6}f^{(5)}(\xi).$$

7.1.6 Effect of round-off error (舍入误差影响)

A particularly important subject in the study of numerical differentiation is the effect that round-off error plays in the approximation.

Let us examine Eq. (7.1.5)*

$$f'(x_0)=\frac{1}{2h}[f(x_0+h)-f(x_0-h)]-\frac{h^2}{6}f^{(3)}(\xi_1),$$

more closely. Suppose that in evaluating $f(x_0+h)$ and $f(x_0-h)$ we encounter round-off errors $e(x_0+h)$ and $e(x_0-h)$. Then our computed values $\tilde{f}(x_0+h)$ and $\tilde{f}(x_0-h)$ are related to the true values $f(x_0+h)$ and $f(x_0-h)$ by the formulas

$$f(x_0+h)=\tilde{f}(x_0+h)+e(x_0+h), \text{ and } f(x_0-h)=\tilde{f}(x_0-h)+e(x_0-h).$$

The total error in the approximation,

$$f'(x_0)-\frac{\tilde{f}(x_0+h)-\tilde{f}(x_0-h)}{2h}=\frac{e(x_0+h)-e(x_0-h)}{2h}-\frac{h^2}{6}f^{(3)}(\xi_1)$$

is due in part to round-off error and in part to truncation error.

If we assume that the round-off errors $e(x_0\pm h)$ are bounded by some number $\varepsilon>0$ and that the third derivative of f is bounded by a number $M>0$, then

$$\left|f'(x_0)-\frac{\tilde{f}(x_0+h)-\tilde{f}(x_0-h)}{2h}\right|\leqslant\frac{\varepsilon}{h}+\frac{h^2}{6}M.$$

To reduce the truncation error, $\frac{h^2}{6}M$, we must reduce h. However, on the other hand, as

h is reduced, the round-off error $\frac{\varepsilon}{h}$ grows. Therefore it is not a good practice to choose h to be too small since the round-off error will dominate the calculations.

7.1.7 Examples (例题)

Example 7.1 Derive an $O(h^4)$ five-point formula to approximate $f'(x_0)$ that uses $f(x_0-h)$, $f(x_0)$, $f(x_0+h)$, $f(x_0+2h)$ and $f(x_0+3h)$.

Solution 1. Let

$$t_0 = x_0 - h, t_1 = x_0, t_2 = x_0 + h, t_3 = x_0 + 2h, t_4 = x_0 + 3h.$$

Then

$$\begin{aligned} f(x) = & f(t_0) + (x-t_0)f[t_0,t_1] + (x-t_0)(x-t_1)f[t_0,t_1,t_2] \\ & + (x-t_0)(x-t_1)(x-t_2)f[t_0,t_1,t_2,t_3] + (x-t_0)\cdots(x-t_3)f[t_0,t_1,t_2,t_3,t_4] \\ & + (x-t_0)\cdots(x-t_4)f[t_0,t_1,t_2,t_3,t_4,x] \end{aligned}$$

so

$$\begin{aligned} f'(t_1) = & f[t_0,t_1] + (t_1-t_0)f[t_0,t_1,t_2] + (t_1-t_0)(t_1-t_2)f[t_0,t_1,t_2,t_3] \\ & + (t_1-t_0)(t_1-t_2)(t_1-t_3)f[t_0,t_1,t_2,t_3,t_4] + (t_1-t_0)\cdots(t_1-t_4)f[t_0,t_1,t_2,t_3,t_4,t_1] \\ = & \frac{1}{h}[f(t_1)-f(t_0)] + \frac{1}{h}\left[\frac{1}{2}f(t_0) - f(t_1) + \frac{1}{2}f(t_2)\right] \\ & - \frac{1}{h}\left[-\frac{1}{6}f(t_0) + \frac{1}{2}f(t_1) - \frac{1}{2}f(t_2) + \frac{1}{6}f(t_3)\right] \\ & + \frac{1}{h}\left[\frac{1}{12}f(t_0) - \frac{1}{3}f(t_1) + \frac{1}{2}f(t_2) - \frac{1}{3}f(t_3) + \frac{1}{12}f(t_4)\right] - 6h^4 \cdot \frac{f^{(5)}(\xi)}{5!}. \end{aligned}$$

Then

$$f'(t_1) = \frac{1}{12h}[-3f(t_0) - 10f(t_1) + 18f(t_2) - 6f(t_3) + f(t_4)] - \frac{h^4}{20}f^{(5)}(\xi).$$

Consequently

$$f'(x_0) = \frac{1}{12h}[-3f(x_0-h) - 10f(x_0) + 18f(x_0+h) - 6f(x_0+2h) + f(x_0+3h)] - \frac{h^4}{20}f^{(5)}(\xi).$$

Solution 2. Expand $f(x)$ in a fourth Taylor polynomial about the point x_0 and evaluate at $x_0-h, x_0+h, x_0+2h, x_0+3h$. Then

$$f(x_0-h) = f(x_0) - hf'(x_0) + \frac{h^2}{2}f''(x_0) - \frac{h^3}{6}f'''(x_0) + \frac{h^4}{24}f^{(4)}(x_0) - \frac{h^5}{120}f^{(5)}(\xi_1), \xi_1 \in (x_0-h, x_0),$$

$$f(x_0+h) = f(x_0) + hf'(x_0) + \frac{h^2}{2}f''(x_0) + \frac{h^3}{6}f'''(x_0) + \frac{h^4}{24}f^{(4)}(x_0) + \frac{h^5}{120}f^{(5)}(\xi_2), \xi_2 \in (x_0, x_0+h),$$

$$f(x_0+2h) = f(x_0) + 2hf'(x_0) + 2h^2f''(x_0) + \frac{4}{3}h^3f'''(x_0) + \frac{2}{3}h^4f^{(4)}(x_0) + \frac{4}{15}h^5f^{(5)}(\xi_3),$$

$\xi_3 \in (x_0, x_0+2h)$,

$$f(x_0+3h)=f(x_0)+3hf'(x_0)+\frac{9}{2}h^2f''(x_0)+\frac{27}{6}h^3f'''(x_0)+\frac{81}{24}h^4f^{(4)}(x_0)+\frac{243}{120}h^5f^{(5)}(\xi_4),$$

$\xi_4\in(x_0,x_0+3h)$,

and

$$\begin{aligned}&Af(x_0-h)+Bf(x_0+h)+Cf(x_0+2h)+Df(x_0+3h)\\&=(A+B+C+D)f(x_0)+(-A+B+2C+3D)hf'(x_0)+(A+B+4C+9D)\frac{h^2}{2}f''(x_0)\\&\quad+\frac{h^3}{6}(-A+B+8C+27D)f'''(x_0)+\frac{h^4}{24}(A+B+16C+81D)f^{(4)}(x_0)\\&\quad+\frac{h^5}{120}[-Af^{(5)}(\xi_1)+Bf^{(5)}(\xi_2)+32Cf^{(5)}(\xi_3)+243Df^{(5)}(\xi_4)]\end{aligned}$$

Now let

$$\begin{cases}A+B+4C+9D=0\\-A+B+8C+27D=0\\A+B+16C+81D=0.\end{cases}$$

Solving these equations for A,B,C in terms of D gives

$$A=-3D,B=18D,C=-6D.$$

Hence

$$\begin{aligned}&-3Df(x_0-h)+18Df(x_0+h)-6Df(x_0+2h)+Df(x_0+3h)\\&=10Df(x_0)+12Dhf'(x_0)+\frac{h^5}{120}[3Df^{(5)}(\xi_1)+18Df^{(5)}(\xi_2)-192Df^{(5)}(\xi_3)+243Df^{(5)}(\xi_4)].\end{aligned}$$

Further simplifying the equation gives

$$\begin{aligned}f'(x_0)=&\frac{1}{12h}[-3f(x_0-h)-10f(x_0)+18f(x_0+h)-6f(x_0+2h)+f(x_0+3h)]\\&-\frac{h^4}{480}[f^{(5)}(\xi_1)+6f^{(5)}(\xi_2)-64f^{(5)}(\xi_3)+81f^{(5)}(\xi_4)].\end{aligned}$$

If $f^{(5)}$ is continuous, then for small h, we can find a ξ such that

$$f^{(5)}(\xi_1)+6f^{(5)}(\xi_2)-64f^{(5)}(\xi_3)+81f^{(5)}(\xi_4)=24f^{(5)}(\xi).$$

As a result,

$$f'(x_0)=\frac{1}{12h}[-3f(x_0-h)-10f(x_0)+18f(x_0+h)-6f(x_0+2h)+f(x_0+3h)]-\frac{h^4}{20}f^{(5)}(\xi).$$

7.2 Richardson's extrapolation (Richardson 外推)

7.2.1 Basic idea (基本思想)

Richardson's extrapolation is used to generate high-accuracy results while using low-order formulas. Although the name attached to the method refers to a paper written by L. F. Richardson and J. A. Gaunt in 1927, the idea behind the technique is much older.

Extrapolation can be applied whenever it is known that an approximation technique has an error term with a predictable form, depending on a parameter, usually the step size h. Suppose that for each number $h \neq 0$ we have a formula $N_1(h)$ that approximates an unknown constant M, and that the truncation error involved with the approximation has the form

$$M - N_1(h) = K_1 h + K_2 h^2 + K_3 h^3 + \cdots,$$

for some collection of (unknown) constants $K_1, K_2, K_3, \cdots$.

The truncation error is $O(h)$. Our purpose is to seek for a new approximation to M of higher orders. The object of extrapolation is to find an easy way to combine these rather inaccurate $O(h)$ approximations in an appropriate way to produce formulas with a higher-order truncation error.

To see specifically how we can generate the extrapolation formulas, consider the $O(h)$ formula for approximating M

$$M = N_1(h) + K_1 h + K_2 h^2 + K_3 h^3 + \cdots. \tag{7.2.1}$$

The formula is assumed to hold for all positive h, so we replace the parameter h by half its value. Then we have a second $O(h)$ approximation formula

$$M = N_1\left(\frac{h}{2}\right) + K_1 \frac{h}{2} + K_2\left(\frac{h}{2}\right)^2 + K_3\left(\frac{h}{2}\right)^3 + \cdots. \tag{7.2.2}$$

Subtracting Eq. (7.2.1) from twice Eq. (7.2.2) eliminates the term involving K_1 and gives

$$M = \left[2N_1\left(\frac{h}{2}\right) - N_1(h)\right] - \frac{1}{2}K_2 h^2 - \frac{3}{4}K_3 h^3 - \cdots. \tag{7.2.3}$$

Define

$$N_2(h) = \frac{2N_1\left(\frac{h}{2}\right) - N_1(h)}{2-1}.$$

Then Eq. (7.2.3) is an $O(h^2)$ approximation formula for M:

$$M = N_2(h) - \frac{1}{2}K_2 h^2 - \frac{3}{4}K_3 h^3 - \cdots. \tag{7.2.4}$$

Now replace h in Eq. (7.2.4) with $\frac{h}{2}$ to produce a second $O(h^2)$ formula

$$M = N_2\left(\frac{h}{2}\right) - \frac{1}{2}K_2\left(\frac{h}{2}\right)^2 - \frac{3}{4}K_3\left(\frac{h}{2}\right)^3 - \cdots. \tag{7.2.5}$$

Subtracting Eq. (7.2.4) from 4 times this equation eliminates the h^2 term and gives

$$(2^2 - 1)M = \left[2^2 N_2\left(\frac{h}{2}\right) - N_2(h)\right] + \frac{3}{8}K_3 h^3 + \cdots.$$

Dividing this equation by 3 produces the new $O(h^3)$ formula

$$M = \frac{2^2 N_2\left(\frac{h}{2}\right) - N_2(h)}{2^2 - 1} + \frac{1}{8}K_3 h^3 + \cdots.$$

We now have the $O(h^3)$ approximation formula

$$N_3(h) = \frac{2^2 N_2\left(\frac{h}{2}\right) - N_2(h)}{2^2 - 1}.$$

Continuing this procedure gives, for each $j = 2,3,\cdots$, the $O(h^j)$ approximation

$$N_j(h) = \frac{2^{j-1} N_{j-1}\left(\frac{h}{2}\right) - N_{j-1}(h)}{2^{j-1} - 1}.$$

Table 7.1 shows the order in which the approximations are generated when

$$M = N_1(h) + K_1 h + K_2 h^2 + K_3 h^3 + \cdots.$$

It is conservatively assumed that the true result is accurate at least to within the agreement of the bottom two results in the diagonal, in this case, to within $|N_3(h) - N_4(h)|$.

Table 7.1 Approximation order

$O(h)$	$O(h^2)$	$O(h^3)$	$O(h^4)$
$N_1(h)$			
$N_1\left(\frac{h}{2}\right)$	$N_2(h)$		
$N_1\left(\frac{h}{2^2}\right)$	$N_2\left(\frac{h}{2}\right)$	$N_3(h)$	
$N_1\left(\frac{h}{2^3}\right)$	$N_2\left(\frac{h}{2^2}\right)$	$N_3\left(\frac{h}{2}\right)$	$N_4(h)$

Extrapolation can be applied whenever the truncation error for a formula has the form

$$K_1 h^{\alpha_1} + K_2 h^{\alpha_2} + \cdots + K_{m-1} h^{\alpha_{m-1}} + O(h^{\alpha_m}),$$

for a collection of constants K_j and when $\alpha_1 < \alpha_1 < \cdots < \alpha_{m-1} < \alpha_m$. Many formulas used for extrapolation have truncation errors that contain only even powers of h, that is, have the form

$$M = N_1(h) + K_1 h^2 + K_2 h^4 + K_3 h^6 + \cdots.$$

The extrapolation is much more effective than when all powers of h are present because the averaging process produces results with errors $O(h^2), O(h^4), O(h^6), \cdots$, with essentially no increase in computation, over the results with errors, $O(h), O(h^2), O(h^3), \cdots$.

7.2.2 Examples (例题)

Example 7.2 Suppose we expand the function f in a fourth Taylor polynomial about x_0. Then

$$f(x)=f(x_0)+f'(x_0)(x-x_0)+\frac{1}{2}f''(x_0)(x-x_0)^2$$
$$+\frac{1}{6}f'''(x_0)(x-x_0)^3+\frac{1}{24}f^{(4)}(x-x_0)^4+\frac{1}{120}f^{(5)}(x-x_0)^5.$$

Evaluating $f(x)$ at x_0+h and x_0-h gives

$$f(x_0+h)=f(x_0)+hf'(x_0)+\frac{h^2}{2}f''(x_0)+\frac{h^3}{6}f'''(x_0)+\frac{h^4}{24}f^{(4)}(x_0)+\frac{h^5}{120}f^{(5)}(\xi_1), \quad \text{(a)}$$

$$f(x_0-h)=f(x_0)-hf'(x_0)+\frac{h^2}{2}f''(x_0)-\frac{h^3}{6}f'''(x_0)+\frac{h^4}{24}f^{(4)}(x_0)-\frac{h^5}{120}f^{(5)}(\xi_2), \quad \text{(b)}$$

where

$$x_0-h<\xi_2<x_0<\xi_1<x_0+h.$$

Subtracting (b) from (a) produces

$$f(x_0+h)-f(x_0-h)=2hf'(x_0)+\frac{h^3}{3}f'''(x_0)+\frac{h^5}{120}\left[f^{(5)}(\xi_1)+f^{(5)}(\xi_2)\right]. \quad \text{(c)}$$

If $f^{(5)}$ is continuous on $[x_0-h,x_0+h]$, the Intermediate Value Theorem implies that a number ξ in (x_0-h, x_0+h) exists with

$$f^{(5)}(\xi)=\frac{1}{2}\left[f^{(5)}(\xi_1)+f^{(5)}(\xi_2)\right].$$

Solving (c) for $f'(x_0)$ gives the $O(h^2)$ approximation

$$f'(x_0)=\frac{1}{2h}\left[f(x_0+h)-f(x_0-h)\right]-\frac{h^2}{6}f'''(x_0)-\frac{h^4}{120}f^{(5)}(\xi), \quad \text{(d)}$$

Let us now make use of the Richardson's extrapolation to develop the $O(h^4)$ approximation to $f'(x_0)$ based on (d).

Replacing h in (d) with $2h$ yields

$$f'(x_0)=\frac{1}{4h}\left[f(x_0+2h)-f(x_0-2h)\right]-\frac{4h^2}{6}f'''(x_0)-\frac{16h^4}{120}f^{(5)}(\eta), \quad \text{(e)}$$

Multiplying (d) by 4 and subtracting Eq. (e) produces

$$3f'(x_0)=\frac{2}{h}\left[f(x_0+h)-f(x_0-h)\right]-\frac{1}{4h}\left[f(x_0+2h)-f(x_0-2h)\right]$$
$$-\frac{h^4}{30}f^{(5)}(\xi)+\frac{4h^4}{30}f^{(5)}(\eta).$$

Namely,

$$f'(x_0)=\frac{2}{3h}\left[f(x_0+h)-f(x_0-h)\right]-\frac{1}{12h}\left[f(x_0+2h)-f(x_0-2h)\right]+\frac{h^4}{30}\frac{4f^{(5)}(\eta)-f^{(5)}(\xi)}{3}.$$

If $f^{(5)}$ is continuous on $[x_0-2h,x_0+2h]$, an alternative method can be used to show $f^{(5)}(\xi)$ and $f^{(5)}(\eta)$ can be replaced by a common value $f^{(5)}(\xi)$. Using this result and dividing by 3 produces the five-point formula

$$f'(x_0)=\frac{1}{12h}\left[f(x_0-2h)-8f(x_0-h)+8f(x_0+h)-f(x_0+2h)\right]+\frac{h^4}{30}f^{(5)}(\xi).$$

Example 7.3 The following data give approximation to the integral

$$M=\int_0^{\pi}\sin x\mathrm{d}x$$

$$N_1(h)=1.570796,\quad N_1\left(\frac{h}{2}\right)=1.896119,$$

$$N_1\left(\frac{h}{4}\right)=1.974232,\quad N_1\left(\frac{h}{8}\right)=1.993570.$$

Assuming

$$M=N_1(h)+k_1h^2+k_2h^4+k_3h^6+k_4h^8+O\left(h^{10}\right),$$

construct an extrapolation table to determine $N_4(h)$.

Solution. From

$$M=N_1(h)+k_1h^2+k_2h^4+k_3h^6+\cdots,$$

$$M=N_1\left(\frac{h}{2}\right)+\frac{k_1}{4}h^2+\frac{k_2}{16}h^4+\frac{k_3}{64}h^6+\cdots,$$

we see that

$$3M=4N_1\left(\frac{h}{2}\right)-N_1(h)-\frac{3}{4}k_2h^4-\frac{15}{16}k_3h^6-\cdots,$$

which leads to

$$M=\frac{4N_1\left(\frac{h}{2}\right)-N_1(h)}{3}-\frac{k_2}{4}h^4-\frac{5}{16}k_3h^6-\cdots$$
$$=N_2(h)+\alpha_1h^4+\alpha_2h^6+\alpha_3h^8+\cdots.$$

Using

$$M=N_2\left(\frac{h}{2}\right)+\frac{\alpha_1}{16}h^4+\frac{\alpha_2}{64}h^6+\frac{\alpha_3}{256}h^8+\cdots$$

gives

$$15M=16N_2\left(\frac{h}{2}\right)-N_2(h)-\frac{3}{4}\alpha_2h^6-\frac{15}{16}\alpha_3h^8-\cdots,$$

which leads to

$$
\begin{aligned}
M &= \frac{16N_2\left(\dfrac{h}{2}\right)-N_2(h)}{15}-\frac{\alpha_2}{20}h^6-\frac{\alpha_3}{16}h^8-\cdots \\
&= N_3(h)+\beta_1 h^6+\beta_2 h^8+\cdots.
\end{aligned}
$$

And

$$
M = N_3\left(\frac{h}{2}\right)+\frac{\beta_1}{64}h^6+\frac{\beta_2}{256}h^8+\cdots.
$$

So

$$
\begin{aligned}
64M &= 64N_3\left(\frac{h}{2}\right)+\beta_1 h^6+\frac{\beta_2}{4}h^8+\cdots, \\
63M &= 64N_3\left(\frac{h}{2}\right)-N_3(h)-\frac{3}{4}\beta_2 h^8-\cdots.
\end{aligned}
$$

As a result,

$$
\begin{aligned}
M &= \frac{64N_3\left(\dfrac{h}{2}\right)-N_3(h)}{63}+O\left(h^8\right) \\
&= N_4(h)+O\left(h^8\right).
\end{aligned}
$$

We can put the relations

$$
\begin{aligned}
N_2(h) &= \frac{4N_1\left(\dfrac{h}{2}\right)-N_1(h)}{3}=\frac{4N_1\left(\dfrac{h}{2}\right)-N_1(h)}{4-1}, \\
N_3(h) &= \frac{16N_2\left(\dfrac{h}{2}\right)-N_2(h)}{15}=\frac{4^2N_2\left(\dfrac{h}{2}\right)-N_2(h)}{4^2-1}, \\
N_4(h) &= \frac{64N_3\left(\dfrac{h}{2}\right)-N_3(h)}{63}=\frac{4^3N_3\left(\dfrac{h}{2}\right)-N_3(h)}{4^3-1}
\end{aligned}
$$

into the table

$$
\begin{array}{llll}
N_1(h) & & & \\
N_1\left(\dfrac{h}{2}\right) & N_2(h) & & \\
N_1\left(\dfrac{h}{4}\right) & N_2\left(\dfrac{h}{2}\right) & N_3(h) & \\
N_1\left(\dfrac{h}{8}\right) & N_2\left(\dfrac{h}{4}\right) & N_3\left(\dfrac{h}{2}\right) & N_4(h)
\end{array}
$$

where

$$N_2(h)=\frac{4N_1\left(\frac{h}{2}\right)-N_1(h)}{3}=\frac{4\times 1.896119-1.570796}{3}=2.004560000,$$

$$N_2\left(\frac{h}{2}\right)=\frac{4N_1\left(\frac{h}{4}\right)-N_1\left(\frac{h}{2}\right)}{3}=\frac{4\times 1.974232-1.896119}{3}=2.000269667,$$

$$N_2\left(\frac{h}{4}\right)=\frac{4N_1\left(\frac{h}{8}\right)-N_1\left(\frac{h}{4}\right)}{3}=\frac{4\times 1.993570-1.974232}{3}=2.000016000,$$

$$N_3(h)=\frac{16N_2\left(\frac{h}{2}\right)-N_2(h)}{15}=\frac{16\times 2.000269667-2.00456}{15}=2.000257245,$$

$$N_3\left(\frac{h}{2}\right)=\frac{16N_2\left(\frac{h}{4}\right)-N_2\left(\frac{h}{2}\right)}{15}=\frac{16\times 2.000016000-2.000269667}{15}=1.999999089,$$

$$N_4(h)=\frac{64N_3\left(\frac{h}{2}\right)-N_3(h)}{63}=\frac{64\times 1.999999089-2.000257245}{63}=1.999994991.$$

Example 7.4 The forward-difference formula can be expressed as

$$f'(x_0)=\frac{1}{h}[f(x_0+h)-f(x_0)]-\frac{h}{2}f''(x_0)-\frac{h^2}{6}f'''(x_0)-\frac{h^3}{24}f^{(4)}(\xi).$$

Note

it follows from

$$f(x_0+h)=f(x_0)+hf'(x_0)+\frac{h^2}{2}f''(x_0)+\frac{h^3}{6}f'''(x_0)+\frac{h^4}{24}f^{(4)}(\xi).$$

Use extrapolation to derive an $O(h^3)$ formula for $f'(x_0)$.

Solution. Define

$$N_1(h)=\frac{1}{h}[f(x_0+h)-f(x_0)].$$

Then

$$f'(x_0)=N_1(h)-\frac{h}{2}f''(x_0)-\frac{h^2}{6}f'''(x_0)-\frac{h^3}{24}f^{(4)}(\xi).$$

Replacing h with $2h$ in the above expression produces

$$f'(x_0)=N_1(2h)-hf''(x_0)-\frac{4h^2}{6}f'''(x_0)-\frac{8h^3}{24}f^{(4)}(\xi_1).$$

Subtracting this equation from 2 times the foregoing one gives

$$f'(x_0)=2N_1(h)-N_1(2h)+\frac{h^2}{3}f'''(x_0)+\frac{h^3}{4}f^{(4)}(\xi_2).$$

where $f^{(4)}(x)$ is supposed to be continuous and

$$f^{(4)}(\xi_2)=\frac{4f^{(4)}(\xi_1)-f^{(4)}(\xi)}{3}.$$

Define

$$N_2(h)=2N_1(h)-N_1(2h).$$

Then

$$f'(x_0)=N_2(h)+\frac{h^2}{3}f'''(x_0)+\frac{h^3}{4}f^{(4)}(\xi_2).$$

To eliminate the power of h^2 in the above formula, we replace h with $2h$ and get

$$f'(x_0)=N_2(2h)+\frac{4h^2}{3}f'''(x_0)+2h^3f^{(4)}(\xi_3).$$

Subtracting the above equation from 4 times the foregoing one yields

$$3f'(x_0)=4N_2(h)-N_2(2h)-h^3f^{(4)}(\xi_4),$$

where

$$f^{(4)}(\xi_4)=2f^{(4)}(\xi_3)-f^{(4)}(\xi_2).$$

Therefore

$$f'(x_0)=\frac{4N_2(h)-N_2(2h)}{3}-\frac{h^3}{3}f^{(4)}(\xi_4).$$

Define

$$N_3(h)=\frac{4N_2(h)-N_2(2h)}{3}.$$

Then

$$f'(x_0)=N_3(h)-\frac{h^3}{3}f^{(4)}(\xi_4).$$

which means that $N_3(h)$ is an $O(h^3)$ approximation to $f'(x_0)$, where $N_3(h)$ can be expressed as follows:

$$\begin{aligned}
N_3(h)&=\frac{1}{3}[4N_2(h)-N_2(2h)]\\
&=\frac{1}{3}\{4[2N_1(h)-N_1(2h)]-[2N_1(2h)-N_1(4h)]\}\\
&=\frac{1}{3}[8N_1(h)-6N_1(2h)+N_1(4h)]\\
&=\frac{1}{3}\left\{8\cdot\frac{1}{h}[f(x_0+h)-f(x_0)]-6\cdot\frac{1}{2h}[f(x_0+2h)-f(x_0)]\right.\\
&\quad\left.+\frac{1}{4h}[f(x_0+4h)-f(x_0)]\right\}
\end{aligned}$$

$$=\frac{1}{12h}\left[-21f\left(x_0\right)+32f\left(x_0+h\right)-12f\left(x_0+2h\right)+f\left(x_0+4h\right)\right].$$

One can also use Newton's interpolatory divided difference formula to solve this problem. To this end, let

$$t_1=t_0+h,\ t_2=t_0+2h,\ t_3=t_0+4h.$$

Then

$$f(x)=f\left(t_0\right)+\left(x-t_0\right)f\left[t_0,t_1\right]+\left(x-t_0\right)\left(x-t_1\right)f\left[t_0,t_1,t_2\right]$$
$$+\left(x-t_0\right)\left(x-t_1\right)\left(x-t_2\right)f\left[t_0,t_1,t_2,t_3\right]+\left(x-t_0\right)\cdots\left(x-t_3\right)f\left[t_0,t_1,t_2,t_3,x\right].$$

Therefore

$$f'\left(t_0\right)=f\left[t_0,t_1\right]+\left(t_0-t_1\right)f\left[t_0,t_1,t_2\right]+\left(t_0-t_1\right)\left(t_0-t_2\right)f\left[t_0,t_1,t_2,t_3\right]$$
$$+\left(t_0-t_1\right)\left(t_0-t_2\right)\left(t_0-t_3\right)f\left[t_0,t_1,t_2,t_3,t_0\right]$$
$$=f\left[t_0,t_1\right]-hf\left[t_0,t_1,t_2\right]+2h^2f\left[t_0,t_1,t_2,t_3\right]-8h^3\frac{f^{(4)}(\xi)}{4!}.$$

Since

$$f\left[t_0,t_1\right]=\frac{1}{h}\left[f\left(t_1\right)-f\left(t_0\right)\right],$$

$$f\left[t_0,t_1,t_2\right]=\frac{f\left(t_0\right)}{\left(t_0-t_1\right)\left(t_0-t_2\right)}+\frac{f\left(t_1\right)}{\left(t_1-t_0\right)\left(t_1-t_2\right)}+\frac{f\left(t_2\right)}{\left(t_2-t_0\right)\left(t_2-t_1\right)}=\frac{f\left(t_0\right)}{2h^2}-\frac{f\left(t_1\right)}{h^2}+\frac{f\left(t_2\right)}{2h^2},$$

$$f\left[t_0,t_1,t_2,t_3\right]=-\frac{1}{8h^3}f\left(t_0\right)+\frac{1}{3h^3}f\left(t_1\right)-\frac{1}{4h^3}f\left(t_2\right)+\frac{1}{24h^3}f\left(t_3\right),$$

we have

$$f'\left(t_0\right)=\frac{1}{12h}\left[-21f\left(t_0\right)+32f\left(t_1\right)-12f\left(t_2\right)+f\left(t_3\right)\right]-\frac{h^3}{3}f^{(4)}(\xi).$$

Setting

$$t_0=x_0,\ t_1=x_0+h,\ t_2=x_0+2h,\ \text{and}\ \ t_3=x_0+4h$$

gives

$$f'\left(x_0\right)=\frac{1}{12h}\left[-21f\left(x_0\right)+32f\left(x_0+h\right)-12f\left(x_0+2h\right)+f\left(x_0+4h\right)\right]-\frac{h^3}{3}f^{(4)}(\xi).$$

7.3 Elements of numerical integration (数值积分)

- Basic idea (基本思想)
- Midpoint rule (中点公式)
- Trapezoidal rule (梯形公式)
- Simpson's rule (Simpson 公式)
- Newton-Cotes formulas (Newton-Cotes 公式)
- Degree of accuracy (代数精度)

● Examples (例题)

It is well known that the area of the ellipse (椭圆)

$$\frac{x^2}{a^2}+\frac{y^2}{b^2}=1$$

can be calculated by the formula

$$A=4\int_0^a y\mathrm{d}x=4ab\int_0^{\pi/2}\sin^2 t\mathrm{d}t=\pi ab.$$

However it is a challenging problem to calculate its perimeter or length. From calculus we know that this perimeter is

$$P=4\int_0^{\pi/2}\sqrt{a^2\sin^2 t+b^2\cos^2 t}\mathrm{d}t,$$

so the problem reduces to evaluating this integral, an elliptic integral of the second kind, which cannot be evaluated explicitly since the integrand has no explicit antiderivative. Methods are developed in this chapter to approximate the solution to problems of this type.

7.3.1 Basic idea (基本思想)

If $f(x)$ has no explicit antiderivative (say, $f(x)$ is given by its tabulated data) or its antiderivative is not easy to obtain. Then $f(x)$ can be replaced with an approximating function $g(x)$ that is easy to integrate

$$I=\int_a^b f(x)\mathrm{d}x\approx\int_a^b g(x)\mathrm{d}x.$$

The basic method involved in approximating $\int_a^b f(x)\mathrm{d}x$ is called **numerical quadrature** (数值求积). It uses a sum $\sum_{i=0}^{n}A_i f(x_i)$ to approximate $\int_a^b f(x)\mathrm{d}x$.

The methods of quadrature in this section are based on the interpolation polynomials given in Chapter 5. The basic idea is to select a set of distinct nodes $\{x_0,x_1,\cdots,x_n\}$ from the interval $[a,b]$, then integrate the Lagrange interpolating polynomial

$$P_n(x)=\sum_{i=0}^{n}L_i(x)f(x_i)$$

and its truncation error term over $[a,b]$ to obtain

$$\int_a^b f(x)\mathrm{d}x=\int_a^b[P_n(x)+R_n(x)]\mathrm{d}x=\int_a^b\sum_{i=0}^{n}f(x_i)L_i(x)\mathrm{d}x+\int_a^b\prod_{i=0}^{n}(x-x_i)\frac{f^{(n+1)}(\xi)}{(n+1)!}\mathrm{d}x$$

$$=\sum_{i=0}^{n}A_i f(x_i)+\frac{1}{(n+1)!}\int_a^b\prod_{i=0}^{n}(x-x_i)f^{(n+1)}(\xi)\mathrm{d}x.$$

where ξ is in $[a,b]$ for each x and

$$A_i=\int_a^b L_i(x)\mathrm{d}x,\ \text{for each}\ i=0,1,\cdots,n.$$

The quadrature formula is, therefore,

$$\int_a^b f(x)\mathrm{d}x \approx \sum_{i=0}^{n} A_i f(x_i),$$

with error given by

$$E(f) = \frac{1}{(n+1)!}\int_a^b \prod_{i=0}^{n}(x - x_i) f^{(n+1)}(\xi)\mathrm{d}x.$$

Before discussing the general situation of quadrature formulas, let us consider a few well known formulas.

7.3.2 Midpoint rule (中点公式)

Let $x_1 = \dfrac{a+b}{2}$. Then

$$f(x) = f(x_1) + (x - x_1) f[x_1, x_1] + (x - x_1)^2 f[x_1, x_1, x],$$

implies

$$\int_a^b f(x)\mathrm{d}x = \int_a^b (f(x_1) + (x - x_1) f[x_1, x_1])\mathrm{d}x + \int_a^b (x - x_1)^2 f[x_1, x_1, x]\mathrm{d}x,$$

then

$$\begin{aligned}\int_a^b f(x)\mathrm{d}x &= (b-a) f\left(\frac{a+b}{2}\right) + f[x_1, x_1, \eta]\int_a^b (x - x_1)^2 \mathrm{d}x \\ &= (b-a) f\left(\frac{a+b}{2}\right) + \frac{f''(\xi)}{2}\cdot\frac{2}{3}\left(b - \frac{a+b}{2}\right)^3 \\ &= (b-a) f\left(\frac{a+b}{2}\right) + \frac{(b-a)^3}{24} f''(\xi).\end{aligned}$$

The above formula is called the **Midpoint rule**.

7.3.3 Trapezoidal rule (梯形公式)

Let

$$x_0 = a, x_1 = b, h = b - a,$$

and let

$$P_1(x) = \frac{x - x_1}{x_0 - x_1} f(x_0) + \frac{x - x_0}{x_1 - x_0} f(x_1) = \frac{x-b}{a-b} f(a) + \frac{x-a}{b-a} f(b),$$

be the linear Lagrange polynomial. Then from

$$f(x) = P_1(x) + f[x_0, x_1, x](x - x_0)(x - x_1) = P_1(x) + f[a, b, x](x-a)(x-b),$$

it follows

$$\begin{aligned}\int_a^b f(x)\mathrm{d}x &= \int_a^b \left[\frac{x-b}{a-b} f(a) + \frac{x-a}{b-a} f(b)\right]\mathrm{d}x + \int_a^b f[a,b,x](x-a)(x-b)\mathrm{d}x \\ &= \frac{1}{2}(b-a)[f(a) + f(b)] + \int_a^b f[a,b,x](x-a)(x-b)\mathrm{d}x.\end{aligned}$$

Since $(x-a)(x-b)$ does not change sign on $[a,b]$, the weighted Mean Value Theorem for Integrals (积分中值定理) can be applied to the error term to give

$$\int_a^b f[a,b,x](x-a)(x-b)\mathrm{d}x = f[a,b,\eta]\int_a^b (x-a)(x-b)\mathrm{d}x, \quad \eta\in[a,b]$$

$$= f[a,b,\eta]\int_a^b (x-a)[(x-a)-(b-a)]\mathrm{d}x$$

$$= f[a,b,\eta]\left[\frac{1}{3}(x-a)^3 - \frac{b-a}{2}(x-a)^2\right]\Big|_a^b = f[a,b,\eta]\left[\frac{1}{3}(b-a)^3 - \frac{(b-a)^3}{2}\right]$$

$$= -\frac{1}{6}(b-a)^3\frac{f''(\xi)}{2} = -\frac{(b-a)^3}{12}f''(\xi), \quad \xi\in[a,b].$$

Hence we have the following **Trapezoidal Rule**:

$$\int_a^b f(x)\mathrm{d}x = \frac{h}{2}[f(a)+f(b)] - \frac{h^3}{12}f''(\xi),$$

in comparison with the **Midpoint Rule**:

$$\int_a^b f(x)\mathrm{d}x = hf\left(\frac{a+b}{2}\right) + \frac{h^3}{24}f''(\xi).$$

7.3.4 Simpson's rule (Simpson 公式)

Let

$$x_0 = a,\ x_1 = \frac{a+b}{2},\ x_2 = b,\ h = \frac{b-a}{2},$$

and

$$P_2(x) = f(x_0) + (x-x_0)f[x_0,x_1] + (x-x_0)(x-x_1)f[x_0,x_1,x_2].$$

be the quadratic polynomial interpolating $f(x)$ at x_0, x_1 and x_2.

Since

$$f(x) = P_2(x) + (x-x_0)(x-x_1)(x-x_2)f[x_0,x_1,x_2,x],$$

integrating the above equation over $[a,b]$ produces

$$\int_a^b f(x)\mathrm{d}x = I_1 + I_2,$$

where

$$I_1 = \int_a^b P_2(x)\mathrm{d}x = \int_a^b\left(f(a) + (x-a)f\left[a,\frac{a+b}{2}\right] + (x-a)\left(x-\frac{a+b}{2}\right)f\left[a,\frac{a+b}{2},b\right]\right)\mathrm{d}x$$

$$= (b-a)f(a) + \frac{1}{2}(b-a)^2 f\left[a,\frac{a+b}{2}\right] + \left[\frac{1}{3}(b-a)^3 - \frac{1}{4}(b-a)^3\right]f\left[a,\frac{a+b}{2},b\right]$$

$$= \frac{b-a}{6}\left[f(a) + 4f\left(\frac{a+b}{2}\right) + f(b)\right] = \frac{h}{3}\left[f(a) + 4f\left(\frac{a+b}{2}\right) + f(b)\right].$$

Since

$$d[(x-a)(x-b)] = [2x-(a+b)]dx,$$

namely,

$$\left(x-\frac{a+b}{2}\right)dx = \frac{1}{2}d[(x-a)(x-b)].$$

we have

$$\begin{aligned}
I_2 &= \int_a^b (x-x_0)(x-x_1)(x-x_2) f[x_0,x_1,x_2,x]dx \\
&= \int_a^b (x-a)(x-b)\left(x-\frac{a+b}{2}\right) f\left[a,\frac{a+b}{2},b,x\right]dx \\
&= \frac{1}{2}\int_a^b (x-a)(x-b) f\left[a,\frac{a+b}{2},b,x\right]\cdot d[(x-a)(x-b)] \\
&= \frac{1}{4}\int_a^b f\left[a,\frac{a+b}{2},b,x\right]d[(x-a)(x-b)]^2.
\end{aligned}$$

Using integration by parts yields

$$\begin{aligned}
I_2 &= \frac{1}{4}\left\{[(x-a)(x-b)]^2 f\left[a,\frac{a+b}{2},b,x\right]\Big|_a^b - \int_a^b [(x-a)(x-b)]^2 df\left[a,\frac{a+b}{2},b,x\right]\right\} \\
&= -\frac{1}{4}\int_a^b (x-a)^2[(x-a)-(b-a)]^2 \cdot f\left[a,\frac{a+b}{2},b,x,x\right]dx.
\end{aligned}$$

Using weighted Mean Value Theorem for Integrals gives

$$\begin{aligned}
I_2 &= -\frac{1}{4} f\left[a,\frac{a+b}{2},b,\eta,\eta\right]\int_a^b (x-a)^2\cdot\left[(x-a)^2-2(b-a)(x-a)+(b-a)^2\right]dx \\
&= -\frac{f^{(4)}(\xi)}{96}\left[\frac{1}{5}(x-a)^5 - 2(b-a)\cdot\frac{1}{4}(x-a)^4 + (b-a)^2\cdot\frac{1}{3}(x-a)^3\right]\Bigg|_a^b \\
&= -\frac{(b-a)^5}{2880} f^{(4)}(\xi) = -\frac{h^5}{90} f^{(4)}(\xi).
\end{aligned}$$

This gives **Simpson's Rule**

$$\int_a^b f(x)dx = \frac{h}{3}\left[f(a)+4f\left(\frac{a+b}{2}\right)+f(b)\right] - \frac{h^5}{90} f^{(4)}(\xi).$$

Since the error term involves the fourth derivative of f, Simpson's Rule gives exact results when applied to any polynomial of degree three or less, which means that for any polynomial $g(x)$ of degree three or less

$$\int_a^b g(x)dx = \frac{h}{3}\left[g(a)+4g\left(\frac{a+b}{2}\right)+g(b)\right].$$

7.3.5 Newton-Cotes formulas (Newton-Cotes 公式)

1. $n+1$-point closed Newton-Cotes formula

For $i=0,1,\cdots,n$, let

$$x_0=a,\ x_n=b,\ h=\frac{b-a}{n},\ x_i=x_0+ih,$$

$$P_n(x)=\sum_{i=0}^{n} f(x_i)L_i(x),$$

where

$$L_i(x)=\prod_{j=0,j\neq i}^{n}\frac{x-x_j}{x_i-x_j}$$

is the Lagrange interpolating polynomial.

Then

$$f(x)=P_n(x)+\frac{f^{(n+1)}(\xi)}{(n+1)!}(x-x_0)(x-x_1)\cdots(x-x_n)$$

suggests

$$\int_a^b f(x)\mathrm{d}x\approx\int_a^b P_n(x)\mathrm{d}x=\int_a^b\sum_{i=0}^{n} f(x_i)L_i(x)\mathrm{d}x$$

$$=\sum_{i=0}^{n} f(x_i)\int_a^b L_i(x)\,\mathrm{d}x=\sum_{i=0}^{n} A_i f(x_i)$$

with

$$A_i=\int_a^b L_i(x)\mathrm{d}x$$

The above formula is called $n+1$-**point closed Newton-Cotes formula.**

It is easy to know that

$$\sum_{i=0}^{n} A_i=b-a,$$

Since

$$\sum_{i=0}^{n} L_i(x)=1.$$

2. $n+1$-point open Newton-Cotes formula

Let

$$h=\frac{b-a}{n+2},\ x_0=a+h,\ x_n=b-h,\ x_i=a+(i+1)h,$$

$$L_i(x)=\prod_{j=0,j\neq i}^{n}\frac{x-x_j}{x_i-x_j},\ A_i=\int_a^b L_i(x)\mathrm{d}x.$$

Then

$$\int_a^b f(x)\,dx \approx \sum_{i=0}^{n} A_i f(x_i)$$

is called $n+1$**-point open Newton-Cotes formula.**

The closed Newton-Cotes formula is different from the open Newton-Cotes formula in that the former uses endpoints as the interpolation nodes and the latter only uses the nodes within the open interval (a,b).

7.3.6 Degree of accuracy (代数精度)

If n is the largest positive integer to ensure that the quadrature formula

$$\int_a^b f(x)\,dx \approx \sum_{i=0}^{n} A_i f(x_i)$$

is exact for x^k, for each $k=0,1,\cdots,n$, then n is called the degree of accuracy of the quadrature formula.

Namely

$$\int_a^b x^k dx = \sum_{i=0}^{n} A_i x_i^k,\ k=0,1,\cdots,n,\ \int_a^b x^{n+1} dx \neq \sum_{i=0}^{n} A_i x_i^{n+1}.$$

It is easy to see that the degrees of accuracy of Midpoint formula, Trapezoidal formula and Simpson's formula are 1, 1, 3 respectively.

7.3.7 Examples (例题)

Example 7.5 Determine the unknowns A_{-1}, A_0 and A_1 so that the degree of accuracy of the quadrature formula

$$\int_{-h}^{h} f(x)\,dx \approx A_{-1}f(-h) + A_0 f(0) + A_1 f(h)$$

is as high as possible.

Solution. To make the degree of accuracy as large as possible, at least we require that the following equation

$$\int_{-h}^{h} f(x)\,dx = A_{-1}f(-h) + A_0 f(0) + A_1 f(h)$$

is true for $f(x)=1,x,x^2$, which results in the following system of equations

$$\begin{cases} A_{-1} + A_0 + A_1 = 2h, \\ (A_{-1} - A_1)h = 0, \\ (A_{-1} + A_1)h^2 = \frac{2}{3}h^3. \end{cases}$$

Solving for A_{-1}, A_0 and A_1 gives

$$A_{-1}=A_1=\frac{1}{3}h,\ A_0=\frac{4}{3}h.$$

Therefore the quadrature formula turns out to be

$$\int_{-h}^{h} f(x)\,dx \approx \frac{1}{3}hf(-h)+\frac{4}{3}hf(0)+\frac{1}{3}hf(h).$$

Now

$$\int_{-h}^{h} x^3\,dx = 0 = \frac{1}{3}h(-h)^3+\frac{4}{3}h\cdot 0+\frac{1}{3}hh^3.$$

and

$$\int_{-h}^{h} x^4\,dx = \frac{2}{5}h^5 \neq \frac{1}{3}h(-h)^4+\frac{4}{3}h\cdot 0+\frac{1}{3}hh^4=\frac{2}{3}h^5$$

implies that the degree of accuracy of the quadrature formula is 3.

Example 7.6 Let

$$h=\frac{b-a}{3}, x_0=a, x_1=a+h, \text{ and } x_2=b.$$

Find the degree of precision of the quadrature formula

$$\int_a^b f(x)\,dx \approx \frac{9}{4}hf(x_1)+\frac{3}{4}hf(x_2).$$

Solution. For $f(x)=1$,

$$\int_a^b 1\,dx = b-a=\frac{9}{4}h+\frac{3}{4}h.$$

For $f(x)=x$,

$$\int_a^b x\,dx=\frac{1}{2}(b^2-a^2),$$

$$\begin{aligned}\frac{9}{4}hx_1+\frac{3}{4}hx_2 &= \frac{3h}{4}(3x_1+x_2)\\ &=\frac{3h}{4}(3a+3h+b)\\ &=\frac{3}{4}\cdot\frac{b-a}{3}(3a+b+b-a)\\ &=\frac{1}{2}(b^2-a^2).\end{aligned}$$

For $f(x)=x^2$,

$$\int_a^b x^2\,dx=\frac{1}{3}(b^3-a^3),$$

$$\begin{aligned}\frac{9}{4}hx_1^2+\frac{3}{4}hx_2^2 &= \frac{3h}{4}(3x_1^2+x_2^2)\\ &=\frac{3}{4}h[3(a+h)^2+b^2]\\ &=\frac{3}{4}\cdot\frac{b-a}{3}\left[\frac{1}{3}(b^2+4a^2+4ab)+b^2\right]\\ &=\frac{1}{4}\cdot\frac{b-a}{3}(4b^2+4a^2+4ab)\\ &=\frac{1}{3}(b^3-a^3).\end{aligned}$$

For $f(x)=x^3$,
$$\int_a^b x^3 \mathrm{d}x=\frac{1}{3}(b^4-a^4),$$
$$\begin{aligned}\frac{9}{4}hx_1^3+\frac{3}{4}hx_2^3&=\frac{3h}{4}(3x_1^3+x_2^3)\\&=\frac{3}{4}h[3(a+h)^3+b^3]\\&=\frac{h}{12}[(2a+b)^3+9b^3]\\&=\frac{h}{12}(8a^3+12a^2\mathrm{b}+6ab^2+10b^3)\\&=\frac{b-a}{18}(4a^3+6a^2b+3ab^2+5b^3).\end{aligned}$$

As a result, the degree of precision is 2.

Example 7.7 Find the constants c_0, c_1 and x_1 so that the quadrature formula
$$\int_0^1 f(x)\,\mathrm{d}x\approx c_0 f(0)+c_1 f(x_1)$$
has the highest possible degree of precision.

Solution. Let
$$f(x)=x^i, i=0,1,2.$$
Then
$$\begin{cases}c_0+c_1=1,\\ c_1x_1=\dfrac{1}{2},\\ c_1x_1^2=\dfrac{1}{3}.\end{cases}$$
Solving for c_0, c_1 and x_1 gives
$$c_0=\frac{1}{4}, c_1=\frac{3}{4}, x_1=\frac{2}{3}.$$
Since
$$\int_0^1 x^3\mathrm{d}x=\frac{1}{4},$$
$$c_0\cdot 0^3+c_1\cdot x_1^3=c_1x_1^3=\frac{3}{4}\cdot\left(\frac{2}{3}\right)^3=\frac{2}{9}\neq\frac{1}{4},$$
the highest possible degree of precision of the quadrature formula is 2.

7.4 Composite numerical integration (复化数值积分)

- Composite midpoint rule (复化中点公式)

- Composite trapezoidal rule (复化梯形公式)
- Composite simpson's rule (复化 Simpson 公式)
- Stability (稳定性)

The Newton-Cotes formulas are generally unsuitable for use over large integration intervals. In this case, decreasing the step-size means increasing the number of interpolation nodes, which results in high-degree formulas. As we know, high-degree polynomials are disadvantageous not only because of their oscillatory nature, but also because of the difficulties in finding out the coefficients in quadrature formulas

$$A_i = \int_a^b L_i(x)\,\mathrm{d}x.$$

In this section, we discuss a piecewise approach (分段逼近) to numerical integration that uses the low-order Newton-Cotes formulas. These are the techniques most often applied.

7.4.1 Composite midpoint rule (复化中点公式)

Let

$$h = \frac{b-a}{2n},\ x_0 = a,\ x_i = a + ih,\ i = 0, 1, \cdots, 2n.$$

And suppose $f(x) \in C^2[a,b]$. Applying the Midpoint Rule to each subinterval $[x_{2i}, x_{2i+2}]$ gives

$$\begin{aligned}\int_a^b f(x)\mathrm{d}x &= \sum_{i=0}^{n-1}\int_{x_{2i}}^{x_{2i+2}} f(x)\mathrm{d}x \\ &= \sum_{i=0}^{n-1}\left[(x_{2i+2} - x_{2i})f(x_{2i+1}) + \frac{(x_{2i+2} - x_{2i})^3}{24} f''(\xi_i)\right] \\ &= \frac{b-a}{n}\sum_{i=0}^{n-1} f(x_{2i+1}) + \frac{h^3}{3}\sum_{i=0}^{n-1} f''(\xi_i).\end{aligned}$$

$f(x) \in C^2[a,b]$ implies that there exists a number $\xi \in (a,b)$ such that

$$\frac{1}{n}\sum_{i=0}^{n-1} f''(\xi_i) = f''(\xi).$$

As a result,

$$\int_a^b f(x)\mathrm{d}x = 2h\sum_{i=0}^{n-1} f(x_{2i+1}) + \frac{b-a}{6}h^2 f''(\xi). \tag{7.4.1}$$

Eq.(7.4.1) is called the **Composite Midpoint Rule**.

7.4.2 Composite trapezoidal rule (复化梯形公式)

Let

$$h = \frac{b-a}{n},\ x_0 = a,\ x_i = a + ih,\ i = 0, 1, \cdots, n.$$

And suppose $f(x) \in C^2[a,b]$. Applying the Trapezoidal Rule to each subinterval $[x_i, x_{i+1}]$

gives

$$\begin{aligned}\int_a^b f(x)\mathrm{d}x &= \sum_{i=0}^{n-1}\int_{x_i}^{x_{i+1}} f(x)\mathrm{d}x \\ &= \sum_{i=0}^{n-1}\left\{\frac{x_{i+1}-x_i}{2}[f(x_i)+f(x_{i+1})]-\frac{(x_{i+1}-x_i)^3}{12}f''(\xi_i)\right\} \\ &= \frac{h}{2}\sum_{i=0}^{n-1}[f(x_i)+f(x_{i+1})]-\frac{h^3}{12}\sum_{i=0}^{n-1}f''(\xi_i).\end{aligned}$$

$f(x)\in C^2[a,b]$ implies that there exists a number $\xi\in(a,b)$ such that

$$\frac{1}{n}\sum_{i=0}^{n-1}f''(\xi_i)=f''(\xi).$$

As a result,

$$\int_a^b f(x)\mathrm{d}x=\frac{h}{2}[f(a)+f(b)+2\sum_{i=1}^{n-1}f(x_i)]-\frac{b-a}{12}h^2f''(\xi). \tag{7.4.2}$$

Eq.(7.4.2) is called the **Composite Trapezoidal Rule.**

7.4.3 Composite Simpson's rule (复化 Simpson 公式)

Let

$$h=\frac{b-a}{2n},\ x_0=a,\ x_i=a+ih,\ i=0,1,\cdots,2n.$$

And suppose $f(x)\in C^4[a,b]$. Applying the Simpson's Rule to each subinterval $[x_{2i},x_{2i+2}]$ gives

$$\begin{aligned}\int_a^b f(x)\mathrm{d}x &= \sum_{i=0}^{n-1}\int_{x_{2i}}^{x_{2i+2}} f(x)\mathrm{d}x \\ &= \sum_{i=0}^{n-1}\left\{\frac{x_{2i+2}-x_{2i}}{6}[f(x_{2i})+4f(x_{2i+1})+f(x_{2i+2})]-\frac{h^5}{90}f^{(4)}(\xi_i)\right\} \\ &= \frac{h}{3}\sum_{i=0}^{n-1}[f(x_{2i})+4f(x_{2i+1})+f(x_{2i+2})]-\frac{h^5}{90}\sum_{i=0}^{n-1}f^{(4)}(\xi_i) \\ &= \frac{h}{3}\left[f(a)+f(b)+2\sum_{i=1}^{n-1}f(x_{2i})+4\sum_{i=0}^{n-1}f(x_{2i+1})\right]-\frac{h^5}{90}\sum_{i=0}^{n-1}f^{(4)}(\xi_i).\end{aligned}$$

$f(x)\in C^4[a,b]$ implies that there exists a number $\xi\in(a,b)$ such that

$$\frac{1}{n}\sum_{i=0}^{n-1}f^{(4)}(\xi_i)=f^{(4)}(\xi).$$

As a result,

$$\int_a^b f(x)\mathrm{d}x=\frac{h}{3}\left[f(a)+f(b)+2\sum_{i=1}^{n-1}f(x_{2i})+4\sum_{i=0}^{n-1}f(x_{2i+1})\right]-\frac{b-a}{180}h^4f^{(4)}(\xi). \tag{7.4.3}$$

Eq. (7.4.3) is called the **Composite Simpson's Rule**.

7.4.4 Stability (稳定性)

All the composite integration techniques share an important property: Stability with respect to round-off error. To demonstrate this, we take the Composite Simpson's Rule as example.

Assume that $\tilde{f}(x_i)$ is the approximation to $f(x_i)$ and

$$f(x_i)=\tilde{f}(x_i)+\mathrm{e}_i$$

where for each $i=0,1,\cdots,n$, e_i denotes the round-off error associated with approximating $f(x_i)$ by $\tilde{f}(x_i)$. Suppose $\varepsilon=\max\limits_{0\leqslant i\leqslant n}|\mathrm{e}_i|$. Then the accumulated error, $\mathrm{e}(h)$, generated by the Composite Simpson's Rule is given by

$$\begin{aligned}\mathrm{e}(h)&=\left|\frac{h}{3}\left(\mathrm{e}_0+\mathrm{e}_{2n}+2\sum_{i=1}^{n-1}\mathrm{e}_{2i}+4\sum_{i=0}^{n-1}\mathrm{e}_{2i+1}\right)\right|\\&\leqslant\frac{h}{3}\left(|\mathrm{e}_0|+|\mathrm{e}_{2n}|+2\sum_{i=1}^{n-1}|\mathrm{e}_{2i}|+4\sum_{i=0}^{n-1}|\mathrm{e}_{2i+1}|\right)\\&\leqslant\frac{h}{3}(\varepsilon+\varepsilon+2(n-1)\varepsilon+4n\varepsilon)\\&=2nh\varepsilon\\&=(b-a)\varepsilon.\end{aligned}$$

Therefore the accumulated error is independent of the number of subintervals, even if an interval is to be divided into as many parts as we want to ensure accuracy, the increased computation that is required does not increase the round-off error, which implies that the procedure is stable as the step size h approaches zero.

Example 7.8 The Midpoint rule for approximating $\int_{-1}^{1}f(x)\mathrm{d}x$ gives the value 12, the Composite Midpoint rule with $n=2$ gives 5, and Composite Simpson's rule with $n=2$ gives 6. Use the fact that $f(-1)=f(1)$ and $f(-0.5)=f(0.5)-1$ to determine $f(-1)$, $f(-0.5)$, $f(0)$, $f(0.5)$, and $f(1)$.

Solution. The given conditions imply the following equations:

(1) (Midpoint rule)

$$2f(0)=12.$$

(2) (Composite Midpoint rule)

$$f(-0.5)+f(0.5)=5.$$

(3) (Composite Simpson's rule)

$$\frac{1}{6}[f(-1)+4f(-0.5)+f(0)]+\frac{1}{6}[f(0)+4f(0.5)+f(1)]=6.$$

Solving these equations based on the fact $f(-1)=f(1)$ and $f(-0.5)=f(0.5)-1$ gives

$f(0)=6, f(0.5)=3, f(-0.5)=2, f(-1)=f(1)=2.$

7.5 Romberg integration (Romberg 积分)

Romberg Integration uses the Composite Trapezoidal Rule to give preliminary approximations and then applies the Richardson extrapolation process to improve the approximations. First of all, we point out that the Trapezoidal Rule

$$\int_a^b f(x)\mathrm{d}x = \frac{h}{2}[f(a)+f(b)] - \frac{h^3}{12}f''(\xi)$$

can be rewritten as

$$\int_a^b f(x)\mathrm{d}x = \frac{h}{2}[f(a)+f(b)] + \alpha_1 h^3 + \alpha_2 h^5 + \alpha_3 h^7 + \cdots. \tag{7.5.1}$$

Let

$$m_k = 2^{k-1},\ h_k = \frac{b-a}{m_k},\ x_i = a + ih_k,\ i = 0, 1, \cdots, m_k;\ k = 1, 2, \cdots.$$

Applying Eq. (7.5.1) to each subinterval $[x_i, x_{i+1}]$ gives

$$\begin{aligned}\int_a^b f(x)\mathrm{d}x &= \sum_{i=0}^{m_k-1}\int_{x_i}^{x_{i+1}} f(x)\mathrm{d}x \\ &= \sum_{i=0}^{m_k-1}\left\{\frac{x_{i+1}-x_i}{2}[f(x_i)+f(x_{i+1})] + \sum_{i=0}^{m_k-1}\alpha_{i1}h_k^3 + \sum_{i=0}^{m_k-1}\alpha_{i2}h_k^5 + \sum_{i=0}^{m_k-1}\alpha_{i3}h_k^7 + \cdots\right. \\ &= \frac{h_k}{2}\left[f(a)+f(b)+2\sum_{i=1}^{m_k-1}f(x_i)\right] + K_1 h_k^2 + K_2 h_k^4 + K_3 h_k^6 + \cdots\end{aligned} \tag{7.5.2}$$

where

$$K_j = \frac{(b-a)\sum_{i=0}^{m_k-1}\alpha_{ij}}{m_k},\quad j = 1, 2, \cdots$$

Let

$$R_{k,1} = \frac{h_k}{2}\left[f(a)+f(b)+2\sum_{i=1}^{m_k-1}f(x_i)\right]. \tag{7.5.3}$$

Then we have

$$\int_a^b f(x)\mathrm{d}x = R_{k,1} + K_1 h_k^2 + K_2 h_k^4 + K_3 h_k^6 + \cdots. \tag{7.5.4}$$

Now the so called Romberg Integration is nothing but applying the Richardson extrapolation to Eq. (7.5.4). Replacing k with $k+1$ in (7.5.4) and noting $h_{k+1} = h_k / 2$ yields

$$\int_a^b f(x)\mathrm{d}x = R_{k+1,1} + K_1 h_{k+1}^2 + K_2 h_{k+1}^4 + K_3 h_{k+1}^6 + \cdots. \tag{7.5.5}$$

Subtracting Eq. (7.5.4) from 4 times Eq. (7.5.5) and simplifying gives the following $O(h_k^4)$ formula

$$\int_a^b f(x)\mathrm{d}x = R_{k+1,2} + K_2^{(1)} h_{k+1}^4 + K_3^{(1)} h_{k+1}^6 + \cdots \tag{7.5.6}$$

where

$$R_{k+1,2} = \frac{4R_{k+1,1} - R_{k,1}}{3}, \tag{7.5.7}$$

$$K_l^{(1)} = \frac{4-4^l}{3} K_l, \quad l = 2,3,\cdots.$$

Replacing k with $k+1$ in (7.5.6) and noting $h_{k+1} = h_k/2$ yields

$$\int_a^b f(x)\mathrm{d}x = R_{k+2,2} + K_2^{(1)} h_{k+2}^4 + K_3^{(1)} h_{k+2}^6 + \cdots. \tag{7.5.8}$$

Subtracting Eq. (7.5.6) from 16 times Eq. (7.5.8) and simplifying gives the following $O(h_k^6)$ formula

$$\int_a^b f(x)\mathrm{d}x = R_{k+2,3} + K_3^{(2)} h_{k+2}^6 + K_4^{(2)} h_{k+2}^8 + \cdots \tag{7.5.9}$$

where

$$R_{k+2,3} = \frac{4^2 R_{k+2,2} - R_{k+1,2}}{4^2 - 1}, \tag{7.5.10}$$

$$K_l^{(2)} = \frac{4^2 - 4^l}{4^2 - 1} K_l^{(1)}, \quad l = 3, 4, \cdots.$$

In general, for $k = 1, 2, \cdots$

$$R_{k+p,p+1} = \frac{4^p R_{k+p,p} - R_{k+p-1,p}}{4^p - 1}, \tag{7.5.11}$$

or

$$R_{k,j} = \frac{4^{j-1} R_{k,j-1} - R_{k-1,j-1}}{4^{j-1} - 1}, \tag{7.5.12}$$

which provides $O(h_k^{2j})$ approximations, namely,

$$\int_a^b f(x)\mathrm{d}x = R_{k,j} + O(h_k^{2j}). \tag{7.5.13}$$

The Romberg Integration can be conducted as in the following table.

4	4^2	4^3	4^4	
$R_{1,1}$				
$R_{2,1}$	$R_{2,2}$			
$R_{3,1}$	$R_{3,2}$	$R_{3,3}$		
$\vdots$	$\vdots$	$\vdots$	$\ddots$	
$R_{n,1}$	$R_{n,2}$	$R_{n,3}$	$\cdots$	$R_{n,n}$

Remark Instead of using the expansion (7.5.1) to get expansion (7.5.2), an expansion based on the Bernoulli polynomials, known as the Euler-Maclaurin expansion (Euler-Maclaurin 展开), is widely used to elaborate the Romberg integration.

Euler-Maclaurin Expansion (Euler- Maclaurin 展开) Let

$$x_0 = a, \quad h = \frac{b-a}{n}, \quad x_i = a + ih, \quad i = 0, 1, \cdots, n.$$

Suppose $f(x) \in C^{2m+2}[a,b]$, then there exists the following Euler-Maclaurin expansion for the integration:

$$\int_a^b f(x)\mathrm{d}x = \frac{h}{2}\left[f(a) + 2\sum_{i=1}^{n-1} f(a+ih) + f(b)\right] + \sum_{k=1}^{m} \frac{B_{2k}}{(2k)!} h^{2k}[f^{(2k-1)}(a) - f^{(2k-1)}(b)]$$
$$-\frac{B_{2m+2}}{(2m+2)!}(b-a)h^{2m+2} f^{(2m+2)}(\xi),$$

where, B_{2k}, $k = 1, 2, \cdots, m+1$, are Bernoulli numbers, and $\xi \in (0,1)$.

We refer to Theorem 7.24 in the Extending Reading part for details. (关于 Euler-Maclaurin 展开的详细证明，参见 7.11 延伸阅读，定理 7.24)

Example 7.9 Use the following data to approximate $\int_1^5 f(x)\mathrm{d}x$ as accurately as possible.

x	1	2	3	4	5
$f(x)$	2.4142	2.6734	2.8974	3.0976	3.2804

Solution. Use Romberg integration

$$\begin{array}{lll} R_{1,1} & & \\ R_{2,1} & R_{2,2} & \\ R_{3,1} & R_{3,2} & R_{3,3} \end{array}$$

$$R_{1,1} = \frac{5-1}{2}[f(1) + f(5)] = 2f(1) + 2f(5),$$

$$R_{2,1} = \frac{3-1}{2}[f(1) + f(3)] + \frac{5-3}{2}[f(3) + f(5)] = f(1) + 2f(3) + f(5),$$

$$R_{3,1} = \frac{1}{2}[f(1) + 2f(2) + 2f(3) + 2f(4) + f(5)],$$

$$R_{2,2} = \frac{4R_{2,1} - R_{1,1}}{3} = \frac{1}{3}[2f(1) + 8f(3) + 2f(5)],$$

$$R_{3,2} = \frac{4R_{3,1} - R_{2,1}}{3} = \frac{1}{3}[f(1) + 4f(2) + 2f(3) + 4f(4) + f(5)],$$

$$R_{3,3} = \frac{4^2 R_{3,2} - R_{2,2}}{4^2 - 1} = \frac{2}{45}[7f(1) + 32f(2) + 12f(3) + 32f(4) + 7f(5)] = 11.5246.$$

Therefore

$$\int_1^5 f(x)\mathrm{d}x \approx 11.5246.$$

Example 7.10 Romberg integration is used to approximate $\int_2^3 f(x)\mathrm{d}x$. If

$$f(2) = 0.51342, f(3) = 0.36788, R_{3,1} = 0.43687, R_{3,3} = 0.43622,$$

find $f(2.5)$.

Solution. By Romberg integration we have

$$R_{1,1} = \frac{1}{2}[f(2) + f(3)],$$

$$R_{2,1} = \frac{1}{4}[f(2) + 2f(2.5) + f(3)],$$

$$R_{3,1} = \frac{1}{8}[f(2) + 2f(2.25) + 2f(2.5) + 2f(2.75) + f(3)],$$

$$R_{2,2} = \frac{4R_{2,1} - R_{1,1}}{3} = \frac{1}{6}[f(2) + f(3) + 4f(2.5)],$$

$$R_{3,2} = \frac{4R_{3,1} - R_{2,1}}{3} = \frac{1}{12}[f(2) + f(3) + 4f(2.25) + 4f(2.75) + 2f(2.5)],$$

$$R_{3,3} = \frac{4^2 R_{3,2} - R_{2,2}}{4^2 - 1} = \frac{1}{90}[7f(2) + 7f(3) + 32f(2.25) + 32f(2.75) + 12f(2.5)].$$

Let

$$a = f(2.25) + f(2.75), b = f(2.5).$$

Then

$$R_{3,1} = \frac{1}{8}[2a + 2b + f(2) + f(3)],$$

$$R_{3,3} = \frac{1}{90}[32a + 12b + 7f(2) + 7f(3)].$$

Solving for b gives

$$f(2.5) = b = \frac{128R_{3,1} - 90R_{3,3} - 9f(2) - 9f(3)}{20} \approx 0.434593.$$

7.6 Gaussian quadrature (Gauss 求积)

- Basic idea (基本思想)
- Two-point Gaussian quadrature (两点 Gauss 求积公式)
- Gaussian nodes (Gauss 结点)
- The error estimation for the Gaussian quadrature (Gauss 求积的误差估计)
- Method to get Gaussian quadrature (Gauss 求积的方法)
- Examples (例题)

7.6.1 Basic idea (基本思想)

All the Newton-Cotes formulas use values of the function at equally-spaced points (等距结点). This restriction brings convenience to the composite integration. However, it can significantly decrease the accuracy of the approximation (逼近精度).
As we know, the Lagrange interpolating polynomial to $f(x)$ with remainder term

$$f(x)=P_n(x)+R_n(x)=\sum_{i=0}^{n} f(x_i)L_i(x)+\frac{f^{(n+1)}(\xi_x)}{(n+1)!}\prod_{i=0}^{n}(x-x_i)$$

results in the $(n+1)$ -point Newton-Cotes formula

$$\int_a^b f(x)\mathrm{d}x=\sum_{i=0}^{n} A_i f(x_i)+\frac{1}{(n+1)!}\int_a^b \prod_{i=0}^{n}(x-x_i)f^{(n+1)}(\xi_x)\mathrm{d}x,$$

where ξ_x is in $[a,b]$ for each x and

$$A_i=\int_a^b L_i(x)\mathrm{d}x\text{, for each } i=0,1,\cdots,n\text{ .}$$

The quadrature formula

$$\int_a^b f(x)\mathrm{d}x\approx\sum_{i=0}^{n} A_i f(x_i),$$

holds true for

$$f(x)=x^k, k=0,1,\cdots,n$$

since

$$(x^k)^{(n+1)}=0, k=0,1,\cdots,n$$

which means the $(n+1)$-point Newton-Cotes formula has the degree of accuracy at least n.

If we regard both the coefficients A_i and the nodes x_i as parameters to be determined other than choosing x_i as the equally-spaced nodes as in the case of Newton-Cotes formulas, then to make the quadrature formula

$$\int_a^b f(x)\mathrm{d}x\approx\sum_{i=0}^{n} A_i f(x_i)$$

have the highest possible degree of accuracy is to make the following equations

$$\int_a^b x^k\mathrm{d}x=\sum_{i=0}^{n} A_i x_i^k,\quad k=0,1,\cdots,m$$

hold true for the largest m.

Since $2n+2$ parameters $A_i, x_i, i=0,1,\cdots,n$, need $2n+2$ equations to determine, we know that the quadrature formula can reach at least the degree of accuracy $2n+1$. However we want to point out that the quadrature formula

$$\int_a^b f(x)\mathrm{d}x\approx\sum_{i=0}^{n} A_i f(x_i)$$

cannot reach the degree of accuracy $2n+2$, since if $f(x)$ is chosen to be a polynomial of degree $2n+2$ of the following form

$$f(x)=p_{2n+2}(x)=\prod_{i=0}^{n}(x-x_i)^2,$$

then $\sum_{i=0}^{n}A_i p_{2n+2}(x_i)=0$, and $\int_a^b p_{2n+2}(x)\mathrm{d}x>0$, which implies

$$\int_a^b p_{2n+2}(x)\mathrm{d}x\neq\sum_{i=0}^{n}A_i p_{2n+2}(x_i).$$

7.6.2 Two-point Gaussian quadrature (两点 Gauss 求积公式)

Let us consider the simplest case $n=1$ in the interval $[-1,1]$. We want to determine the nodes x_0 and x_1, and the coefficients A_0, A_1 so that the integration formula

$$\int_{-1}^{1}f(x)\mathrm{d}x\approx A_0f(x_0)+A_1f(x_1)$$

has the degree of precision at least three.

Making the following equalities

$$\int_{-1}^{1}x^k\mathrm{d}x=A_0x_0^k+A_1x_1^k,\quad k=0,1,2,3$$

hold true produces the equations below

$$\begin{aligned}A_0+A_1&=2,\\A_0x_0+A_1x_1&=0,\\A_0x_0^2+A_1x_1^2&=\frac{2}{3},\\A_0x_0^3+A_1x_1^3&=0.\end{aligned}$$

Solving for A_0, A_1, x_0, x_1 gives

$$A_0=A_1=1,\ x_0=-\frac{\sqrt{3}}{3},\ x_1=\frac{\sqrt{3}}{3}.$$

Therefore the two-point quadrature

$$\int_{-1}^{1}f(x)\mathrm{d}x\approx f\left(-\frac{\sqrt{3}}{3}\right)+f\left(\frac{\sqrt{3}}{3}\right)$$

has the degree of precision at least three. This formula can be extended to general interval $[a,b]$.

Let

$$x=\frac{b-a}{2}t+\frac{a+b}{2}.$$

Then

$$\int_a^b f(x)\mathrm{d}x = \int_{-1}^1 f\left(\frac{b-a}{2}t + \frac{a+b}{2}\right)\frac{b-a}{2}\mathrm{d}t$$
$$\approx \frac{b-a}{2}\left[f\left(-\frac{\sqrt{3}(b-a)}{6} + \frac{a+b}{2}\right) + f\left(\frac{\sqrt{3}(b-a)}{6} + \frac{a+b}{2}\right)\right].$$

7.6.3 Gaussian nodes (Gauss 结点)

If the quadrature formula

$$\int_a^b f(x)\,\mathrm{d}x \approx \sum_{i=0}^n A_i f(x_i)$$

can reach the degree of accuracy $2n+1$ by properly choosing coefficients $A_i, i = 0, 1, \cdots, n,$ and nodes $x_i, i = 0, 1, \cdots, n,$ then the formula is called Gaussian formula and the nodes $x_i, i = 0, 1, \cdots, n,$ are called Gaussian nodes (Gauss 结点).

Let us now explore how the Gaussian nodes are generated. Suppose $x_i, i = 0, 1, \cdots, n,$ are the Gaussian nodes, and define

$$\omega(x) = \prod_{k=0}^n (x - x_k).$$

Then for any polynomial $p(x)$ of degree not exceeding n, $p(x)\omega(x)$ is a polynomial of degree $\leqslant 2n+1$. Therefore, Gaussian formula holds true for $f(x) = p(x)\omega(x)$, i.e.

$$\int_a^b p(x)\omega(x)\,\mathrm{d}x = \sum_{i=0}^n A_i p(x_i)\omega(x_i) = 0,$$

which shows that the polynomial $\omega(x)$ of degree $n+1$ is orthogonal to all the polynomials of degree not exceeding n.

On the other hand, if

$$\omega(x) = \prod_{k=0}^n (x - x_k)$$

is orthogonal to any polynomial $p(x)$ of degree $\leqslant n$, i.e.

$$\int_a^b x^k \omega(x)\,\mathrm{d}x = 0, \quad k = 0,1,\cdots,n.$$

Then, we can verify that $x_i, i = 0, 1, \cdots, n,$ are the Gaussian nodes. In fact, suppose $p_{2n+1}(x)$ is any polynomial of degree $2n+1$, then there exist polynomials $q(x)$ and $r(x)$ of degree $\leqslant n$ such that

$$p_{2n+1}(x) = \omega(x)q(x) + r(x).$$

From

$$\int_a^b p_{2n+1}(x)\,\mathrm{d}x = \int_a^b \omega(x)q(x)\mathrm{d}x + \int_a^b r(x)\mathrm{d}x$$
$$= \int_a^b r(x)\mathrm{d}x$$

$$= \sum_{i=0}^{n} A_i r(x_i)$$

$$= \sum_{i=0}^{n} A_i p_{2n+1}(x_i),$$

it follows that $x_i, i = 0,1,\cdots,n,$ are the Gaussian nodes.

7.6.4 The error estimation for the Gaussian quadrature (Gauss 求积的误差估计)

Theorem 7.1 Suppose $f(x)$ has a continuous derivative of order $2n+2$ on $[a,b]$. Then the Gaussian quadrature formula

$$\int_a^b f(x)\,\mathrm{d}x \approx \sum_{i=0}^{n} A_i f(x_i)$$

with the Gaussian nodes $x_i, i = 0, 1, \cdots, n,$ has the following error estimation

$$\int_a^b f(x)\mathrm{d}x - \sum_{i=0}^{n} A_i f(x_i) = \frac{f^{(2n+2)}(\xi)}{(2n+2)!}\int_a^b \left[\prod_{i=0}^{n}(x - x_i)\right]^2 \mathrm{d}x,$$

where ξ is a number in (a,b).

Proof. Let $H_{2n+1}(x)$ denote the Hermite polynomial of degree $2n+1$ satisfying

$$H_{2n+1}(x_i)=f(x_i),\ H'_{2n+1}(x_i)=f'(x_i),\ i = 0, 1, \cdots, n.$$

Then using the Newton's interpolatory divided difference formula with remainder term gives

$$f(x) = H_{2n+1}(x) + \left[\prod_{i=0}^{n}(x - x_i)\right]^2 f[x_0, x_0, x_1, x_1, \cdots, x_{n-1}, x_{n-1}, x_n, x_n, x].$$

By the weighted Mean Value Theorem for Integrals (积分中值定理), we have

$$\int_a^b f(x)\,\mathrm{d}x = \int_a^b H_{2n+1}(x)\mathrm{d}x + \int_a^b \left[\prod_{i=0}^{n}(x - x_i)\right]^2 f[x_0, x_0, x_1, x_1, \cdots, x_{n-1}, x_{n-1}, x_n, x_n, x]\mathrm{d}x$$

$$= \int_a^b H_{2n+1}(x)\mathrm{d}x + f[x_0, x_0, x_1, x_1, \cdots, x_{n-1}, x_{n-1}, x_n, x_n, \eta]\int_a^b \left[\prod_{i=0}^{n}(x - x_i)\right]^2 \mathrm{d}x.$$

Since the Gaussian quadrature formula

$$\int_a^b f(x)\mathrm{d}x \approx \sum_{i=0}^{n} A_i f(x_i)$$

is exact for any polynomial of degree not exceeding $2n+1$, it follows

$$\int_a^b H_{2n+1}(x)\mathrm{d}x = \sum_{i=0}^{n} A_i H_{2n+1}(x_i) = \sum_{i=0}^{n} A_i f(x_i).$$

Therefore

$$\int_a^b f(x)\mathrm{d}x-\sum_{i=0}^{n}A_i f(x_i)=f[x_0,x_0,x_1,x_1,\cdots,x_{n-1},x_{n-1},x_n,x_n,\eta]\int_a^b\left[\prod_{i=0}^{n}(x-x_i)\right]^2\mathrm{d}x$$

$$=\frac{f^{(2n+2)}(\xi)}{(2n+2)!}\int_a^b\left[\prod_{i=0}^{n}(x-x_i)\right]^2\mathrm{d}x,$$

where ξ is a number in (a,b).

The proof is completed.

Define the Legendre polynomials: $p_0(x)=1,\ p_n(x)=\frac{1}{2^n n!}\frac{\mathrm{d}^n}{\mathrm{d}x^n}(x^2-1)^n,\ n=1,2,\cdots$. Then it is not difficult to show that $p_n(x)$ is a polynomial of degree n with leading coefficient $\frac{(2n)!}{2^n(n!)^2}$, and

$$\int_{-1}^{1}p_n(x)p_m(x)\mathrm{d}x=\begin{cases}0, & m\neq n;\\ \dfrac{2}{2n+1}, & m=n.\end{cases}$$

Thus the Gaussian nodes $x_i, i=0,1,\cdots,n,$ in the Gaussian quadrature formula

$$\int_{-1}^{1}f(x)\mathrm{d}x\approx\sum_{i=0}^{n}A_i f(x_i)$$

are nothing but the zeros of the Legendre polynomial $p_n(x)$ of degree n. Therefore the error estimation for the Gaussian quadrature on the interval $[-1,1]$ can be further reformulated as follows.

Theorem 7.2 Suppose $f(x)$ has a continuous derivative of order $2n+2$ on $[-1,1]$. Then the Gaussian quadrature formula

$$\int_{-1}^{1}f(x)\mathrm{d}x\approx\sum_{i=0}^{n}A_i f(x_i)$$

with the Gaussian nodes $x_i, i=0,1,\cdots,n,$ has the following error estimation

$$\int_{-1}^{1}f(x)\mathrm{d}x-\sum_{i=0}^{n}A_i f(x_i)=\frac{2^{2n+3}[(n+1)!]^4}{(2n+3)[(2n+2)!]^3}f^{(2n+2)}(\xi),$$

where ξ is a number in (a,b).

7.6.5 Method to get Gaussian quadrature (Gauss 求积的方法)

Step1. Solving equations for $x_i, i=0,1,\cdots,n,$

$$\int_a^b x^k\omega(x)\mathrm{d}x=0,\quad k=0,1,\cdots,n.$$

gives the Gaussian nodes.

Step2. Solving the following equations for $A_i, i=0,1,\cdots,n,$

$$\int_a^b x^k \mathrm{d}x = \sum_{i=0}^{n} A_i x_i^k, \quad k = 0, 1, \cdots, n$$

7.6.6 Examples (例题)

Example 7.11 Given two-point Gaussian formula

$$\int_{-1}^{1} f(x)\,\mathrm{d}x \approx A_0 f(x_0) + A_1 f(x_1),$$

Determine A_0, A_1, x_0, x_1.

Solution (method 1). Firstly, we can make the following equalities hold

$$\int_{-1}^{1} x^k (x - x_0)(x - x_1)\mathrm{d}x = 0, \quad k = 0, 1,$$

i.e.

$$\int_{-1}^{1} (x - x_0)(x - x_1)\mathrm{d}x = 0,$$

$$\int_{-1}^{1} x(x - x_0)(x - x_1)\mathrm{d}x = 0,$$

namely,

$$\frac{2}{3} + 2x_0 x_1 = 0,$$

$$-\frac{2}{3}(x_0 + x_1) = 0.$$

Solving for x_0 and x_1 gives

$$x_0 = \frac{1}{\sqrt{3}}, \quad x_1 = -\frac{1}{\sqrt{3}}.$$

Then solving the equations

$$\int_{-1}^{1} x^k \mathrm{d}x = A_0 \left(\frac{1}{\sqrt{3}}\right)^k + A_1 \left(-\frac{1}{\sqrt{3}}\right)^k, \quad k = 0,1$$

for A_0 and A_1 gives

$$A_0 = A_1 = 1.$$

As a result, Gaussian quadrature formula turns out to be

$$\int_{-1}^{1} f(x)\,\mathrm{d}x \approx f\left(-\frac{1}{\sqrt{3}}\right) + f\left(\frac{1}{\sqrt{3}}\right).$$

Solution (method 2). As we know, that the two-point formula

$$\int_{-1}^{1} f(x)\,\mathrm{d}x \approx A_0 f(x_0) + A_1 f(x_1),$$

is the Gaussian quadrature formula means that the nodes x_0 and x_1 must be the zeros of the quadratic orthogonal polynomial $p_2(x)$ on $[-1,1]$ with respect to the weight function $\omega(x)\equiv 1$. We can make use of the Gram-Schmidt process to find out $p_2(x)$. Let

$$p_0(x)\equiv 1,\ w(x)\equiv 1.$$

Suppose

$$p_1(x)=x-B_1,$$

then from

$$\int_{-1}^{1} p_0(x)p_1(x)\,\mathrm{d}x=0,$$

it follows

$$B_1=\frac{\int_{-1}^{1} x\,\mathrm{d}x}{\int_{-1}^{1}\mathrm{d}x}=0.$$

Therefore

$$p_1(x)=x.$$

Now suppose

$$p_2(x)=(x-B_2)p_1(x)-C_2p_0(x)=(x-B_2)x-C_2.$$

Then from

$$\int_{-1}^{1} p_2(x)p_1(x)\,\mathrm{d}x=0,$$

it follows

$$\int_{-1}^{1}(x-B_2)x^2\,\mathrm{d}x=0,$$

namely

$$B_2=0.$$

From

$$\int_{-1}^{1} p_2(x)p_0(x)\,\mathrm{d}x=0,$$

it follows

$$\int_{-1}^{1}[(x-B_2)p_1(x)-C_2]\mathrm{d}x=\int_{-1}^{1}(x^2-C_2)\mathrm{d}x=0,$$

namely

$$C_2=\frac{1}{3}.$$

Hence

$$p_2(x)=x^2-\frac{1}{3}.$$

As a result, $x_0=\frac{1}{\sqrt{3}}$, $x_1=-\frac{1}{\sqrt{3}}$. And

$$A_0=\int_{-1}^{1}L_0(x)\,\mathrm{d}x=\int_{-1}^{1}\frac{x-x_1}{x_0-x_1}\mathrm{d}x=\int_{-1}^{1}\frac{x-\frac{1}{\sqrt{3}}}{-\frac{2}{\sqrt{3}}}\mathrm{d}x=1,$$

$$A_1=\int_{-1}^{1}L_1(x)\,\mathrm{d}x=\int_{-1}^{1}\frac{x-x_0}{x_1-x_0}\mathrm{d}x=\int_{-1}^{1}\frac{x+\frac{1}{\sqrt{3}}}{\frac{2}{\sqrt{3}}}\mathrm{d}x=1.$$

Consequently, Gaussian quadrature formula turns out to be

$$\int_{-1}^{1}f(x)\,\mathrm{d}x\approx f\left(-\frac{1}{\sqrt{3}}\right)+f\left(\frac{1}{\sqrt{3}}\right).$$

7.7 本章算法程序及实例 (Algorithms and examples)

7.7.1 数值微分 (Numerical differentiation)

1. 三点公式 (Three-point formula)

例 7.12 已知数据点

x	0	0.2	0.4	0.6	0.8	1.0
y	0	0.04	0.16	0.36	0.64	1

利用三点公式计算以上各点的一阶导数和二阶导数近似值.

```
%% 一阶导数
clear all; clc
x=[0 0.2 0.4 0.6 0.8 1.0];  %等距结点
y=[0 0.04 0.16 0.36 0.64 1]; %y=x^2
n=length(x);
m=length(y);
if n~=m
    error('x和y长度不一致')
end
if n<3
    error('x和y长度不够，长度至少要有3')
end
dx=diff(x);
```

```
h=mean(dx);
if max(abs(h-dx))>1e-12
    error('x不是等间距的')
end
dy=zeros(size(y));
dy(2:end-1)=(y(3:end)-y(1:end-2))/(2*h);%中间点
dy(1)=(-3*y(1)+4*y(2)-y(3))/(2*h);%首端点
dy(n)=(y(n-2)-4*y(n-1)+3*y(n))/(2*h);%末端点
disp(dy)
```

结果

```
  dy=[ 0    0.4000    0.8000    1.2000    1.6000    2.0000]
```

```
%% 二阶导数
clear all; clc
x=[0 0.2 0.4 0.6 0.8 1.0];  %等距结点
y=[0 0.04 0.16 0.36 0.64 1]; %y=x^2
n=length(x);
m=length(y);
if n~=m
    error('x和y长度不一致')
end
if n<3
    error('x和y长度不够，长度至少要有3')
end
dx=diff(x);
h=mean(dx);
if max(abs(h-dx))>1e-12
    error('x不是等间距的')
end
hd=1/h/h;
dy2=zeros(size(y));
dy2(1:end-2)=(y(1:end-2)-2*y(2:end-1)+y(3:end))*hd;%1~n-2点
dy2(n-1)=(y(n-2)-2*y(n-1)+y(n))*hd;%n-1点
dy2(n)=(y(n-2)-2*y(n-1)+y(n))*hd;%末端点
disp(dy2)
```

结果

```
  dy2=[2.0000    2.0000    2.0000    2.0000    2.0000    2.0000]
```

2. 五点公式 (Five-point formula)

例 7.13 已知数据点

x	0	0.2	0.4	0.6	0.8	1.0
y	0	0.04	0.16	0.36	0.64	1

利用五点公式计算以上各点的一阶导数和二阶导数近似值.

```
%% 一阶导数
clear all; clc
x=0:0.2:1;
y=x.^2;
n=length(x);
m=length(y);
if n~=m
   error('x和y长度不一致')
end
if n<3
   error('x和y长度不够，长度至少要有3')
end
dx=diff(x);
h=mean(dx);
if max(abs(h-dx))>1e-12
   error('x不是等间距的')
end
dy=zeros(size(y));
dy(3:end-2)=(y(1:end-4)-8*y(2:end-3)+8*y(4:end-1)-y(5:end))/(12*h);
dy(1)=(-25*y(1)+48*y(2)-36*y(3)+16*y(4)-3*y(5))/(12*h);%首端点
dy(2)=(-3*y(1)-10*y(2)+18*y(3)-6*y(4)+y(5))/(12*h);%第二点
dy(n-1)=(-y(n-4)+6*y(n-3)-18*y(n-2)+10*y(n-1)+3*y(n))/(12*h);%倒数第二点
dy(n)=(3*y(n-4)-16*y(n-3)+36*y(n-2)-48*y(n-1)+25*y(n))/(12*h);%末端点
disp(dy)
```

结果

```
   dy=[ 0     0.4000     0.8000     1.2000     1.6000     2.0000]
```

```
%% 二阶导数
clear all; clc
x=0:0.2:1;
y=x.^2;
```

```
n=length(x);
m=ength(y);
if n~=m
    error('x和y长度不一致')
end
if n<3
    error('x和y长度不够，长度至少要有3')
end
dx=diff(x);
h=mean(dx);
if max(abs(h-dx))>1e-12
    error('x不是等间距的')
end
dy2=zeros(size(y));
dy2(3:end-2)=(-y(1:end-4)+16*y(2:end-3)-30*y(3:end-2)+16*y(4:end-1)-y(5:end))/(12*h.^2);%中间点
dy2(1)=(35*y(1)-104*y(2)+114*y(3)-56*y(4)+11*y(5))/(12*h.^2);%首端点
dy2(2)=(11*y(1)-20*y(2)+6*y(3)+4*y(4)-y(5))/(12*h.^2);%第二点
dy2(n-1)=(-y(n-4)+4*y(n-3)+6*y(n-2)-20*y(n-1)+11*y(n))/(12*h.^2);
dy2(n)=(11*y(n-4)-56*y(n-3)+114*y(n-2)-104*y(n-1)+35*y(n))/(12*h.^2);
disp(dy2)
```

结果

```
dy=[ 2.0000    2.0000    2.0000    2.0000    2.0000    2.0000]
```

3. 理查德森外推法计算一阶导数 (Evaluating derivatives by Richardson's extrapolation)

```
function g=Richardson(fun,x,h,e)
if nargin<4
    e=1e-16;
end
i=1;
j=1;
R(1,1)=(feval(fun,x+h)-feval(fun,x-h))/(2*h);
R(i+1,1)=(feval(fun,x+h/2)-feval(fun,x-h/2))/h;
R(i+1,j+1)=(4^j*R(i+1,j)-R(i,j))/(4^j-1);
while abs(R(i+1,j+1)-R(i+1,i))>e
    i=i+1;
    R(i+1,1)=(feval(fun,x+h/2^i)-feval(fun,x-h/2^i))/(2*h/2^i);
for j=1:i
```

```
    R(i+1,j+1)=(4^j*R(i+1,j)-R(i,j))/(4^j-1);
end
end
R
g=R(i+1,j+1)
```

例 7.14 利用理查德森外推法计算 $f(x)=x^2e^{-x}, x=0.5$ 处，初始步长 $h=0.1$，并使得误差不超过10^{-7}.

运行程序

```
clc;clear all
format long
Richardson(inline('x^2*exp(-x)'),0.5,0.1,1e-7); %函数，点，步长，误差
```

结果

```
R =
[0.45160491,          0,          0,          0]
[0.45407617, 0.45489992,          0,          0]
[0.45469263, 0.45489812, 0.45489799,          0]
[0.45484666, 0.45489800, 0.45489799, 0.45489799]
g = 0.454897994784476
```

7.7.2 数值积分 (Numerical integration)

1. Newton-Cotes 公式 (Newton-Cotes formula)

```
% fun为被积函数，n为求积结点的个数， a为积分下限， b为积分上限
% k为有效数字个数
function y=Newton_Cotes(fun,a,b,n,k)
syms t;
sum=0;
h=(b-a)/n;%h为步长
for i=1:n+1
   s=sym(1);
   for j=1:n+1
     if j~=i
         s=s*(t-j+1)/(i-j);    %计算连乘
     end
   end
   s(i)=int(s,0,n);     %计算科特斯系数保存在s
   y(i)=feval(fun,(a+(i-1)*h));     %计算函数在结点上的函数值
   sum=sum+y(i)*h*s(i);     %计算Newton-Cotes数值积分
```

```
end
sum=vpa(sum,k)  %显示计算结果，有效数字位数为k
```

例 7.15 利用 Newton-Cotes 公式计算 $\int_0^1 e^{x^2} dx$，结点数为 5，保留 8 位有效数字.

```
Newton_Cotes(inline('exp(x^2)'),0,1,5,8);
```

结果 sum=1.4627995.

2. 复化梯形公式 (Composite trapezoidal rule)

```
%fun为被积函数，n为求积结点的个数，a为积分下限，b为积分上限
function I=Com_Trap(fun,a,b,n,k)
sum=0;
h=(b-a)/n;
for i=1:n-1;
   sum=sum+feval(fun,a+i*h);
end
I=(feval(fun,a)+2*sum+feval(fun,b))*h/2;
I=vpa(I,k)
```

例 7.16 利用复化梯形公式计算 $\int_0^1 e^{x^2} dx$，结点数为 5，保留 8 位有效数字.

```
Com_Trap(inline('exp(x^2)'),0,1,5,8)
```

结果 I=1.4806546.

3. 复化 Simpson 公式 (Composite Simpson's rule)

```
%fun为被积函数，n为求积结点的个数，a为积分下限，b为积分上限
%k为有效数字个数
function I=Com_Simp(fun,a,b,n,k)
sum1=0;
sum2=0;
h=(b-a)/n;
for i=0:n-1
    sum1=sum1+feval(fun,a+(i+1/2)*h);
end
for i=1:n-1
   sum2=sum2+feval(fun,a+i*h);
end
I=(feval(fun,a)+4*sum1+2*sum2+feval(fun,b))*h/6;
I=vpa(I,k)
```

例 7.17 利用复化 Simpson 公式计算 $\int_0^1 e^{x^2}dx$，节点数为 5，保留 8 位有效数字.

```
Com_Simp(inline('exp(x^2)'),0,1,5,8)
```

结果 I=1.4626814.

4. Romberg 公式 (Romberg formula)

```
% fun为被积函数,a为积分下限,b为积分上限,n区间个数
function []=Romberg(fun,a,b,n)
format long
R=zeros(n+1,n+1);
R(1,1)=(b-a)/2*(feval(fun,a)+feval(fun,b));
for i=1:n
    sum=0;
    h=(b-a)/2^i;
    for j=1:2^i-1
        sum=sum+feval(fun,a+j*h);
    end
    R(i+1,1)=(feval(fun,a)+2*sum+feval(fun,b))*h/2;
end
for k=2:n+1
    for m=k:n+1
        R(m,k)=4^(k-1)/(4^(k-1)-1)*R(m,k-1)-1/(4^(k-1)-1)*R(m-1,k-1);
    end
end
R
I=R(n+1,n+1)
```

例 7.18 利用 Romberg 公式计算 $\int_0^1 e^{x^2}dx$，初始分割区间个数为 3.

```
Romberg(inline('exp(x^2)'),0,1,3)
```

结果

```
R =
   [1.8591409,            0,            0,            0]
   [1.5715832,    1.4757306,            0,            0]
   [1.4906789,    1.4637108,    1.4629094,            0]
   [1.4697123,    1.4627234,    1.4626576,    1.4626536]
I=1.4626536.
```

5. 高斯求积公式 (Gaussian quadrature)

```
% fun:函数,wx:权函数,a,b:积分下,上限, n:结点个数
function I=Gauss_Int(fun,wx,a,b,n)
%% 第一步构造正交多项式, 求出高斯点
syms x;
p=sym('p',[1,n+1]); %% 定义一个初始多项式函数向量
for i=1:n+1  %% 计算前k个正交多项式函数
   p(i)= x^(i-1);
end
q=sym('q',[1,n+1]); %% 定义一个正交后函数向量
q(1)=1;
for i=2:n+1
   for j=2:i
      p(i)=p(i)-(int(wx*q(j-1)*p(i),x,a,b))/(int(wx*q(j-1)*q(j-1),...
      x,a,b))*q(j-1);
   end
   q(i)=p(i);
end
phi=q(n+1);
x1=solve(phi);
x1=double(x1);
x0=sort(x1)'  %显示高斯点
%% 第二步利用高斯点构造Lagrange基函数
L=ones(1,n);
L=sym(L);
for i=1:n
   for j=1:n
      if j~=i
         L(i)=L(i)*(x-x0(j))/(x0(i)-x0(j));
      end
   end
end
%% 第三步利用积分算出积分系数
A=[];
for i=1:n
   A(i)=int(wx*(L(i))^2,x,a,b);
end
```

```
A  %显示高斯求积系数
%% 第四步求和，得到高斯求积公式
for i=1:n
   B(i)=feval(fun,x0(i));
end
I=A*B' %显示积分值
```

例 7.19 利用三点高斯求积公式计算 $\int_0^1 e^{x^2}dx$.

解 先利用换元法 $x=\frac{1}{2}(1+t)$，将积分转化为权系数为 1 的古典高斯公式

$$\int_0^1 e^{x^2}dx=\int_{-1}^1 \frac{1}{2}e^{\frac{(1+t)^2}{4}}dt$$

调用程序

```
Gauss_Int(inline('exp((x+1)^2/4)/2'),1,-1,1,3)
```

结果依次为

高斯点：–0.774596669241483，0，0.774596669241483

高斯求积系数：0.555555555555556，0.888888888888889，0.555555555555556

积分值：I=1.462409711477322.

7.8 本章要点 (Highlights)

1. 一阶及高阶导数的三点公式、五点公式不仅可以通过插值方法推出，也可以借助 Taylor 展开通过求解线性方程组的方法得到.
2. Richardson 外推法是在步长为 h 的低阶公式基础上，通过将步长加倍为 $2h$ 或减半成 $h/2$，再利用线性组合的方式得到高阶公式.
3. Richardson 外推法可构造指定精度的数值微分公式.
4. Newton-Cotes 公式中 $n=1$ 时为梯形公式，$n=2$ 时为 Simpson 公式.
5. 数值微分的近似程度是通过误差阶来衡量的，而数值求积公式的近似程度则采用代数精度来衡量.
6. 复化求积方法就是将积分区间分成若干个子区间，对每个子区间利用低阶求积公式，求得每个子区间的近似值，再将其累加求和作为原积分的近似值.
7. 利用累积舍入误差可判别复化求积公式的稳定性.
8. Romberg 积分的实质就是在复化梯形求积公式基础上，利用 Richardson 外推提高数值积分计算精度.
9. Gauss 求积是基于非等距点的求积公式，具有 $n+1$ 个 Gauss 结点的 Gauss 求积公式其代数精度为 $2n+1$.
10. 所谓 n 点 Gauss 求积公式的 Gauss 点实际上是带权 n 次正交多项式 (Legendre 多项式) 的 n 个根.

7.9 问题讨论 (Questions for discussion)

1. 设已知一系列具体值 $(x_i, y_i), i=1,2,\cdots,n$, 为求 $f''(x_i)$, 可以选择以下两种方法. 方法一：先用向前或向后差分公式计算一阶导数，再用一阶导数的三点公式计算二阶导数；方法二：直接利用二阶导数的三点公式计算二阶导数. 试问哪种方法得到的二阶导数具有更高的精度？
2. 试分析 Richardson 外推法中 $N_n(h)$ 与 $f'(x_0)$ 的关系，并证明 $\lim\limits_{n\to\infty} N_n(h)=f'(x_0)$.
3. 试推出一阶导数的 $(2n+1)$ 点公式.
4. 讨论 $(n+1)$ 点 Newton-Cotes 公式的代数精度.
5. 设 $f(x)\in C^4[a,b]$, 证明当 $n\to\infty$ 时，$\int_a^b f(x)\,\mathrm{d}x$ 的复化梯形公式和复化 Simpson 公式收敛于积分 $\int_a^b f(x)\,\mathrm{d}x$.
6. 证明梯形求积公式

$$\int_a^b f(x)\mathrm{d}x=\frac{h}{2}[f(a)+f(b)]-\frac{h^3}{12}f''(\xi)$$

可以写成级数形式

$$\int_a^b f(x)\mathrm{d}x=\frac{h}{2}[f(a)+f(b)]+\alpha_1 h^3+\alpha_2 h^5+\alpha_3 h^7+\cdots.$$

7. Simpson 求积公式能否写成如下级数形式？

$$\int_a^b f(x)\mathrm{d}x=\frac{h}{3}\left[f(a)+4f\left(\frac{a+b}{2}\right)+f(b)\right]+\beta_2 h^5+\beta_3 h^7+\beta_4 h^9+\cdots$$

能否在上述公式基础上，利用复化 Simpson 求积公式建立 Richardson 外推？这种外推算法与 Romberg 积分有何关系？
8. 简述一般插值型求积公式的积分原理，Newton-Cotes 求积公式为什么没有 Gauss 求积公式代数精度高？
9. Legendre 多项式定义为：$p_0(x)=1,\ p_n(x)=\dfrac{1}{2^n n!}\dfrac{\mathrm{d}^n}{\mathrm{d}x^n}(x^2-1)^n$，$n=1,2,\cdots$. 证明 $p_n(x)$ 是首项系数为 $\dfrac{(2n)!}{2^n(n!)^2}$ 的 n 次多项式，且有如下性质：

$$\int_{-1}^1 p_n(x)p_m(x)\,\mathrm{d}x=\begin{cases}0, & m\neq n;\\ \dfrac{2}{2n+1}, & m=n.\end{cases}$$

因此，n 次 Legendre 多项式 $p_n(x)$ 的零点即是 Gauss 点.
10. 设 $f(x)$ 在区间 $[-1,1]$ 上有 $2n+2$ 阶连续导数，则 Gauss 求积公式

$$\int_{-1}^{1} f(x)\,dx \approx \sum_{i=0}^{n} A_i f(x_i)$$

有如下误差估计:

$$\int_{-1}^{1} f(x)dx - \sum_{i=0}^{n} A_i f(x_i) = \frac{2^{2n+3}[(n+1)!]^4}{(2n+3)[(2n+2)!]^3} f^{(2n+2)}(\xi),$$

其中，$x_i = 0, 1, \cdots, n,$ 为 Gauss 点，$\xi \in (-1,1)$.

11. 令

$$\omega(x) = \frac{2^{n+1}[(n+1)!]^2}{[2(n+1)]!} p_{n+1}(x),$$

其中，$p_n(x) = \dfrac{1}{2^n n!}\dfrac{d^n}{dx^n}(x^2-1)^n$，是 n 次 Legendre 多项式，则 Gauss 求积公式

$$\int_{-1}^{1} f(x)\,dx \approx \sum_{i-0}^{n} A_i f(x_i)$$

中的 $A_i, i = 0, 1, \cdots, n,$ 可表示为

$$A_i = \int_{-1}^{1} L_i(x)\,dx = \int_{-1}^{1} \frac{\omega(x)}{(x-x_i)\omega'(x_i)} dx,$$

$$= \int_{-1}^{1} \frac{p_{n+1}(x)}{(x-x_i)p'_{n+1}(x_i)} dx, \quad i = 0, 1, \cdots, n,$$

令

$$q_{2n,k}(x) = \frac{p_{n+1}(x)p'_{n+1}(x)}{x-x_k}, \quad B_k = \int_{-1}^{1} q_{2n,k}(x)\,dx, \quad k = 0, 1, \cdots, n.$$

则由 $q_{2n,k}(x) \in \Pi_{2n}, k = 0, 1, \cdots, n,$ 知

$$B_k = \int_{-1}^{1} q_{2n,k}(x)\,dx = \sum_{i=0}^{n} A_i q_{2n,k}(x_i) = \sum_{i=0}^{n} A_i \frac{p_{n+1}(x_i)}{x_i - x_k} p'_{n+1}(x_i) = A_k [p'_{n+1}(x_k)]^2.$$

因此，

$$A_k = \frac{B_k}{[p'_{n+1}(x_k)]^2}.$$

另外，由分部积分可得

$$B_k = \int_{-1}^{1} \frac{p_{n+1}(x)p'_{n+1}(x)}{x-x_k} dx = \int_{-1}^{1} \frac{p_{n+1}(x)}{x-x_k} dp_{n+1}(x)$$

$$= \left.\frac{[p_{n+1}(x)]^2}{x-x_k}\right|_{-1}^{1} - \int_{-1}^{1} p_{n+1}(x)\left(\frac{p_{n+1}(x)}{x-x_k}\right)' dx.$$

根据 Legendre 多项式的定义，容易证明：$p_{n+1}(1) = p_{n+1}(-1) = 1$ (请证明). 由于 $\left(\dfrac{p_{n+1}(x)}{x-x_k}\right)'$

是 $n-1$ 次多项式，因此，$\int_{-1}^{1} p_{n+1}(x)\left(\frac{p_{n+1}(x)}{x-x_k}\right)'\mathrm{d}x=0$ (请说出理由)，于是，

$$B_k=\frac{1}{1-x_k}+\frac{1}{1+x_k}=\frac{2}{1-x_k^2},\quad k=0,1,\cdots,n.$$

因此，最后得到

$$A_k=\frac{2}{(1-x_k^2)[p_{n+1}'(x_k)]^2},\quad k=0,1,\cdots,n.$$

上式提供了计算 Gauss 求积公式系数 $A_k, k=0,1,\cdots,n,$ 的另一种途径.

12. 设 $f(x)$ 在区间 $[-1,1]$ 上有 $2n+2$ 阶连续导数，$w(x)=1/\sqrt{1-x^2}$，考虑 n 次 Chebyshev 多项式 $T_n(x)=\cos(n\arccos x)$，由第 6 章第 3 节内容，有

$$\int_{-1}^{1}\frac{1}{\sqrt{1-x^2}}T_n(x)T_m(x)\mathrm{d}x\ \ (x=\cos\theta)=\begin{cases}0, & n\neq m;\\ \frac{\pi}{2}, & n=m\neq 0;\\ \pi, & n=m=0.\end{cases}$$

即，Chebyshev 多项式 $\{T_n(x)\}$ 是区间 $[-1,1]$ 上关于权函数 $w(x)=1/\sqrt{1-x^2}$ 正交的多项式系. $n+1$ 次 Chebyshev 多项式 $T_{n+1}(x)=\cos((n+1)\arccos x)$ 的 $n+1$ 个零点为

$$x_k=\cos\frac{2k+1}{2n+2}\pi,\quad k=0,1,\cdots,n,$$

利用上述结点构造如下加权求积公式

$$\int_{-1}^{1}w(x)f(x)\mathrm{d}x\approx\sum_{i=0}^{n}A_if(x_i),\quad w(x)=1/\sqrt{1-x^2},$$

则有

$$\int_{-1}^{1}\frac{x^k}{\sqrt{1-x^2}}\mathrm{d}x=\sum_{i=0}^{n}A_ix_i^k,\quad k=0,1,\cdots,2n+1,$$

且

$$\int_{-1}^{1}\frac{x^{2n+2}}{\sqrt{1-x^2}}\mathrm{d}x\neq\sum_{i=0}^{n}A_ix_i^{2n+2}.$$

类似于 Gauss 求积公式的误差估计，可以证明如下误差估计 (试证明！)：

$$\int_{-1}^{1}\frac{f(x)}{\sqrt{1-x^2}}\mathrm{d}x-\sum_{i=0}^{n}A_if(x_i)=\frac{\pi}{2^{2n+1}(2n+2)!}f^{(2n+2)}(\xi),$$

其中，$x_i, i=0,1,\cdots,n,$ 为 $n+1$ 次 Chebyshev 多项式 $T_{n+1}(x)=\cos[(n+1)\arccos x]$ 的 $n+1$ 个零点，$\xi\in(-1,1)$.

13. 证明：下述求积公式

$$\int_{-1}^{1}w(x)f(x)\mathrm{d}x\approx\sum_{i=0}^{n}A_if(x_i),\quad w(x)=1/\sqrt{1-x^2},$$

代数精度为 $2n+1$, 当且仅当 $x_i, i=0,1,\cdots,n$, 为 $n+1$ 次 Chebyshev 多项式 $T_{n+1}(x)=\cos[(n+1)\arccos x]$ 的 $n+1$ 个零点.

14. 设 $f(x)$ 在区间 $[-1,1]$ 上有 $2n+2$ 阶连续导数, $x_i, i=0,1,\cdots,n$, 为 $n+1$ 次 Chebyshe 多项式 $T_{n+1}(x)=\cos[(n+1)\arccos x]$ 的 $n+1$ 个零点, $\tilde{T}_n(x)=T_n(x)/2^{n-1}$ 是首项系数为 1 的 n 次 Chebyshev 多项式. 在求积公式

$$\int_{-1}^{1} w(x)f(x)\,\mathrm{d}x \approx \sum_{i=0}^{n} A_i f(x_i), \quad w(x)=1/\sqrt{1-x^2},$$

中, 令 $f(x)=\dfrac{\tilde{T}_{n+1}(x)}{(x-x_k)\tilde{T}'_{n+1}(x_k)}$, 得

$$A_k=\int_{-1}^{1} w(x)\frac{\tilde{T}_{n+1}(x)}{(x-x_k)\tilde{T}'_{n+1}(x_k)}\,\mathrm{d}x=\int_{-1}^{1} w(x)\frac{T_{n+1}(x)}{(x-x_k)T'_{n+1}(x_k)}\,\mathrm{d}x.$$

试证明: $A_k=\dfrac{\pi}{n+1}$, $k=0,1,\cdots,n$. 从而

$$\int_{-1}^{1}\frac{1}{\sqrt{1-x^2}}f(x)\,\mathrm{d}x=\frac{\pi}{n+1}\sum_{i=0}^{n} f(x_i)+\frac{\pi}{2^{2n+1}(2n+2)!}f^{(2n+2)}(\xi).$$

上述求积公式称为 Mehler 求积公式.

7.10 关键术语 (Key terms)

forward-difference (向前差分)
remainder term (余项)
backward-difference (向后差分)
higher derivatives (高阶导数)
round-off error (舍入误差)
Richardson's extrapolation (Richardson 外推)
approximation (逼近)
Taylor polynomial (泰勒多项式)
integrate (积分)
numerical quadrature (数值求积)
midpoint rule (中点公式)
trapezoidal rule (梯形公式)
Simpson's rule (Simpson 公式)
the weighted mean value theorem for integrals (积分中值定理)
degree of accuracy (代数精度)
piecewise approach (分段逼近)
composite midpoint rule (复化中点公式)
composite trapezoidal rule (复化梯形公式)

composite Simpson's rule (复化 Simpson 公式)
stability (稳定性)
accumulated error (累积误差)
subinterval (子区间)
Romberg integration (Romberg 积分)
equally-spaced points (等距结点)
accuracy of the approximation (逼近精度).
Gaussian formula (Gauss 求积公式)
Gaussian nodes (Gauss 结点)

7.11 延伸阅读 (Extending reading)

7.11.1 Gauss 求积公式的收敛性(The convergence of Gaussian quadrature)

定理 7.3 设

$$\int_a^b f(x)\mathrm{d}x \approx \sum_{i=0}^{n} A_i f(x_i)$$

是高斯求积公式，令

$$G_n(f) = \sum_{i=0}^{n} A_i f(x_i)$$

如果 $f(x)$ 在 $[a,b]$ 上连续，则有

$$\lim_{n\to\infty} G_n(f) = \int_a^b f(x)\mathrm{d}x.$$

为证明定理 7.3，需要如下引理.

引理 7.4 (Weierstrass 逼近定理) 设 $f(x)$ 在 $[a,b]$ 上连续，$\forall \varepsilon > 0$, 则存在多项式 $p(x)$，使得

$$\|f - p\|_\infty < \varepsilon,$$

其中，$\|f\|_\infty = \max\limits_{x\in[a,b]} |f(x)|$.

引理 7.4 的证明详见第 6 章定理 6.9(延伸阅读部分).

引理 7.5 设

$$\int_a^b f(x)\mathrm{d}x \approx \sum_{i=0}^{n} A_i f(x_i)$$

是高斯求积公式，则 $A_i > 0, i = 0, 1, \cdots, n$, 且 $\sum\limits_{i=0}^{n} A_i = b - a$.

证明 因高斯求积公式的代数精度为 $2n+1$，即对次数不高于 $2n+1$ 的任意多项式 $p(x)$，恒有

$$\int_a^b p(x)\mathrm{d}x = \sum_{i=0}^{n} A_i p(x_i).$$

对 $k = 0, 1, \cdots, n$, 分别令

$$p(x) = \left[L_k(x)\right]^2 = \left[\frac{\prod\limits_{j=0, j\neq k}^{n} (x - x_j)}{\prod\limits_{j=0, j\neq k}^{n} (x_k - x_j)}\right]^2,$$

则 $p(x)$ 是 $2n$ 次多项式，且 $p(x) > 0, p(x_k) = 1, p(x_i) = 0, i \in \{0, 1, \cdots, n\} - \{k\}$，因此有

$$A_k = \sum_{i=0}^{n} A_i \left[L_k(x_i)\right]^2 = \int_a^b \left[L_k(x)\right]^2 \mathrm{d}x > 0, \quad k = 0, 1, \cdots, n.$$

特别的，令 $p(x) = 1$, 则有

$$\sum_{i=0}^{n} A_i = \int_a^b 1\mathrm{d}x = b - a.$$

引理 7.5 证毕.

定理 7.3 的证明 $\forall \varepsilon > 0$, 由引理 7.4 (Weierstrass 逼近定理) 知，存在 N 次多项式 $p_N(x)$，使得

$$\left\|f - p_N\right\|_\infty < \frac{\varepsilon}{2(b-a)}.$$

由于

$$\begin{aligned}\int_a^b f(x)\mathrm{d}x - G_n(f) &= \int_a^b f(x)\mathrm{d}x - \int_a^b p_N(x)\mathrm{d}x + \int_a^b p_N(x)\mathrm{d}x - G_n(p_N) + G_n(p_N) - G_n(f) \\ &\triangleq I_1 + I_2 + I_3,\end{aligned}$$

其中，

$$\begin{aligned} I_1 &= \int_a^b f(x)\mathrm{d}x - \int_a^b p_N(x)\mathrm{d}x, \\ I_2 &= \int_a^b p_N(x)\mathrm{d}x - G_n(p_N), \\ I_3 &= G_n(p_N) - G_n(f). \end{aligned}$$

易知，

$$\begin{aligned} |I_1| &= \left|\int_a^b f(x)\mathrm{d}x - \int_a^b p_N(x)\mathrm{d}x\right| \\ &\leqslant \int_a^b \left|f(x) - p_N(x)\right| \mathrm{d}x \\ &\leqslant (b-a)\left\|f - p_N\right\|_\infty \\ &< (b-a)\frac{\varepsilon}{2(b-a)} = \frac{\varepsilon}{2}, \end{aligned}$$

当$n>\left[\dfrac{N-1}{2}\right]$时，有$N<2n+1$，此时，$I_2=\int_a^b p_N(x)\mathrm{d}x-G_n(p_N)=0$.

由引理 7.5 知，

$$\begin{aligned}|I_3|&=|G_n(p_N)-G_n(f)|\\&=\left|\sum_{i=0}^{n}A_i[p_N(x_i)-f(x_i)]\right|\\&\leqslant\sum_{i=0}^{n}A_i|p_N(x_i)-f(x_i)|\\&\leqslant\sum_{i=0}^{n}A_i\|p_N-f\|_\infty\\&<\frac{\varepsilon}{2(b-a)}(b-a)=\frac{\varepsilon}{2}.\end{aligned}$$

因此，当$n>\left[\dfrac{N-1}{2}\right]$时，有

$$\left|\int_a^b f(x)\mathrm{d}x-G_n(f)\right|=|I_1+I_2+I_3|\leqslant|I_1|+|I_2|+|I_3|<\frac{\varepsilon}{2}+\frac{\varepsilon}{2}=\varepsilon.$$

从而，$\lim\limits_{n\to\infty}G_n(f)=\int_a^b f(x)\mathrm{d}x$.

定理 7.3 证毕.

如果对$f(x)$施加更强的条件，则有如下结果：

定理 7.6 设

$$\int_a^b f(x)\mathrm{d}x\approx\sum_{i=0}^{n}A_if(x_i)$$

是高斯求积公式，令

$$G_n(f)=\sum_{i=0}^{n}A_if(x_i),$$

如果$f(x)$在$[a,b]$上有直到$2n+2$阶连续导数，则有

$$\lim_{n\to\infty}G_n(f)=\int_a^b f(x)\mathrm{d}x.$$

证明 由 Gauss 求积公式的误差估计，有

$$\int_a^b f(x)\mathrm{d}x-\sum_{i=0}^{n}A_if(x_i)=\frac{f^{(2n+2)}(\xi)}{(2n+2)!}\int_a^b\left[\prod_{i=0}^{n}(x-x_i)\right]^2\mathrm{d}x,$$

因此，

$$\left|\int_a^b f(x)\mathrm{d}x-G_n(f)\right|\leqslant\frac{M}{(2n+2)!}(b-a)^{2n+3},$$

其中，$M=\max\limits_{x\in[a,b]}\left|f^{(2n+2)}(x)\right|$. 由

$$\lim_{n\to\infty}\frac{M}{(2n+2)!}(b-a)^{2n+3}=0,$$

推得

$$\lim_{n\to\infty}G_n(f)=\int_a^b f(x)\mathrm{d}x.$$

定理 7.6 证毕.

7.11.2 Radau 求积公式与 Lobatto 求积公式(Radau quadrature and Lobatto quadrature)

在正文中,我们介绍了两种不同类型的求积公式,即 Newton-Cotes 求积公式和 Gauss 求积公式. 就 $n+1$ 点求积公式

$$\int_a^b f(x)\mathrm{d}x\approx\sum_{i=0}^{n}A_i f(x_i) \tag{7.11.1}$$

而言,如果所有结点 $x_i, i=0,1,\cdots,n,$ 事先给定且等距分布,则上述公式就是 Newton-Cotes 求积公式,其代数精度一般为 n 或 $n+1$ (当 n 为偶数时),梯形公式和 Simpson 公式是 Newton-Cotes 求积公式当 n 分别等于 1 和 2 时的特例. 如果所有结点 $x_i, i=0,1,\cdots,n,$ 与 $A_i, i=0,1,\cdots,n,$ 一起均作为待定参数,从而可将公式(7.11.1)的代数精度提高到 $2n+1$,此时(7.11.1)即为 Gauss 求积公式.

如果不是所有结点 $x_i, i=0,1,\cdots,n,$ 而只有部分结点与 $A_i, i=0,1,\cdots,n,$ 一起作为待定参数,则此时的求积公式既不属于 Newton-Cotes 求积公式,也不是 Gauss 求积公式,在这类求积公式中,有两种情况值得特别关注.

情形一 积分区间的两个端点有一个预先选定.

(1) **左端点 $x_0=a$ 固定**,而其余结点 $x_i, i=1,2,\cdots,n,$ 待定,此时的求积公式可写成

$$\int_a^b f(x)\mathrm{d}x\approx A_0 f(a)+\sum_{i=1}^{n}A_i f(x_i) \tag{7.11.2}$$

公式(7.11.2)中的未知结点 $x_i, i=1,2,\cdots,n,$ 和未知系数 $A_i, i=0,1,\cdots,n,$ 可通过下述 $2n+1$ 个方程联立求解得

$$\int_a^b x^k\mathrm{d}x=A_0a^k+\sum_{i=1}^{n}A_i x_i^k,\quad k=0,1,\cdots,2n. \tag{7.11.3}$$

由(7.11.3)求得的 $x_i, i=1,2,\cdots,n,$ 和 $A_i, i=0,1,\cdots,n,$ 可保证求积公式(7.11.2)达到 $2n$ 次代数精度. 公式(7.11.2)称作 **Radau (左端点) 求积公式.** 关于 Radau (左端点) 求积公式,有如下结果.

定理 7.7 设

$$\int_a^b f(x)\mathrm{d}x\approx A_0 f(a)+\sum_{i=1}^{n}A_i f(x_i)$$

是 Radau (左端点) 求积公式,令 $\omega_n(x)=\prod_{i=1}^{n}(x-x_i)$, 则 Radau (左端点) 求积公式的代数精度为 $2n$ 的充分必要条件是,$\omega_n(x)$ 在 $[a,b]$ 上与所有不超过 $n-1$ 次的多项式关于权函数

$w(x)=x-a$ 正交.

证明 必要性. 设 Radau (左端点) 求积公式的代数精度为 $2n$，若 $q(x)$ 是任何不超过 $n-1$ 次的多项式，则 $(x-a)q(x)\omega_n(x)$ 是一个不超过 $2n$ 次的多项式，因此有

$$\int_a^b (x-a)q(x)\omega_n(x)\mathrm{d}x = A_0(x-a)q(x)\omega_n(x)|_{x=a} + \sum_{i=1}^n A_i(x_i-a)q(x_i)\omega_n(x_i)=0,$$

故 $\omega_n(x)$ 在 $[a,b]$ 上与所有不超过 $n-1$ 次的多项式关于权函数 $w(x)=x-a$ 正交.

下面证明充分性. 设 $\omega_n(x)$ 在 $[a,b]$ 上与所有不超过 $n-1$ 次的多项式关于权函数 $w(x)=x-a$ 正交. 设 $p_{2n}(x)$ 是不超过 $2n$ 次的多项式，则有不超过 $n-1$ 次的多项式 $q(x)$ 和不超过 n 次的多项式 $r(x)$，使得

$$p_{2n}(x)=(x-a)\omega_n(x)q(x)+r(x),$$

由 $\omega_n(x)$ 在 $[a,b]$ 上与所有不超过 $n-1$ 次的多项式关于权函数 $w(x)=x-a$ 正交，推得

$$\int_a^b (x-a)\omega_n(x)q(x)\,\mathrm{d}x=0.$$

拥有 $n+1$ 个系数 $A_i, i=0,1,\cdots,n,$ 的求积公式

$$\int_a^b f(x)\mathrm{d}x \approx A_0 f(a)+\sum_{i=1}^n A_i f(x_i)$$

其代数精度至少为 $n+1$，故对不超过 n 次的多项式 $r(x)$ 有

$$\int_a^b r(x)\mathrm{d}x = A_0 r(a)+\sum_{i=1}^n A_i r(x_i).$$

因此，

$$\begin{aligned}\int_a^b p_{2n}(x)\mathrm{d}x &= \int_a^b (x-a)\omega_n(x)q(x)\mathrm{d}x+\int_a^b r(x)\mathrm{d}x=\int_a^b r(x)\mathrm{d}x\\ &= A_0 r(a)+\sum_{i=1}^n A_i r(x_i)\\ &= A_0 p_{2n}(a)+\sum_{i=1}^n A_i p_{2n}(x_i).\end{aligned}$$

于是，Radau (左端点) 求积公式的代数精度为 $2n$. 定理 7.7 证毕.

定理 7.8 设

$$\int_a^b f(x)\mathrm{d}x \approx A_0 f(a)+\sum_{i=1}^n A_i f(x_i)$$

是 Radau (左端点) 求积公式，则 $A_i>0, i=0,1,\cdots,n,$ 且 $\sum_{i=0}^n A_i=b-a$.

证明 因求积公式

$$\int_a^b f(x)\mathrm{d}x \approx A_0 f(a)+\sum_{i=1}^n A_i f(x_i)$$

具有 $2n$ 次代数精度，故对任意不高于 $2n$ 次的多项式 $p(x)$, 有

$$\int_a^b p(x)\mathrm{d}x = A_0 p(a) + \sum_{i=1}^{n} A_i p(x_i).$$

令

$$p(x) = \left(\prod_{i=0,i\neq k}^{n} \frac{x - x_i}{x_k - x_i} \right)^2,$$

显然 $p(x)$ 是 $2n$ 次的多项式，且 $p(x) > 0, p(x_k) = 1, p(x_i) = 0, i = 0,1,\cdots,n, i \neq k$. 从而，

$$A_k = \sum_{i=0}^{n} A_i p(x_i) = \int_a^b p(x)\mathrm{d}x > 0, k = 0,1,\cdots,n.$$

特别令 $p(x) = 1$, 则有

$$\sum_{i=0}^{n} A_i = \int_a^b 1\mathrm{d}x = b - a.$$

定理 7.8 证毕.

定理 7.9 (收敛性) 设

$$\int_a^b f(x)\mathrm{d}x \approx A_0 f(a) + \sum_{i=1}^{n} A_i f(x_i)$$

是 Radau (左端点) 求积公式，令

$$R_n^L(f) = A_0 f(a) + \sum_{i=1}^{n} A_i f(x_i)$$

如果 $f(x)$ 在 $[a,b]$ 上连续，则有

$$\lim_{n\to\infty} R_n^L(f) = \int_a^b f(x)\mathrm{d}x.$$

定理 7.9 的证明完全类似于定理 7.3 的证明，只需在定理 7.3 的证明过程中，将 $n > \left[\frac{N-1}{2}\right]$ 改为 $n > \left[\frac{N}{2}\right]$ 即可.

定理 7.10 (误差估计) 设 $f(x)$ 在 $[a,b]$ 上有直到 $2n$+1 阶的连续导数，则 Radau (左端点) 求积公式有如下误差估计

$$\int_a^b f(x)\mathrm{d}x = A_0 f(a) + \sum_{i=1}^{n} A_i f(x_i) + \frac{f^{(2n+1)}(\xi)}{(2n+1)!}\int_a^b (x-a)\prod_{i=1}^{n}(x-x_i)^2\mathrm{d}x,$$

其中，$\xi \in (a,b)$.

证明 设 $2n$ 次多项式 $p_{2n}(x)$ 满足如下插值条件：

$$p_{2n}(a) = f(a),\ p_{2n}(x_i) = f(x_i),\ i = 1,2,\cdots,n;\ p'_{2n}(x_i) = f'(x_i),\ i = 1,2,\cdots,n,$$

利用带误差余项的重结点 Newton 插值公式，得到

$$\begin{aligned} p_{2n}(x) = {} & f(a) + (x-a)f[a,x_1] + (x-a)(x-x_1)f[a,x_1,x_1] \\ & +(x-a)(x-x_1)^2 f[a,x_1,x_1,x_2] + (x-a)(x-x_1)^2(x-x_2)f[a,x_1,x_1,x_2,x_2] \\ & +(x-a)(x-x_1)^2(x-x_2)^2 f[a,x_1,x_1,x_2,x_2,x_3] \end{aligned}$$

$$+\cdots$$
$$+(x-a)(x-x_1)^2\cdots(x-x_{n-1})^2(x-x_n)f[a,x_1,x_1,\cdots,x_{n-1},x_{n-1},x_n,x_n],$$

$$f(x)=p_{2n}(x)+(x-a)(x-x_1)^2\cdots(x-x_n)^2f[a,x_1,x_1,\cdots,x_n,x_n,x].$$

上式两边在 $[a,b]$ 上求积分，得到

$$\int_a^b f(x)\mathrm{d}x=\int_a^b p_{2n}(x)\mathrm{d}x+\int_a^b(x-a)(x-x_1)^2\cdots(x-x_n)^2f[a,x_1,x_1,\cdots,x_n,x_n,x]\mathrm{d}x.$$

由于 Radau (左端点) 求积公式对不高于 $2n$ 次多项式精确成立，故有

$$\int_a^b p_{2n}(x)\mathrm{d}x=A_0p_{2n}(a)+\sum_{i=1}^n A_ip_{2n}(x_i).$$

由 $p_{2n}(a)=f(a),\ p_{2n}(x_i)=f(x_i),\ i=1,2,\cdots,n$,知

$$\int_a^b p_{2n}(x)\mathrm{d}x=A_0f(a)+\sum_{i=1}^n A_if(x_i).$$

因 $(x-a)(x-x_1)^2\cdots(x-x_n)^2$ 在 $[a,b]$ 上非负，由积分中值定理得

$$\begin{aligned}&\int_a^b(x-a)(x-x_1)^2\cdots(x-x_n)^2f[a,x_1,x_1,\cdots,x_n,x_n,x]\mathrm{d}x\\&=f[a,x_1,x_1,\cdots,x_n,x_n,\eta]\int_a^b(x-a)(x-x_1)^2\cdots(x-x_n)^2\mathrm{d}x\\&=\frac{f^{(2n+1)}(\xi)}{(2n+1)!}\int_a^b(x-a)(x-x_1)^2\cdots(x-x_n)^2\mathrm{d}x.\end{aligned}$$

其中， $\eta\in[a,b],\ \xi\in(a,b)$.

因此，

$$\int_a^b f(x)\mathrm{d}x=A_0f(a)+\sum_{i=1}^n A_if(x_i)+\frac{f^{(2n+1)}(\xi)}{(2n+1)!}\int_a^b(x-a)\prod_{i=1}^n(x-x_i)^2\mathrm{d}x.$$

定理 7.10 证毕.

(2) **右端点 $x_n=b$ 固定**，而其余结点 $x_i,i=0,1,\cdots,n-1$, 待定，此时的求积公式可写成

$$\int_a^b f(x)\mathrm{d}x\approx\sum_{i=0}^{n-1}A_if(x_i)+A_nf(b)\tag{7.11.4}$$

公式 (7.11.4) 中的未知结点 $x_i,i=0,1,\cdots,n-1$, 和未知系数 $A_i,i=0,1,\cdots,n$, 可通过下述 $2n+1$ 个方程联立求解得到

$$\int_a^b x^k\mathrm{d}x=\sum_{i=0}^{n-1}A_ix_i^k+A_nb^k,\quad k=0,1,\cdots,2n.\tag{7.11.5}$$

由(7.11.5)求得的 $x_i,i=0,1,\cdots,n-1$, 和 $A_i,i=0,1,\cdots,n$, 可保证求积公式(7.11.4)达到 $2n$ 次代数精度. 公式(7.11.4)称作 **Radau (右端点) 求积公式.** 关于 Radau (右端点) 求积公式，完全类似于定理 7.7.

定理 7.11 设

$$\int_a^b f(x)\mathrm{d}x\approx\sum_{i=0}^{n-1}A_if(x_i)+A_nf(b)$$

是 Radau (右端点) 求积公式，令 $\omega_n(x)=\prod_{i=0}^{n-1}(x-x_i)$，则 Radau (右端点) 求积公式的代数精度为 $2n$ 的充分必要条件是，$\omega_n(x)$ 在 $[a,b]$ 上与所有不超过 $n-1$ 次的多项式关于权函数 $w(x)=b-x$ 正交.

证明 必要性. 设 Radau (右端点) 求积公式的代数精度为 $2n$，若 $q(x)$ 是任何不超过 $n-1$ 次的多项式，则 $(b-x)q(x)\omega_n(x)$ 是一个不超过 $2n$ 次的多项式，因此有

$$\int_a^b (b-x)q(x)\omega_n(x)\mathrm{d}x=\sum_{i=0}^{n-1}A_i(b-x_i)q(x_i)\omega_n(x_i)+A_n(b-x)q(x)\omega_n(x)|_{x=b}=0,$$

故 $\omega_n(x)$ 在 $[a,b]$ 上与所有不超过 $n-1$ 次的多项式关于权函数 $w(x)=b-x$ 正交.

下面证明充分性. 设 $\omega_n(x)$ 在 $[a,b]$ 上与所有不超过 $n-1$ 次的多项式关于权函数 $w(x)=b-x$ 正交. 设 $p_{2n}(x)$ 是不超过 $2n$ 次的多项式，则有不超过 $n-1$ 次的多项式 $q(x)$ 和不超过 n 次的多项式 $r(x)$，使得

$$p_{2n}(x)=(b-x)\omega_n(x)q(x)+r(x),$$

由 $\omega_n(x)$ 在 $[a,b]$ 上与所有不超过 $n-1$ 次的多项式关于权函数 $w(x)=b-x$ 正交，推得

$$\int_a^b (b-x)\omega_n(x)q(x)\,\mathrm{d}x=0.$$

拥有 $n+1$ 个系数 $A_i, i=0,1,\cdots,n,$ 的求积公式

$$\int_a^b f(x)\mathrm{d}x\approx\sum_{i=0}^{n-1}A_if(x_i)+A_nf(b)$$

其代数精度至少为 $n+1$，故对不超过 n 次的多项式 $r(x)$ 有

$$\int_a^b r(x)\mathrm{d}x=\sum_{i=0}^{n-1}A_ir(x_i)+A_nr(b).$$

因此，

$$\begin{aligned}\int_a^b p_{2n}(x)\mathrm{d}x&=\int_a^b(b-x)\omega_n(x)q(x)\mathrm{d}x+\int_a^b r(x)\mathrm{d}x\\&=\int_a^b r(x)\mathrm{d}x\\&=\sum_{i=0}^{n-1}A_ir(x_i)+A_nr(b)\\&=\sum_{i=0}^{n-1}A_ip_{2n}(x_i)+A_np_{2n}(b).\end{aligned}$$

于是，Radau (右端点) 求积公式的代数精度为 $2n$. 定理 7.11 证毕.

定理 7.12 设

$$\int_a^b f(x)\mathrm{d}x\approx\sum_{i=0}^{n-1}A_if(x_i)+A_nf(b)$$

是 Radau (右端点) 求积公式，则 $A_i > 0, i = 0,1,\cdots,n$, 且 $\sum_{i=0}^{n} A_i = b - a$.

定理 7.12 的证明完全类似于定理 7.8 的证明.

定理 7.13 (收敛性) 设

$$\int_a^b f(x)\mathrm{d}x \approx \sum_{i=0}^{n-1} A_i f(x_i) + A_n f(b)$$

是 Radau (右端点) 求积公式，令

$$\mathrm{R}_n^R(f) = \sum_{i=0}^{n-1} A_i f(x_i) + A_n f(b)$$

如果 $f(x)$ 在 $[a,b]$ 上连续，则有

$$\lim_{n\to\infty} R_n^{\mathrm{R}}(f) = \int_a^b f(x)\mathrm{d}x.$$

定理 7.13 的证明完全类似于定理 7.3 的证明，只需在定理 7.3 的证明过程中，将 $n > \left[\dfrac{N-1}{2}\right]$ 改为 $n > \left[\dfrac{N}{2}\right]$ 即可.

定理 7.14 (误差估计) 设 $f(x)$ 在 $[a,b]$ 上有直到 $2n+1$ 阶的连续导数，则 Radau (右端点) 求积公式有如下误差估计

$$\int_a^b f(x)\mathrm{d}x = \sum_{i=0}^{n-1} A_i f(x_i) + A_{\mathrm{n}} f(b) + \frac{f^{(2n+1)}(\xi)}{(2n+1)!}\int_a^b \prod_{i=0}^{n-1}(x-x_i)^2(x-b)\,\mathrm{d}x,$$

其中，$\xi \in (a,b)$.

证明 设 $2n$ 次多项式 $p_{2n}(x)$ 满足如下插值条件:

$$p_{2n}(x_i) = f(x_i), i = 0,1,\cdots,n-1,\ p_{2n}(b) = f(b);\ p_{2n}'(x_i) = f'(x_i), i = 0,1,\cdots,n-1,$$

利用带误差余项的重结点 Newton 插值公式，得到

$$\begin{aligned} p_{2n}(x) &= f(x_0) + (x-x_0)f[x_0,x_0] + (x-x_0)^2 f[x_0,x_0,x_1] \\ &\quad + (x-x_0)^2(x-x_1)f[x_0,x_0,x_1,x_1] + (x-x_0)^2(x-x_1)^2 f[x_0,x_0,x_1,x_1,x_2] \\ &\quad + (x-x_0)^2(x-x_1)^2(x-x_2)f[x_0,x_0,x_1,x_1,x_2,x_2] \\ &\quad + \cdots \\ &\quad + (x-x_0)^2(x-x_1)^2\cdots(x-x_{n-1})^2 f[x_0,x_0,x_1,x_1,\cdots,x_{n-1},x_{n-1},x_n], \end{aligned}$$

$$f(x) = p_{2n}(x) + (x-x_0)^2\cdots(x-x_{n-1})^2(x-b)f[x_0,x_0,x_1,x_1,\cdots,x_{n-1},x_{n-1},b,x].$$

上式两边在 $[a,b]$ 上求积分，得到

$$\int_a^b f(x)\mathrm{d}x = \int_a^b p_{2n}(x)\mathrm{d}x + \int_a^b (x-x_0)^2\cdots(x-x_{n-1})^2(x-b)f[x_0,x_0,\cdots,x_{n-1},x_{n-1},b,x]\mathrm{d}x.$$

由于 Radau (右端点) 求积公式对不高于 $2n$ 次多项式精确成立，故有

$$\int_a^b p_{2n}(x)\mathrm{d}x = \sum_{i=0}^{n-1} A_i p_{2n}(x_i) + A_n p_{2n}(b).$$

由 $p_{2n}(x_i) = f(x_i), i = 0,1,\cdots,n-1, p_{2n}(b) = f(b)$,知

$$\int_a^b p_{2n}(x)\mathrm{d}x = \sum_{i=0}^{n-1} A_i f(x_i) + A_n f(b).$$

因 $(x-x_0)^2\cdots(x-x_{n-1})^2(x-b)$ 在 $[a,b]$ 上不变号，由积分中值定理得

$$\int_a^b (x-x_0)^2\cdots(x-x_{n-1})^2(x-b)f[x_0,x_0,\cdots,x_{n-1},x_{n-1},b,x]\mathrm{d}x$$

$$= f[x_0,x_0,\cdots,x_{n-1},x_{n-1},b,\eta]\int_a^b (x-x_0)^2\cdots(x-x_{n-1})^2(x-b)\mathrm{d}x$$

$$= \frac{f^{(2n+1)}(\xi)}{(2n+1)!}\int_a^b (x-x_0)^2\cdots(x-x_{n-1})^2(x-b)\mathrm{d}x.$$

其中，$\eta\in[a,b], \xi\in(a,b)$.

因此，

$$\int_a^b f(x)\mathrm{d}x = \sum_{i=0}^{n-1} A_i f(x_i) + A_\mathrm{n} f(b) + \frac{f^{(2n+1)}(\xi)}{(2n+1)!}\int_a^b \prod_{i=0}^{n-1}(x-x_i)^2(x-b)\,\mathrm{d}x,$$

定理 7.14 证毕.

情形二　积分区间的两个端点均预先选定.

设左端点 $x_0=a$ 固定，右端点 $x_n=b$ 亦固定. 其余结点 $x_i, i=1,2,\cdots,n-1,$ 待定，此时的求积公式可写成

$$\int_a^b f(x)\mathrm{d}x \approx A_0 f(a) + \sum_{i=1}^{n-1} A_i f(x_i) + A_n f(b) \tag{7.11.6}$$

公式 (7.11.6) 中的未知结点 $x_i, i=1,2,\cdots,n-1,$ 和未知系数 $A_i, i=0,1,\cdots,n,$ 可通过下述 $2n$ 个方程联立求解得到

$$\int_a^b x^k\mathrm{d}x = A_0 a^k + \sum_{i=1}^{n-1} A_i x_i^k + A_n b^k,\quad k=0,1,\cdots,2n-1. \tag{7.11.7}$$

由 (7.11.7) 求得的 $x_i, i=1,2,\cdots,n-1$ 和 $A_i, i=0,1,\cdots,n,$ 可保证求积公式 (7.11.6) 达到 $2n-1$ 次代数精度. 公式 (7.11.6) 称作 **Lobatto 求积公式**.

定理 7.15　设

$$\int_a^b f(x)\mathrm{d}x \approx A_0 f(a) + \sum_{i=1}^{n-1} A_i f(x_i) + A_n f(b)$$

是 Lobatto 求积公式，令 $\omega_{n-1}(x)=\prod_{i=1}^{n-1}(x-x_i)$，则 Lobatto 求积公式的代数精度为 $2n-1$ 的充分必要条件是，$\omega_{n-1}(x)$ 在 $[a,b]$ 上与所有不超过 $n-2$ 次的多项式关于权函数 $w(x)=(x-a)(b-x)$ 正交.

证明　必要性. 设 Lobatto 求积公式的代数精度为 $2n-1$，若 $q(x)$ 是任何不超过 $n-2$ 次的多项式，则 $(x-a)(b-x)q(x)\omega_{n-1}(x)$ 是一个不超过 $2n-1$ 次的多项式，因此有

$$\begin{aligned}&\int_a^b (x-a)(b-x)q(x)\omega_{n-1}(x)\mathrm{d}x\\&=A_0(x-a)(b-x)q(x)\omega_{n-1}(x)|_{x=a}+\sum_{i=1}^{n-1}A_i(x_i-a)(b-x_i)q(x_i)\omega_{n-1}(x_i)\\&\quad+A_0(x-a)(b-x)q(x)\omega_{n-1}(x)|_{x=b}\\&=0,\end{aligned}$$

故 $\omega_{n-1}(x)$ 在 $[a,b]$ 上与所有不超过 $n-2$ 次的多项式关于权函数 $w(x)=(x-a)(b-x)$ 正交.

下面证明充分性. 设 $\omega_{n-1}(x)$ 在 $[a,b]$ 上与所有不超过 $n-2$ 次的多项式关于权函数 $w(x)=(x-a)(b-x)$ 正交. 设 $p_{2n-1}(x)$ 是不超过 $2n-1$ 次的多项式，则有不超过 $n-2$ 次的多项式 $q(x)$ 和不超过 n 次的多项式 $r(x)$，使得

$$p_{2n-1}(x)=(x-a)(b-x)\omega_{n-1}(x)q(x)+r(x),$$

由 $\omega_{n-1}(x)$ 在 $[a,b]$ 上与所有不超过 $n-2$ 次的多项式关于权函数 $w(x)=(x-a)(b-x)$ 正交，推得

$$\int_a^b (x-a)(b-x)\omega_{n-1}(x)q(x)\,\mathrm{d}x=0.$$

拥有 $n+1$ 个系数 $A_i, i=0,1,\cdots,n,$ 的求积公式

$$\int_a^b f(x)\mathrm{d}x\approx A_0f(a)+\sum_{i=1}^{n-1}A_if(x_i)+A_nf(b)$$

其代数精度至少为 $n+1$，故对不超过 n 次的多项式 $r(x)$ 有

$$\int_a^b r(x)\mathrm{d}x=A_0r(a)+\sum_{i=1}^{n-1}A_ir(x_i)+A_nr(b).$$

因此，

$$\begin{aligned}\int_a^b p_{2n-1}(x)\mathrm{d}x&=\int_a^b (x-a)(b-x)\omega_{n-1}(x)q(x)\mathrm{d}x+\int_a^b r(x)\mathrm{d}x\\&=\int_a^b r(x)\mathrm{d}x\\&=A_0r(a)+\sum_{i=1}^{n-1}A_ir(x_i)+A_nr(b)\\&=A_0p_{2n-1}(a)+\sum_{i=1}^{n-1}A_ip_{2n-1}(x_i)+A_np_{2n-1}(b).\end{aligned}$$

于是，Lobatto 求积公式的代数精度为 $2n-1$. 定理 7.15 证毕.

定理 7.16 设

$$\int_a^b f(x)\mathrm{d}x\approx A_0f(a)+\sum_{i=1}^{n-1}A_if(x_i)+A_nf(b)$$

是 Lobatto 求积公式，则 $A_i>0, i=0,1,\cdots,n,$ 且 $\sum_{i=0}^{n}A_i=b-a$.

证明　因求积公式

$$\int_a^b f(x)\mathrm{d}x \approx A_0 f(a) + \sum_{i=1}^{n-1} A_i f(x_i) + A_n f(b)$$

具有 $2n-1$ 次代数精度，故对任意不高于 $2n-1$ 次的多项式 $p(x)$, 有

$$\int_a^b p(x)\mathrm{d}x = A_0 p(a) + \sum_{i=1}^{n-1} A_i p(x_i) + A_n p(b).$$

令

$$p_a(x) = \left(\prod_{i=1}^{n-1} \frac{x - x_i}{a - x_i}\right)^2 \cdot \frac{b - x}{b - a},$$

$$p_k(x) = \frac{x - a}{x_k - a} \cdot \left(\prod_{i=1, i\neq k}^{n-1} \frac{x - x_i}{x_k - x_i}\right)^2 \cdot \frac{b - x}{b - x_k}, \quad k = 1, 2, \cdots, n-1,$$

$$p_b(x) = \frac{x - a}{b - a} \cdot \left(\prod_{i=1}^{n-1} \frac{x - x_i}{b - x_i}\right)^2,$$

则 $p_a(x)$ 和 $p_b(x)$ 是 $2n-1$ 次的多项式， $p_k(x), k = 1, 2, \cdots, n-1$, 是 $2n-2$ 次的多项式，且

$$\forall x \in [a,b], p_a(x) \geqslant 0,\ p_a(a) = 1,\ p_a(x_i) = 0, i = 1, \cdots, n-1, p_a(b) = 0;$$

$$\forall x \in [a,b], p_k(x) \geqslant 0,\ p_k(x_k) = 1,\ p_k(a) = 0,\ p_k(x_j) = 0,\ j = 1, \cdots, n-1, j \neq k, p_k(b) = 0;$$

$$\forall x \in [a,b], p_b(x) \geqslant 0,\ p_b(b) = 1,\ p_b(a) = 0, p_b(x_i) = 0, i = 1, \cdots, n-1.$$

从而，

$$A_0 = A_0 p_a(a) + \sum_{i=1}^{n-1} A_i p_a(x_i) + A_n p_a(b) = \int_a^b p_a(x)\mathrm{d}x > 0,$$

$$A_k = A_0 p_k(a) + \sum_{i=1}^{n-1} A_i p_k(x_i) + A_n p_k(b) = \int_a^b p_k(x)\mathrm{d}x > 0, \quad k = 1, 2, \cdots, n-1;$$

$$A_n = A_0 p_b(a) + \sum_{i=1}^{n-1} A_i p_b(x_i) + A_n p_b(b) = \int_a^b p_b(x)\mathrm{d}x > 0.$$

特别令 $p(x) = 1$, 则有

$$\sum_{i=0}^{n} A_i = \int_a^b 1\mathrm{d}x = b - a.$$

定理 7.16 证毕.

定理 7.17　(收敛性) 设

$$\int_a^b f(x)\mathrm{d}x \approx A_0 f(a) + \sum_{i=1}^{n-1} A_i f(x_i) + A_n f(b)$$

是 Lobatto 求积公式，令

$$L_n(f) = A_0 f(a) + \sum_{i=1}^{n-1} A_i f(x_i) + A_n f(b)$$

如果 $f(x)$ 在 $[a,b]$ 上连续，则有

$$\lim_{n\to\infty} L_n(f) = \int_a^b f(x)\mathrm{d}x.$$

证明 $\forall \varepsilon > 0$, 由引理 7.4 (Weierstrass 逼近定理) 知，存在 N 次多项式 $p_N(x)$，使得

$$\|f - p_N\|_\infty < \frac{\varepsilon}{2(b-a)}.$$

由于

$$\int_a^b f(x)\mathrm{d}x - L_n(f) = \int_a^b f(x)\mathrm{d}x - \int_a^b p_N(x)\mathrm{d}x + \int_a^b p_N(x)\mathrm{d}x - L_n(p_N) + L_n(p_N) - L_n(f)$$
$$\overset{\Delta}{=} I_1 + I_2 + I_3,$$

其中，

$$I_1 = \int_a^b f(x)\mathrm{d}x - \int_a^b p_N(x)\mathrm{d}x,$$
$$I_2 = \int_a^b p_N(x)\mathrm{d}x - L_n(p_N),$$
$$I_3 = L_n(p_N) - L_n(f).$$

易知，

$$\begin{aligned}|I_1| &= \left|\int_a^b f(x)\mathrm{d}x - \int_a^b p_N(x)\mathrm{d}x\right| \\ &\leqslant \int_a^b |f(x) - p_N(x)|\mathrm{d}x \\ &\leqslant (b-a)\|f - p_N\|_\infty \\ &< (b-a)\frac{\varepsilon}{2(b-a)} = \frac{\varepsilon}{2},\end{aligned}$$

当 $n > \left[\dfrac{N+1}{2}\right]$ 时，有 $N < 2n-1$，此时，$I_2 = \int_a^b p_N(x)\mathrm{d}x - L_n(p_N) = 0$. 由定理 7.16 知，

$$\begin{aligned}|I_3| &= |L_n(p_N) - L_n(f)| \\ &= \left|A_0[p_N(a) - f(a)] + \sum_{i=1}^{n-1} A_i[p_N(x_i) - f(x_i)] + A_n[p_N(b) - f(b)]\right| \\ &\leqslant A_0|p_N(a) - f(a)| + \sum_{i=1}^{n-1} A_i|p_N(x_i) - f(x_i)| + A_n|p_N(b) - f(b)| \\ &\leqslant A_0\|p_N - f\|_\infty + \sum_{i=1}^{n-1} A_i\|p_N - f\|_\infty + A_n\|p_N - f\|_\infty \\ &= \|p_N - f\|_\infty\left(A_0 + \sum_{i=1}^{n-1} A_i + A_n\right) \\ &< \frac{\varepsilon}{2(b-a)}(b-a) = \frac{\varepsilon}{2}.\end{aligned}$$

因此，当 $n > \left[\dfrac{N+1}{2}\right]$ 时，有

$$\left|\int_a^b f(x)\mathrm{d}x - L_n(f)\right| = \left|I_1 + I_2 + I_3\right| \leqslant \left|I_1\right| + \left|I_2\right| + \left|I_3\right| < \frac{\varepsilon}{2} + \frac{\varepsilon}{2} = \varepsilon.$$

从而，$\lim\limits_{n\to\infty} L_n(f) = \int_a^b f(x)\mathrm{d}x$. 定理 7.17 证毕.

定理 7.18 (误差估计) 设 $f(x)$ 在 $[a,b]$ 上有直到 $2n$ 阶的连续导数，则 Lobatto 求积公式有如下误差估计

$$\int_a^b f(x)\mathrm{d}x = A_0 f(a) + \sum_{i=1}^{n-1} A_i f(x_i) + A_n f(b) + \frac{f^{(2n)}(\xi)}{(2n)!}\int_a^b (x-a)\prod_{i=1}^{n-1}(x-x_i)^2(x-b)\,\mathrm{d}x,$$

其中，$\xi \in (a,b)$.

证明 设 $2n-1$ 次多项式 $p_{2n-1}(x)$ 满足如下插值条件：

$$p_{2n-1}(a) = f(a),\ p_{2n-1}(b) = f(b),\ p_{2n-1}(x_i) = f(x_i),\ p'_{2n-1}(x_i) = f'(x_i),\ i = 1,2,\cdots,n-1,$$

利用带误差余项的重结点 Newton 插值公式，得到

$$\begin{aligned}
p_{2n-1}(x) = {} & f(a) + (x-a)f[a,x_1] + (x-a)(x-x_1)f[a,x_1,x_1] \\
& + (x-a)(x-x_1)^2 f[a,x_1,x_1,x_2] + (x-a)(x-x_1)^2(x-x_2)f[a,x_1,x_1,x_2,x_2] \\
& + (x-a)(x-x_1)^2(x-x_2)^2 f[a,x_1,x_1,x_2,x_2,x_3] \\
& + \cdots \\
& + (x-a)(x-x_1)^2\cdots(x-x_{n-1})^2 f[a,x_1,x_1,\cdots,x_{n-1},x_{n-1},b],
\end{aligned}$$

$$f(x) = p_{2n-1}(x) + (x-a)(x-x_1)^2\cdots(x-x_{n-1})^2(x-b)f[a,x_1,x_1,\cdots,x_{n-1},x_{n-1},b,x].$$

上式两边在 $[a,b]$ 上求积分，得到

$$\int_a^b f(x)\mathrm{d}x = \int_a^b p_{2n-1}(x)\mathrm{d}x + \int_a^b (x-a)(x-x_1)^2\cdots(x-x_{n-1})^2(x-b)f[a,x_1,x_1,\cdots,x_{n-1},x_{n-1},b,x]\mathrm{d}x.$$

由于 Lobatto 求积公式对不高于 $2n-1$ 次多项式精确成立，故有

$$\int_a^b p_{2n-1}(x)\mathrm{d}x = A_0 p_{2n-1}(a) + \sum_{i=1}^{n-1} A_i p_{2n-1}(x_i) + A_n p_{2n-1}(b).$$

由 $p_{2n-1}(a) = f(a),\ p_{2n-1}(b) = f(b),\ p_{2n-1}(x_i) = f(x_i),\ i = 1,2,\cdots,n-1$,知

$$\int_a^b p_{2n-1}(x)\mathrm{d}x = A_0 f(a) + \sum_{i=1}^{n-1} A_i f(x_i) + A_n f(b).$$

因 $(x-a)(x-x_1)^2\cdots(x-x_n)^2(x-b)$ 在 $[a,b]$ 上不变号，由积分中值定理得

$$\begin{aligned}
& \int_a^b (x-a)(x-x_1)^2\cdots(x-x_{n-1})^2(x-b)f[a,x_1,x_1,\cdots,x_{n-1},x_{n-1},b,x]\mathrm{d}x \\
& = f[a,x_1,x_1,\cdots,x_{n-1},x_{n-1},b,\eta]\int_a^b (x-a)(x-x_1)^2\cdots(x-x_{n-1})^2(x-b)\mathrm{d}x \\
& = \frac{f^{(2n)}(\xi)}{(2n)!}\int_a^b (x-a)(x-x_1)^2\cdots(x-x_{n-1})^2(x-b)\mathrm{d}x.
\end{aligned}$$

其中，$\eta \in [a,b],\ \xi \in (a,b)$.

因此，

$$\int_a^b f(x)\mathrm{d}x = A_0 f(a) + \sum_{i=1}^{n-1} A_i f(x_i) + A_n f(b) + \frac{f^{(2n)}(\xi)}{(2n)!}\int_a^b (x-a)\prod_{i=1}^{n-1}(x-x_i)^2(x-b)\,\mathrm{d}x,$$

定理 7.18 证毕.

7.11.3 Euler-Maclaurin 展开(Euler-Maclaurin expansion)

Romberg 积分是在复化梯形求积公式基础上，采用 Richard 外推算法得到的. 复化梯形求积公式的一般形式如下：

$$\int_a^b f(x)\mathrm{d}x = \frac{h}{2}\left[f(a) + 2\sum_{i=1}^{n-1} f(x_i) + f(b)\right] - \frac{b-a}{12}h^2 f''(\xi).$$

上述形式显然无法应用 Richard 外推算法，为此，需要将误差项进一步拓展，Euler-Maclaurin 展开正是服务于这个目的. 在阐述 Euler- Maclaurin 展开公式之前，需要了解 Bernoulli 多项式和 Bernoulli 数的相关性质.

定义 7.1 如下生成函数定义的多项式 $B_n(x), n=0,1,\cdots$, 称为 Bernoulli 多项式

$$\frac{te^{tx}}{\mathrm{e}^t - 1} = \sum_{n=0}^{\infty} B_n(x)\frac{t^n}{n!}. \tag{7.11.8}$$

定理 7.19 Bernoulli 多项式 $B_n(x), n=0,1,\cdots$, 有如下性质：

(1) $B_0(x) = 1$;

(2) $B_n'(x) = nB_{n-1}(x), n = 1,2,\cdots$;

(3) $B_n(1-x) = (-1)^n B_n(x), n = 0,1,\cdots$;

(4) $\int_0^1 B_n(x)\,\mathrm{d}x = 0, n = 1,2,\cdots$.

证明 在等式(7.11.8)两边关于 $t \to 0$ 取极限，得

$$B_0(x) = \lim_{t\to 0}\sum_{n=0}^{\infty} B_n(x)\frac{t^n}{n!} = \lim_{t\to 0}\frac{te^{tx}}{\mathrm{e}^t - 1} = \lim_{t\to 0}\frac{\mathrm{e}^{tx} + tx\mathrm{e}^{tx}}{\mathrm{e}^t} = 1.$$

性质(1) 得证.

在等式(7.11.8)两边关于 x 求导，得

$$\frac{t^2\mathrm{e}^{tx}}{\mathrm{e}^t - 1} = \sum_{n=0}^{\infty} B_n'(x)\frac{t^n}{n!},$$

由(7.11.8)知，

$$\begin{aligned}\frac{t^2\mathrm{e}^{tx}}{\mathrm{e}^t - 1} &= t\frac{te^{tx}}{\mathrm{e}^t - 1}\\ &= t\sum_{n=0}^{\infty} B_n(x)\frac{t^n}{n!}\\ &= \sum_{n=0}^{\infty} B_n(x)\frac{t^{n+1}}{n!}\\ &= \sum_{n=0}^{\infty}(n+1)B_n(x)\frac{t^{n+1}}{(n+1)!}\end{aligned}$$

$$=\sum_{n=1}^{\infty} nB_{n-1}(x)\frac{t^n}{n!}.$$

因此，有

$$\sum_{n=0}^{\infty} B_n'(x)\frac{t^n}{n!}=\sum_{n=1}^{\infty} nB_{n-1}(x)\frac{t^n}{n!}.$$

比较上式两端t^n系数，得：$B_n'(x)=nB_{n-1}(x),\quad n=1,2,\ldots$. 性质 (2) 得证.

将(7.11.8)式两端的变量x换成$1-x$，得

$$\begin{aligned}\sum_{n=0}^{\infty} B_n(1-x)\frac{t^n}{n!}&=\frac{te^{t(1-x)}}{e^t-1}\\&=\frac{te^te^{-tx}}{e^t-1}\\&=\frac{te^{-tx}}{1-e^{-t}}\\&=\frac{-te^{-tx}}{e^{-t}-1}.\end{aligned}$$

将(7.11.8)两端的变量t换成$-t$，得

$$\begin{aligned}\frac{-te^{-tx}}{e^{-t}-1}&=\sum_{n=0}^{\infty} B_n(x)\frac{(-t)^n}{n!}\\&=\sum_{n=0}^{\infty}(-1)^n B_n(x)\frac{t^n}{n!}.\end{aligned}$$

由此推得

$$\sum_{n=0}^{\infty} B_n(1-x)\frac{t^n}{n!}=\sum_{n=0}^{\infty}(-1)^n B_n(x)\frac{t^n}{n!}.$$

比较上式两端t^n系数，得：$B_n(1-x)=(-1)^n B_n(x), n=0,1,\cdots$, 性质(3) 得证.

等式(7.11.8)两边在$[0,1]$上求积分，得

$$\sum_{n=0}^{\infty}\left[\int_0^1 B_n(x)\mathrm{d}x\right]\frac{t^n}{n!}=\int_0^1\sum_{n=0}^{\infty} B_n(x)\frac{t^n}{n!}\mathrm{d}x=\int_0^1\frac{te^{tx}}{e^t-1}\mathrm{d}x=\frac{e^{tx}\big|_{x=0}^{x=1}}{e^t-1}=\frac{e^t-1}{e^t-1}=1.$$

上式两边关于t求k阶导数，得

$$\left(\sum_{n=0}^{\infty}\left[\int_0^1 B_n(x)\mathrm{d}x\right]\frac{t^n}{n!}\right)^{(k)}=0,\quad k=1,2,\cdots$$

即

$$\sum_{n=k}^{\infty}\left[\int_0^1 B_n(x)\mathrm{d}x\right]\frac{t^{n-k}}{(n-k)!}=0,\quad k=1,2,\cdots$$

上式改写为

$$\int_0^1 B_k(x)\mathrm{d}x+\sum_{n=1}^{\infty}\left[\int_0^1 B_{k+n}(x)\mathrm{d}x\right]\frac{t^n}{n!}=0,\quad k=1,2,\cdots$$

令 $t=0$, 得： $\int_0^1 B_k(x)\mathrm{d}x=0,\quad k=1,2,\cdots$ 性质 (4) 得证. 定理 7.19 证毕.

定理 7.20 Bernoulli 多项式 $B_n(x), n=0,1,\cdots$, 是首项系数为 1 的 n 次多项式，且当 n 为偶数时，$B_n(x)$ 是关于 $x-\frac{1}{2}$ 的偶函数，当 n 为奇数时，$B_n(x)$ 是关于 $x-\frac{1}{2}$ 的奇函数.

证明 由定理 7.19 的性质 (1) 和 (2) 知， $B_0(x)=1$, 从 $B_1'(x)=B_0(x)=1$ 容易推出 $B_1(x)=x+c$, 由定理 7.19 的性质 (4) 知

$$\int_0^1 B_1(x)\mathrm{d}x=0,$$

即

$$\int_0^1 (x+c)\mathrm{d}x=0,$$

解得, $c=-\frac{1}{2}$, 从而，

$$B_1(x)=x-\frac{1}{2}.$$

从 $B_2'(x)=2B_1(x)=2\left(x-\frac{1}{2}\right)$, 容易推出 $B_2(x)=\left(x-\frac{1}{2}\right)^2+c$, 由定理 7.19 的性质 (4) 知

$$\int_0^1 B_2(x)\mathrm{d}x=0,$$

即

$$\int_0^1\left[\left(x-\frac{1}{2}\right)^2+c\right]\mathrm{d}x=0,$$

$$\int_0^{1/2}\left(t^2+c\right)\mathrm{d}x=0,$$

解得， $c=-\frac{1}{12}$, 从而，

$$B_2(x)=\left(x-\frac{1}{2}\right)^2-\frac{1}{12},$$

从 $B_3'(x)=3B_2(x)=3\left(x-\frac{1}{2}\right)^2-\frac{1}{4}$, 推出 $B_3(x)=\left(x-\frac{1}{2}\right)^3-\frac{1}{4}\left(x-\frac{1}{2}\right)+c$, 由定理 7.19 的性质 (4) 知

$$\int_0^1 B_3(x)\mathrm{d}x=0,$$

即

$$\int_0^1\left[\left(x-\frac{1}{2}\right)^3-\frac{1}{4}\left(x-\frac{1}{2}\right)+c\right]\mathrm{d}x=0,$$

$$\int_{-1/2}^{1/2}\left(t^3-\frac{1}{4}t+c\right)\mathrm{d}t=0,$$

由此解得，$c=0$, 从而，

$$B_3(x)=\left(x-\frac{1}{2}\right)^3-\frac{1}{4}\left(x-\frac{1}{2}\right),$$

从 $B_4'(x)=4B_3(x)=4\left(x-\frac{1}{2}\right)^3-\left(x-\frac{1}{2}\right)$, 推出 $B_4(x)=\left(x-\frac{1}{2}\right)^4-\frac{1}{2}\left(x-\frac{1}{2}\right)^2+c$, 由定理 7.19 的性质 (4) 知

$$\int_0^1 B_4(x)\mathrm{d}x=0,$$

即

$$\int_0^1\left[\left(x-\frac{1}{2}\right)^4-\frac{1}{2}\left(x-\frac{1}{2}\right)^2+c\right]\mathrm{d}x=0,$$

$$\int_0^{1/2}\left(t^4-\frac{1}{2}t^2+c\right)\mathrm{d}t=0,$$

由此解得，$c=\frac{7}{240}$, 从而，

$$B_4(x)=\left(x-\frac{1}{2}\right)^4-\frac{1}{2}\left(x-\frac{1}{2}\right)^2+\frac{7}{240},$$

从 $B_5'(x)=5B_4(x)=5\left(x-\frac{1}{2}\right)^4-\frac{5}{2}\left(x-\frac{1}{2}\right)^2+\frac{7}{48}$, 推出

$$B_5(x)=\left(x-\frac{1}{2}\right)^5-\frac{5}{6}\left(x-\frac{1}{2}\right)^3+\frac{7}{48}\left(x-\frac{1}{2}\right)+c,$$

由定理 7.19 的性质 (4) 知

$$\int_0^1 B_5(x)\mathrm{d}x=0,$$

即

$$\int_0^1\left[\left(x-\frac{1}{2}\right)^5-\frac{5}{6}\left(x-\frac{1}{2}\right)^3+\frac{7}{48}\left(x-\frac{1}{2}\right)+c\right]\mathrm{d}x=0,$$

$$\int_{-1}^1\left(t^5-\frac{5}{6}t^3+\frac{7}{48}t+c\right)\mathrm{d}t=0,$$

由此解得，$c=0$, 从而，

$$B_5(x)=\left(x-\frac{1}{2}\right)^5-\frac{5}{6}\left(x-\frac{1}{2}\right)^3+\frac{7}{48}\left(x-\frac{1}{2}\right),$$

反复利用恒等式

$$B_n'(x)=nB_{n-1}(x),\quad n=1,2,\cdots,$$

用归纳法不难证明，$B_n(x), n=0,1,\cdots,$ 是首项系数为 1 的 n 次多项式.

由定理 7.19 的性质 (3)，即

$$B_n(1-x)=(-1)^n B_n(x), \quad n=0,1,\cdots$$

推得

$$B_{2k+1}\left(\frac{1}{2}+x\right)=-B_{2k+1}\left(\frac{1}{2}-x\right), \quad k=0,1,\cdots$$

于是，$B_{2k+1}(x)$ 是关于 $x-\frac{1}{2}$ 的奇函数. 由 $B_{2k}(x)=\frac{1}{2k+1}B'_{2k+1}(x)$, $k=0,1,\cdots$, 推得

$$B_{2k}\left(\frac{1}{2}+x\right)=\frac{1}{2k+1}B'_{2k+1}\left(\frac{1}{2}+x\right)=\frac{1}{2k+1}B'_{2k+1}\left(\frac{1}{2}-x\right)=B_{2k}\left(\frac{1}{2}-x\right), \quad k=0,1,\cdots,$$

于是，$B_{2k}(x)$ 是关于 $x-\frac{1}{2}$ 的偶函数. 定理 7.20 证毕.

定理 7.21 当 n 为奇数时，$B_n(0)=B_n(1)=0$, $n=3,5,7,\cdots$.

证明 由定理 7.19 的性质 (4) 和 (2)，知 $B_n(0)=B_n(1), n=2,3,\cdots$,

$$B_n(1)-B_n(0)=\int_0^1 B'_n(x)\,\mathrm{d}x=\int_0^1 nB_{n-1}(x)\,\mathrm{d}x=0, \quad n=2,3,\cdots.$$

因此，

$$B_n(0)=B_n(1), n=2,3,\cdots$$

在定理 7.19 的性质 (3) 中令 $x=0$，得，$B_n(1)=(-1)^n B_n(0), n=0,1,\cdots$ 从而，

$$B_{2k+1}(1)=-B_{2k+1}(0), k=1,2,\cdots$$

又

$$B_{2k+1}(1)=B_{2k+1}(0), k=1,2,\cdots$$

因此，当 n 为奇数时，$B_n(0)=B_n(1)=0$, $n=3,5,7,\cdots$. 定理 7.21 证毕.

定义 7.2 称 $B_n=B_n(0)=B_n(1), n=2,3,\cdots$, 为 Bernoulli 数.

由定理 7.20 的证明过程，易知，$B_2=\frac{1}{6}, B_3=0, B_4=-\frac{1}{30}, B_5=0.$

定理 7.22 设 $C_{2m}(x)=B_{2m}(x)-B_{2m}, m=1,2,\cdots$, 则 $C_{2m}(x)$ 在 $[0,1]$ 上不变号.

证明 用反证法. 假设 $C_{2m}(x)$ 在 $[0,1]$ 上改变符号，则 $C_{2m}(x)$ 在 $(0,1)$ 内一定有零点，又 $C_{2m}(0)=B_{2m}(0)-B_{2m}=B_{2m}-B_{2m}=0$, $C_{2m}(1)=B_{2m}(1)-B_{2m}=B_{2m}-B_{2m}=0$, 由 Rolle 定理，$C_{2m}(x)$ 在 $[0,1]$ 上有 3 个零点，于是 $C'_{2m}(x)=B'_{2m}(x)=2mB_{2m-1}(x)$ 在 $(0,1)$ 上有两个零点，即，$B_{2m-1}(x)$ 在 $(0,1)$ 上有两个零点，又，$B_{2m-1}(0)=0$, $B_{2m-1}(1)=0$, 故 $B_{2m-1}(x)$ 在 $[0,1]$ 上有 4 个零点，由 Rolle 定理，$B'_{2m-1}(x)=(2m-1)B_{2m-2}(x)$ 在 $(0,1)$ 上有 3 个零点，即，$B_{2m-2}(x)$ 在 $(0,1)$ 上有 3 个零点. 由 Rolle 定理，$B'_{2m-2}(x)=(2m-2)B_{2m-3}(x)$ 在 $(0,1)$ 上有两个零点，即，$B_{2m-3}(x)$ 在 $(0,1)$ 上有两个零点，又，$B_{2m-3}(0)=0$, $B_{2m-3}(1)=0$, 故 $B_{2m-3}(x)$ 在 $[0,1]$ 上有 4 个零点. 不断重复上述过程，最后推出，三次多项式 $B_3(x)$ 在 $[0,1]$ 上有 4 个零点. 矛盾. 故 $C_{2m}(x)$ 在 $[0,1]$ 上不变号. 定理 7.22 证毕.

定理 7.23 **(Euler-Maclaurin 展开)** 设 $f(x)\in C^{2m+2}[0,1]$, 则有如下展开

$$\int_0^1 f(x)\,\mathrm{d}x=\frac{1}{2}[f(0)+f(1)]+\sum_{k=1}^{m}\frac{B_{2k}}{(2k)!}[f^{(2k-1)}(0)-f^{(2k-1)}(1)]-\frac{B_{2m+2}}{(2m+2)!}f^{(2m+2)}(\eta),$$

其中，$B_{2k}, k=1,2,\cdots,m+1$, 是 Bernoulli 数，$\eta\in(0,1)$.

证明　应用定理 7.19 的性质 (1) 和 (2) ： $B_0(x)=1,\ B_n'(x)=nB_{n-1}(x),\ n=1,2,\cdots$，以及 $B_1(x)=x-\frac{1}{2}$，和定理 7.21： $B_n=B_n(0)=B_n(1), n=2,3,\cdots$， $B_{2k+1}=0, k=1,2,\cdots$，有

$$\begin{aligned}
\int_0^1 f(x)\,\mathrm{d}x &= \int_0^1 f(x)B_0(x)\,\mathrm{d}x \\
&= \int_0^1 f(x)B_1'(x)\,\mathrm{d}x \\
&= \int_0^1 f(x)\,\mathrm{d}B_1(x) \\
&= f(x)B_1(x)\Big|_0^1 - \int_0^1 f'(x)B_1(x)\mathrm{d}x \\
&= \frac{1}{2}[f(0)+f(1)] - \frac{1}{2}\int_0^1 f'(x)\,\mathrm{d}B_2(x) \\
&= \frac{1}{2}[f(0)+f(1)] + \frac{1}{2}B_2[f'(0)-f'(1)] + \frac{1}{2}\int_0^1 f''(x)B_2(x)\mathrm{d}x \\
&= \frac{1}{2}[f(0)+f(1)] + \frac{1}{2}B_2[f'(0)-f'(1)] + \frac{1}{2}\cdot\frac{1}{3}\int_0^1 f''(x)\,\mathrm{d}B_3(x) \\
&= \frac{1}{2}[f(0)+f(1)] + \frac{1}{2}B_2[f'(0)-f'(1)] - \frac{1}{3!}\int_0^1 f'''(x)B_3(x)\mathrm{d}x \\
&= \frac{1}{2}[f(0)+f(1)] + \frac{1}{2}B_2[f'(0)-f'(1)] - \frac{1}{4!}\int_0^1 f'''(x)\,\mathrm{d}B_4(x) \\
&= \frac{1}{2}[f(0)+f(1)] + \frac{B_2}{2!}[f'(0)-f'(1)] + \frac{B_4}{4!}[f'''(0)-f'''(1)] \\
&\quad + \frac{1}{4!}\int_0^1 f^{(4)}(x)B_4(x)\mathrm{d}x \\
&= \frac{1}{2}[f(0)+f(1)] + \frac{B_2}{2!}[f'(0)-f'(1)] + \frac{B_4}{4!}[f'''(0)-f'''(1)] + \frac{1}{5!}\int_0^1 f^{(4)}(x)\,\mathrm{d}B_5(x) \\
&= \frac{1}{2}[f(0)+f(1)] + \frac{B_2}{2!}[f'(0)-f'(1)] + \frac{B_4}{4!}[f'''(0)-f'''(1)] - \frac{1}{5!}\int_0^1 f^{(5)}(x)B_5(x)\,\mathrm{d}x \\
&= \frac{1}{2}[f(0)+f(1)] + \frac{B_2}{2!}[f'(0)-f'(1)] + \frac{B_4}{4!}[f'''(0)-f'''(1)] - \frac{1}{6!}\int_0^1 f^{(5)}(x)\,\mathrm{d}B_6(x) \\
&= \frac{1}{2}[f(0)+f(1)] + \frac{B_2}{2!}[f'(0)-f'(1)] + \frac{B_4}{4!}[f'''(0)-f'''(1)] \\
&\quad + \frac{B_6}{6!}\left[f^{(5)}(0)-f^{(5)}(1)\right] + \frac{1}{6!}\int_0^1 f^{(6)}(x)B_6(x)\mathrm{d}x \\
&= \cdots\cdots \\
&= \frac{1}{2}[f(0)+f(1)] + \frac{B_2}{2!}[f'(0)-f'(1)] + \frac{B_4}{4!}[f'''(0)-f'''(1)]
\end{aligned}$$

$$+\frac{B_6}{6!}\left[f^{(5)}(0)-f^{(5)}(1)\right]+\cdots+\frac{B_{2m+2}}{(2m+2)!}\left[f^{(2m+1)}(0)-f^{(2m+1)}(1)\right]$$
$$+\frac{1}{(2m+2)!}\int_0^1 f^{(2m+2)}(x)B_{2m+2}(x)\mathrm{d}x$$
$$=\frac{1}{2}\left[f(0)+f(1)\right]+\sum_{k=1}^{m+1}\frac{B_{2k}}{(2k)!}\left[f^{(2k-1)}(0)-f^{(2k-1)}(1)\right]$$
$$+\frac{1}{(2m+2)!}\int_0^1 f^{(2m+2)}(x)\left(B_{2m+2}(x)-B_{2m+2}\right)\mathrm{d}x+\frac{B_{2m+2}}{(2m+2)!}\int_0^1 f^{(2m+2)}(x)\,\mathrm{d}x$$
$$=\frac{1}{2}\left[f(0)+f(1)\right]+\sum_{k=1}^{m+1}\frac{B_{2k}}{(2k)!}\left[f^{(2k-1)}(0)-f^{(2k-1)}(1)\right]$$
$$+\frac{f^{(2m+2)}(\eta)}{(2m+2)!}\int_0^1\left(B_{2m+2}(x)-B_{2m+2}\right)\mathrm{d}x-\frac{B_{2m+2}}{(2m+2)!}\left[f^{(2m+1)}(0)-f^{(2m+1)}(1)\right]$$
$$=\frac{1}{2}\left[f(0)+f(1)\right]+\sum_{k=1}^{m}\frac{B_{2k}}{(2k)!}\left[f^{(2k-1)}(0)-f^{(2k-1)}(1)\right]$$
$$-\frac{B_{2m+2}}{(2m+2)!}f^{(2m+2)}(\eta),$$

其中，用到了定理 7.22：$C_{2m}(x)=B_{2m}(x)-B_{2m}, m=1,2,\cdots$，在$[0,1]$上不变号、积分中值定理、定理 7.19 的性质 (4) ：$\int_0^1 B_n(x)\mathrm{d}x=0, n=1,2,\cdots$ 等结论.

定理 7.23 证毕.

定理 7.24 (Euler-Maclaurin 展开) 令 $x_0=a, h=\dfrac{b-a}{n}, x_i=a+ih, i=0,1,\cdots,n$, 设 $f(x)\in C^{2m+2}[a,b]$, 则有如下展开

$$\int_a^b f(x)\mathrm{d}x=\frac{h}{2}\left[f(a)+2\sum_{i=1}^{n-1}f(a+ih)+f(b)\right]+\sum_{k=1}^{m}\frac{B_{2k}}{(2k)!}h^{2k}[f^{(2k-1)}(a)-f^{(2k-1)}(b)]$$
$$-\frac{B_{2m+2}}{(2m+2)!}(b-a)h^{2m+2}f^{(2m+2)}(\xi),$$

其中，$B_{2k}, k=1,2,\cdots,m+1$, 是 Bernoulli 数，$\xi\in(0,1)$.

证明 由定积分性质，有

$$\int_a^b f(x)\mathrm{d}x=\sum_{i=0}^{n-1}\int_{x_i}^{x_{i+1}}f(x)\mathrm{d}x=\sum_{i=0}^{n-1}\int_{a+ih}^{a+(i+1)h}f(x)\mathrm{d}x$$

作变量替换：$t=\dfrac{1}{h}(x-a-ih)$, 得

$$\int_a^b f(x)\mathrm{d}x=\sum_{i=0}^{n-1}\int_{a+ih}^{a+(i+1)h}f(x)\mathrm{d}x=h\sum_{i=0}^{n-1}\int_0^1 f[a+(t+i)h])\mathrm{d}t$$

由定理 7.23，知

$$\int_0^1 f[a+(t+i)h]\mathrm{d}t=\frac{1}{2}[f(a+ih)+f(a+(i+1)h)]$$
$$+\sum_{k=1}^{m}\frac{B_{2k}}{(2k)!}h^{2k-1}[f^{(2k-1)}(a+ih)-f^{(2k-1)}(a+(i+1)h)]$$
$$-\frac{B_{2m+2}}{(2m+2)!}h^{2m+2}f^{(2m+2)}(\eta_i),$$

其中，$\eta_i\in[a+ih,a+(i+1)h]$. 因此，

$$\int_a^b f(x)\,\mathrm{d}x=h\sum_{i=0}^{n-1}\int_0^1 f[a+(t+i)h])\mathrm{d}t=\frac{1}{2}h\sum_{i=0}^{n-1}[f(a+ih)+f(a+(i+1)h)]$$
$$+h\sum_{i=0}^{n-1}\sum_{k=1}^{m}\frac{B_{2k}}{(2k)!}h^{2k-1}[f^{(2k-1)}(a+ih)-f^{(2k-1)}(a+(i+1)h)]$$
$$-h\sum_{i=0}^{n-1}\frac{B_{2m+2}}{(2m+2)!}h^{2m+2}f^{(2m+2)}(\eta_i)$$
$$=\frac{h}{2}\left[f(a)+2\sum_{i=1}^{n-1}f(a+ih)+f(b)\right]+\sum_{k=1}^{m}\frac{B_{2k}}{(2k)!}h^{2k}[f^{(2k-1)}(a)-f^{(2k-1)}(b)]$$
$$-\frac{B_{2m+2}}{(2m+2)!}(b-a)h^{2m+2}f^{(2m+2)}(\xi),$$

其中，$\xi\in[a,b]$, $f(\xi)=\frac{1}{n}\sum_{i=0}^{n-1}f^{(2m+2)}(\eta_i)$. 定理 7.24 证毕.

7.12 习题 (Exercises)

1. Suppose $f(x)\in C^5(x_0,x_0+4h)$. Show that there exists $\xi\in(x_0,x_0+4h)$ such that

$$f'(x_0)=\frac{1}{12h}[-25f(x_0)+48f(x_0+h)-36f(x_0+2h)+16f(x_0+3h)-3f(x_0+4h)]$$
$$+\frac{h^4}{5}f^{(5)}(\xi_1).$$

2. Suppose $f(x)\in C^5(x_0-3h,x_0+h)$. Show that there exists $\xi\in(x_0-3h,x_0+h)$ such that

$$f'(x_0)=\frac{1}{12h}[-f(x_0-3h)+6f(x_0-2h)-18f(x_0-h)+10f(x_0)+3f(x_0+h)]$$
$$-\frac{h^4}{20}f^{(5)}(\xi_3).$$

3. Suppose $f(x)\in C^5(x_0-4h,x_0)$. Show that there exists $\xi\in(x_0-4h,x_0)$ such that

$$f'(x_0)=\frac{1}{12h}[3f(x_0-4h)-16f(x_0-3h)+36f(x_0-2h)-48f(x_0-h)+25f(x_0)]$$
$$+\frac{h^4}{5}f^{(5)}(\xi_4).$$

4. Suppose $f(x)\in C^3(x_0,x_0+2h)$. Show that there exists $\xi\in(x_0,x_0+2h)$ such that

$$f''(x_0)=\frac{1}{h^2}[f(x_0)-2f(x_0+h)+f(x_0+2h)]+hf'''(\xi).$$

5. Suppose $f(x)\in C^3(x_0-2h,x_0)$. Show that there exists $\xi\in(x_0-2h,x_0)$ such that

$$f''(x_0)=\frac{1}{h^2}[f(x_0-2h)-2f(x_0-h)+f(x_0)]-hf'''(\xi).$$

6. Suppose $f(x)\in C^5(x_0,x_0+4h)$. Show that there exists $\xi\in(x_0,x_0+4h)$ such that

$$f''(x_0)=\frac{1}{12h^2}[35f(x_0)-104f(x_0+h)+114f(x_0+2h)-56f(x_0+3h)+11f(x_0+4h)]$$
$$-\frac{5h^3}{6}f^{(5)}(\xi).$$

7. Suppose $f(x)\in C^5(x_0-h,x_0+3h)$. Show that there exists $\xi\in(x_0-h,x_0+3h)$ such that

$$f''(x_0)=\frac{1}{12h^2}[11f(x_0-h)-20f(x_0)+6f(x_0+h)+4f(x_0+2h)-f(x_0+3h)]+\frac{h^3}{12}f^{(5)}(\xi).$$

8. Suppose $f(x)\in C^6(x_0-2h,x_0+2h)$. Show that there exists $\xi\in(x_0-2h,x_0+2h)$ such that

$$f''(x_0)=-\frac{1}{12h^2}[f(x_0-2h)-16f(x_0-h)+30f(x_0)-16f(x_0+h)+f(x_0+2h)]$$
$$+\frac{h^4}{90}f^{(6)}(\xi).$$

9. Suppose $f(x)\in C^5(x_0-3h,x_0+h)$. Show that there exists $\xi\in(x_0-3h,x_0+h)$ such that

$$f''(x_0)=\frac{1}{12h^2}[-f(x_0-3h)+4f(x_0-2h)+6f(x_0-h)-20f(x_0)+11f(x_0+h)]$$
$$-\frac{h^3}{12}f^{(5)}(\xi).$$

10. Suppose $f(x)\in C^5(x_0-4h,x_0)$ Show that there exists $\xi\in(x_0-4h,x_0)$ such that

$$f''(x_0)=\frac{1}{12h^2}[11f(x_0-4h)-56f(x_0-3h)+114f(x_0-2h)-104f(x_0-h)+35f(x_0)]$$
$$+\frac{5h^3}{6}f^{(5)}(\xi).$$

11. Suppose that the following data $N_1(h)=2.356194$, $N_1(h/2)=-0.4879837$, $N_1(h/4)=-0.8815732$, $N_1(h/8)=-0.9709157$, can be used to approximate the integral

$$M=\int_0^{3\pi/2}\cos x dx.$$

Assuming

$$M=N_1(h)+K_1h^2+K_2h^4+K_3h^6+K_4h^8+O(h^{10}),$$

construct an extrapolation table to determine $N_4(h)$.

12. Suppose that $N(h)$ is an approximation to M for every $h>0$ and that

$$M=N(h)+K_1h+K_2h^2+K_3h^3+\cdots,$$

for some constants $K_1, K_2, K_3, \cdots$. Use the values $N(h)$, $N(h/3)$ and $N(h/9)$ to produce an $O(h^3)$ approximation to M.

13. Suppose that $N(h)$ is an approximation to M for every $h > 0$ and that

$$M = N(h) + K_1 h^2 + K_2 h^4 + K_3 h^6 + \cdots,$$

for some constants $K_1, K_2, K_3, \cdots$. Use the values $N(h)$, $N(h/3)$, and $N(h/9)$ to produce an $O(h^6)$ approximation to M.

14. Approximate the integral $\int_{0.5}^{1} x^4 \mathrm{d}x$ using the Trapezoidal rule (梯形求积公式).

15. Approximate the integral $\int_{0.5}^{1} x^4 \mathrm{d}x$ using Simpson's rule (Simpson 求积公式)

16. Approximate the integral $\int_{1}^{1.5} x^2 \ln x \, \mathrm{d}x$ using the Trapezoidal rule (梯形求积公式).

17. Approximate the integral $\int_{1}^{1.5} x^2 \ln x \, \mathrm{d}x$ using Simpson's rule (Simpson 求积公式).

18. Approximate the integral $\int_{1}^{1.6} \frac{2x}{x^2 - 4} \mathrm{d}x$ using the Trapezoidal rule (梯形求积公式).

19. Approximate the integral $\int_{0}^{\pi/4} x \sin x \, \mathrm{d}x$ using the Trapezoidal rule (梯形求积公式).

20. The Trapezoidal rule applied to $\int_{0}^{2} f(x) \, \mathrm{d}x$ gives the value 4, and Simpson's rule gives the value 2. What is $f(1)$?

21. Show that the error $(b-a)h^2 f''(\xi)/6$ in the composite midpoint rule

$$\int_a^b f(x)\mathrm{d}x = 2h\sum_{i=0}^{n-1} f(x_{2i+1}) + \frac{b-a}{6} h^2 f''(\xi)$$

can be approximated by $h^2[f'(b) - f'(a)]/6$.

22. Show that the error $-(b-a)h^2 f''(\xi)/12$ in the composite trapezoidal rule

$$\int_a^b f(x)\mathrm{d}x = \frac{h}{2}\left[f(a) + f(b) + 2\sum_{i=1}^{n-1} f(x_i)\right] - \frac{b-a}{12} h^2 f''(\xi)$$

can be approximated by $-h^2[f'(b) - f'(a)]/12$.

23. Show that the error $-(b-a)h^4 f^{(4)}(\xi)/180$ in the composite Simpson's rule

$$\int_a^b f(x)\mathrm{d}x = \frac{h}{3}\left[f(a) + f(b) + 2\sum_{i=1}^{n-1} f(x_{2i}) + 4\sum_{i=0}^{n-1} f(x_{2i+1})\right] - \frac{b-a}{180} h^4 f^{(4)}(\xi)$$

can be approximated by $-h^4[f'''(b) - f'''(a)]/180$.

24. Find the degree of precision (代数精度) of the quadrature formula (求积公式)

$$\int_{-1}^{1} f(x)\mathrm{d}x \approx f\left(-\frac{\sqrt{3}}{3}\right) + f\left(\frac{\sqrt{3}}{3}\right).$$

25. Let $h = (b-a)/3, x_0 = a, x_1 = a + h,$ and $x_2 = b$. Find the degree of precision of the

quadrature formula

$$\int_a^b f(x)\mathrm{d}x \approx \frac{9}{4}hf(x_1) + \frac{3}{4}hf(x_2).$$

26. The quadrature formula $\int_{-1}^{1} f(x)\mathrm{d}x \approx c_0 f(-1) + c_1 f(0) + c_2 f(1)$ is exact for all polynomials of degree less than or equal to 2. Determine c_0, c_1 and c_2.

27. The quadrature formula $\int_0^2 f(x)\mathrm{d}x \approx c_0 f(0) + c_1 f(1) + c_2 f(2)$ is exact for all polynomials of degree less than or equal to 2. Determine c_0, c_1 and c_2.

28. Determine constants a, b and x_0 such that the following quadrature formula

$$\int_0^1 f(x)\mathrm{d}x \approx af(x_0) + bf(1)$$

has degree of precision as high as possible and give the error.

29. Determine constants a, b and x_1 such that the following quadrature formula

$$\int_0^1 f(x)\mathrm{d}x \approx af(0) + bf(x_1)$$

has degree of precision as high as possible and give the error.

30. Determine constants a, b, x_1 and x_2 such that the following quadrature formula

$$\int_0^1 f(x)\mathrm{d}x \approx af(x_1) + bf(x_2)$$

has degree of precision as high as possible and give the error.

31. Find the constants x_0, x_1 and c_1 so that the quadrature formula

$$\int_0^1 f(x)\mathrm{d}x \approx \frac{1}{2} f(x_0) + c_1 f(x_1)$$

has the highest possible degree of precision (代数精度).

32. Derive Simpson's rule with error term by using

$$\int_{x_0}^{x_2} f(x)\mathrm{d}x \approx a_0 f(x_0) + a_1 f(x_1) + a_2 f(x_2) + kf^{(4)}(\xi).$$

Find a_0, a_1 and a_2 from the fact that Simpson's rule is exact for $f(x) = x^n$ when $n = 1, 2$, and 3. Then find k by applying the integration formula $f(x) = x^4$.

33. Show that a quadrature formula has degree of precision n if and only if the error $E(P(x)) = \int_a^b P(x)\mathrm{d}x - \sum_{i=0}^{n} a_i P(x_i) = 0$ for all polynomials $P(x)$ of degree $k = 0, 1, \cdots, n$, but $E(P(x)) \neq 0$ for some polynomial $P(x)$ of degree $n+1$.

34. Show that the degree of accuracy of the following quadrature formula

$$\int_a^b f(x)\mathrm{d}x \approx \sum_{i=0}^{n} A_i f(x_i)$$

is n if and only if

$$A_i = \int_a^b L_i(x)\mathrm{d}x,\ i = 0, 1, \cdots, n$$

where $L_i(x), i = 0, 1, \cdots, n$, are the Lagrange basis functions.

35. Show that the $(n+1)$-point Newton-Cotes formula

$$\int_a^b f(x)\mathrm{d}x \approx \sum_{i=0}^{n} A_i f(x_i)$$

has the degree of accuracy at least $n+1$ when n is even.

36. Suppose that $f(0) = 1, f(0.5) = 2.5, f(1) = 2$ and $f(0.25) = f(0.75) = \alpha$. Find α if the Composite Trapezoidal rule with $n = 4$ gives the value 1.75 for $\int_0^1 f(x)\mathrm{d}x$.

37. The Midpoint rule for approximating $\int_{-1}^1 f(x)\mathrm{d}x$ gives the value 12, the Composite Midpoint rule with $n = 2$ gives 5, and Composite Simpson's rule gives 6. Use the fact that $f(-1) = f(1)$ and $f(-0.5) = f(0.5) - 1$ to determine $f(-1), f(-0.5), f(0), f(0.5)$ and $f(1)$.

38. Romberg integration is used to approximate $\int_2^3 f(x)\mathrm{d}x$. If $f(2) = 0.51342$, $f(3) = 0.36788$, $R_{31} = 0.43687$ and $R_{33} = 0.43662$, find $f(2.5)$.

39. Romberg integration for approximating $\int_0^1 f(x)\mathrm{d}x$ gives $R_{11} = 4$ and $R_{22} = 5$. Find $f(0.5)$.

40. Romberg integration for approximating $\int_a^b f(x)\mathrm{d}x$ gives $R_{11} = 8$ and $R_{22} = 16/3$ and $R_{33} = 208/45$. Find R_{31}.

41. Romberg integration is used to approximate

$$\int_0^1 \frac{x^2}{1+x^3}\mathrm{d}x.$$

If $R_{11} = 0.250$ and $R_{22} = 0.2315$, what is R_{21}?

42. Approximate the integral $\int_0^{0.35} \frac{2}{x^2-4}\mathrm{d}x$ using Gaussian quadrature with $n = 2$, and compare your result with the exact value of the integral.

43. Approximate the integral $\int_1^{1.5} x^2 \ln x\,\mathrm{d}x$ using Gaussian quadrature with $n = 2$, and compare your result with the exact value of the integral.

44. Approximate the integral $\int_0^1 x^2 \mathrm{e}^{-x}\mathrm{d}x$ using Gaussian quadrature with $n = 2$, and compare your result with the exact value of the integral.

45. Approximate the integral $\int_0^{\pi/4} x^2 \sin x\mathrm{d}x$ using Gaussian quadrature with $n = 2$, and compare your result with the exact value of the integral.

46. Determine constants a, b, c and d that will produce a quadrature formula

$$\int_{-1}^{1} f(x)\mathrm{d}x \approx af(-1) + bf(1) + cf'(-1) + \mathrm{d}f'(1)$$

that has degree of precision 3.

47. Determine constants a, b, c, d and e that will produce a quadrature formula

$$\int_{-1}^{1} f(x)\mathrm{d}x \approx af(-1) + bf(0) + cf(1) + df'(-1) + ef'(1)$$

that has degree of precision 4.

48. Determine constants a, b, c and γ that will produce a quadrature formula

$$\int_{0}^{1} f(x)\mathrm{d}x \approx af(0) + bf(1) + cf''(\gamma)$$

that has degree of precision 3.

49. Suppose $\int_a^b f(x)\mathrm{d}x \approx \sum_{i=0}^{n} A_i f(x_i)$ is the Gaussian quadrature, $x_i, i = 0, 1, \cdots, n$, are the Gaussian nodes and

$$L_k(x) = \frac{\prod_{j=0, j\neq k}^{n} (x - x_j)}{\prod_{j=0, j\neq k}^{n} (x_k - x_j)}, \quad k = 0, 1, \cdots, n.$$

Show $A_i = \int_a^b L_i(x)\mathrm{d}x = \int_a^b [L_i(x)]^2\, \mathrm{d}x, \quad i = 0, 1, \cdots, n.$

50. Consider a quadrature formula of the form

$$\int_0^1 f(x)\,\mathrm{d}x \approx a_0 f(0) + a_1 f'(0) + \sum_{k=1}^{n} w_k f(x_k) + b_0 f(1).$$

a. Call the formula "Hermite-interpolatory" if the right-hand side is obtained by integrating on the left instead of the interpolation polynomial $p(x)$ satisfying

$$p(0) = f(0),\ p'(0) = f'(0),\ p(1) = f(1),$$
$$p(x_k) = f(x_k), \quad k = 1, 2, \cdots, n.$$

What degree of accuracy does the formula have in this case (regardless of how the nodes x_k are chosen, as long as they are mutually distinct and strictly inside the interval $[0,1]$?

b. What is the maximum degree of accuracy expected to be if all coefficients and nodes x_k are allowed to be freely chosen?

c. Show that for the maximum degree of accuracy to be achieved, it is necessary that $\{x_k\}$ are the zeros of the polynomial π_n of degree n which is orthogonal on $[0,1]$ with respect to the weight function $w(x) = x^2(1 - x)$.

d. Show that the choice of the x_k in (c) together with the requirement of the quadrature formula to be Hermite-interpolatory is sufficient for the maximum degree of accuracy to be attained.

51. (Turán quadrature formula) Let $w(t)$ be a nonnegative weight function on $[a,b]$. Prove that the quadrature formula

$$\int_a^b f(t)w(t)\mathrm{d}t = \sum_{k=1}^{n}[w_k f(t_k) + w_k' f'(t_k) + w_k'' f''(t_k)] + E_n(f)$$

has degree of accuracy $d = 4n-1$ if and only if the following conditions are satisfied:
a. The formula is (Hermite-) interpolatory; that is, $E_n(f) = 0$ if $f \in \Pi_{3n-1}$.
b. The node polynomial $\omega_n(t) = \prod_{k=1}^{n}(t-t_k)$ satisfies

$$\int_a^b [\omega_n(t)]^3 p(t)w(t)\mathrm{d}t = 0 \quad \text{for all } p \in \Pi_{n-1}.$$

52. Consider a quadrature formula of the type

$$\int_0^\infty \mathrm{e}^{-x} f(x)\,\mathrm{d}x = af(0) + bf(c) + E(f).$$

a. Find a,b,c such that the formula has degree of accuracy 2. (Point of information: $\int_0^\infty \mathrm{e}^{-x}x^r\,\mathrm{d}x = r!$)

b. Let $p_2(x)$ be the quadratic Hermite interpolation polynomial interpolating $f(x)$ at the simple point $x=0$ and the double point $x=2$. Namely, $p_2(0) = f(0)$, $p_2(2) = f(2)$, $p_2'(2) = f'(2)$. Determine $\int_0^\infty \mathrm{e}^{-x} p_2(x)\,\mathrm{d}x$ and compare with the result in (a).

c. Obtain the remainder $E(f)$.

53. a. Determine a, b, c such that the following formula has degree of accuracy 2

$$\int_0^1 f(x)\,\mathrm{d}x \approx af(0) + bf(1) - c[f'(1) - f'(0)].$$

b. Transform the rule in (a) into one appropriate for approximating $\int_x^{x+h} f(t)\mathrm{d}t$.

c. Obtain a composite integration rule based on the formula in (b) for approximating $\int_a^b f(t)\mathrm{d}t$.

54. Consider a quadrature formula of the type

$$\int_0^1 f(x)\,\mathrm{d}x = af(x_1) + b[f(1) - f(0)] + E(f)$$

and determine a, b, x_1 such that the degree of accuracy is as large as possible. What is the maximum degree attainable?

Chapter 8 Initial-Value Problems for Ordinary Differential Equations (常微分方程初值问题)

本章内容 (Contents)

8.1 The elementary theory of initial- value problem (初值问题基本理论)

8.2 Euler's method (Euler 方法)

8.3 Runge-Kutta methods (Runge-Kutta 方法)

8.4 Multistep methods (多步方法)

学习目标 (Learning Objectives)

- 理解李普希茨条件以及微分方程的适定问题
- 洞悉欧拉方法的基本思想和截断误差
- 理解微分方程数值解法阶的定义
- 深谙欧拉方法与隐式欧拉方法的内涵
- 掌握两步欧拉方法以及方法的阶
- 掌握梯形公式、改进的欧拉公式
- 明辨梯形公式和改进的欧拉公式的区别
- 会用欧拉公式、改进的欧拉公式求解微分方程的数值解
- 理解 Runge-Kutta 方法的基本思想
- 掌握二阶 Runge-Kutta 方法
- 知悉三阶、四阶 Runge-Kutta 方法
- 会用二阶、三阶 Runge-Kutta 方法求微分方程的数值解
- 掌握 Adams 显示格式和 Adams 隐式格式

8.1 The elementary theory of initial- value problem (初值问题基本理论)

- Convex set (凸集)
- Lipschitz condition (李普希茨条件)

- Picard's Theorem (Picard 定理)
- Well-posed problem (适定问题)

Differential equations have been widely used to model the problems in science, technology and engineering and in most cases, solutions to these differential equations are subject to the conditions of initial values. A differential equation appearing in practice is usually so complicated that it is hardly possible to find out its analytical solution. Therefore it is of significance to study how to approximate the true solution. This chapter is mainly concerned with approximating the solution $y(t)$ to a problem of the form

$$\frac{\mathrm{d}y}{\mathrm{d}t} = f(t,y), \quad \text{for} \quad a \leqslant t \leqslant b$$

subject to an initial condition

$$y(a) = \alpha.$$

8.1.1 Convex set (凸集)

Definition 8.1 A set D is said to be convex if any two points (t_1, y_1) and (t_2, y_2) in D and $\lambda \in [0,1]$ implies

$$(\lambda t_1 + (1-\lambda)t_2, \lambda y_1 + (1-\lambda)y_2) \in D.$$

Example 8.1 The set $D = \{(t,y) \mid t^2 + y^2 \leqslant 1\}$ is a convex set, since for any two points $(t_1, y_1) \in D, (t_2, y_2) \in D$, and $\lambda \in [0,1]$, we have

$$\begin{aligned}
&[\lambda t_1 + (1-\lambda)t_2]^2 + [\lambda y_1 + (1-\lambda)y_2]^2 \\
&= \lambda^2(t_1^2 + y_1^2) + (1-\lambda)^2(t_2^2 + y_2^2) + 2\lambda(1-\lambda)(t_1 t_2 + y_1 y_2) \\
&\leqslant \lambda^2 + (1-\lambda)^2 + \lambda(1-\lambda)(t_1^2 + t_2^2 + y_1^2 + y_2^2) \\
&\leqslant \lambda^2 + (1-\lambda)^2 + 2\lambda(1-\lambda) \\
&= [\lambda + (1-\lambda)]^2 = 1,
\end{aligned}$$

which implies $(\lambda t_1 + (1-\lambda)t_2, \lambda y_1 + (1-\lambda)y_2) \in D$. Therefore the set

$$D = \{(t,y) \mid t^2 + y^2 \leqslant 1\}$$

is a convex set.

Example 8.2 The set $D = \{(t,y) \mid 0 \leqslant y \leqslant t^2, -1 \leqslant t \leqslant 1\}$ is not a convex set, since for $(-1,1) \in D$, $(1,1) \in D$, taking $\lambda = \frac{1}{2}$ yields $(\frac{1}{2}\cdot(-1) + \frac{1}{2}\cdot 1, \frac{1}{2}\cdot 1 + \frac{1}{2}\cdot 1) = (0,1) \notin D$.

8.1.2 Lipschitz condition (李普希茨条件)

Definition 8.2 A function $f(t,y)$ is said to satisfy a Lipschitz condition in the variable y on a set $D \subset \mathbf{R}^2$ if there exists a constant $L > 0$ such that

$$|f(t,y_1) - f(t,y_2)| \leqslant L|y_1 - y_2|.$$

Example 8.3 If

$$D=\{(t,y)\mid 1\leqslant t\leqslant 2,-3\leqslant y\leqslant 4\}$$

and

$$f(t,y)=t|y|,$$

then for each pair of points (t,y_1) and (t,y_2) in D, we have

$$|f(t,y_1)-f(t,y_2)|=|t|\big||y_1|-|y_2|\big|\leqslant 2|y_1-y_2|.$$

Thus $f(t,y)$ satisfies a Lipschitz condition on D in the variable y with Lipschitz constant 2.

Theorem 8.1 Suppose $f(t,y)$ is defined on a convex set $D\subset\mathbf{R}^2$. If $f(t,y)$ is differentiable with respect to the variable y and a positive constant L exists with

$$\left|\frac{\partial f(t,y)}{\partial y}\right|\leqslant L,\quad \forall(t,y)\in D,$$

then $f(t,y)$ satisfies a Lipschitz condition on D in the variable y with Lipschitz constant L.

Proof. For each pair of points (t,y_1) and (t,y_2) in D, we have by the Mean Value Theorem that

$$f(t,y_2)-f(t,y_1)=f'_y(t,\xi)(y_2-y_1),$$

where (t,ξ) is located between (t,y_1) and (t,y_2), and hence lies in D since D is a convex set. Therefore

$$|f(t,y_2)-f(t,y_1)|=|f'_y(t,\xi)||y_2-y_1|\leqslant L|y_2-y_1|,$$

which indicates that $f(t,y)$ satisfies a Lipschitz condition on D in the variable y with Lipschitz constant L.

The proof of theorem 8.1 is completed.

8.1.3 Picard's theorem (Picard 定理)

Theorem 8.2 **(Picard 定理)** Suppose the real valued function $f(t,y)$ is continuous in the rectangular region $D=\{(t,y)\mid t_0\leqslant t\leqslant t_M, y_0-C\leqslant y\leqslant y_0+C\}$, and satisfies the Lipschitz condition in the variable y on D with Lipschitz constant $L>0$, i.e.,

$$|f(t,y_1)-f(t,y_2)|\leqslant L|y_1-y_2|. \tag{8.1.1}$$

Let

$$K=\max_{t_0\leqslant t\leqslant t_M}|f(t,y_0)|. \tag{8.1.2}$$

If

$$C\geqslant\frac{K}{L}(\mathrm{e}^{L(t_M-t_0)}-1), \tag{8.1.3}$$

then there exists a unique function $y(t)\in C[t_0,t_M]$ such that $y(t_0)=y_0$ and $y'(t)=f(t,y)$ for $(t,y)\in D$.

Proof. Let

$$y_0(t) = y_0,$$
$$y_n(t) = y_0 + \int_{t_0}^{t} f(s, y_{n-1}(s))\mathrm{d}s, \quad n = 1, 2, \cdots \tag{8.1.4}$$

Then we get a sequence of functions $\{y_n(t)\}_{n=0}^{\infty}$. It is obvious that each function $y_n(t)$ is continuous on $[t_0, t_M]$ since $f(t, y)$ is continuous in the rectangular region $D = \{(t, y) \mid t_0 \leqslant t \leqslant t_M, y_0 - C \leqslant y \leqslant y_0 + C\}$. For $t \in [t_0, t_M]$ and $n = 1$ it follows from (8.1.4) and (8.1.2)

$$|y_1(t) - y_0| \leqslant \int_{t_0}^{t} |f(s, y_0(s))|\mathrm{d}s \leqslant K(t - t_0). \tag{8.1.5}$$

Using (8.1.4), (8.1.1), (8.1.2) and (8.1.5) gives

$$\begin{aligned}
|y_2(t) - y_1(t)| &\leqslant \int_{t_0}^{t} |f(s, y_1(s)) - f(s, y_0(s))|\mathrm{d}s \\
&\leqslant L\int_{t_0}^{t} |y_1(s) - y_0(s)|\mathrm{d}s \\
&= L\int_{t_0}^{t} |y_1(s) - y_0|\mathrm{d}s \\
&\leqslant L\int_{t_0}^{t} K(s - t_0)\mathrm{d}s \\
&= LK\frac{(t - t_0)^2}{2} \\
&= \frac{K}{L}\frac{[L(t - t_0)]^2}{2}.
\end{aligned} \tag{8.1.6}$$

Assume that the following inequality

$$|y_k(t) - y_{k-1}(t)| \leqslant \frac{K}{L}\frac{[L(t - t_0)]^k}{k!} \tag{8.1.7}$$

holds true for $1 \leqslant k \leqslant n$, then using (8.1.4), (8.1.1), (8.1.2) and (8.1.7) yields

$$\begin{aligned}
|y_{n+1}(t) - y_n(t)| &\leqslant \int_{t_0}^{t} |f(s, y_n(s)) - f(s, y_{n-1}(s))|\mathrm{d}s \\
&\leqslant L\int_{t_0}^{t} |y_n(s) - y_{n-1}(s)|\mathrm{d}s \\
&\leqslant L\int_{t_0}^{t} \frac{K}{L}\frac{[L(s - t_0)]^n}{n!}\mathrm{d}s \\
&= \frac{K}{L}\frac{[L(t - t_0)]^{n+1}}{(n+1)!},
\end{aligned} \tag{8.1.8}$$

which means that the inequality (8.1.7) is true for $k = n + 1$, and we get by induction

$$|y_k(t) - y_{k-1}(t)| \leqslant \frac{K}{L}\frac{[L(t - t_0)]^k}{k!}, \quad k = 1, 2, \cdots; t \in [t_0, t_M]. \tag{8.1.9}$$

As a result,

$$\sum_{k=1}^{\infty}\left|y_k(t)-y_{k-1}(t)\right| \leqslant \frac{K}{L}\sum_{k=1}^{\infty}\frac{[L(t-t_0)]^k}{k!} < \frac{K}{L}\mathrm{e}^{L(t-t_0)} \leqslant \frac{K}{L}\mathrm{e}^{L(t_M-t_0)}, \tag{8.1.10}$$

which shows that the infinite series

$$\sum_{k=1}^{\infty}[y_k(t)-y_{k-1}(t)]$$

converges absolutely and uniformly for $t\in[t_0,t_M]$. Since

$$y_n(t)=y_0+\sum_{k=1}^{n}[y_k(t)-y_{k-1}(t)],$$

we see that the sequence of functions $\{y_n(t)\}_{n=0}^{\infty}$ converges to a limit function, say, $y(t)$, uniformly on $[t_0,t_M]$ and $y(t)\in C[t_0,t_M]$ because of $y_n(t)\in C[t_0,t_M]$ for any n.

Moreover for any positive integer n, we have

$$\begin{aligned}\left|y_n(t)-y_0\right| &\leqslant \sum_{k=1}^{n}\left|y_k(t)-y_{k-1}(t)\right| \\ &\leqslant \frac{K}{L}\sum_{k=1}^{n}\frac{[L(t-t_0)]^k}{k!}, \quad t\in[t_0,t_M] \\ &\leqslant \frac{K}{L}(\mathrm{e}^{L(t_M-t_0)}-1) \\ &\leqslant C,\end{aligned} \tag{8.1.11}$$

which shows that $(t,y_n(t))\in D=\{(t,y)\,|\,t_0\leqslant t\leqslant t_M, y_0-C\leqslant y\leqslant y_0+C\}$. From (8.1.4) it follows

$$y(t)=\lim_{n\to\infty}y_n(t)=y_0+\int_{t_0}^{t}\lim_{n\to\infty}f(s,y_{n-1}(s))\mathrm{d}s=y_0+\int_{t_0}^{t}f(s,y(s))\mathrm{d}s. \tag{8.1.12}$$

Clearly $y(t_0)=y_0$. Differentiating both sides of (8.1.12) results in $y'(t)=f(t,y)$ for $(t,y)\in D$.

Next we show that the solution is unique. To this end, suppose that $y(t)\in C[t_0,t_M]$ and $z(t)\in C[t_0,t_M]$ are two solutions such that

$$y(t)=y_0+\int_{t_0}^{t}f(s,y(s))\mathrm{d}s,$$

$$z(t)=z_0+\int_{t_0}^{t}f(s,z(s))\mathrm{d}s,$$

where $y_0=z_0$. Then we have

$$\begin{aligned}\left|y(t)-z(t)\right| &\leqslant \int_{t_0}^{t}\left|f(s,y(s))-f(s,z(s))\right|\mathrm{d}s \\ &\leqslant L\int_{t_0}^{t}\left|y(s)-z(s)\right|\mathrm{d}s \\ &\leqslant LM(t-t_0),\end{aligned} \tag{8.1.13}$$

where $M = \max_{t_0 \leqslant t \leqslant t_M} |y(t) - z(t)|$. Substituting the inequality

$$|y(s) - z(s)| \leqslant LM(s - t_0)$$

into $|y(t) - z(t)| \leqslant L\int_{t_0}^{t} |y(s) - z(s)| \mathrm{d}s$ gives

$$|y(t) - z(t)| \leqslant ML^2 \int_{t_0}^{t} (s - t_0) \mathrm{d}s = ML^2 \frac{(t - t_0)^2}{2}.$$

Substituting the inequality

$$|y(s) - z(s)| \leqslant ML^2 \frac{(s - t_0)^2}{2}$$

into $|y(t) - z(t)| \leqslant L\int_{t_0}^{t} |y(s) - z(s)| \mathrm{d}s$ gives

$$|y(t) - z(t)| \leqslant ML^3 \int_{t_0}^{t} \frac{(s - t_0)^2}{2} \mathrm{d}s = ML^3 \frac{(t - t_0)^3}{3!}.$$

Proceeding with the above arguments as many times as required, one gets for any positive integer n

$$|y(t) - z(t)| \leqslant ML^n \int_{t_0}^{t} \frac{(s - t_0)^{n-1}}{(n-1)!} \mathrm{d}s = M \frac{[L(t - t_0)]^n}{n!}.$$

Taking limits on the both sides of the above inequality as $n \to \infty$ leads to

$$0 \leqslant |y(t) - z(t)| \leqslant \lim_{n \to \infty} M \frac{[L(t - t_0)]^n}{n!} = 0,$$

which implies $y(t) = z(t)$, as asserted. The proof of Theorem 8.2 is completed.

In particular, as $C \to +\infty$ we have the following corollary of Picard's theorem.

Corollary 8.3 Suppose the real valued function $f(t, y)$ is continuous in the region $D = \{(t, y) \mid a \leqslant t \leqslant b, -\infty < y < +\infty\}$, and satisfies the Lipschitz condition in the variable y on D, then the initial-value problem

$$y'(t) = f(t, y), \quad a \leqslant t \leqslant b, \quad y(a) = \alpha,$$

has a unique solution $y(t)$ for $a \leqslant t \leqslant b$.

Example 8.4 Show that the following initial-value problem

$$y'(t) = 1 + t\sin(ty), \quad 0 \leqslant t \leqslant 2, \quad y(0) = 0,$$

has a unique solution.

Proof. Applying the Mean Value Theorem to the function $f(t, y) = 1 + t\sin(ty)$ with respect to the variable y gives

$$f(t, y_2) - f(t, y_1) = f_y'(t, \xi)(y_2 - y_1) = t^2 \cos(t\xi)(y_2 - y_1),$$

which implies that

$$|f(t, y_2) - f(t, y_1)| = |t^2 \cos(t\xi)||y_2 - y_1| \leqslant 4|y_2 - y_1|,$$

where ξ is located between y_1 and y_2 . Therefore $f(t,y)=1+t\sin(ty)$ satisfies a Lipschitz condition in the variable y with Lipschitz constant $L=4$. In addition, $f(t,y)$ is clearly continuous in the region $D=\{(t,y)\,|\,0\leqslant t\leqslant 2,-\infty<y<+\infty\}$, by the Corollary of Picard's Theorem there exists a unique solution to this initial-value problem. The proof is completed.

8.1.4 Well-posed problem (适定问题)

Definition 8.3 The initial-value problem

$$\begin{cases}\dfrac{\mathrm{d}y}{\mathrm{d}t}=f(t,y), \quad a\leqslant t\leqslant b;\\ y(a)=\alpha\end{cases}$$

is said to be a well-posed problem if:

(1) A unique solution $y(t)$ to the problem exists;

(2) For any $\varepsilon>0$, there is a positive constant k (which may depend on ε) such that whenever $|\varepsilon_0|\leqslant\varepsilon$ and $\delta(t)$ is continuous with $|\delta(t)|<\varepsilon$ on $[a,b]$, a unique solution $z(t)$ to the perturbed problem

$$z_0=z(t_0)=z(a)=\alpha+\varepsilon_0,$$

exists with

$$|z(t)-y(t)|<k\varepsilon, \forall t\in[a,b].$$

The following theorem tells when an initial-value problem is well-posed.

Theorem 8.4 Let $D=\{(t,y)\,|\,a\leqslant t\leqslant b,-\infty<y<+\infty\}$. If the real valued function $f(t,y)$ is continuous and satisfies a Lipschitz condition in the variable y on D, then the initial-value problem

$$y'(t)=f(t,y), \quad a\leqslant t\leqslant b, \quad y(a)=\alpha$$

is well posed.

Proof. Since the real valued function $f(t,y)$ is continuous in the region $D=\{(t,y)\,|\,a\leqslant t\leqslant b,-\infty<y<+\infty\}$, and satisfies a Lipschitz condition in the variable y on D, we might as well suppose

$$|f(t,y_1)-f(t,y_2)|\leqslant L|y_1-y_2|, \quad \forall y_1,y_2,$$

by Corollary 8.3 the initial-value problem

$$y'(t)=f(t,y), \quad a\leqslant t\leqslant b, \quad y(a)=\alpha,$$

has a unique solution $y(t)\in C[a,b]$.

For any $\varepsilon>0$, consider the following perturbed problem

$$\frac{\mathrm{d}z}{\mathrm{d}t}=f(t,z)+\delta(t), a\leqslant t\leqslant b, z(a)=\alpha+\varepsilon_0$$

where $|\varepsilon_0|\leqslant\varepsilon$ and $\delta(t)$ is continuous with $|\delta(t)|<\varepsilon$ on $[a,b]$. Let

$$\tilde{f}(t,z)=f(t,z)+\delta(t), a\leqslant t\leqslant b.$$

Then $\tilde{f}(t,z)$ is continuous in the region $D=\{(t,z)\,|\,a\leqslant t\leqslant b,-\infty<z<+\infty\}$, and

$$\left|\tilde{f}(t,z_1)-\tilde{f}(t,z_2)\right|=\left|f(t,z_1)-f(t,z_2)\right|\leqslant L\left|z_1-z_2\right|,\quad \forall z_1,z_2.$$

Thus $\tilde{f}(t,z)$ satisfies the same Lipschitz condition as $f(t,y)$. By Corollary 8.3 the initial-value problem

$$z'(t)=\tilde{f}(t,z),\quad a\leqslant t\leqslant b,\quad z(a)=\alpha+\varepsilon_0,$$

has a unique solution $z(t)\in C[a,b]$. Let

$$y_0=y(t_0)=y(a)=\alpha,\ z_0=z(t_0)=z(a)=\alpha+\varepsilon_0.$$

Then

$$y(t)=y_0+\int_{t_0}^{t} f(s,y(s))\mathrm{d}s,$$

$$z(t)=z_0+\int_{t_0}^{t} \tilde{f}(s,z(s))\mathrm{d}s,$$

we have

$$\begin{aligned}\left|y(t)-z(t)\right|&\leqslant\left|y_0-z_0\right|+\int_{t_0}^{t}\left|f(s,y(s))-\tilde{f}(s,z(s))\right|\mathrm{d}s\\&\leqslant\left|\varepsilon_0\right|+\int_{t_0}^{t}\left[\left|f(s,y(s))-f(s,z(s))\right|+\left|\delta(s)\right|\right]\mathrm{d}s\\&\leqslant\varepsilon+\int_{t_0}^{t}\left(L\left|y(s)-z(s)\right|+\varepsilon\right)\mathrm{d}s\\&\leqslant\varepsilon+\varepsilon(t-t_0)+LM(t-t_0),\end{aligned}$$

where $M=\max\limits_{a\leqslant t\leqslant b}\left|y(t)-z(t)\right|$. Substituting the inequality

$$\left|y(s)-z(s)\right|\leqslant\varepsilon+\varepsilon(s-t_0)+LM(s-t_0)$$

into $\left|y(t)-z(t)\right|\leqslant\varepsilon+\int_{t_0}^{t}\left(L\left|y(s)-z(s)\right|+\varepsilon\right)\mathrm{d}s$ gives

$$\begin{aligned}\left|y(t)-z(t)\right|&\leqslant\varepsilon+\varepsilon(L+1)(t-t_0)+\left(ML^2+L\varepsilon\right)\int_{t_0}^{t}(s-t_0)\mathrm{d}s\\&=\varepsilon+\varepsilon(L+1)(t-t_0)+\left(ML^2+L\varepsilon\right)\frac{(t-t_0)^2}{2}\\&<\varepsilon\left\{1+(L+1)(t-t_0)+\frac{\left[(L+1)(t-t_0)\right]^2}{2}\right\}+ML^2\frac{(t-t_0)^2}{2}.\end{aligned}$$

Substituting the inequality

$$\left|y(s)-z(s)\right|<\varepsilon\left\{1+(L+1)(s-t_0)+\frac{\left[(L+1)(s-t_0)\right]^2}{2}\right\}+ML^2\frac{(s-t_0)^2}{2}$$

into $\left|y(t)-z(t)\right|\leqslant\varepsilon+\int_{t_0}^{t}\left(L\left|y(s)-z(s)\right|+\varepsilon\right)\mathrm{d}s$ gives

$$|y(t)-z(t)|<\varepsilon\left\{1+(L+1)(t-t_0)+\frac{\left[(L+1)(t-t_0)\right]^2}{2}+\frac{\left[(L+1)(t-t_0)\right]^3}{3!}\right\}$$
$$+M\frac{[L(t-t_0)]^3}{3!}.$$

Proceeding with the above arguments as many times as required, one gets for any positive integer n

$$|y(t)-z(t)|<\varepsilon\left\{1+(L+1)(t-t_0)+\frac{\left[(L+1)(t-t_0)\right]^2}{2}+\cdots+\frac{\left[(L+1)(t-t_0)\right]^n}{n!}\right\}$$
$$+M\frac{[L(t-t_0)]^n}{n!}.$$

Taking limits on the both sides of the above inequality as $n\to\infty$ and using

$$\lim_{n\to\infty}\left\{1+(L+1)(t-t_0)+\frac{\left[(L+1)(t-t_0)\right]^2}{2}+\cdots+\frac{\left[(L+1)(t-t_0)\right]^n}{n!}\right\}$$
$$=\mathrm{e}^{(L+1)(t-t_0)}\leqslant\mathrm{e}^{(L+1)(b-a)},$$

$$\lim_{n\to\infty}M\frac{[L(t-t_0)]^n}{n!}=0,$$

we have

$$|y(t)-z(t)|<\mathrm{e}^{(L+1)(b-a)}\varepsilon,\quad\forall t\in[a,b].$$

Let $k=\mathrm{e}^{(L+1)(b-a)}$. Then we derive $|y(t)-z(t)|<k\varepsilon, \forall t\in[a,b]$.

The proof of Theorem 8.4 is completed.

Example 8.5 Show that the initial-value problem

$$\frac{\mathrm{d}y}{\mathrm{d}t}=y-t^2+1,\quad 0\leqslant t\leqslant 2,\quad y(0)=0.5$$

is well posed on $D=\{(t,y)\,|\,0\leqslant t\leqslant 2,-\infty<y<\infty\}$.

Solution. Since

$$\left|\frac{\partial(y-t^2+1)}{\partial y}\right|=1,$$

it follows from Theorem 8.1 that $f(t,y)=y-t^2+1$ satisfies a Lipschitz condition in variable y on $D=\{(t,y)\,|\,0\leqslant t\leqslant 2,-\infty<y<\infty\}$ with Lipschitz constant 1. Clearly $f(t,y)=y-t^2+1$ is continuous on D. By Theorem 8.4 the given initial-value problem is well posed.

8.2 Euler's method (Euler 方法)

- Basic ideas of Euler's method (Euler 方法的基本思想)
- Order of a method (方法的阶)
- Implicit Euler's scheme (隐式 Euler 格式)
- Two-step Euler's scheme (两步 Euler 格式)
- Trapezoidal scheme (梯形格式)
- Modified Euler scheme (改进的 Euler 格式)
- Error bounds for the Euler's scheme (Euler 格式的误差界)

8.2.1 Basic ideas of Euler's method (Euler 方法的基本思想)

As we mentioned, our purpose is to find out an approximation to the well-posed initial-value problem

$$\frac{\mathrm{d}y}{\mathrm{d}t} = f(t,y), \quad a \leqslant t \leqslant b, \quad y(a) = \alpha. \tag{8.2.1}$$

The basic idea is to design an algorithm or a scheme so that the approximate values of the true solution $y(t)$ to the problem (8.2.1) at the equally spaced points in the interval $[a,b]$ can be recursively computed. The Euler's method is conducted in the following steps.

Step 1. Choose a positive integer N, and select the mesh points

$$t_i = a + ih, \quad i = 0,1,\cdots,N$$

with $h = (b-a)/N$.

Step 2. Write out the following equation at the mesh point t_i,

$$y'(t_i) = f(t_i, y(t_i)). \tag{8.2.2}$$

Step 3. Replacing $y'(t_i)$ with forward divided difference

$$\frac{y(t_{i+1}) - y(t_i)}{h} \approx y'(t_i)$$

in Eq.(8.2.2) gives

$$y(t_{i+1}) \approx y(t_i) + hf(t_i, y(t_i)). \tag{8.2.3}$$

Step 4. Denote by w_i the approximation to $y(t_i)$ for each $i = 1,2,\cdots,N$.

And for each $i = 0,1,\cdots,N-1$ we call the following scheme

$$\begin{cases} w_0 = \alpha, \\ w_{i+1} = w_i + hf(t_i, w_i) \end{cases} \tag{8.2.4}$$

the Euler's scheme (欧拉格式).

Example 8.6 Use Euler's scheme with $N=10$ to approximate the solution to the initial-value problem

$$y' = y - t^2 + 1, \quad 0 \leqslant t \leqslant 2, \quad y(0) = 0.5.$$

Solution. With $N=10$ we have $h=2/10=0.2$, $t_i=0.2i$, and $w_0=y(0)=0.5$. Since $f(t,y)=y-t^2+1$, applying the Euler's scheme (8.2.4) to the problem gives

$$\begin{aligned} w_0 &= 0.5, \\ w_{i+1} &= w_i + hf(t_i, w_i) \\ &= w_i + 0.2(w_i - 0.04i^2 + 1) \\ &= 1.2w_i - 0.008i^2 + 0.2, \quad i = 0,1,\cdots,9. \end{aligned}$$

Starting with $w_0 = 0.5$, one can obtain

$$\begin{aligned} w_1 &= 1.2w_0 - 0.008\cdot 0^2 + 0.2 = 0.8, \\ w_2 &= 1.2w_1 - 0.008\cdot 1^2 + 0.2 = 1.152, \\ &\vdots \\ w_{10} &= 1.2w_9 - 0.008\cdot 9^2 + 0.2 = 4.8657845. \end{aligned}$$

It is not difficult to verify that the exact solution to the initial-value problem

$$y' = y - t^2 + 1, \quad 0 \leqslant t \leqslant 2, \quad y(0) = 0.5$$

is given by $y(t)=(t+1)^2-0.5\mathrm{e}^t$. The following Table 8.1 shows the comparison between the approximate values generated by the Euler's scheme and the actual values at $t_i=0.2i, i=0,1,\cdots,10$.

Table 8.1 Comparison between the approximations by Euler's scheme and the actual values

i	t_i	w_i	$y_i = y(t_i)$	$\lvert y_i - w_i \rvert$
0	0.0	0.5000000	0.5000000	0.0000000
1	0.2	0.8000000	0.8292986	0.0292986
2	0.4	1.1520000	1.2140877	0.0620877
3	0.6	1.5505000	1.6489406	0.0985406
4	0.8	1.9884800	2.1272295	0.1387495
5	1.0	2.4581760	2.6408591	0.1826831
6	1.2	2.9498112	3.1799415	0.2301303
7	1.4	3.4517734	3.7324000	0.2806266
8	1.6	3.9501281	4.2834838	0.3333557
9	1.8	4.4281538	4.8151763	0.3870225
10	2.0	4.8657845	5.3054720	0.4396874

8.2.2 Order of a method (方法的阶)

Definition 8.4 If $w_i = y(t_i)$, then $y(t_{i+1}) - w_{i+1}$ is called the local truncation error (局部截断误差).

Definition 8.5 If a scheme produces the local truncation error $O(h^{p+1})$, then the scheme is said to be of order p.

Theorem 8.5 Euler's scheme has order one (Euler 方法是一阶方法)

Proof. Suppose $w_i = y(t_i)$, then

$$w_{i+1} = w_i + hf(t_i, w_i) = y(t_i) + hf(t_i, y(t_i)).$$

Using the first Taylor's polynomial about t_i gives

$$\begin{aligned} y(t_{i+1}) &= y(t_i) + (t_{i+1} - t_i)y'(t_i) + \frac{(t_{i+1} - t_i)^2}{2} y''(\xi_i) \\ &= y(t_i) + hy'(t_i) + \frac{h^2}{2} y''(\xi_i), \quad \xi_i \in (t_i, t_{i+1}). \end{aligned}$$

Therefore

$$y(t_{i+1}) - w_{i+1} = \frac{h^2}{2} y''(\xi_i) = O(h^2).$$

As a result, Euler's scheme is of order one.

8.2.3 Implicit Euler's scheme (隐式 Euler 格式)

If we use the backward divided difference to replace the derivative

$$y'(t_{i+1}) \approx \frac{y(t_{i+1}) - y(t_i)}{h},$$

then from

$$y'(t_{i+1}) = f(t_{i+1}, y(t_{i+1}))$$

it follows

$$y(t_{i+1}) \approx y(t_i) + hf(t_{i+1}, y(t_{i+1}))$$

and for $i = 1, 2, \cdots, N-1$, we call the scheme

$$\begin{cases} w_0 = \alpha, \\ w_{i+1} = w_i + hf(t_{i+1}, w_{i+1}) \end{cases} \tag{8.2.5}$$

the implicit Euler's scheme.

Similarly to the proof of Theorem 8.5, it is not difficult to prove the following.

Theorem 8.6 Implicit Euler's scheme has order one.

The implicit Euler's scheme may be solved by iteration:

$$\begin{aligned} w_{i+1}^{(0)} &= w_i + hf(t_i, w_i), \\ w_{i+1}^{(k+1)} &= w_i + hf(t_i, w_{i+1}^{(k)}), k = 0, 1, \cdots. \end{aligned}$$

8.2.4 Two-step Euler's scheme (两步 Euler 格式)

If we use the three-point symmetrical formula

$$y'(t_i) \approx \frac{y(t_{i+1}) - y(t_{i-1})}{2h},$$

then we get the following scheme

$$\begin{cases} w_0 = \alpha, \\ w_{i+1} = w_{i-1} + 2hf(t_i, w_i). \end{cases} \tag{8.2.6}$$

This scheme is called two-step Euler's scheme.

When we talk about the two-step scheme, the local truncation error (局部截断误差) described in Definition 8.4 has to be reformulated as follows: if $w_{i-1} = y(t_{i-1})$, $w_i = y(t_i)$, then $y(t_{i+1}) - w_{i+1}$ is called the local truncation error for a two-step scheme.

Theorem 8.7 The two step Euler's scheme has order two.

Proof. Using second Taylor's polynomial expansion about t_i and evaluating at t_{i-1} and t_{i+1}, respectively, gives

$$y(t_{i+1}) = y(t_i) + hy'(t_i) + \frac{h^2}{2}y''(t_i) + \frac{h^3}{6}y'''(\xi_i),$$

$$y(t_{i-1}) = y(t_i) - hy'(t_i) + \frac{h^2}{2}y''(t_i) - \frac{h^3}{6}y'''(\overline{\xi}_i).$$

If y''' is continuous on (a,b), then

$$y(t_{i+1}) - y(t_{i-1}) = 2hy'(t_i) + \frac{h^3}{3}y'''(\eta_i),$$

i.e.

$$y(t_{i+1}) = y(t_{i-1}) + 2hy'(t_i) + \frac{h^3}{3}y'''(\eta_i).$$

Suppose

$$w_{i-1} = y(t_{i-1}),\ w_i = y(t_i).$$

Then

$$\begin{aligned} y(t_{i+1}) &= w_{i-1} + 2hy'(t_i) + \frac{h^3}{3}y'''(\eta_i) \\ &= w_{i-1} + 2hf(t_i, w_i) + \frac{h^3}{3}y'''(\eta_i). \end{aligned} \tag{8.2.7}$$

From Eq. (8.2.6) and (8.2.7) it follows

$$y(t_{i+1}) - w_{i+1} = \frac{h^3}{3}y'''(\eta_i) = O(h^3). \tag{8.2.8}$$

8.2.5 Trapezoidal scheme (梯形格式)

Integrating the differential equation

$$y' = f(t, y)$$

over $[t_i, t_{i+1}]$ yields

$$y(t_{i+1}) = y(t_i) + \int_{t_i}^{t_{i+1}} f(t, y)\,\mathrm{d}t. \tag{8.2.9}$$

It is not difficult to observe that Euler's (explicit) scheme is obtained from Eq.(8.2.9) by using the numerical integration

$$\int_{t_i}^{t_{i+1}} f(t,y)\,\mathrm{d}t \approx \int_{t_i}^{t_{i+1}} f(t_i, y(t_i))\,\mathrm{d}t = hf(t_i, y(t_i))$$

while Euler's implicit scheme is obtained from (8.2.9) by using the numerical integration

$$\int_{t_i}^{t_{i+1}} f(t,y)\,\mathrm{d}t \approx \int_{t_i}^{t_{i+1}} f(t_{i+1}, y(t_{i+1}))\,\mathrm{d}t = hf(t_{i+1}, y(t_{i+1})).$$

Integrating the differential equation

$$y' = f(t,y)$$

over $[t_{i-1}, t_{i+1}]$ yields

$$y(t_{i+1}) = y(t_{i-1}) + \int_{t_{i-1}}^{t_{i+1}} f(t,y)\,\mathrm{d}t. \tag{8.2.10}$$

Applying the Midpoint formula to the integral $\int_{t_{i-1}}^{t_{i+1}} f(t,y)\,\mathrm{d}t$ gives

$$\int_{t_{i-1}}^{t_{i+1}} f(t,y)\,\mathrm{d}t \approx 2hf(t_i, y(t_i)). \tag{8.2.11}$$

Substituting Eq. (8.2.11) into Eq. (8.2.10) leads to

$$y(t_{i+1}) \approx y(t_{i-1}) + 2hf(t_i, y(t_i)) \tag{8.2.12}$$

which results in the two-step Euler's scheme

$$\begin{cases} w_0 = \alpha, \\ w_{i+1} = w_{i-1} + 2hf(t_i, w_i). \end{cases}$$

Now let us apply the Trapezoidal formula to the integral $\int_{t_i}^{t_{i+1}} f(t,y)\,\mathrm{d}t$ to obtain

$$\int_{t_i}^{t_{i+1}} f(t,y)\,\mathrm{d}t \approx \frac{h}{2}[f(t_i, y(t_i)) + f(t_{i+1}, y(t_{i+1}))]. \tag{8.2.13}$$

Substituting Eq. (8.2.13) into Eq. (8.2.9) results in

$$y(t_{i+1}) \approx y(t_i) + \frac{h}{2}[f(t_i, y(t_i)) + f(t_{i+1}, y(t_{i+1}))] \tag{8.2.14}$$

which produces the following scheme:

$$\begin{aligned} &w_0 = \alpha, \\ &w_{i+1} = w_i + \frac{h}{2}[f(t_i, w_i) + f(t_{i+1}, w_{i+1})], \end{aligned} \tag{8.2.15}$$

known as the Trapezoidal scheme (梯形格式). It is easy to see that the Trapezoidal scheme is just the average of explicit and implicit Euler's schemes.

Theorem 8.8 The trapezoidal scheme has order two.

Proof. Since using the error estimation of trapezoidal rule gives

$$\int_{t_i}^{t_{i+1}} f(t,y)\,\mathrm{d}t - \frac{h}{2}[f(t_i, y(t_i)) + f(t_{i+1}, y(t_{i+1}))] = O(h^3), \tag{8.2.16}$$

it follows from (8.2.15), (8.2.16) and Mean Value Theorem

$$\begin{aligned}
& y(t_{i+1}) - w_{i+1} \\
&= y(t_i) + \int_{t_i}^{t_{i+1}} f(t,y)\,\mathrm{d}t - \left\{ w_i + \frac{h}{2}[f(t_i, w_i) + f(t_{i+1}, w_{i+1})] \right\} \\
&= \int_{t_i}^{t_{i+1}} f(t,y)\,\mathrm{d}t - \frac{h}{2}[f(t_i, w_i) + f(t_{i+1}, w_{i+1})] \\
&= \int_{t_i}^{t_{i+1}} f(t,y)\,\mathrm{d}t - \frac{h}{2}[f(t_i, y(t_i)) + f(t_{i+1}, y(t_{i+1}))] \\
&\quad + \frac{h}{2}[f(t_i, y(t_i)) + f(t_{i+1}, y(t_{i+1}))] - \frac{h}{2}[f(t_i, w_i) + f(t_{i+1}, w_{i+1})] \\
&= O(h^3) + \frac{h}{2}[f(t_{i+1}, y(t_{i+1})) - f(t_{i+1}, w_{i+1})] \\
&= O(h^3) + \frac{h}{2}[y(t_{i+1}) - w_{i+1}] f'_y(t_{i+1}, \xi_{i+1}).
\end{aligned}$$

As a result,

$$y(t_{i+1}) - w_{i+1} = O(h^3).$$

Therefore trapezoidal scheme is of order two.

8.2.6 Modified Euler scheme (改进的 Euler 格式)

We know that Euler's (explicit) scheme has only order one, and Trapezoidal scheme is of order two, but it is implicit.

A new scheme can be established based on Euler's scheme and Trapezoidal scheme.

$$\begin{cases}
\mathit{Predictor}: & \overline{w}_{i+1} = w_i + hf(t_i, w_i), \\
\mathit{Corrector}: & w_{i+1} = w_i + \dfrac{h}{2}[f(t_i, y(t_i)) + f(t_{i+1}, \overline{w}_{i+1})].
\end{cases}$$

This scheme is called modified Euler's scheme and can be rewritten as

$$\begin{cases}
w_p = w_i + hf(t_i, w_i), \\
w_c = w_i + hf(t_{i+1}, w_p), \\
w_{i+1} = \dfrac{1}{2}(w_p + w_c).
\end{cases}$$

Theorem 8.9 The modified Euler's scheme has order two.

Proof. Suppose

$$\overline{w}_{i+1} = w_i + hf(t_i, w_i),$$

$$w_{i+1} = w_i + \frac{h}{2}[f(t_i, w_i) + f(t_{i+1}, \overline{w}_{i+1})].$$

Using the error estimation of trapezoidal rule

$$\int_{t_i}^{t_{i+1}} f(t,y)\,\mathrm{d}t - \frac{h}{2}[f(t_i, y(t_i)) + f(t_{i+1}, y(t_{i+1}))] = O(h^3),$$

gives

$$\begin{aligned}
& y(t_{i+1}) - w_{i+1} \\
&= y(t_i) + \int_{t_i}^{t_{i+1}} f(t,y)\,\mathrm{d}t - \left\{ w_i + \frac{h}{2}[f(t_i, w_i) + f(t_{i+1}, \overline{w}_{i+1})] \right\} \\
&= \int_{t_i}^{t_{i+1}} f(t,y)\,\mathrm{d}t - \frac{h}{2}[f(t_i, w_i) + f(t_{i+1}, \overline{w}_{i+1})] \\
&= \int_{t_i}^{t_{i+1}} f(t,y)\,\mathrm{d}t - \frac{h}{2}[f(t_i, y(t_i)) + f(t_{i+1}, y(t_{i+1}))] + \frac{h}{2}[f(t_{i+1}, y(t_{i+1})) - f(t_{i+1}, \overline{w}_{i+1})] \\
&= O(h^3) + \frac{h}{2}[f(t_{i+1}, y(t_{i+1})) - f(t_{i+1}, \overline{w}_{i+1})] \\
&= O(h^3) + \frac{h}{2} f_y'(t_{i+1}, \xi_{i+1})(y(t_{i+1}) - \overline{w}_{i+1}) \\
&= O(h^3) + \frac{h}{2} f_y'(t_{i+1}, \xi_{i+1})[y(t_{i+1}) - y(t_i) - hf(t_i, w_i)] \\
&= O(h^3) + \frac{h}{2} f_y'(t_{i+1}, \xi_{i+1})[hy'(t_i) + O(h^2) - hf(t_i, w_i)] \\
&= O(h^3) + \frac{h}{2} f_y'(t_{i+1}, \xi_{i+1}) \cdot O(h^2) \\
&= O(h^3).
\end{aligned}$$

Therefore the modified Euler's scheme has order two. The proof of Theorem 8.9 is completed.

8.2.7 Error bounds for the Euler's scheme (Euler 格式的误差界)

To derive an error bound generated by the Euler's scheme, we need the following lemmas.

Lemma 8.10 The following inequalities

$$0 \leqslant (1+x)^m \leqslant \mathrm{e}^{mx}$$

hold true for $x \geqslant 1$ and any positive m.

Proof. Using the Maclaurin expansion with remainder term gives

$$\mathrm{e}^x = 1 + x + \frac{x^2}{2}\mathrm{e}^{\xi},$$

where ξ is located between x and zero. Therefore

$$1 + x \leqslant \mathrm{e}^x = 1 + x + \frac{x^2}{2}\mathrm{e}^{\xi},$$

which implies $0 \leqslant (1+x)^m \leqslant \mathrm{e}^{mx}$ for $x \geqslant -1$ and any positive m.

The proof is completed.

Lemma 8.11 Suppose that the sequence $\{a_i\}_{i=0}^{N}$ satisfies

$$a_0 \geqslant -\frac{\beta}{\alpha}, \quad a_{i+1} \leqslant (1+\alpha)a_i + \beta, \quad i = 0,1,\cdots,N-1,$$

where α and β are positive real numbers, then we have

$$a_{i+1} \leqslant e^{(i+1)\alpha}\left(a_0 + \frac{\beta}{\alpha}\right) - \frac{\beta}{\alpha}.$$

Proof. From $a_{i+1} \leqslant (1+\alpha)a_i + \beta$, it follows

$$\begin{aligned}
a_{i+1} &\leqslant (1+\alpha)a_i + \beta \\
&\leqslant (1+\alpha)[(1+\alpha)a_{i-1} + \beta] + \beta \\
&= (1+\alpha)^2 a_{i-1} + [1 + (1+\alpha)]\beta \\
&\leqslant (1+\alpha)^2\left[(1+\alpha)a_{i-2} + \beta\right] + \left[1 + (1+\alpha)\right]\beta \\
&= (1+\alpha)^3 a_{i-2} + \left[1 + (1+\alpha) + (1+\alpha)^2\right]\beta \\
&\cdots \\
&\leqslant (1+\alpha)^{i+1} a_0 + \left[1 + (1+\alpha) + (1+\alpha)^2 + \cdots + (1+\alpha)^i\right]\beta \\
&= (1+\alpha)^{i+1} a_0 + \frac{(1+\alpha)^{i+1} - 1}{\alpha}\beta \\
&= (1+\alpha)^{i+1}\left[a_0 + \frac{\beta}{\alpha}\right] - \frac{\beta}{\alpha} \\
&\leqslant e^{(i+1)\alpha}\left[a_0 + \frac{\beta}{\alpha}\right] - \frac{\beta}{\alpha},
\end{aligned}$$

where Lemma 8.10 is applied to get $(1+\alpha)^{i+1} \leqslant e^{(i+1)\alpha}$ and $a_0 \geqslant -\frac{\beta}{\alpha}$ is used to obtain $(1+\alpha)^{i+1}\left[a_0 + \frac{\beta}{\alpha}\right] \leqslant e^{(i+1)\alpha}\left[a_0 + \frac{\beta}{\alpha}\right]$. The proof is completed.

Theorem 8.12 Suppose the real valued function $f(t,y)$ is continuous in the region $D = \{(t,y) \mid a \leqslant t \leqslant b, -\infty < y < +\infty\}$, and satisfies the Lipschitz condition with constant L in the variable y on D, and there exists a constant M such that

$$\max_{a \leqslant t \leqslant b} |y''(t)| \leqslant M,$$

where $y(t)$ denotes the unique solution to the the initial-value problem

$$y'(t) = f(t,y), \quad a \leqslant t \leqslant b, \quad y(a) = \alpha.$$

Then the error bound between the approximations w_i generated by Euler's scheme and the actual values $y(t_i)$ has the following estimation:

$$|y(t_i) - w_i| \leqslant \frac{hM}{2L}\left[e^{L(t_i - a)} - 1\right], \quad i = 0,1,\cdots,N.$$

Proof. Since $w_0 = y(t_0) = y(a) = \alpha$, the result is clear true for $i=0$. By Taylor's expansion with remainder term, we have

$$y(t_{i+1})=y(t_i+h)=y(t_i)+hy'(t_i)+\frac{h^2}{2}y''(\xi_i)$$
$$=y(t_i)+hf(t_i,y(t_i))+\frac{h^2}{2}y''(\xi_i),\quad i=0,1,\cdots,N-1.$$

And the Euler's scheme is

$$w_{i+1}=w_i+hf(t_i,w_i),\quad i=0,1,\cdots,N-1.$$

Subtracting the above two systems of equations gives

$$y(t_{i+1})-w_{i+1}=y(t_i)-w_i+h\left[f(t_i,y(t_i))-f(t_i,w_i)\right]+\frac{h^2}{2}y''(\xi_i),\quad i=0,1,\cdots,N-1.$$

Since $f(t,y)$ satisfies the Lipschitz condition with constant L in the variable y on D, and there exists a constant M such that

$$\max_{a\leqslant t\leqslant b}\left|y''(t)\right|\leqslant M,$$

we have for $i=0,1,\cdots,N-1$,

$$\left|y(t_{i+1})-w_{i+1}\right|\leqslant\left|y(t_i)-w_i\right|+h\left|f(t_i,y(t_i))-f(t_i,w_i)\right|+\frac{h^2}{2}M$$
$$\leqslant\left|y(t_i)-w_i\right|+hL\left|y(t_i)-w_i\right|+\frac{h^2M}{2}$$
$$\leqslant(1+hL)\left|y(t_i)-w_i\right|+\frac{h^2M}{2}.$$

Let

$$a_i=\left|y(t_i)-w_i\right|,\quad \alpha=hL,\quad \beta=\frac{Mh^2}{2}.$$

Then Lemma 8.11 implies

$$a_i\leqslant e^{i\alpha}\left(a_0+\frac{\beta}{\alpha}\right)-\frac{\beta}{\alpha},$$

namely

$$\left|y(t_i)-w_i\right|\leqslant\left(e^{ihL}-1\right)\frac{Mh^2/2}{hL}=\frac{Mh}{2L}\left(e^{ihL}-1\right)=\frac{Mh}{2L}\left(e^{(t_i-a)L}-1\right),$$

where $ih=t_i-t_0=t_i-a$ is used. The proof is completed.

Example 8.7 As in Example 8.6, Euler's scheme with $N=10$ was adopted to approximate the solution to the initial-value problem

$$y'=y-t^2+1,\quad 0\leqslant t\leqslant 2,\quad y(0)=0.5.$$

Now use the inequality in Theorem 8.12 to find a bound for the approximation errors and compare these with the actual ones.

Solution. With $N=10$ we have $h=2/10=0.2$, $t_i=0.2i$, $i=0,1,\cdots,10$, and $w_0=y(0)=0.5$. Since $f(t,y)=y-t^2+1$, we have $\partial f(t,y)/\partial y=1$ for all y, which implies the Lipschitz constant L is 1. It is easy to verify that the exact solution to the given initial-value problem is $y(t)=(t+1)^2-0.5e^t$, so we have $y''(t)=2-0.5e^t$, and $|y''(t)|\leqslant 0.5e^2-2, \forall t\in[0,2]$. Using the inequality in Theorem 8.12

$$|y(t_i)-w_i|\leqslant \frac{hM}{2L}\left[e^{L(t_i-a)}-1\right],\quad i=0,1,\cdots,N,$$

for the error bounds with $h=0.2, L=1$, and $M=0.5e^2-2$, gives

$$|y(t_i)-w_i|\leqslant 0.1(0.5e^2-2)\ (e^{t_i}-1),\quad i=0,1,\cdots,10.$$

Consequently

$$|y(0.2)-w_1|\leqslant 0.1(0.5e^2-2)\ (e^{0.2}-1)=0.03752;$$

$$|y(0.4)-w_2|\leqslant 0.1(0.5e^2-2)\ (e^{0.4}-1)=0.08334;$$

$$\vdots$$

$$|y(2)-w_{10}|\leqslant 0.1(0.5e^2-2)\ (e^2-1)=1.08264.$$

All the involved actual errors and error bounds are listed in Table 8.2.

Table 8.2 Error bounds

i	t_i	w_i	$y_i=y(t_i)$	Actual error $\lvert y_i-w_i\rvert$	Error bounds $\lvert y_i-w_i\rvert$
1	0.2	0.8000000	0.8292986	0.0292986	0.03752
2	0.4	1.1520000	1.2140877	0.0620877	0.08344
3	0.6	1.5505000	1.6489406	0.0985406	0.13931
4	0.8	1.9884800	2.1272295	0.1387495	0.20767
5	1.0	2.4581760	2.6408591	0.1826831	0.29117
6	1.2	2.9498112	3.1799415	0.2301303	0.39315
7	1.4	3.4517734	3.7324000	0.2806266	0.51771
8	1.6	3.9501281	4.2834838	0.3333557	0.66985
9	1.8	4.4281538	4.8151763	0.3870225	0.85568
10	2.0	4.8657845	5.3054720	0.4396874	1.08264

8.3 Runge-Kutta method (Runge-Kutta 方法)

- Taylor polynomial in two variables (二元 Taylor 多项式)
- Basic ideas of Runge-Kutta method (Runge-Kutta 方法的基本思想)
- Runge-Kutta method of order two (二阶 Runge-Kutta 方法)
- Runge-Kutta scheme of order three (三阶 Runge-Kutta 方法)
- Runge-Kutta method of order four (四阶 Runge-Kutta 方法)

8.3.1 Taylor polynomial in two variables (二元 Taylor 多项式)

Suppose that $f(t, y)$ and all its partial derivatives of order less than or equal to $n+1$

are continuous on $D=\{(t, y) \mid a \leqslant t \leqslant b, c \leqslant y \leqslant d\}$ and let $(t_0, y_0) \in D$. For every $(t, y) \in D$, there exists ξ between t and t_0 and μ between y and y_0 with

$$f(t, y) = P_n(t, y) + R_n(t, y)$$

where

$$P_n(t, y) = f(t_0, y_0) + \sum_{k=0}^{n} \frac{D^k f(t_0, y_0)}{k!},$$

$$R_n(t, y) = \frac{D^{n+1} f(\xi, \mu)}{(n+1)!},$$

with

$$D^k f = \left[(t-t_0)\frac{\partial}{\partial t} + (y-y_0)\frac{\partial}{\partial y}\right]^k f = \sum_{j=0}^{k} \binom{k}{j} \frac{\partial^k f}{\partial t^j \partial y^{k-j}} (t-t_0)^j (y-y_0)^{k-j}.$$

8.3.2 Basic ideas of Runge-Kutta method (Runge-Kutta 方法的基本思想)

Mean-value Theorem tells us that if $y(t) \in C[a, b]$ and $y'(t)$ exists on (a, b), then a number $\xi_i \in (t_i, t_{i+1})$ exists with

$$\frac{y(t_{i+1}) - y(t_i)}{h} = y'(\xi_i)$$

or

$$y(t_{i+1}) = y(t_i) + hy'(\xi_i) = y(t_i) + hf(\xi_i, y(\xi_i)) = y(t_i) + K_i^* \cdot h,$$

where $K_i^* = f(\xi_i, y(\xi_i))$ is called the average slope on $[t_i, t_{i+1}]$.

It is not difficult to understand that every approximation to the average slope results in a new scheme. As we know, in Euler's scheme

$$w_{i+1} = w_i + hf(t_i, w_i),$$

we have $K_i^* = f(t_i, w_i)$.

In Euler's implicit scheme

$$w_{i+1} = w_i + hf(t_{i+1}, w_{i+1}),$$

we have $K_i^* = f(t_{i+1}, w_{i+1})$.

And in modified Euler's scheme

$$\begin{cases} K_1 = f(t_i, w_i), \\ K_2 = f(t_{i+1}, w_i + hK_1), \\ w_{i+1} = w_i + \dfrac{1}{2}h(K_1 + K_2), \end{cases}$$

we have $K_i^* = \dfrac{1}{2}(K_1 + K_2)$.

This implies us to construct the scheme of high orders by considering the linear combination of more slopes over $[t_i, t_{i+1}]$.

8.3.3 Runge-Kutta method of order two (二阶 Runge-Kutta 方法)

As we know, every point in $[t_i, t_{i+1}]$ can be expressed as

$$t_{i+p} = t_i + ph, 0 \leqslant p \leqslant 1.$$

Let

$$K_1 = f(t_i, w_i),\ w_{i+p} = w_i + phK_1,$$

$$K_2 = f(t_{i+p}, w_{i+p}),\ K_i^* = (1-\lambda)K_1 + \lambda K_2.$$

we set up the following scheme

$$\begin{cases} w_{i+1} = w_i + hK_i^*, \\ K_i^* = (1-\lambda)K_1 + \lambda K_2, \\ K_1 = f(t_i, w_i), \\ K_2 = f(t_i + ph, w_i + phK_1). \end{cases} \tag{8.3.1}$$

Since

$$\begin{aligned} y''(t_i) &= \frac{\mathrm{d}}{\mathrm{d}t} y'(t)\bigg|_{t=t_i} \\ &= \frac{\mathrm{d}}{\mathrm{d}t} f(t, y(t))\bigg|_{t=t_i} \\ &= [f_t(t, y(t)) + f_y(t, y(t))y'(t)]_{t=t_i} \\ &= f_t(t_i, w_i) + f_y(t_i, w_i)y'(t_i) \\ &= f_t(t_i, w_i) + f(t_i, w_i)f_y(t_i, w_i), \end{aligned}$$

to determine λ, p, we suppose $w_i = y(t_i)$, then

$$\begin{aligned} K_1 &= f(t_i, w_i), \\ K_2 &= f(t_{i+p}, w_i + phK_1) \\ &= f(t_i, w_i) + ph\frac{\partial f}{\partial t}(t_i, w_i) + phK_1\frac{\partial f}{\partial y}(t_i, w_i) + O(h^2) \\ &= f(t_i, w_i) + ph[f_t(t_i, w_i) + f(t_i, w_i)f_y(t_i, w_i)] + O(h^2) \\ &= y'(t_i) + phy''(t_i) + O(h^2). \end{aligned}$$

Substituting into (8.3.1) gives

$$\begin{aligned} w_{i+1} &= w_i + h[(1-\lambda)y'(t_i) + \lambda y'(t_i) + \lambda phy''(t_i) + O(h^2)] \\ &= w_i + hy'(t_i) + \lambda ph^2 y''(t_i) + O(h^3). \end{aligned} \tag{8.3.2}$$

By Taylor's expansion

$$y(t_{i+1}) = y(t_i) + hy'(t_i) + \frac{h^2}{2}y''(t_i) + O(h^3). \tag{8.3.3}$$

To ensure that the truncation error is $O(h^3)$, we must get

$$\lambda p = \frac{1}{2}.$$

In particular, for $p=1$, $\lambda=\frac{1}{2}$, the Runge-Kutta scheme is the modified Euler's scheme. If we choose $p=\frac{1}{2}$, $\lambda=1$, then Runge-Kutta scheme reads as follows

$$\begin{aligned}
K_1 &= f(t_i, w_i), \\
K_2 &= f(t_{i+1/2}, w_i + \frac{1}{2}hK_1), \\
w_{i+1} &= w_i + hK_2.
\end{aligned}$$

8.3.4 Runge-Kutta method of order three (三阶 Runge-Kutta 方法)

For $0 \leqslant p \leqslant q \leqslant 1$, let

$$\begin{aligned}
& t_{i+p} = t_i + ph, t_{i+q} = t_i + qh, \\
& K_1 = f(t_i, w_i), \\
& K_2 = f(t_i + ph, w_i + phK_1), \\
& K_3 = f(t_i + qh, w_i + qh[(1-\alpha)K_1 + \alpha K_2]).
\end{aligned}$$

Consider the scheme

$$w_{i+1} = w_i + h\left[(1-\lambda-\mu)K_1 + \lambda K_2 + \mu K_3\right]$$

and determine p, q, λ, μ, α so that the above scheme has order three. What follows is just one of the scheme with order three.

$$\begin{aligned}
w_{i+1} &= w_i + \frac{h}{6}(K_1 + 4K_2 + K_3), \\
K_1 &= f(t_i, w_i), \\
K_2 &= f(t_i + \frac{h}{2}, w_i + \frac{h}{2}K_1), \\
K_3 &= f(t_i + h, w_i + h(-K_1 + 2K_2)).
\end{aligned}$$

8.3.5 Runge-Kutta method of order four (四阶 Runge-Kutta 方法)

For $0 \leqslant p \leqslant q \leqslant r \leqslant 1$, let

$$\begin{aligned}
& t_{i+p} = t_i + ph, \quad t_{i+q} = t_i + qh, \quad t_{i+r} = t_i + rh, \\
& K_1 = f(t_i, w_i), \\
& K_2 = f(t_i + ph, w_i + phK_1), \\
& K_3 = f(t_i + qh, w_i + qhK_2), \\
& K_4 = f(t_i + rh, w_i + rh[\alpha K_1 + \beta K_2 + (1-\alpha-\beta)K_3]),
\end{aligned}$$

Consider the scheme

$$w_{i+1} = w_i + h\left[\lambda K_1 + \mu K_2 + \nu K_3 + (1-\lambda-\mu-\nu)K_4\right]$$

and determine $p,q,r,\alpha,\beta,\lambda,\mu,\nu$ so that the above scheme has order four. What follows is the well known Runge-Kutta scheme of order four, just one of the above family of schemes of order four.

$$\begin{aligned}
w_{i+1} &= w_i + \frac{h}{6}(K_1 + 2K_2 + 2K_3 + K_4),\\
K_1 &= f(t_i, w_i),\\
K_2 &= f(t_i + \frac{h}{2}, w_i + \frac{h}{2}K_1),\\
K_3 &= f(t_i + \frac{h}{2}, w_i + \frac{h}{2}K_2),\\
K_4 &= f(t_i + h, w_i + hK_3).
\end{aligned}$$

8.4 Multistep methods (多步方法)

- Adams explicit schemes (Adams 显式格式)
- Adams implicit schemes (Adams 隐式格式)

Except for the two-step Euler's scheme, almost all the methods introduced to this point belong to one-step methods since the scheme can be expressed as the form:

$$w_{i+1} = w_i + hF(t_i, t_{i+1}, w_i, w_{i+1}),$$

which includes Euler's scheme when $F(t_i,t_{i+1},w_i,w_{i+1}) = f(t_i,w_i)$, Euler's implicit scheme when $F(t_i,t_{i+1},w_i,w_{i+1}) = f(t_{i+1},w_{i+1})$, and trapezoidal scheme when $F(t_i,t_{i+1},w_i,w_{i+1}) = \frac{1}{2}[f(t_i,w_i) + f(t_{i+1},w_{i+1})]$. All the information used by these methods is obtained within the subinterval $[t_i, t_{i+1}]$ over which the solution is being approximated. There is no reason not to believe that more mesh points are involved in a scheme, better approximation can be obtained. If a method makes use of the approximations at more than one previous mesh point to determine the approximation at the next point, then the method is called multistep methods. In this section we introduce two kinds of multistep methods: Adams explicit schemes or Adams-Bashforth techniques, and Adams implicit schemes or Adams-Moulton techniques.

8.4.1 Adams explicit schemes (Adams 显式格式)

Adams explicit schemes are also known as Adams-Bashforth techniques.

(1) **Adams-Bashforth two-step explicit scheme** (Adams 两步显式格式)

Let

$$w_i = y(t_i),\ w_{i-1} = y(t_{i-1}),$$

$$w_i' = y'(t_i) = f(t_i, y(t_i)) = f(t_i, w_i),$$
$$w_{i-1}' = y'(t_{i-1}) = f(t_{i-1}, y(t_{i-1})) = f(t_{i-1}, w_{i-1}).$$

Choose parameter λ so that the following scheme

$$\begin{cases} w_{i+1} = w_i + h\left[(1-\lambda)w_i' + \lambda w_{i-1}'\right] \\ w_i' = f(t_i, w_i) \\ w_{i-1}' = f(t_{i-1}, w_{i-1}) \end{cases} \tag{8.4.1}$$

is of order 2.

Since

$$w_i' = y'(t_i),$$
$$w_{i-1}' = y'(t_{i-1}) = y'(t_i) - hy''(t_i) + O(h^2),$$

substituting into (8.4.1) yields,

$$\begin{aligned} w_{i+1} &= w_i + h\left[(1-\lambda)y'(t_i) + \lambda y'(t_i) - \lambda hy''(t_i) + O(h^2)\right] \\ &= w_i + hy'(t_i) - \lambda h^2 y''(t_i) + O(h^3) \\ &= y(t_i) + hy'(t_i) - \lambda h^2 y''(t_i) + O(h^3). \end{aligned}$$

On the other hand,

$$y(t_{i+1}) = y(t_i) + hy'(t_i) + \frac{h^2}{2}y''(t_i) + O(h^3).$$

To make (8.4.1) be of order two is to make

$$y(t_{i+1}) - w_{i+1} = O(h^3).$$

As a consequence, $\lambda = -\dfrac{1}{2}$. And we get Adams two-step explicit scheme

$$w_{i+1} = w_i + \frac{h}{2}(3w_i' - w_{i-1}') = w_i + \frac{h}{2}\left[3f(t_i, w_i) - f(t_{i-1}, w_{i-1})\right].$$

(2) **Adams-Bashforth three-step explicit scheme** (Adams 三步显式格式)

Adams three-step explicit scheme is a scheme of the following form:

$$w_{i+1} = w_i + h(\lambda w_i' + \mu w_{i-1}' + \gamma w_{i-2}'), \quad \lambda + \mu + \gamma = 1;$$
$$w_k' = f(t_k, w_k) = f(t_k, y(t_k)), \quad k = i, i-1, i-2.$$

The parameters λ, μ and γ are chosen so that the scheme is of order three.

Note

$$w_i' = f(t_i, w_i) = f(t_i, y(t_i)) = y'(t_i),$$

$$w_{i-1}' = y'(t_{i-1}) = y'(t_i) - hy''(t_i) + \frac{h^2}{2}y'''(t_i) + O(h^3),$$

$$w_{i-2}' = y'(t_{i-2}) = y'(t_i) - 2hy''(t_i) + 2h^2 y'''(t_i) + O(h^3),$$

we have

$$\begin{aligned} w_{i+1} &= w_i + h\Big[\lambda y'(t_i) + \mu y'(t_i) + \gamma y'(t_i) - (\mu h + 2\gamma h)y''(t_i) \\ &\quad + \left(\frac{\mu}{2} + 2\gamma\right)h^2 y'''(t_i)\Big] + O(h^4) \\ &= y(t_i) + hy'(t_i) + h^2(-\mu - 2\gamma)y''(t_i) + h^3\left(\frac{\mu}{2} + 2\gamma\right)y'''(t_i) + O(h^4). \end{aligned}$$

On the other hand,

$$y(t_{i+1}) = y(t_i) + hy'(t_i) + \frac{h^2}{2}y''(t_i) + \frac{h^3}{3!}y'''(t_i) + O(h^4),$$

$$y(t_{i+1}) - w_{i+1} = O(h^4)$$

implies:

$$\begin{cases} -\mu - 2\gamma = \dfrac{1}{2} \\ \dfrac{\mu}{2} + 2\gamma = \dfrac{1}{6} \end{cases} \Rightarrow \begin{cases} \mu = -\dfrac{4}{3} \\ \gamma = \dfrac{5}{12} \end{cases} \Rightarrow \lambda = 1 - \mu - \gamma = \frac{23}{12}.$$

So Adams three-step explicit scheme is

$$\begin{aligned} w_{i+1} &= w_i + \frac{h}{12}(23w_i' - 16w_{i-1}' + 5w_{i-2}') \\ &= w_i + \frac{h}{12}\left[23f(t_i, w_i) - 16f(t_{i-1}, w_{i-1}) + 5f(t_{i-2}, w_{i-2})\right], \end{aligned}$$

which is a method of order three.

(3) **Adams-Bashforth four-step explicit scheme** (Adams 四步显式格式)

Adams four-step explicit scheme is a scheme of the following form

$$w_{i+1} = w_i + h(\lambda_1 w_i' + \lambda_2 w_{i-1}' + \lambda_3 w_{i-2}' + \lambda_4 w_{i-3}')$$

$$w_k' = f(t_k, w_k) = f(t_k, y(t_k)) = y'(t_k), \quad k = i, i-1, i-2, i-3$$

where

$$\lambda_1 + \lambda_2 + \lambda_3 + \lambda_4 = 1.$$

The parameters $\lambda_1, \lambda_2, \lambda_3, \lambda_4$ are chosen so that the scheme is of order four.

Applying Taylor's expansion to w_{i-1}', w_{i-2}' and w_{i-3}' gives

$$
\begin{aligned}
w_{i+1} = w_i + h\Big\{&\lambda_1 y'(t_i) \\
&+\lambda_2\left[y'(t_i) - hy''(t_i) + \frac{h^2}{2}y'''(t_i) - \frac{h^3}{3!}y^{(4)}(t_i) + O(h^4)\right] \\
&+\lambda_3\left[y'(t_i) - 2hy''(t_i) + 2h^2y'''(t_i) - \frac{4}{3}h^3y^{(4)}(t_i) + O(h^4)\right] \\
&+\lambda_4\left[y'(t_i) - 3hy''(t_i) + \frac{9}{2}h^2y'''(t_i) - \frac{9}{2}h^3y^{(4)}(t_i) + O(h^4)\right]\Big\} \\
= w_i + hy'(t_i) &+ h^2(-\lambda_2 - 2\lambda_3 - 3\lambda_4)y''(t_i) \\
&+ h^3\left(\frac{1}{2}\lambda_2 + 2\lambda_3 + \frac{9}{2}\lambda_4\right)y'''(t_i) + h^4\left(-\frac{1}{6}\lambda_2 - \frac{4}{3}\lambda_3 - \frac{9}{2}\lambda_4\right)y^{(4)}(t_i) + O(h^5)
\end{aligned}
$$

$$
y(t_{i+1}) = y(t_i) + hy'(t_i) + \frac{1}{2!}h^2y''(t_i) + \frac{1}{3!}h^3y'''(t_i) + \frac{1}{4!}h^4y^{(4)}(t_i) + O(h^5)
$$

$$
y(t_{i+1}) - w_{i+1} = O(h^5)
$$

which implies

$$
\begin{cases}
\lambda_1 + \lambda_2 + \lambda_3 + \lambda_4 = 1, \\
-\lambda_2 - 2\lambda_3 - 3\lambda_4 = \dfrac{1}{2}, \\
\dfrac{1}{2}\lambda_2 + 2\lambda_3 + \dfrac{9}{2}\lambda_4 = \dfrac{1}{6}, \\
-\dfrac{1}{6}\lambda_2 - \dfrac{4}{3}\lambda_3 - \dfrac{9}{2}\lambda_4 = \dfrac{1}{24}.
\end{cases}
\Rightarrow
\begin{cases}
\lambda_1 = \dfrac{55}{24}, \\
\lambda_2 = -\dfrac{59}{24}, \\
\lambda_3 = \dfrac{37}{24}, \\
\lambda_4 = -\dfrac{9}{24}.
\end{cases}
$$

So we have the Adams four-step explicit scheme

$$
\begin{aligned}
w_{i+1} &= w_i + \frac{h}{24}(55w_i' - 59w_{i-1}' + 37w_{i-2}' - 9w_{i-3}') \\
&= w_i + \frac{h}{24}[55f(t_i, w_i) - 59f(t_{i-1}, w_{i-1}) + 37f(t_{i-2}, w_{i-2}) - 9f(t_{i-3}, w_{i-3})],
\end{aligned}
$$

which is a method of order four.

8.4.2 Adams implicit schemes (Adams 隐式格式)

Adams implicit schemes are also known as Adams-Moulton techniques.

(1) **Adams implicit scheme** (Adams 隐式格式)

Adams implicit scheme has the following general form

$$
\begin{aligned}
&w_0 = \alpha, \\
&w_{i+1} = w_i + h[(1-\lambda)w_{i+1}' + \lambda w_i'],
\end{aligned}
$$

where $w_k' = f(t_k, w_k),\quad k = i, i+1$.

Since

$$w'_{i+1} = y'(t_{i+1}) = y'(t_i) + hy''(t_i) + O(h^2),$$
$$w'_i = y'(t_i),$$

we have

$$w_{i+1} = w_i + h\Big[(1-\lambda)\big(y'(t_i) + hy''(t_i) + O(h^2)\big) + \lambda y'(t_i)\Big]$$
$$= y(t_i) + hy'(t_i) + (1-\lambda)h^2 y''(t_i) + O(h^3).$$

Using

$$y(t_{i+1}) = y(t_i) + hy'(t_i) + \frac{h^2}{2}y''(t_i) + O(h^3),$$

and

$$y(t_{i+1}) - w_{i+1} = O(h^3),$$

derives $\lambda = \frac{1}{2}$.

So we obtain the Adams implicit scheme

$$w_{i+1} = w_i + \frac{h}{2}(w'_i + w'_{i+1}) = w_i + \frac{h}{2}[f(t_i, w_i) + f(t_{i+1}, w_{i+1})].$$

(2) **Adams-Moulton two-step implicit scheme** (Adams-Moulton 两步隐式格式)

$$w_0 = \alpha, \quad w_1 = \alpha_1,$$
$$w_{i+1} = w_i + \frac{h}{12}[5f(t_{i+1}, w_{i+1}) + 8f(t_i, w_i) - f(t_{i-1}, w_{i-1})],$$

where $i = 1, 2, \cdots, N-1$. The local truncation error is $O(h^3)$.

(3) **Adams-Moulton three-step implicit scheme** (Adams-Moulton 三步隐式格式)

$$w_0 = \alpha,\ w_1 = \alpha_1,\ w_2 = \alpha_2,$$
$$w_{i+1} = w_i + \frac{h}{24}[9f(t_{i+1}, w_{i+1}) + 19f(t_i, w_i) - 5f(t_{i-1}, w_{i-1}) + f(t_{i-2}, w_{i-2})],$$

where $i = 1, 2, \cdots, N-1$. The local truncation error is $O(h^4)$.

(4) **Adams-Moulton four-step implicit scheme** (Adams- Moulton 四步隐式格式)

$$w_0 = \alpha,\ w_1 = \alpha_1,\ w_2 = \alpha_2,\ w_3 = \alpha_3,$$
$$w_{i+1} = w_i + \frac{h}{720}[251f(t_{i+1}, w_{i+1}) + 646f(t_i, w_i)$$
$$- 264f(t_{i-1}, w_{i-1}) + 106f(t_{i-2}, w_{i-2}) - 19f(t_{i-3}, w_{i-3})],$$

where $i = 1, 2, \cdots, N-1$. The local truncation error is $O(h^5)$.

8.5 本章算法程序及实例 (Algorithms and examples)

8.5.1 Euler 方法 (Euler's scheme)

```
%% y'=f(x,y),fun为f(x,y), a,b为数值解区间,y0为a点的初值,n为区间中取点个数
```

```
function y=Euler_1(fun,a,b,y0,h)
n=(b-a)/h;
x=zeros(n+1,1);
y=zeros(n+1,1);
x(1)=a;
y(1)=y0;
for i=2:n
   x(i)=a+(i-1)*h;
   y(i)=y(i-1)+h*feval(fun,x(i-1),y(i-1));
end
x(n+1)=b
y(n+1)=y(n)+h*feval(fun,x(n),y(n));
plot(x,y,'-d')
```

例 8.8 利用 Euler 方法求常微分方程

$$\begin{cases} \dfrac{\mathrm{d}y}{\mathrm{d}x}=2\dfrac{y}{x}+x^2\mathrm{e}^x, & 1\leqslant x\leqslant 2; \\ y(1)=0. \end{cases}$$

步长 $h=0.1$.

调用程序

```
Euler_1(inline('2/x*y+x^2*exp(x)'),1,2,0,0.1)
```

可得结果如图 8.1.

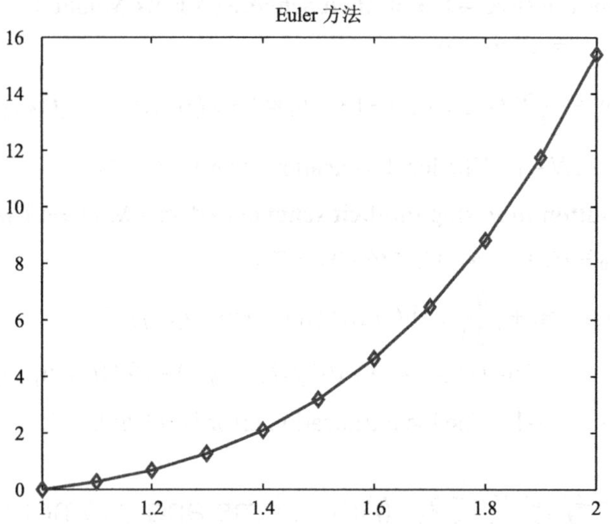

图 8.1 Euler 方法

8.5.2 隐式 Euler 方法 (Euler implicit scheme)

```
% y'=f(x,y),fun为f(x,y), a,b为数值解区间,y0为a点的初值, h为步长
% 利用迭代法求y_(n+1),迭代次数为k
function y=Implicit_Euler(fun,a,b,y0,h,k)
n=(b-a)/h;
x=zeros(n+1,1);
y=zeros(n+1,1);
x(1)=a;
y(1)=y0;
for i=2:n
    x(i)=a+(i-1)*h;
    y(i)=y(i-1)+h*feval(fun,x(i-1),y(i-1));
end
x(n+1)=b
y(n+1)=y(n)+h*feval(fun,x(n),y(n));
y0=y;
for i=2:n+1
    for j=1:k
        y(i)=y0(i-1)+h*feval(fun,x(i),y(i));
    end
end
hold on
plot(x,y,'-d')
title('隐式 Euler 方法')
```

例 8.9 利用 (两步) 隐式 Eular 方法求常微分方程

$$\begin{cases}\dfrac{\mathrm{d}y}{\mathrm{d}x}=2\dfrac{y}{x}+x^2\mathrm{e}^x, & 1\leqslant x\leqslant 2;\\ y(1)=0.\end{cases}$$

步长 $h=0.1$.

分别调用程序

```
Implicit_Euler(inline('2/x*y+x^2*exp(x)'),1,2,0,0.1,1)
Implicit_Euler(inline('2/x*y+x^2*exp(x)'),1,2,0,0.1,2)
```

可得结果如图 8.2.

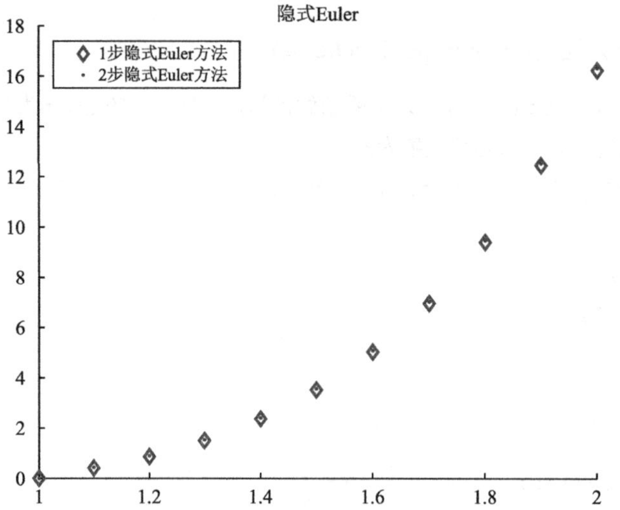

图 8.2 隐式 Euler 方法

8.5.3 梯形方法 (Trapezoidal scheme)

```
% y'=f(x,y),fun为f(x,y), a,b为数值解区间,y0为a点的初值, h为步长
% 利用迭代法求y_(n+1),迭代次数为k
% y_(n+1)=y_n+h/2*(f(x_n,y_n)+f(x_n+1),f(y_n+1))
function y=Trap_Euler(fun,a,b,y0,h,k)
n=(b-a)/h;
x=zeros(n+1,1);
y=zeros(n+1,1);
x(1)=a;y(1)=y0;
for i=2:n
    x(i)=a+(i-1)*h;
    y(i)=y(i-1)+h*feval(fun,x(i-1),y(i-1));
end
x(n+1)=b
y(n+1)=y(n)+h*feval(fun,x(n),y(n));
y0=y;
for i=2:n+1
    for j=1:k
        y(i)=y0(i-1)+h/2*(feval(fun,x(i-1),y0(i-1))+feval(fun,x(i),y(i)));
    end
end
hold on
plot(x,y,'-o')
```

```
title('梯形方法')
```

例 8.10 利用梯形方法求常微分方程

$$\begin{cases} \dfrac{dy}{dx} = 2\dfrac{y}{x} + x^2 e^x, & 1 \leqslant x \leqslant 2; \\ y(1) = 0. \end{cases}$$

步长 $h=0.1$.

调用程序

```
Trap_Euler(inline('2/x*y+x^2*exp(x)'),1,2,0,0.1,2)
```

运行结果如图 8.3.

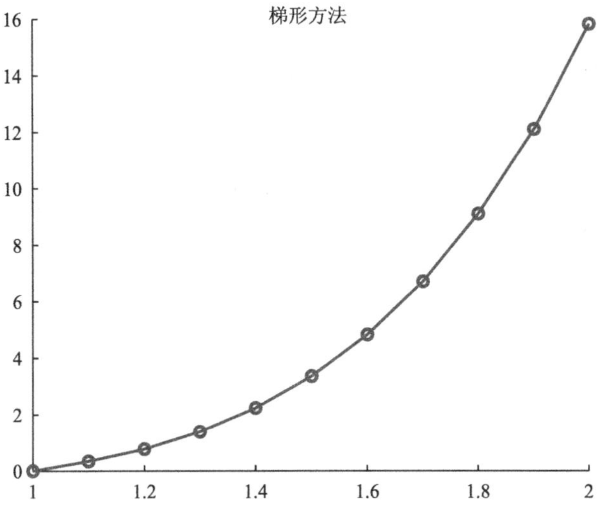

图 8.3 梯形方法

8.5.4 改进的 Euler 公式 (Modified Euler's scheme)

```
% dy/dx=fun,a,b为区间,y0为a点的初值,h为步长
% yp=yn+h*f(xn,yn);yc=yn+h*f(xn+1,yp),yn+1=1/2*(yp+yc)
function y=Modi_Euler(fun,a,b,y0,h)
n=(b-a)/h;
x=zeros(1,n+1);
y=zeros(1,n+1);
x(1)=a;
y(1)=y0;
for i=2:n+1
    x(i)=a+(i-1)*h;
    yp(i)=y(i-1)+h*feval(fun,x(i-1),y(i-1));
    yc(i)=y(i-1)+h*feval(fun,x(i),yp(i));
```

```
    y(i)=1/2*(yp(i)+yc(i));
end
hold on
plot(x,y,'-s')
title('改进Euler方法')
```

例 8.11　利用改进的 Euler 方法求常微分方程

$$\begin{cases}\dfrac{dy}{dx}=2\dfrac{y}{x}+x^2e^x, & 1\leqslant x\leqslant 2;\\ y(1)=0.\end{cases}$$

步长 $h=0.1$.

调用程序

```
Modi_Euler(inline('2/x*y+x^2*exp(x)'),1,2,0,0.1);
```

运行结果见图 8.4.

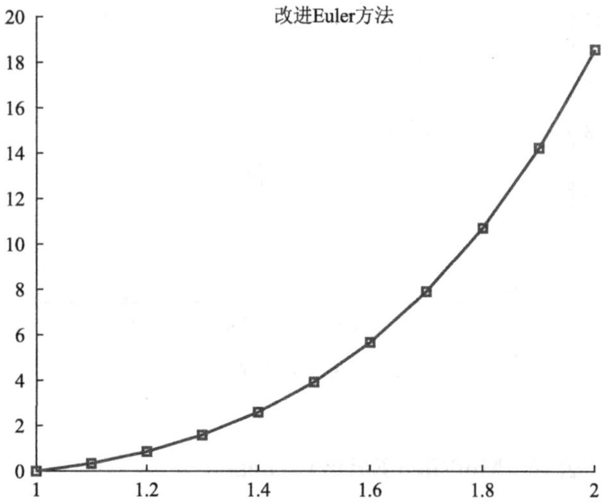

图 8.4　改进的 Euler 方法

例 8.12　利用 Euler 方法、隐式 Euler 方法、梯形方法、改进 Euler 方法分别求常微分方程

$$\begin{cases}\dfrac{dy}{dx}=2\dfrac{y}{x}+x^2e^x, & 1\leqslant x\leqslant 2;\\ y(1)=0.\end{cases}$$

步长 $h=0.1$. 并与精确解 $y=x^2(e^x-1)$ 进行对比.

运行程序

```
%% 修改几个函数中plot线型表示
clc,clear
Euler_1(inline('2/x*y+x^2*exp(x)'),1,2,0,0.1);
```

```
hold on
Implicit_Euler(inline('2/x*y+x^2*exp(x)'),1,2,0,0.1,1);
hold on
Trap_Euler(inline('2/x*y+x^2*exp(x)'),1,2,0,0.1,1);
hold on
Modi_Euler(inline('2/x*y+x^2*exp(x)'),1,2,0,0.1);
hold on
x1=1:0.1:2;
y1=x1.^2.*(exp(x1)-exp(1));
hold on
plot(x1,y1,'-s')
title('四种方法与精确解对比图')
legend('Euler','隐式Euler','梯形方法','改进Euler','精确解')
```

结果如图 8.5 所示.

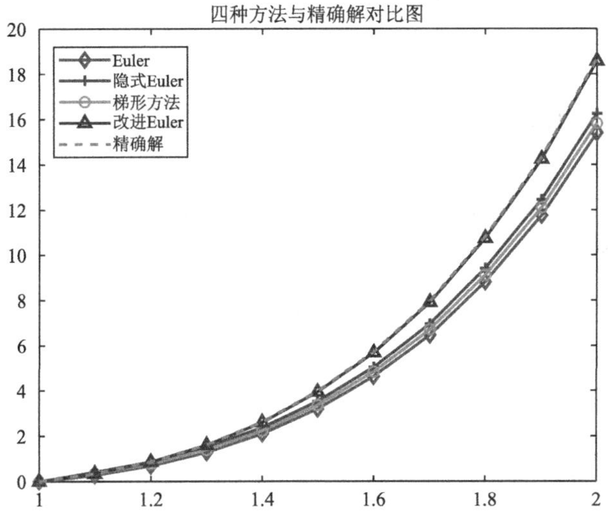

图 8.5 四种方法对比

8.5.5 二阶龙格-库塔方法 (Runge-Kutta order two)

```
% dy/dx=fun;a,b为数值解区间; h为步长; d为参数
% d=1/2时龙格库塔方法为改进Euler方法
function y=RK_2(fun,a,b,y0,h,d)
n=(b-a)/h;
c1=1-d;
c2=d;
lambda=1/(2*d);
```

```
mu=1/(2*d);
x=zeros(1,n+1);
y=zeros(1,n+1);
x(1)=a;
y(1)=y0;
for i=2:n+1
    x(i)=a+(i-1)*h;
    k1(i)=feval(fun,x(i-1),y(i-1));
    k2(i)=feval(fun,x(i-1)+lambda*h,y(i-1)+mu*h*k1(i));
    y(i)=y(i-1)+h*(c1*k1(i)+c2*k2(i));
end
hold on
plot(x,y,'-s')
```

例 8.13　利用二阶龙格-库塔方法求常微分方程

$$\begin{cases}\dfrac{dy}{dx}=2\dfrac{y}{x}+x^2e^x, & 1\leqslant x\leqslant 2;\\ y(1)=0.\end{cases}$$

步长 $h=0.1$. 参数为 1.

调用函数

```
RK_2(inline('2/x*y+x^2*exp(x)'),1,2,0,0.1,1)
```

结果如图 8.6 所示.

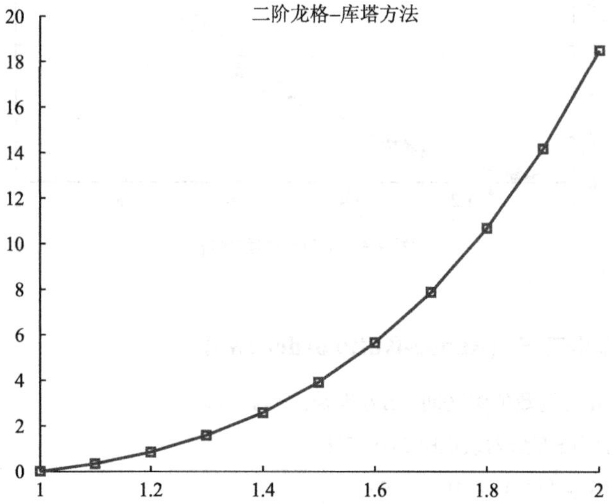

图 8.6　二阶龙格-库塔方法

8.5.6 三阶龙格-库塔方法 (Runge-Kutta order three)

```
% dy/dx=fun;a,b为数值解区间; h为步长; d为参数
% 参数选择: c1=c3=1/6,c2=2/3;lambda2=1/2,lambda3=1,mu21=1/2,mu31=mu32= 1/6;
function y=RK_3(fun,a,b,y0,h)
n=(b-a)/h;
x=zeros(1,n+1);
y=zeros(1,n+1);
x(1)=a;
y(1)=y0;
for i=2:n+1
    x(i)=a+(i-1)*h;
    k1(i)=feval(fun,x(i-1),y(i-1));
    k2(i)=feval(fun,x(i-1)+1/2*h,y(i-1)+1/2*h*k1(i));
    k3(i)=feval(fun,x(i-1)+h,y(i-1)+1/6*h*k1(i)+1/6*h*k2(i));
    y(i)=y(i-1)+h*(1/6*k1(i)+4/6*k2(i)+1/6*k3(i));
end
hold on
plot(x,y,'-^')
title('三阶龙格-库塔方法')
```

例 8.14 利用三阶龙格-库塔方法求常微分方程

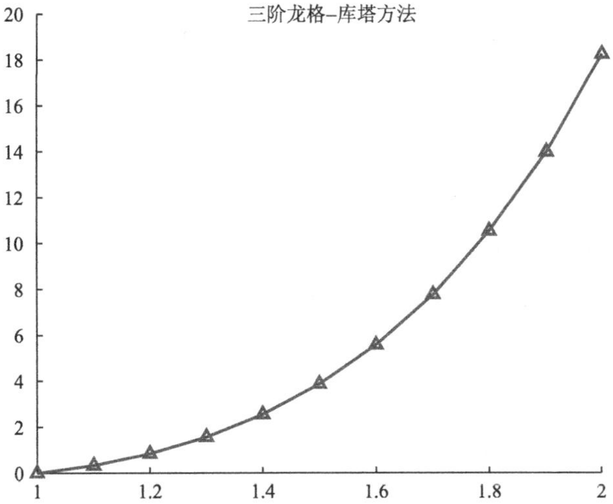

图 8.7 三阶龙格-库塔方法

$$\begin{cases} \dfrac{dy}{dx} = 2\dfrac{y}{x} + x^2 e^x, & 1 \leqslant x \leqslant 2; \\ y(1) = 0. \end{cases}$$

步长 $h = 0.1$.

调用函数

```
RK_3(inline('2/x*y+x^2*exp(x)'),1,2,0,0.1)
```

结果如图 8.7 所示.

8.5.7 四阶龙格-库塔方法 (Runge-Kutta order four)

```
% dy/dx=fun;a,b为数值解区间; h为步长; d为参数
% 经典的四阶龙格-库塔方法
function y=RK_4(fun,a,b,y0,h)
n=(b-a)/h;
x=zeros(1,n+1);
y=zeros(1,n+1);
x(1)=a;
y(1)=y0;
for i=2:n+1
    x(i)=a+(i-1)*h;
    k1(i)=feval(fun,x(i-1),y(i-1));
    k2(i)=feval(fun,x(i-1)+1/2*h,y(i-1)+1/2*h*k1(i));
    k3(i)=feval(fun,x(i-1)+1/2*h,y(i-1)+1/2*h*k2(i));
    k4(i)=feval(fun,x(i-1)+h,y(i-1)+h*k3(i));
    y(i)=y(i-1)+h*(1/6*k1(i)+2/6*k2(i)+2/6*k3(i)+1/6*k4(i));
end
hold on
plot(x,y,'-o')
title('四阶龙格-库塔方法')
```

例 8.15 利用四阶龙格-库塔方法求常微分方程

$$\begin{cases} \dfrac{dy}{dx} = 2\dfrac{y}{x} + x^2 e^x, & 1 \leqslant x \leqslant 2; \\ y(1) = 0. \end{cases}$$

步长 $h = 0.1$.

调用函数

```
RK_4(inline('2/x*y+x^2*exp(x)'),1,2,0,0.1)
```

结果如图 8.8 所示.

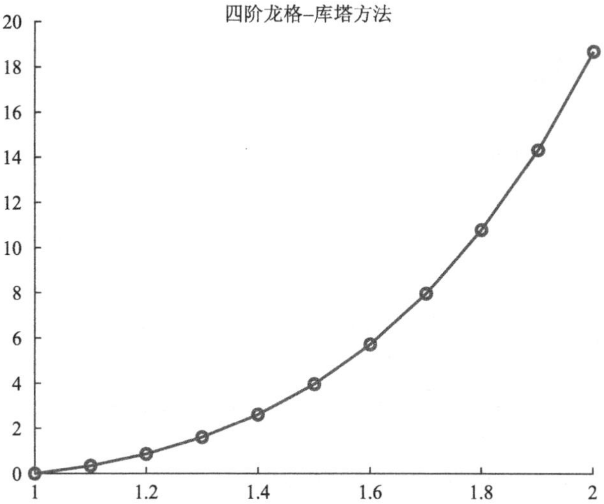

图 8.8 四阶龙格-库塔方法

例 8.16 利用二阶、三阶、四阶龙格-库塔方法求常微分方程

$$\begin{cases}\dfrac{\mathrm{d}y}{\mathrm{d}x}=2\dfrac{y}{x}+x^2\mathrm{e}^x, \quad 1\leqslant x\leqslant 2;\\ y(1)=0.\end{cases}$$

步长 $h=0.1$.

结果见图 8.9.

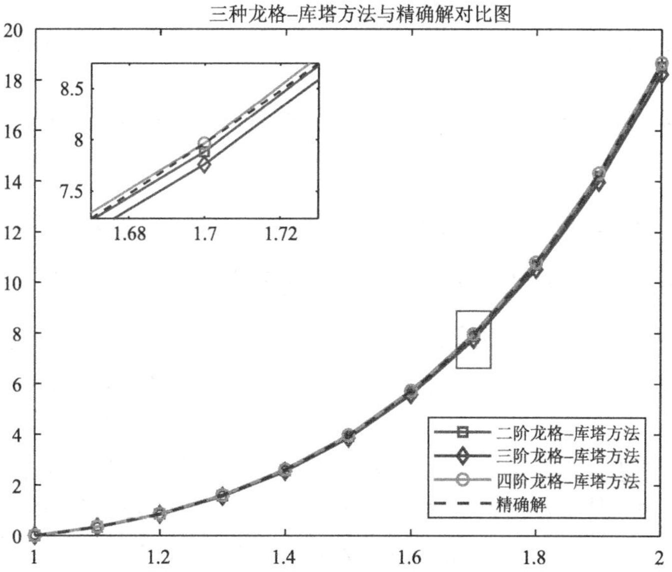

图 8.9 三种龙格-库塔方法对比

8.6 本章要点 (Hightlights)

1. Euler 方法是初值问题数值解的一阶方法，改进的 Euler 方法 (modified Euler's scheme) 是二阶方法.
2. (显式) Euler 格式：$w_{i+1}=w_i+hf(t_i,w_i)$ 和隐式 Euler 格式(implicit Euler's scheme)：$w_{i+1}=w_i+hf(t_{i+1},w_{i+1})$ 既可以分别通过两点向前数值微分公式(差商公式) (forward divided difference)：$y'(t_i)\approx\dfrac{y(t_{i+1})-y(t_i)}{h}$ 和两点向后数值微分公式(backward divided difference)：$y'(t_{i+1})\approx\dfrac{y(t_{i+1})-y(t_i)}{h}$ 导出，也可以首先将常微分方程 $y'=f(t,y)$ 转化为等价的积分方程：$y(t)=y(t_i)+\int_{t_i}^{t}f(t,y)\mathrm{d}t$ 得到 $y(t_{i+1})=y(t_i)+\int_{t_i}^{t_{i+1}}f(t,y)\mathrm{d}t$ 再分别用数值积分中的左矩形求积公式和右矩形求积公式推得.
3. 两步 Euler 格式 (two-step Euler's scheme)：$w_{i+1}=w_{i-1}+2hf(t_i,w_i)$，既可以通过对称形式的三点中心数值微分公式(three-point symmetrical formula)：$y'(t_i)\approx\dfrac{y(t_{i+1})-y(t_{i-1})}{2h}$ 推出，也可以从转换后的积分关系式 $y(t_{i+1})=y(t_{i-1})+\int_{t_{i-1}}^{t_{i+1}}f(t,y)\mathrm{d}t$ 出发，应用数值积分中的中点求积公式 (mid-point rule)：$y(t_{i+1})\approx y(t_{i-1})+2hf(t_i,y(t_i))$ 推得.
4. 梯形格式 (trapezoidal scheme)：$w_{i+1}=w_i+\dfrac{h}{2}[f(t_i,w_i)+f(t_{i+1},w_{i+1})]$ 来源于对转换后的积分关系式：$y(t_{i+1})=y(t_i)+\int_{t_i}^{t_{i+1}}f(t,y)\mathrm{d}t$ 应用梯形求积公式 (trapezoidal rule)：$\int_{t_i}^{t_{i+1}}f(t,y)\mathrm{d}t\approx\dfrac{h}{2}[f(t_i,y(t_i))+f(t_{i+1},y(t_{i+1}))]$ 而得到. 梯形格式是一种二阶隐式格式，是 Euler 格式和隐式 Euler 格式的平均.
5. 改进的 Euler 格式实际上是对隐式的梯形格式进行显式化改造的结果. 主要途径是，将梯形格式：$w_{i+1}=w_i+\dfrac{h}{2}[f(t_i,w_i)+f(t_{i+1},w_{i+1})]$ 右端 $f(t_{i+1},w_{i+1})$ 中的 w_{i+1} 用 (显式) Euler 格式：$w_{i+1}=w_i+hf(t_i,w_i)$ 进行预处理. 梯形格式和改进的 Euler 格式都是二阶方法.
6. 改进的 Euler 公式是二阶 Runge-Kutta 方法的特例，二阶 Runge-Kutta 方法局部截断误差为 $O(h^3)$.
7. Runge-Kutta 方法既不是多步 (multi-step) 方法，也不同于经典的单步 (one-step) 方法，可以理解为一种单步内的加密细分方法，其基本思想是，在每一步 $[t_i,t_{i+1}]$ 之内引进多个细分点 $t_{i+p},t_{i+q},t_{i+r},\cdots,t_{i+\mu}$，$0\leqslant p\leqslant q\leqslant r\leqslant\cdots\leqslant\mu\leqslant 1$, 对这些分点处的斜率 $K_p=f(t_{i+p},w_{i+p})$，$K_q=f(t_{i+q},w_{i+q})$，$K_r=f(t_{i+r},w_{i+r}),\cdots$，$K_\mu=f(t_{i+\mu},w_{i+\mu})$, 作线性组合：$K_i^*=\alpha_pK_p+\alpha_qK_q+\alpha_rK_r+\cdots+\alpha_\mu K_\mu$, 其中，$\alpha_p+\alpha_q+\alpha_r+\cdots+\alpha_\mu=1$, 即可建立 Runge-Kutta 格式：$w_{i+1}=w_i+hK_i^*$. Runge-Kutta 方法是以簇类形式出现的，每一簇类

均含有无穷多种格式，同一簇类的所有格式具有相同的阶.

8. Adams 方法属于多步数值求解方法，其基本格式为

$$w_{i+1} = w_i + hK_i^*.$$

对于 Adams m 步显式格式，

$$K_i^* = \alpha_{i-m+1}w'_{i-m+1} + \alpha_{i-m+2}w'_{i-m+2} + \cdots + \alpha_{i-1}w'_{i-1} + \alpha_i w'_i, \text{且} \sum_{k=1}^{m} \alpha_{i-m+k} = 1;$$

对于 Adams m 步隐式格式，

$$K_i^* = \alpha_{i-m+2}w'_{i-m+2} + \alpha_{i-m+1}w'_{i-m+1} + \cdots + \alpha_i w'_i + \alpha_{i+1}w'_{i+1}, \text{且} \sum_{k=1}^{m} \alpha_{i-m+k+1} = 1;$$

其中，$w'_k = f(t_k, w_k) = f(t_k, y(t_k)) = y'(t_k), k = i-m+1, i-m+2, \cdots, i, i+1.$

通过 Taylor 展开建立方程组的方式，可使

$$y(t_{i+1}) - w_{i+1} = O(h^{m+1}),$$

从而 Adams m 步显 (隐) 式格式达到 m 阶.

8.7 问题讨论 (Questions for discussion)

1. 常微分方程初值问题的数值求解公式中局部截断误差是什么？各种求解公式的特点是什么？
2. 常微分方程初值问题的单步求解方法与多步求解方法的区别在哪？如果采用多步方法，如何提供初始值？并保证初始值的精度？
3. 如所知，常微分方程 $y' = f(t, y)$ 等价于积分方程：

$$y(t) = y(t_i) + \int_{t_i}^{t} f(t, y)\,\mathrm{d}t,$$

特别有

$$y(t_{i+1}) = y(t_i) + \int_{t_i}^{t_{i+1}} f(t, y)\,\mathrm{d}t.$$

令

$$g(t) = f(t, y(t)).$$

由带误差余项的线性插值公式，有

$$\begin{aligned} g(t) &= g(t_{i-1})\frac{t - t_i}{t_{i-1} - t_i} + g(t_i)\frac{t - t_{i-1}}{t_i - t_{i-1}} + (t - t_{i-1})(t - t_i)g[t_{i-1}, t_i, t] \\ &= -g(t_{i-1})\frac{t - t_i}{h} + g(t_i)\frac{t - t_{i-1}}{h} + (t - t_{i-1})(t - t_i)g[t_{i-1}, t_i, t]. \end{aligned}$$

从而，

$$\begin{aligned}\int_{t_i}^{t_{i+1}} g(t)\mathrm{d}t &= -\frac{1}{h}g(t_{i-1})\int_{t_i}^{t_{i+1}}(t-t_i)\mathrm{d}t + \frac{1}{h}g(t_i)\int_{t_i}^{t_{i+1}}(t-t_{i-1})\mathrm{d}t + \int_{t_i}^{t_{i+1}}(t-t_{i-1})(t-t_i)g[t_{i-1},t_i,t]\mathrm{d}t \\ &= -\frac{h}{2}g(t_{i-1}) + \frac{3h}{2}g(t_i) + g[t_{i-1},t_i,\eta]\int_{t_i}^{t_{i+1}}(t-t_{i-1})(t-t_i)\mathrm{d}t \\ &= -\frac{h}{2}g(t_{i-1}) + \frac{3h}{2}g(t_i) + \frac{5}{12}h^3 g''(\xi) \\ &= -\frac{h}{2}f(t_{i-1},y(t_{i-1})) + \frac{3h}{2}f(t_i,y(t_i)) + O(h^3)\end{aligned}$$

因此推得

$$y(t_{i+1}) = y(t_i) + \frac{h}{2}[3f(t_i,y(t_i)) - f(t_{i-1},y(t_{i-1}))] + O(h^3).$$

由此推出二阶 Adams 两步显式格式

$$w_{i+1} = w_i + \frac{h}{2}[3f(t_i,w_i) - f(t_{i-1},w_{i-1})].$$

受上述论证过程启发，能否推出下述三阶 Adams 三步显式格式：

$$w_{i+1} = w_i + \frac{h}{12}[23f(t_i,w_i) - 16f(t_{i-1},w_{i-1}) + 5f(t_{i-2},w_{i-2})]$$

和四阶 Adams 四步显式格式：

$$w_{i+1} = w_i + \frac{h}{24}[55f(t_i,w_i) - 59f(t_{i-1},w_{i-1}) + 37f(t_{i-2},w_{i-2}) - 9f(t_{i-3},w_{i-3})]?$$

对 Adams 各阶隐式格式，是否可以采用类似方式推导？

4. 四步显式 Adams 方法与四步隐式 Adams 方法有什么区别？

5. 四阶 Runge-Kutta 格式可以写成如下形式：

$$\begin{aligned} w_0 &= \alpha, \\ w_{i+1} &= w_i + \frac{h}{6}f(t_i,w_i) + \frac{h}{3}f(t_i+\alpha_1 h, w_i+\delta_1 hf(t_i,w_i)) \\ &\quad + \frac{h}{3}f(t_i+\alpha_2 h, w_i+\delta_2 hf(t_i+\gamma_2 h, w_i+\gamma_3 hf(t_i,w_i))) \\ &\quad + \frac{h}{6}f(t_i+\alpha_3 h, w_i+\delta_3 hf(t_i+\gamma_4 h, w_i+\gamma_5 hf(t_i+\gamma_6 h, w_i+\gamma_7 hf(t_i,w_i)))). \end{aligned}$$

求常数 $\alpha_1,\alpha_2,\alpha_3,\delta_1,\delta_2,\delta_3,\gamma_2,\gamma_3,\gamma_4,\gamma_5,\gamma_6$ 和 γ_7.

8.8 关键术语 (Key terms)

convex set (凸集)
Lipschitz condition (Lipschitz 条件)
initial-value problem (初值问题)
well-posed problem (适定问题)
local truncation error (局部截断误差)

order (阶)
Euler's scheme (Euler 格式)
implicit Euler's scheme (隐式 Euler 格式)
trapezoidal scheme (梯形格式)
modified Euler's scheme (改进的 Euler 格式)
error bounds (误差界)
Taylor polynomial (Taylor 多项式)
Runge-Kutta method (Runge-Kutta 方法)
one-step (单步)
Taylor's expansion (Taylor 展开)
multistep methods (多步方法)
Adams explicit scheme (Adams 显式格式)
Adams implicit schemes (Adams 隐式格式)
two-step (两步)
truncation error (截断误差)

8.9 延伸阅读 (Extending reading)

常微分方程边值问题 (Boundary value problems for ODEs)

在前述常微分方程初值问题中，我们考虑的是一阶常微分方程，只涉及一个初始条件，即已知函数在求解区间左端点的函数值. 工程实际中遇到的常微分方程可能是高阶微分方程，已知条件可能涉及区间上多点函数值，此时的常微分方程不再是初值问题，而是所谓的边值问题. 我们考虑如下最简单的二阶常微分方程两点边值问题：

$$-y''(t)+r(t)y(t)=f(t),\quad a\leqslant t\leqslant b, \tag{8.9.1}$$

$$y(a)=\alpha,\quad y(b)=\beta, \tag{8.9.2}$$

其中，α 和 β 是给定的实数. 设 $r(t)$ 和 $f(t)$ 是定义在 $[a,b]$ 上的连续函数，且

$$r(t)\geqslant 0,\quad a\leqslant t\leqslant b.$$

设

$$h=(b-a)/N,\quad t_i=a+ih,\quad i=0,1,\cdots,N,$$

则在每一网格结点 t_i 处有

$$-y''(t_i)+r(t_i)y(t_i)=f(t_i), \tag{8.9.1*}$$

$$y(t_0)=\alpha,\quad y(t_N)=\beta. \tag{8.9.2*}$$

定义 8.6 定义 y 的一阶中心差分 δy 如下:

$$\delta y(t_j)=y\left(t_j+\frac{h}{2}\right)-y\left(t_j-\frac{h}{2}\right).$$

而高阶中心差分可用下述方式递归定义：

$$\delta^{m+1}y(t_j)=\delta[\delta^m y(t_j)]=\delta^m y\left(t_j+\frac{h}{2}\right)-\delta^m y\left(t_j-\frac{h}{2}\right).$$

特别，二阶中心差分可写成

$$\begin{aligned}\delta^2 y(t_j)&=\delta y\left(t_j+\frac{h}{2}\right)-\delta y\left(t_j-\frac{h}{2}\right)\\&=y(t_j+h)-2y(t_j)+y(t_j-h).\end{aligned}$$

定理 8.13　设 $y\in C^4[t-h,t+h]$, 则 $\exists\xi\in(t-h,t+h)$, 使得

$$\frac{\delta^2 y(t)}{h^2}=y''(t)+\frac{h^2}{12}y^{(4)}(\xi). \tag{8.9.3}$$

证明　由 Taylor 展开得

$$y(t-h)=y(t)-hy'(t)+\frac{h^2}{2}y''(t)-\frac{h^3}{6}y^{(3)}(t)+\frac{h^4}{24}y^{(4)}(\xi_1),$$

$$y(t+h)=y(t)+hy'(t)+\frac{h^2}{2}y''(t)+\frac{h^3}{6}y^{(3)}(t)+\frac{h^4}{24}y^{(4)}(\xi_2),$$

其中，$\xi_1\in(t-h,t)$, $\xi_2\in(t,t+h)$. 于是有

$$\delta^2 y(t)=y(t-h)-2y(t)+y(t+h)=h^2y''(t)+\frac{h^4}{24}[y^{(4)}(\xi_1)+y^{(4)}(\xi_2)].$$

由 $y\in C^4[t-h,t+h]$, 知 $\exists\xi\in(t-h,t+h)$, 使得

$$y^{(4)}(\xi_1)+y^{(4)}(\xi_2)=2y^{(4)}(\xi),$$

于是，$\dfrac{\delta^2 y(t)}{h^2}=y''(t)+\dfrac{h^2}{12}y^{(4)}(\xi)$. 定理 8.13 得证.

定理 8.14　设 $y\in C^6[t-h,t+h]$, 则 $\exists\eta\in(t-h,t+h)$, 使得

$$\frac{\delta^2 y(t)}{h^2}=y''(t)+\frac{h^2}{12}y^{(4)}(t)+\frac{h^4}{360}y^{(6)}(\eta). \tag{8.9.4}$$

证明　由 Taylor 展开得

$$y(t-h)=y(t)-hy'(t)+\frac{h^2}{2}y''(t)-\frac{h^3}{6}y^{(3)}(t)+\frac{h^4}{24}y^{(4)}(t)-\frac{h^5}{120}y^{(5)}(t)+\frac{h^6}{720}y^{(6)}(\eta_1),$$

$$y(t+h)=y(t)+hy'(t)+\frac{h^2}{2}y''(t)+\frac{h^3}{6}y^{(3)}(t)+\frac{h^4}{24}y^{(4)}(t)+\frac{h^5}{120}y^{(5)}(t)+\frac{h^6}{720}y^{(6)}(\eta_2),$$

其中，$\eta_1\in(t-h,t)$, $\eta_2\in(t,t+h)$. 于是有

$$\delta^2 y(t)=y(t-h)-2y(t)+y(t+h)=h^2y''(t)+\frac{h^4}{12}y^{(4)}(t)+\frac{h^6}{720}[y^{(6)}(\eta_1)+y^{(6)}(\eta_2)].$$

由 $y\in C^6[t-h,t+h]$, 知 $\exists\eta\in(t-h,t+h)$, 使得

$$y^{(6)}(\eta_1)+y^{(6)}(\eta_2)=2y^{(6)}(\eta),$$

于是，$\dfrac{\delta^2 y(t)}{h^2}=y''(t)+\dfrac{h^2}{12}y^{(4)}(t)+\dfrac{h^4}{360}y^{(6)}(\eta)$. 定理 8.14 得证.

用 w_i 表示 $y(t_i)$ 的数值解，$i=1,2,\cdots,N$，由(8.9.3)知

$$\frac{\delta^2 y(t)}{h^2}-y''(t)=\frac{h^2}{12}y^{(4)}(\xi).$$

因此，二阶导数 $y''(t_i)$ 可借助如下中心差商进行逼近

$$y''(t_i)\approx\frac{1}{h^2}\delta^2 y(t_i)=\frac{1}{h^2}[y(t_i+h)-2y(t_i)+y(t_i-h)]=\frac{1}{h^2}(w_{i+1}-2w_i+w_{i-1}).$$

于是，边值问题(8.9.1)和(8.9.2)可利用下述线性方程组数值求解：

$$-\frac{1}{h^2}(w_{i+1}-2w_i+w_{i-1})+r_i w_i=f_i,\quad i=1,2,\cdots,N-1,$$

$$w_0=\alpha,\quad w_N=\beta,$$

其中，$r_i=r(t_i)$，$f_i=f(t_i)$. 上述线性方程组可改写为

$$-\frac{1}{h^2}w_{i-1}+\left(\frac{2}{h^2}+r_i\right)w_i-\frac{1}{h^2}w_{i+1}=f_i,\quad i=1,2,\cdots,N-1,\tag{8.9.5}$$

$$w_0=\alpha,\quad w_N=\beta,\tag{8.9.6}$$

即

$$\begin{aligned}
(2/h^2+r_1)w_1-(1/h^2)w_2&=f_1+\alpha/h^2,\\
(-1/h^2)w_1+(2/h^2+r_2)w_2-(1/h^2)w_3&=f_2,\\
(-1/h^2)w_2+(2/h^2+r_3)w_3-(1/h^2)w_4&=f_3,\\
\cdots\quad\cdots\quad\cdots\quad\cdots\quad\cdots&\\
(-1/h^2)w_{N-3}+(2/h^2+r_{N-2})w_{N-2}-(1/h^2)w_{N-1}&=f_{N-2},\\
(-1/h^2)w_{N-2}+(2/h^2+r_{N-1})w_{N-1}&=f_{N-1}+\beta/h^2.
\end{aligned}$$

令

$$M_{ii}=2/h^2+r_i,\ M_{ii-1}=M_{ii+1}=-1/h^2,\ M_{ij}=0,\ j\neq i-1,i,i+1,\ i=1,2,\cdots,N-1;$$

$$g_1=f_1+\alpha/h^2,\ g_i=f_i, i=2,3,\cdots,N-2,\ g_{N-1}=f_{N-1}+\beta/h^2,$$

则线性方程组 (8.9.5)和(8.9.6) 可写成如下矩阵形式：

$$\boldsymbol{M}\boldsymbol{w}=\boldsymbol{g},\tag{8.9.7}$$

其中，$\boldsymbol{M}=(M_{ij})_{(N-1)\times(N-1)}$, $\boldsymbol{w}=(w_1,w_2,\cdots,w_{N-1})^{\mathrm{T}}$, $\boldsymbol{g}=(g_1,g_2,\cdots,g_{N-1})^{\mathrm{T}}$.

由于 $r(t)\geqslant 0, a\leqslant t\leqslant b$, 故有 $r_i\geqslant 0, i=1,2,\cdots,N-1$, 因此 $\boldsymbol{M}$ 是对角占优的三对角对称矩阵，从而 $\boldsymbol{M}$ 非奇异，于是 (8.9.7) 有唯一解.

定义 8.7 用中心差分逼近求解常微分方程边值问题 (8.9.1)和(8.9.2) 产生的截断误差定义为

$$T_i=-\frac{1}{h^2}\delta^2 y(t_i)+r_i y(t_i)-f_i,\quad i=1,2,\cdots,N-1,$$

其中，$y(t)$ 是 (8.9.1)和(8.9.2) 的精确解.

实际上，由 (8.9.1) 知，$r_i y(t_i)-f_i=y''(t_i)$，因此，

$$T_i = -\frac{1}{h^2}\delta^2 y(t_i) + r_i y(t_i) - f_i = y''(t_i) - \frac{1}{h^2}\delta^2 y(t_i), \quad i = 1,2,\cdots,N-1.$$

即，截断误差实际上是用中心差商 $\delta^2 y(t)/h^2$ 逼近二阶导数 $y''(t)$ 所产生的误差. 关于截断误差，有下述定理.

定理 8.15 设边值问题 (8.9.1)、(8.9.2) 的精确解 $y(t) \in C^4[a,b]$, 则有

$$|T_i| \leqslant T = \frac{1}{12}h^2 M_4, \quad i = 1,2,\cdots,N-1,$$

其中，$M_4 = \max\limits_{t\in[a,b]}\left|y^{(4)}(t)\right|$.

证明 由定理 8.13 知，

$$\begin{aligned} T_i &= -\frac{1}{h^2}\delta^2 y(t_i) + r_i y(t_i) - f_i \\ &= -\left[y''(t_i) + \frac{h^2}{12}y^{(4)}(\xi_i)\right] + r_i y(t_i) - f_i \\ &= -y''(t_i) + r_i y(t_i) - f_i + \frac{h^2}{12}y^{(4)}(\xi_i) \\ &= \frac{h^2}{12}y^{(4)}(\xi_i), \qquad \xi_i \in [a,b] \end{aligned}$$

记 $M_4 = \max\limits_{t\in[a,b]}\left|y^{(4)}(t)\right|$，则有

$$|T_i| \leqslant T = \frac{1}{12}h^2 M_4, \quad i = 1,2,\cdots,N-1.$$

定理 8.15 得证.

定义 8.8 设 w_i 表示 $y(t_i)$ 的数值解，令

$$e_i = y(t_i) - w_i, \quad i = 0,1,\cdots,N,$$

称 $E = \max\limits_{0\leqslant i\leqslant N}|e_i|$ 为边值问题 (8.9.1)和(8.9.2) 的全局误差.

关于全局误差，有下述定理：

定理 8.16 设边值问题 (8.9.1)和(8.9.2) 的精确解 $y(t) \in C^4[a,b]$, $w_i, i = 0,1,\cdots,N$, 表示 (8.9.5)和(8.9.6) 的解，即用中心差商 $\delta^2 y(t_i)/h^2$ 代替二阶导数 $y''(t_i)$ 时，$y(t_i)$ 的数值解，则全局误差有下述估计：

$$E \leqslant \frac{1}{96}h^2(b-a)^2 \max_{t\in[a,b]}\left|y^{(4)}(t)\right|.$$

为证明定理 8.16，需要下面的引理.

引理 8.17 (最大原理) 设正实数 $a_i, b_i, c_i, i = 0,1,\cdots,N$, 满足

$$b_i \geqslant a_i + c_i, \quad i = 0,1,\cdots,N,$$

且实数 $u_i, i = 0,1,\cdots,N$, 满足

$$-a_i u_{i-1} + b_i u_i - c_i u_{i+1} \leqslant 0, \quad i = 1,2,\cdots,N-1.$$

令 $K = \max\{u_0, u_N, 0\}$, 则 $u_i \leqslant K, i = 0,1,\cdots,N$.

证明 令 $u_r = \max\{u_0, u_1, \cdots, u_N\}$. 当 $r=0, r=N$ 或 $u_r \leqslant 0$ 时，引理 8.17 的结论显然成立. 因此，接下来只需就 $1 \leqslant r \leqslant N-1$ 以及 $u_r > 0$ 的情形证明引理 8.17 的结论即可. 由 $u_r = \max\{u_0, u_1, \cdots, u_N\}$ 推得 $u_r \geqslant u_{r-1}$, $u_r \geqslant u_{r+1}$. 因此，

$$\begin{aligned} b_r u_r &\leqslant a_r u_{r-1} + c_r u_{r+1} \\ &\leqslant a_r u_r + c_r u_r \\ &= (a_r + c_r) u_r \\ &\leqslant b_r u_r, \end{aligned}$$

由此推得 $a_r u_{r-1} + c_r u_{r+1} = a_r u_r + c_r u_r$, 即

$$a_r(u_r - u_{r-1}) + c_r(u_r - u_{r+1}) = 0.$$

$a_r > 0$, $c_r > 0$, $u_r - u_{r-1} \geqslant 0$ 以及 $u_r - u_{r+1} \geqslant 0$ 蕴含： $u_r = u_{r-1} = u_{r+1}$. 同理，由

$$\begin{aligned} b_{r-1} u_r &= b_{r-1} u_{r-1} \\ &\leqslant a_{r-1} u_{r-2} + c_{r-1} u_r \\ &\leqslant a_{r-1} u_r + c_{r-1} u_r \\ &= (a_{r-1} + c_{r-1}) u_r \\ &\leqslant b_{r-1} u_r, \end{aligned}$$

可推得： $a_{r-1} u_{r-2} + c_{r-1} u_r = a_{r-1} u_r + c_{r-1} u_r$, 从而 $u_r = u_{r-2}$. 由

$$\begin{aligned} b_{r+1} u_r &= b_{r+1} u_{r+1} \\ &\leqslant a_{r+1} u_r + c_{r+1} u_{r+2} \\ &\leqslant a_{r+1} u_r + c_{r+1} u_r \\ &= (a_{r+1} + c_{r+1}) u_r \\ &\leqslant b_{r+1} u_r, \end{aligned}$$

可推得： $a_{r+1} u_r + c_{r+1} u_{r+2} = a_{r+1} u_r + c_{r+1} u_r$, 从而 $u_r = u_{r+2}$. 反复利用上述方法，可分别推得

$$u_r = u_0 = u_N = \max\{u_0, u_1, \cdots, u_N\}.$$

令 $K = \max\{u_0, u_N, 0\}$, 则 $u_i \leqslant K, i = 0,1,\cdots,N$. 引理 8.17 得证.

接下来证明定理 8.16. 为此，定义

$$L(u_i) = -\frac{1}{h^2}\delta^2 u_i + r_i u_i, \quad i = 1,2,\cdots,N-1, \tag{8.9.8}$$

则有

$$L(y(t_i)) = -\frac{1}{h^2}\delta^2 y(t_i) + r_i y(t_i) = f_i + T_i, \quad i = 1,2,\cdots,N-1,$$

$$L(w_i) = -\frac{1}{h^2}\delta^2 w_i + r_i w_i = f_i, \quad i = 1,2,\cdots,N-1,$$

其中， $\delta^2 w_i = w_{i+1} - 2w_i + w_{i-1}$. 由 $e_i = y(t_i) - w_i, i = 0,1,\cdots,N$, 得

$$L(e_i) = L(y(t_i)) - L(w_i) = T_i, \quad i = 1,2,\cdots,N-1.$$

令

$$\varphi_i = C[(2i-N)^2 h^2 - N^2 h^2], \quad i = 0,1,\cdots,N, \tag{8.9.9}$$

其中， $h=(b-a)/N$ ， C是待定常数. 于是，

$$
\begin{aligned}
L(\varphi_i) &= -\frac{1}{h^2}\delta^2\varphi_i + r_i\varphi_i \\
&= -\frac{1}{h^2}(\varphi_{i-1} - 2\varphi_i + \varphi_{i+1}) + r_i\varphi_i \\
&= -C[(2i-2-N)^2 - 2(2i-N)^2 + (2i+2-N)^2] + r_i\varphi_i \\
&= -8C + r_i\varphi_i, \quad i = 1,2,\cdots,N-1.
\end{aligned}
$$

因此，

$$L(e_i + \varphi_i) = T_i - 8C + r_i\varphi_i, \quad i = 1,2,\cdots,N-1.$$

取 $C = T/8$, 其中 $T = \frac{1}{12}h^2 M_4$， $M_4 = \max\limits_{t\in[a,b]}\left|y^{(4)}(t)\right|$， 则由 $|T_i| \leqslant T$， $r_i \geqslant 0$, $\varphi_i \leqslant 0$, 推得

$$L(e_i + \varphi_i) = T_i - 8C + r_i\varphi_i \leqslant 0, \quad i = 1,2,\cdots,N-1.$$

现令

$$u_i = e_i + \varphi_i, \quad i = 0,1,\cdots,N,$$

则

$$L(u_i) \leqslant 0, \quad i = 1,2,\cdots,N-1,$$

即

$$-\frac{1}{h^2}(u_{i-1} - 2u_i + u_{i+1}) + r_i u_i \leqslant 0, \quad i = 1,2,\cdots,N-1.$$

上式可改写为

$$-\frac{1}{h^2}u_{i-1} + \left(\frac{2}{h^2} + r_i\right)u_i - \frac{1}{h^2}u_{i+1} \leqslant 0, \quad i = 1,2,\cdots,N-1.$$

令

$$a_i = c_i = \frac{1}{h^2}, b_i = \frac{2}{h^2} + r_i, \quad i = 1,2,\cdots,N-1,$$

则正实数 $a_i, b_i, c_i, i = 1,2,\cdots,N-1$, 满足 $b_i \geqslant a_i + c_i$， $i = 1,2,\cdots,N-1$, 且

$$-a_i u_{i-1} + b_i u_i - c_i u_{i+1} \leqslant 0, \quad i = 1,2,\cdots,N-1,$$

从而根据引理 8.17 得

$$u_i \leqslant K = \max\{u_0, u_N, 0\}, \quad i = 0,1,\cdots,N.$$

又 $e_0 = 0, e_N = 0, \varphi_0 = 0, \varphi_N = 0$, 故 $u_0 = e_0 + \varphi_0 = 0$, $u_N = e_N + \varphi_N = 0$, 从而推得

$$u_i \leqslant 0, \quad i = 0,1,\cdots,N,$$

于是有

$$e_i + \varphi_i \leqslant 0, \quad i = 0,1,\cdots,N.$$

由 (8.9.9) 得

$$\begin{aligned}\varphi_i &= C[(2i-N)^2h^2 - N^2h^2] \\ &\geqslant -CN^2h^2 \\ &= -\frac{T}{8}N^2h^2 \\ &= -\frac{T}{8}(b-a)^2, \quad i=0,1,\cdots,N.\end{aligned}$$

因此，利用$T=\frac{1}{12}h^2M_4=\frac{1}{12}h^2\max\limits_{t\in[a,b]}\left|y^{(4)}(t)\right|$，得

$$e_i \leqslant -\varphi_i \leqslant \frac{T}{8}(b-a)^2 = \frac{1}{96}h^2(b-a)^2\max_{t\in[a,b]}\left|y^{(4)}(t)\right|, \quad i=0,1,\cdots,N,$$

从而

$$E=\max_{0\leqslant i\leqslant N}\left|e_i\right| \leqslant \frac{1}{96}h^2(b-a)^2\max_{t\in[a,b]}\left|y^{(4)}(t)\right|.$$

定理 8.16 得证.

定理 8.16 表明，如果 $y(t)\in C^4[a,b]$，则有

$$\lim_{N\to\infty}\max_{0\leqslant i\leqslant N}\left|y(t_i)-w_i\right|=0.$$

即，如果 $y(t)\in C^4[a,b]$，则由中心差分逼近而获得的二阶常微分方程边值问题 (8.9.1)和(8.9.2) 求解格式是收敛的.

8.10 习题 (Exercises)

1. Show that each of the following initial-value problem has a unique solution.
   a. $y'=y\cos t, \quad 0\leqslant t\leqslant 1, \quad y(0)=1;$
   b. $y'=\frac{2}{t}y+t^2e^t, \quad 1\leqslant t\leqslant 2, \quad y(1)=0;$
   c. $y'=-\frac{2}{t}y+t^2e^t, \quad 1\leqslant t\leqslant 2, \quad y(1)=\sqrt{2}e;$
   d. $y'=\frac{4t^3y}{1+t^4}, \quad 0\leqslant t\leqslant 1, \quad y(0)=1.$
2. Suppose $f(t,y)=t^2y+1$, is the initial-value problem
$$y'=f(t,y), \quad 0\leqslant t\leqslant 1, \quad y(0)=1,$$
   well-posed?
3. Suppose $f(t,y)=ty$, is the initial-value problem
$$y'=f(t,y), \quad 0\leqslant t\leqslant 1, \quad y(0)=1,$$
   well-posed?
4. Suppose $f(t,y)=1-y$, is the initial-value problem
$$y'=f(t,y), \quad 0\leqslant t\leqslant 1, \quad y(0)=1,$$

well-posed?

5. Suppose $f(t,y)=-ty+\frac{4t}{y}$, is the initial-value problem

$$y'=f(t,y),\quad 0\leqslant t\leqslant 1,\quad y(0)=1,$$

well-posed?

6. Suppose $f(t,y)=\mathrm{e}^{t-y}$, is the initial-value problem

$$y'=f(t,y),\quad 0\leqslant t\leqslant 1,\quad y(0)=1,$$

well-posed?

7. Suppose $f(t,y)=\frac{1+y}{1+t}$, is the initial-value problem

$$y'=f(t,y),\quad 0\leqslant t\leqslant 1,\quad y(0)=1,$$

well-posed?

8. Suppose $f(t,y)=\cos(ty)$, is the initial-value problem

$$y'=f(t,y),\quad 0\leqslant t\leqslant 1,\quad y(0)=1,$$

well-posed?

9. Suppose $f(t,y)=\frac{y^2}{1+t}$, is the initial-value problem

$$y'=f(t,y),\quad 0\leqslant t\leqslant 1,\quad y(0)=1,$$

well-posed?

10. Suppose the perturbation $\delta(t)=\delta t$ is proportional to t for some constant δ. Show that the following initial-value problem is well-posed.

$$y'=\frac{2}{t}y+t^2\mathrm{e}^t,\quad 1\leqslant t\leqslant 2,\quad y(1)=0.$$

11. Use Euler's method to approximate the solution for the initial-value problem

$$\frac{\mathrm{d}y}{\mathrm{d}t}=t\mathrm{e}^{3t}-2y,\ 0\leqslant t\leqslant 1,\quad y(0)=0,\ \text{ with }\ h=0.5.$$

12. Use Euler's method to approximate the solution for the initial-value problem

$$\frac{\mathrm{d}y}{\mathrm{d}t}=1+(t-y)^2,\ 2\leqslant t\leqslant 3,\quad y(2)=1,\ \text{ with }\ h=0.5.$$

13. Use Euler's method to approximate the solution for the initial-value problem

$$\frac{\mathrm{d}y}{\mathrm{d}t}=1+y/t,\ 1\leqslant t\leqslant 2,\quad y(1)=2,\ \text{ with }\ h=0.25.$$

14. Use Euler's method to approximate the solution for the initial-value problem

$$\frac{\mathrm{d}y}{\mathrm{d}t}=\cos 2t+\sin 3t,\ 0\leqslant t\leqslant 1,\quad y(0)=0,\ \text{ with }\ h=0.25.$$

15. Use Euler's method to approximate the solution for the initial-value problem

$$\frac{dy}{dt}=e^{t-y},\ 0\leqslant t\leqslant 1,\quad y(0)=1,\ \text{with}\ h=0.5.$$

16. Use Euler's method to approximate the solution for the initial-value problem

$$\frac{dy}{dt}=\frac{1+t}{1+y},\ 1\leqslant t\leqslant 2,\quad y(1)=2,\ \text{with}\ h=0.5.$$

17. Use Euler's method to approximate the solution for the initial-value problem

$$\frac{dy}{dt}=-y+ty^{1/2},\ 2\leqslant t\leqslant 3,\quad y(2)=2,\ \text{with}\ h=0.25.$$

18. Use Euler's method to approximate the solution for the initial-value problem

$$\frac{dy}{dt}=t^{-2}(\sin 2t-2ty),\ 1\leqslant t\leqslant 2,\quad y(1)=2,\ \text{with}\ h=0.25.$$

19. Given the initial-value problem

$$y'=\frac{2}{t}y+t^2e^t,\quad 1\leqslant t\leqslant 2,\quad y(1)=0,$$

with exact solution $y(t)=t^2(e^t-e)$:

a. Use Euler's method with $h=0.1$ to approximate the solution, and compare it with the actual values of y;

b. Use the answers generated in part (a) and linear interpolation to approximate the following values of y, and compare them to the actual values;

i. $y(1.04)$ ii. $y(1.55)$ iii. $y(1.97)$

c. Use Theorem 8.7 to compute the value of h necessary for $|y(t_i)-w_i|\leqslant 0.1$

20. Given the initial-value problem

$$y'=\frac{2}{t^2}-\frac{y}{t}-y^2,\quad 1\leqslant t\leqslant 2,\quad y(1)=-1,$$

with exact solution $y(t)=-1/t$:

a. Use Euler's method with $h=0.05$ to approximate the solution, and compare it with the actual values of y;

b. Use the answers generated in part (a) and linear interpolation to approximate the following values of y; and compare them to the actual values;

i. $y(1.052)$ ii. $y(1.555)$ iii. $y(1.978)$

c. Use Theorem 8.7 to compute the value of h necessary for $|y(t_i)-w_i|\leqslant 0.5$

21. Use the Modified Euler's method to approximate the solution for the initial-value problem

$$\frac{dy}{dt}=te^{3t}-2y,\ 0\leqslant t\leqslant 1,\quad y(0)=0,\ \text{with}\ h=0.5.$$

22. Use the Modified Euler's method to approximate the solution for the initial-value problem

$$\frac{dy}{dt}=1+(t-y)^2,\ 2\leqslant t\leqslant 3,\quad y(2)=1,\ \text{with}\ h=0.5.$$

23. Use the Modified Euler's method to approximate the solution for the initial-value problem

$$\frac{dy}{dt}=1+y/t,\ 1\leqslant t\leqslant 2,\quad y(1)=2,\ \text{ with }\ h=0.25.$$

24. Use the Modified Euler's method to approximate the solution for the initial-value problem

$$\frac{dy}{dt}=\cos 2t+\sin 3t,\ 0\leqslant t\leqslant 1,\quad y(0)=1,\ \text{ with }\ h=0.25.$$

25. Suppose the following method

$$\begin{aligned}
&w_0=\alpha,\\
&K_1=f(t_i,w_i),\\
&K_2=f(t_i+\frac{h}{4},w_i+\frac{h}{4}K_1),\\
&w_{i+1}=w_i+h[(1-\lambda)K_1+\lambda K_2]
\end{aligned}$$

is used to solve the initial value problem

$$\frac{dy}{dt}=f(t,y),\quad a\leqslant t\leqslant b;\quad y(a)=\alpha$$

where $w_i=y(t_i),h=(b-a)/N,t_i=a+ih,0\leqslant i\leqslant N$. Determine λ so that the above method is of order two.

26. Derive the following implicit Euler's scheme

$$w_{i+1}=w_i+hf(t_{i+1},w_{i+1}),\ w_0=\alpha$$

concerning the initial value problem for the ordinary differential equation

$$\frac{dy}{dt}=f(t,y),\quad a\leqslant t\leqslant b,\quad y(a)=\alpha$$

where $w_i=y(t_i),t_i=a+ih,h=(b-a)/N$, and show it is a method of order one.

27. Show the following scheme

$$\begin{aligned}
&w_0=\alpha,\\
&w_{i+1}=w_i+hf(t_i+h,w_i+hf(t_i,w_i))
\end{aligned}$$

for the initial value problem $\frac{dy}{dt}=f(t,y),\quad a\leqslant t\leqslant b,\quad y(a)=\alpha$ has order one, where $h=\frac{b-a}{n}$.

28. Show the two-step Euler's scheme

$$w_0=\alpha,\ w_{i+1}=w_{i-1}+2hf(t_i,w_i)$$

for the initial value problem $\frac{dy}{dt}=f(t,y),\quad a\leqslant t\leqslant b,\quad y(a)=\alpha$ has order two.

29. Use Adams-Bashforth two-step explicit scheme to approximate the solutions to the following initial-value problem

$$y'=te^{3t}-2y,\quad 0\leqslant t\leqslant 1,\quad y(0)=0,\ \text{ with }\ h=0.2,$$

and compare the results with the actual values (actual solution $y(t)=\frac{1}{5}te^{3t}-\frac{1}{25}e^{3t}$

$+\frac{1}{25}e^{-2t}$).

30. Use Adams-Bashforth three-step explicit scheme to approximate the solutions to the following initial-value problem

$$y' = 1 + (t - y)^2, \quad 2 \leqslant t \leqslant 3, \quad y(2) = 1, \quad \text{with} \quad h = 0.2,$$

and compare the results with the actual values (actual solution $y(t) = t + \frac{1}{1-t}$).

31. Use Adams-Bashforth four-step explicit scheme to approximate the solutions to the following initial-value problem.

$$y' = 1 + y/t, \quad 1 \leqslant t \leqslant 2, \quad y(1) = 2, \quad \text{with} \quad h = 0.2,$$

and compare the results with the actual values (actual solution $y(t) = t \ln t + 2t$).

32. Use Adams-Moulton two-step implicit scheme to approximate the solutions to the following initial-value problem.

$$y' = y/t - (y/t)^2, \quad 1 \leqslant t \leqslant 2, \quad y(1) = 1, \quad \text{with} \quad h = 0.1,$$

and compare the results with the actual values (actual solution $y(t) = \frac{t}{1 + \ln t}$).

33. Use Adams-Moulton three-step implicit scheme to approximate the solutions to the following initial-value problem.

$$y' = -(y+1)(y+3), \quad 0 \leqslant t \leqslant 2, \quad y(0) = -2, \quad \text{with} \quad h = 0.1,$$

and compare the results with the actual values (actual solution $y(t) = -3 + \frac{2}{1 + e^{-2t}}$).

34. Use Adams-Moulton four-step implicit scheme to approximate the solutions to the following initial-value problem.

$$y' = -5y + 5t^2 + 2t, \quad 0 \leqslant t \leqslant 1, \quad y(0) = 1/3, \quad \text{with} \quad h = 0.1,$$

and compare the results with the actual values (actual solution $y(t) = t^2 + \frac{1}{3}e^{-5t}$).

35. Derive the Adams-Bashforth three-step scheme by using the Lagrange form of the interpolating polynomial.

36. Derive the Adams-Bashforth four-step scheme by using Newton's backward- difference form of the interpolating polynomial.

37. Derive the Adams-Moulton two-step method and its local truncation error by using an appropriate form of an interpolating polynomial.

References (参考文献)

蔡大用, 白峰杉. 高等数值分析[M]. 北京: 清华大学出版社, 1997.

邓建中, 葛仁杰, 程正兴. 计算方法[M]. 西安: 西安交通大学出版社, 1985.

关治, 陆金甫. 数值分析基础[M]. 北京: 高等教育出版社, 1998.

李庆扬, 王能超, 易大义. 数值分析[M]. 北京: 清华大学出版社, 2008.

檀结庆等. 连分式理论及其应用[M]. 北京: 科学出版社, 2007.

王仁宏, 朱功勤. 有理函数逼近及其应用[M]. 北京: 科学出版社, 2004.

王仁宏. 数值逼近[M]. 北京: 高等教育出版社, 1999.

徐翠薇, 孙绳武. 计算方法引论[M]. 北京: 高等教育出版社, 2002.

徐利治, 王仁宏, 周蕴时. 函数逼近的理论与方法[M]. 上海: 上海科学技术出版社, 1983.

徐献瑜, 李家楷, 徐国良. Padé 逼近概论[M]. 上海: 上海科学技术出版社, 1990.

周国标, 宋宝瑞, 谢建利. 数值计算[M]. 北京: 高等教育出版社, 2008.

朱晓临. 数值分析[M]. 合肥: 中国科学技术大学出版社, 2010.

ATKINSON K E, HAN W. Elementary numerical analysis[M]. 3rd ed. New York: John Wiley & Sons, Inc. , 2004.

ATKINSON K E. An introduction to numerical analysis[M]. 2nd ed. New York: John Wiley & Sons, Inc. , 1989.

BAKER G A, GRAVES-MORRIS P R. Padé approximants[M]. 2nd ed. Cambridge: Cambridge University Press, 1996.

BURDEN R L, FAIRES J D. Numerical analysis[M]. 影印版. 北京: 高等教育出版社, 2001.

CHENEY E W. Introduction to approximation theory[M]. New York: McGraw-Hill, 1966.

CIARLET P G. Introduction to numerical linear algebra and optimisation[M]. Cambridge: Cambridge University Press, 1989.

DAVIS P J, RABINOWITZ P. Methods of numerical integration[M]. 2nd ed. San Diego: Academic Press, 1984.

ENGELS H. Numerical quadrature and cubature[M]. London: Academic Press, 1980.

EPPERSON J F. An introduction to numerical methods and analysis[M]. 2nd ed. New York: John Wiley & Sons, Inc. , 2013.

GAUTSCHI W. Numerical analysis[M]. 2nd ed. New York: Springer, 2012.

GERALD C F, WHEATLEY P O. Applied numerical analysis[M]. 7th ed. Boston: Pearson Addison Wesley, 2004.

GOLUB G H, VAN LOAN C F. Matrix computations[M]. 3rd ed. Baltimore: Johns Hopkins University Press, 1996.

HIGHAM N J. Accuracy and stability of numerical algorithms[M]. Philadelphia: SIAM, 1996.

ISERLES A. A first course in the numerical analysis of differential equation[M]. 2nd ed. New York:

Cambridge University Press, 2008.

KRESS R. Numerical analysis[M]. New York: Springer, 1998.

LORENTZEN L, WAADELAND H. Continued fractions with applications[M]. Amsterdam: Elsevier Science Publishers B. V. , 1992.

POWELL M J D. Approximation theory and methods[M]. Cambridge: Cambridge University Press, 1996.

SCOTT L R. Numerical analysis[M]. Princeton: Princeton University Press, 2011.

STOER J, BULIRSCH P. Introduction to numerical analysis[M]. 2nd ed. Berlin: Springer, 1992.

SULI E, MAYERS D F. An introduction to numerical analysis[M]. Cambridge: Cambridge University Press, 2003.

TREFETHEN L N, BAU III D. Numerical linear algebra[M]. Philadelphia: SIAM, 1997.